HAZARDOUS MATERIALS CHEMISTRY FOR EMERGENCY RESPONDERS

HAZARDOUS MATERIALS CHEMISTRY FOR EMERGENCY RESPONDERS

ROBERT A. BURKE

CRC Press
Taylor & Francis Group
Boca Raton   London   New York

CRC Press is an imprint of the
Taylor & Francis Group, an **informa** business

CRC Press
Taylor & Francis Group
6000 Broken Sound Parkway NW, Suite 300
Boca Raton, FL 33487-2742

First issued in paperback 2017

© 2013 by Taylor & Francis Group, LLC
CRC Press is an imprint of Taylor & Francis Group, an Informa business

No claim to original U.S. Government works

Version Date: 20130206

ISBN 13: 978-1-4398-4985-9 (hbk)
ISBN 13: 978-1-138-07465-1 (pbk)

Library of Congress Cataloging-in-Publication Data

Burke, Robert (Robert A.)
  Hazardous materials chemistry for emergency responders / author, Robert Burke. -- Third edition.
    pages cm
  Includes bibliographical references and index.
  ISBN 978-1-4398-4985-9 (hardback)
  1. Hazardous substances. I. Title.

T55.3.H3B87 2013
604.7--dc23   2013004218

Visit the Taylor & Francis Web site at
http://www.taylorandfrancis.com

and the CRC Press Web site at
http://www.crcpress.com

This book is affectionately dedicated to my newest granddaughter, Abigale Marie (#21) December 7, 2010. She has brought joy, happiness, and love to my life and has captured my heart. She wrapped me around her little finger from the moment she was born as she held my finger in the incubator in the delivery room. She has lived with my wife and I since birth and we continue to raise her through our retirement. It isn't the way we had planned our retirement and certainly has been a challenge, but I wouldn't change a thing even if I could. While it will be many years before she will be able to read this, in the mean time, I will be kept busy with my grandfatherly duty of loving and spoiling her beyond imagination!

Contents

Preface

Much discussion has taken place concerning the validity of emergency response personnel studying chemistry. Obviously, from the title of this book, my opinion is clear; however, the chemistry presented here is more "street chemistry" than "college chemistry." I have tried to create chemistry subject matter appropriate for response personnel and to express the information in understandable terms.

The material in the book is organized into the U.S. Department of Transportation's (DOT) nine hazard classes, with which emergency response personnel should be intimately familiar. Each hazard class will be covered in its own chapter along with the appropriate chemistry concepts and terminology.

Almost every hazardous material presents more than one hazard; the DOT's placarding and labeling system only identifies the most severe hazards. Along with the hazard classes, typical information is provided about each hazard class. Individual chemicals are discussed, along with their hazards and their physical and chemical characteristics, both as distinct chemicals and within chemical families. Furthermore, the multiple dangers of hazardous materials, including "hidden" dangers, are studied throughout the book.

Common industrial chemicals, along with other hazardous materials, are presented throughout the book. Learning about these chemicals will provide responders with an overview of the varying dangers presented by hazardous materials and will show the similarities and differences among chemical family members, as well as other hazardous materials. Selected reports of various incidents involving hazardous materials are presented to emphasize the effects that chemical and physical characteristics can have on an incident outcome. Chemical terminology will be explained so that response personnel might recognize and understand information that they will encounter when researching reference sources, including books, computer databases, material safety data sheets (MSDS), and shipping papers. Basic chemistry provides emergency responders with a background that will help them not only to understand chemical terminology, but also to talk intelligently with Chemical Transportation Emergency Center (CHEMTREC), the National Response Center (NRC), Occupational Safety and Health Administration (OSHA), Environmental Protection Agency (EPA), other government agencies, shippers, industry representatives, and the media concerning chemicals involved in hazardous materials incidents.

Robert Burke, BA, CFPS
Hazardous Materials
Specialist/Instructor

Acknowledgments

This book would not have been possible if not for the hundreds of emergency response personnel and fellow instructors from various organizations whom I have had the distinct pleasure to know, teach with, and learn from during classes at the National Fire Academy and across the country. It was their inspiration, friendship, and support that gave me the desire and dedication to put together a book of this type. Many of the ideas for material presented here have come from the input of these responders, students, and instructors; for them, and the many lasting friendships that have developed over the years, I am grateful.

Special thanks to Jan Kuzma who was my first instructor at the National Five Academy in 1981 when I took the chemistry of hazardous materials class, which was then called HAZMAT I. Jan and I became friends over the years, and he imparted a great deal of wisdom upon me and helped me at many cross roads of my life. He is the one who encouraged me to write books. So, without his influence this book and my three others would likely not have happened. Thanks Jan, you will always be a special friend to me. I truly recognize and appreciate all you have done for me over the past 30 years.

In memory of Chris Waters

This book would not have been possible without the guidance and patience of Noel P. "Chris" Waters. Chris was the program chair for hazardous materials when I became a chemistry instructor at the National Fire Academy in 1988. He passed away in September 2008 at the age of 67 following a long illness.

Passing of a friend

Noel P. "Chris" Waters retired as New York City Fire Department (FDNY) Acting Lieutenant, Charter Member of Hazmat 1, and former chair of the Hazardous Materials Program at the National Fire Academy. Chris served with the United States Coast Guard for 4 years, and in 1965 he joined FDNY where he served for 22 years. While with FDNY, he was assigned to E-69, L-28, R-3, Battalion 16, and spent most of his time in Harlem and the South Bronx. During his off days, he was the fire marshal for the Town of Woodbury, New York, and a substitute teacher, secondary level. He was program chair for hazardous materials at the National Fire Academy for 7 years. Following his departure from the National Fire Academy, Chris became the Hazmat coordinator for Charleston County, South Carolina, and emergency management director.

Chris' students called him "Dr. Doom." His shaved head, "Fu Manchu" mustache, and physical presence may have precipitated his nickname. He was a big man with a passion for people and teaching. He loved to cook for family and friends and share his recipes with them. Upon meeting him, he seemed a little rough and imposing on the outside; however, he had a heart of gold and you couldn't have a better friend. I met Chris in Cheyenne, Wyoming, in 1988 while attending an instructor class for the National Fire Academy Chemistry of Hazardous Materials. During the 1990s, I had the opportunity to teach with Chris at the National Fire Academy. He was without a doubt the best instructor I have ever known. Teaching with him wasn't easy: How do you possibly follow the best? During mentoring sessions, he always told me to just be yourself and develop your own style, don't try to imitate anyone else. Well there was no way anyone could imitate Chris. He had a wonderful style that was like no other. His teaching was no nonsense, and he had a way of explaining difficult concepts that made them seem easy. He was animated and at times even boisterous, but you could tell he loved teaching and really cared about his students. After class, he would spend hours tutoring students to make sure they were able to pass his classes. He also helped me, and I am sure others, to become better instructors.

While he was the program chair for hazardous materials at the National Fire Academy, I taught a number of classes there in the early 1990s. Chris was always available to help you, giving out teaching pointers for different sections of the courses. I didn't always make his job easy. There was the "Toga" incident following the final exam in one class. Another class was involved in the abduction of a sheep that appeared at the Ott House. Then there was the "field trip" incident and another time the eggs at graduation. Following one of the incidents that caused him some grief with the academy administration, he said "Bobby" come over to my office. Chris always called me "Bobby." I never understood why. It was my actual given name, but I never used it and I don't think he knew it was my real name. Anyway, he sat me down in his office and asked why do these incidents with students only happen in your classes? My response was "you told me to develop my own style!" I knew that the students were pretty keyed up for the final exam in the courses and I just "encouraged" them to let off steam. He just rolled his eyes and gave me one of those stern looks and shook his head. My career would not be where it is today if not for the mentoring, tutoring, and friendship of Chris Waters. I will miss him.

Author

Robert A. Burke graduated high school in Dundee, Illinois. He earned an AA in fire protection technology from Community College of Baltimore County, Catonsville Campus, Maryland, and a BS in fire science from the University of Maryland. He has also completed graduate work at the University of Baltimore in public administration. He has attended numerous classes at the National Fire Academy in Emmitsburg, Maryland, and additional classes on firefighting, hazardous materials, and weapons of mass destruction at Oklahoma State University; Maryland Fire and Rescue Institute; Texas A & M University, College Station, Texas; the Center for Domestic Preparedness in Anniston, Alabama; and others.

Burke has over 30 years experience in the emergency services as a career and volunteer firefighter and has served as a lieutenant for the Anne Arundel County, Maryland Fire Department; an assistant fire chief for the Verdigris Fire Protection District in Claremore, Oklahoma; a deputy state fire marshal in the state of Nebraska; a private fire protection and hazardous materials consultant; an exercise and training officer for the Chemical Stockpile Emergency Preparedness Program (CSEPP) for the Maryland Emergency Management Agency; and retired as the fire marshal for the University of Maryland. He has served on several volunteer fire companies, including West Dundee, Illinois; Carpentersville, Illinois; Sierra Volunteer Fire Department, Chaves County, New Mexico; Ord, Nebraska; and Earleigh Heights Volunteer Fire Company in Severna Park, Maryland, which is a part of the Anne Arundel County, Maryland Fire Department.

Burke is a certified hazardous materials specialist (CFPS) by the National Fire Protection Association (NFPA) and is certified by the National Board on Fire Service Professional Qualifications as a fire instructor III, hazardous materials incident commander, fire inspector III, and plans examiner II. He served on the NFPA technical committee for NFPA 45 Laboratories Using Chemicals for ten years. He has been qualified as an expert witness for arson trials as well.

Burke has been an adjunct instructor at the National Fire Academy in Emmitsburg, Maryland, for hazardous materials, weapons of mass destruction, and fire protection curriculums for 25 years and at the Community College of Baltimore County, Catonsville Campus. He has had articles published in various fire service trade magazines for the past 25 years. He is currently a contributing editor for *Firehouse Magazine*, with a bimonthly column titled "Hazmat Studies." He has had numerous articles published in *Firehouse*, *Fire Chief*, *Fire Engineering*, and *Nebraska Smoke Eater* magazines. He has also been recognized as a subject matter specialist for hazardous materials and has been interviewed by newspapers, radio, and television about incidents that have occurred in local communities, including Fox Television in New York City during a tank farm fire on Staten Island.

Burke has been a presenter at Firehouse Expo in Baltimore on numerous occasions. He has given presentations at the EPA Region III SERC/LEPC Conference in Norfolk, Virginia, in November 1994 and at the 1996 Environmental and Industrial Fire Safety Seminar, Baltimore, Maryland, on North America Emergency Response Guidebook (NAERG). He was a speaker at the 1996 International Hazardous Materials Spills Conference held on June 26, 1996, at New Orleans, Louisiana; at the 5th Annual 1996 Environmental and Industrial Fire Safety Seminar, held at Baltimore, Maryland, sponsored by Baltimore City Fire Department; and at LEPC, Instructor for Hazmat Chemistry, held in August 1999 at Hazmat Expo 2000 in Las Vegas, Nevada. He also gave a keynote presentation at the Western Canadian Hazardous Materials Symposium at Saskatoon, Saskatchewan, Canada, in 2008.

Burke has developed several CD-ROM-based training programs, including the Emergency Response Guidebook, Hazardous Materials and Terrorism Awareness for Dispatchers and 911 Operators, Hazardous Materials and Terrorism Awareness for Law Enforcement, Chemistry of Hazardous Materials Course, Chemistry of Hazardous Materials Refresher, Understanding Ethanol, Understanding Liquefied Petroleum Gases, Understanding Cryogenic Liquids, Understanding Chlorine, and Understanding Anhydrous Ammonia. He has also developed the "Burke Placard Hazard Chart" included in the appendix of this book. He has published three additional books titled *Counterterrorism for Emergency Responders*, *Fire Protection: Systems and Response*, and *Hazmat Teams Across America*. He can be reached via e-mail at robert.burke@windstream.net or through his website www.hazardousmaterialspage.com.

chapter one

Introduction

Mention the word *chemistry* to the average firefighter or other emergency responder and it strikes fear in their heart. Maybe that is a bit overstated, but the idea of chemistry certainly creates a great deal of anxiety to some. Many emergency personnel do not want any part of chemistry, often because they feel it is too difficult or complicated. Difficulty of the subject of chemistry is not the material itself; it is in convincing emergency response personnel that they need to study the concepts to make them better hazardous materials responders at all levels.

Firefighters learn about fire behavior as part of their basic training in order to have a better understanding of fire. Fire behavior is part of the chemistry of fire, which is really a chemical chain reaction. Understanding fire and how it behaves helps firefighters extinguish fires safely and effectively. Emergency medical personnel take courses to learn how to care for sick and injured patients. Sometimes, drugs are used by paramedics as part of the initial treatment. Drugs are chemicals and can be hazardous if they are not handled properly. Emergency medical personnel learn about drugs and treatment techniques in order to effectively treat their patients. Law enforcement personnel take courses in criminal justice to better prepare for their job of enforcing the law. All emergency response personnel receive some type of basic training to better prepare for their jobs. Responders are called upon daily to deal with incidents of all kinds that often involve hazardous chemicals. In order for personnel to better understand the hazardous chemicals they face on a daily basis, an understanding of basic chemistry principles and terminology is as essential as other emergency response training. This book presents emergency response personnel with a view of chemistry as it applies to the hazardous materials that may be encountered in any emergency response. Some of the concepts presented may bend the rules of chemistry a bit. However, the purpose of this book is not to educate chemists, but rather to teach response personnel about basic chemistry concepts in a format that most responders, regardless of educational background, can understand. Concepts taught will work in the street application of chemistry when dealing with hazardous materials. This course is sometimes referred to as "Street Chemistry." This book also presents firefighters, police, and emergency medical services (EMS) personnel with some basic tools to assist them in understanding hazardous materials and their behavior. This information may help keep them from being injured or killed at the scene of a hazardous materials incident.

In an effort to understand chemistry or to make chemistry more understandable, let us look at chemistry as it affects us on a daily basis in life. Chemistry is all around us. Everything on Earth, including the human body (Figure 1.1), is made up of one of the chemical elements or a combination of those elements from the periodic table of elements. In 1869, Russian chemist Dmitri Mendeleev put forth the theory of periodicity. Through his work, the first periodic table of elements was developed. His efforts were the beginning of the understanding of the concept of elements, which led to the understanding

Oxygen	65%	Sulfur	
Carbon	18%	Iron	
Hydrogen	10%	Sodium	
Nitrogen	0.3%	Zinc	1%
Calcium	1.5%	Magnesium	
Phosphorus	0.1%	Silicon	
		Potassium	

Figure 1.1 Chemical makeup of the human body.

of compounds, which make up many of the hazardous materials we face today. Not all elements or compounds are overly hazardous. We use many of the nonhazardous compounds on a daily basis. Plastics, fibers, foods and additives, household goods, recreational equipment, and the list goes on and on. This book will attempt to present those elements and compounds that may be encountered in emergency response that are hazardous and under certain conditions present a hazard to the public and response personnel. Having a better understanding of the physical and chemical characteristics of those materials should help response personnel to better protect themselves and the public.

Chemicals and compounds have been known to exist for centuries. Emergency responders have dealt with them for many years and just referred to them as chemicals or their names if known such as gasoline, chlorine, and ammonia. The modern-day coinage of the term "hazardous material" occurred in the mid-1970s when the U.S. Department of Transportation (DOT) established a definition of hazardous material. DOT began the first major regulation of hazardous materials in transportation, including a hazard class and placard and label system for the identification of hazardous materials. DOT identifies the four modes of transportation that it regulates as rail, highway, air, and water and pipeline. During this same time period, the first DOT *Emergency Response Guidebook* (ERG) was developed (Figure 1.2).

An explosion that occurred in Marshalls Creek, Pennsylvania, on June 25, 1964, that killed three firefighters and injured two others led to the implementation of the DOT placarding and labeling system and the development of the ERG. A truck driver hauling approximately 15 tons of dynamite and the blasting agent nitro-carbo-nitrate (NCN) disconnected the trailer and drove the tractor to a service station. While he was gone, the tires on the trailer caught fire. Another passing tractor-trailer driver reported the fire to the Marshalls Creek Fire Department. The driver reported that there were no markings on the trailer. Three fire engines responded at 4:08 a.m., and an attack line was pulled to fight the blaze. As the firefighters approached the trailer, a detonation occurred creating a crater 10 ft deep and 40 ft wide. The fire had reached the explosive cargo, and the resulting explosion killed six people, including three firefighters, the truck driver who reported the fire, and two bystanders. Property damage was over $600,000, including all three of the fire engines. Responding firefighters had no idea what the cargo of the burning trailer was.

Marshalls Creek firefighters killed in dynamite explosion
F. Earl Miller, 50
Leonard R. Mosier, 48
Edward F. Hines, 48

The ERG is updated and issued approximately every 4 years. It is designed to assist first responders in dealing with hazardous materials releases during the initial phase of an incident (Photo 1.1). Information provided in the book is very generic and does not

Figure 1.2 DOT *Emergency Response Guidebook* 1977–2012.

Photo 1.1 The *Emergency Response Guidebook* along with the DOT placard and label system provides first responders with information to help them identify hazardous materials at an incident scene.

provide the detail necessary to mitigate the incident in most cases. The book is designed to keep response personnel safe and provide them with some initial guidelines to protect themselves and the public from harm.

As other federal agencies began developing regulations dealing with hazardous materials storage and use, different names were also created. The U.S. Occupational Safety and Health Administration (OSHA) and the U.S. Environmental Protection Agency (EPA) both refer to hazardous materials as "hazardous substances." The EPA also regulates chemicals that no longer have a commercial value. When chemicals are no longer useful for the purpose they were intended, they become hazardous waste. Hazardous waste is regulated in the workplaces where it is generated, during transportation to a disposal site, and when it is disposed of. For example, gasoline, when transported, is a *hazardous material* regulated by DOT. When a gasoline tanker off-loads gasoline into an underground storage tank at a gasoline station, it becomes a *hazardous substance* regulated by OSHA and EPA. If some of the gasoline was spilled on the ground during the off-loading, it would become *hazardous waste*, regulated by OSHA, EPA, and DOT. There are three different names for the same gasoline, depending on whether it was transported, in fixed storage, or spilled. For the purposes of this book, we will use the term "hazardous material" interchangeably with all other agency terminology.

On December 3, 1984, Bhopal, India, experienced a release of approximately 40 metric tons of methyl isocyanate (MIC) at the Union Carbide pesticide plant (Photo 1.2). Over 100,000 were injured and 3,000 people were killed, and many more have and continue to die from the long-term effects. The accident occurred around 12:40 a.m. local time,

Photo 1.2 On December 3, 1984, Bhopal, India, experienced a release of approximately 40 metric tons of MIC at the Union Carbide pesticide plant.

when most of the victims were sleeping. The dead included large numbers of infants, children, and older men and women. These age groups are often adversely impacted by toxic exposures. It was this incident that led to the Emergency Planning and Community Right-To-Know Act (EPCRA) of 1986 in the United States. Following the incident in Bhopal, the U.S. Congress was concerned that such an incident could happen here. MIC has been released on several occasions from the Union Carbide plant in Institute, West Virginia, shortly after Bhopal and as recently as 1996. Fortunately, these releases did not affect the surrounding community they certainly could have. Congress was also concerned with the level of preparedness and training available to deal with an incident of the magnitude of Bhopal. In 1986, Congress passed the EPCRA, sometimes referred to as the Superfund Amendments and Reauthorization Act (SARA). With the passage of this important legislation, the federal government for the first time mandated training and competency for emergency responders to hazardous materials releases. Congress mandated that the EPA also create a list of extremely hazardous substances, most of which are poisons, that would require reporting under EPCRA by companies manufacturing, storing, or using them. EPCRA created the State Emergency Response Commission (SERC) and the Local Emergency Planning Committee (LEPC) to facilitate the reporting and oversight process. Also part of the legislation was the requirement that information about extremely hazardous substances be made available to the public so they are aware of what types of hazardous materials may be found in their communities.

According to the EPA, 85 million Americans live, work, and play within a 5 mile radius of 66,000 facilities handling regulated amounts of high-hazard chemicals. Regulations also require that local emergency responders, particularly the fire department, be given access to chemical facilities for the purposes of preplanning for emergencies. The EPCRA also called for the distribution of Material Safety Data Sheets (MSDS) to the local fire department.

The National Fire Protection Association (NFPA) also has developed standards that set forth competencies that emergency responders should be able to display concerning hazardous materials response. Among the competency requirements are certain chemical and physical characteristics of hazardous materials of which responders must have knowledge. From the requirements of EPCRA, and OSHA and NFPA training competencies, the study of the chemistry of hazardous materials, sometimes called "street chemistry" has evolved. You do not need to be a chemist to safely and effectively respond to hazardous materials incidents. The information contained in this book is intended to cover the chemistry requirements of all levels of EPCRA, OSHA, EPA, DOT, and NFPA hazardous materials regulation and response concepts and well beyond.

Hazardous materials statistics

The Chemical Abstract Service (CAS) lists over 295,000 regulated chemicals, and the number increases each year. The DOT regulates over 3600 hazardous materials in transportation, as listed in 49 CFR. Chemicals are also listed by the DOT in the *ERG*. While specific information on hazardous materials in the ERG is limited, the current edition uses chemistry terminology as a heading in each of the orange guide pages. OSHA regulates the occupational exposure of over 600 hazardous substances. Other lists of chemicals are compiled and regulated by various governmental regulatory agencies, such as the EPA. EPA has several listings of hazardous materials, depending on whether they are in storage or hazardous wastes. The Resource Conservation and Recovery Act (RCRA) deals with hazardous wastes exhibiting the characteristics of ignitibility, corrosivity, and reactivity.

Regulations are found in 40 CFR 261.33. The Comprehensive Environmental Response, Compensation, and Liability Act (CERCLA) of 1980 lists hazardous substances in 40 CFR 302, Table 302.4. The Clean Air Act (CAA).

The National Institute for Occupational Safety and Health (NIOSH) also lists hazardous materials that may pose a health hazard to emergency responders. They have published the *NIOSH Pocket Guide to Chemical Hazards*, which can be obtained free from the agency.

According to the Association of American Railroads (AAR) in 2011, 1.7 million carloads of hazardous materials were transported by rail with 99.99% reaching their destination without an accident, over 70% of those were transported in tank cars. Of those, approximately 75,000 carloads are toxic inhalation hazard (TIH) materials like chlorine and anhydrous ammonia (Photo 1.3).

The DOT reports that there were 12 deaths, 162 injuries, and damage costs of over $123 million from 15,007 incidents involving hazardous materials in the year 2011. This is an increase of over 209 incidents from 2010; the number of deaths increased and injuries decreased during the same period. The incidents are broken down by mode of transportation: highway (Photo 1.4): 12,795, air: 1,398, and railway: 743. Deaths were highway 11 and railway 1. Injuries were highway 127, railway 20, air 7, and waterway 8. Most deaths and injuries occurred to transportation workers (employees). Deaths, injuries, and damage occurred most often in incidents involving corrosive and flammable liquids. The top 10 commodities released during the incidents were all flammable or corrosive liquids.

Photo 1.3 Over 70% of all hazardous materials on the rail are shipped in tank cars.

Photo 1.4 Almost 50% of all trucks on the highways are transporting hazardous materials.

DOT/UN hazard classes of hazardous materials

Chemicals regulated by the DOT are listed in the hazardous materials tables in 19 CFR 100–199. This book will use the nine DOT/United Nations (UN) hazard classes to group hazardous materials. The hazard classes are shown in Table 1.1. UN hazard classes recognized by the DOT are as follows: Class 1, Explosives; Class 2, Compressed Gases; Class 3, Flammable Liquids; Class 4, Flammable Solids; Class 5, Oxidizers; Class 6, Poisons; Class 7, Radioactives; Class 8, Corrosives; and Class 9, Miscellaneous Hazardous Materials. Class 1 Explosives are subdivided into six subclasses: 1.1 through 1.6. Class 2 Compressed Gases have three subclasses: 2.1, Flammable; 2.2, Non-flammable; and 2.3, Poison. Class 4 Flammable Solids have three subclasses: 4.1, Flammable solid; 4.2, Spontaneously combustible; and 4.3, Dangerous when wet. Class 5 Oxidizers have two subclasses: 5.1, Oxidizers

Table 1.1 DOT/UN Hazard Classes

Class 1	Explosives
Class 2	Compressed gases
Class 3	Flammable liquids
Class 4	Flammable solids
Class 5	Oxidizers
Class 6	Poisons
Class 7	Radioactives
Class 8	Corrosives
Class 9	Miscellaneous hazmat

and 5.2, Organic peroxide. Class 6 Poisons have two subclasses and some special classifications of placards. Subclass 6.1 is poisons that are liquids or solids. Subclass 6.2 is infectious substances. There are also placards in Class 6 for *keep away from foodstuffs* and *marine
pollutant*. Class 7 Radioactives have no subclasses, but the placard is used only when materials bearing the Radioactive III label are shipped. Class 8 Corrosive materials do not have
any subclasses; however, there are two distinctive chemicals in Class 8: acids and bases.
Class 9 Miscellaneous Hazardous Materials do not have any subclasses.

Each hazard class has associated placards and labels identifying the hazards of the
class during transportation. Each hazard class and associated placard has a color, which
indicates a particular hazard (Figures 1.3–1.4). Hazardous materials may have more than
one hazard. It is important to note that the placard on a transport vehicle depicts only the
most severe hazard of a material as determined by DOT hazard class definitions. When
a material has more than one hazard, the DOT prioritizes the hazard that will be placarded. These hazards are listed by the DOT in 49 CFR 173.2a (Table 1.2) to determine which
hazard will be assigned to a particular material when the material has multiple hazards.
Almost every hazardous material has more than one hazard. As an emergency responder,
you must be familiar with other potential, and often hidden, hazards that chemicals may
present. The Burke placard hazard chart shows the potential hidden hazards of the nine
hazard classes (Figure 1.5). Across the top of the chart are all the potential hazards a chemical could have that would affect emergency responders. Down the left side are all of the
colors representing the UN/DOT hazard classes. An X is used to identify the DOT hazard
designated for the material and the color of the placard, which will be on the shipment.
An asterisk (*) is used to identify all of the other potential hazards of the materials. That
does not mean that a particular chemical has all of the hazards, but until you are able to
obtain additional information, precautions must be taken for each. For example, some corrosive materials are classified as oxidizers, such as perchloric acid above 50% concentration. Perchloric acid above 50% concentration is placarded as an oxidizer. Perchloric acid is
also a strong corrosive material, but will not be placarded corrosive.

Do not focus only on the hazard depicted by the placard. Thoroughly examine all hazardous materials encountered to determine all physical and chemical characteristics and
hazards associated with the chemical. An incident occurred in Kansas City, Missouri, in
which six firefighters were killed fighting a fire involving commercial-grade ammonium
nitrate, a 1.5 Blasting Agent under the DOT hazard class system. The storage containers
on a construction site were not marked as to the type of hazardous materials inside. As a
result of this tragic loss of firefighters' lives, OSHA adopted the DOT placarding and labeling system to be continued in use for fixed storage until the chemicals are used up. OSHA
lists and regulates chemicals that are considered potentially dangerous in the workplace
as part of worker right-to-know regulations.

NFPA 704 marking system

The NFPA has developed a fixed-facility marking system to designate general hazards
of chemicals, referred to as the NFPA 704 marking system (Photo 1.5). It is designed to
warn emergency responders of the presence of hazardous materials in fixed facilities
and to give them some general information about the hazards of the materials. NFPA
704 placards do not identify any specific chemicals or DOT hazard classes other than
flammable. The system utilizes a diamond-shaped placard with four colored sections
(Figure 1.6). Each section indicates a particular hazard: the blue section indicates a health
hazard, the red section flammability, and the yellow section reactivity. The white section

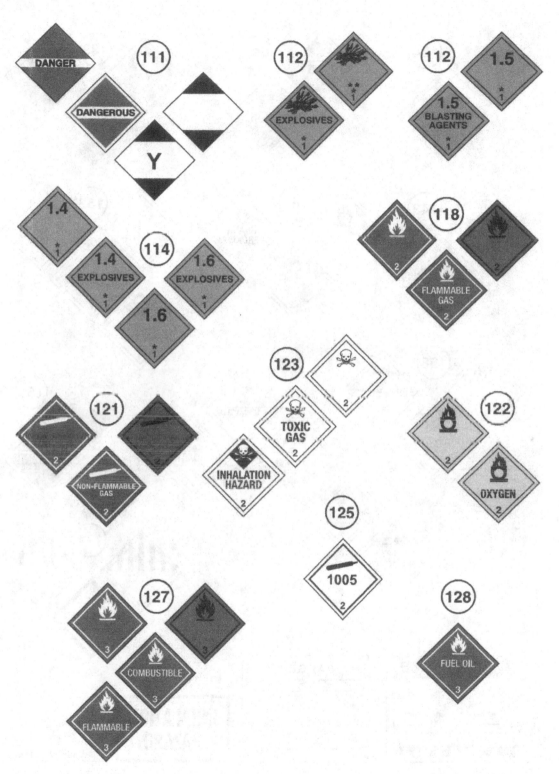

Figure 1.3 DOT placards and labels.

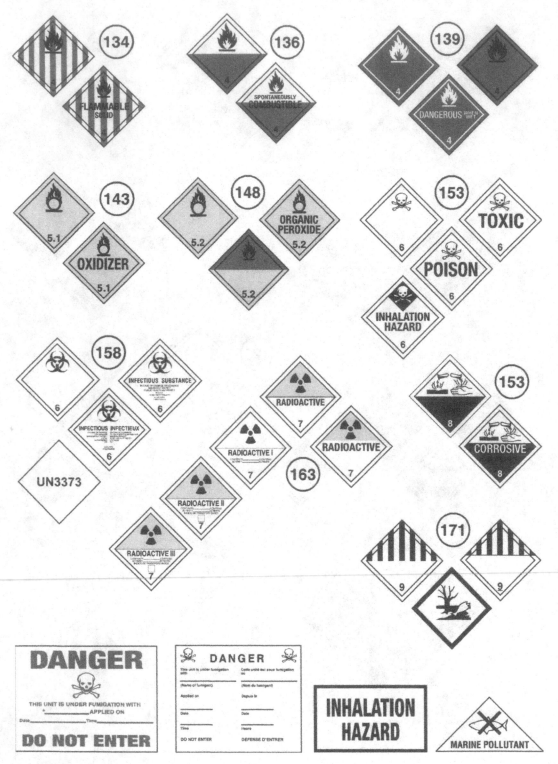

Figure 1.4 DOT placards and labels.

Table 1.2 DOT Classification of Materials with
More Than One Hazard

1. Radioactive
2. Division 2.3: Poison gas
3. Division 2.1: Flammable gas
4. Division 2.2: Nonflammable gas
5. Division 6.1: Poisonous liquid, inhalation hazard

The above chart does not apply to the following:

Class 1 explosives, 5.2 organic peroxides, Division 6.2 infectious substances, and wetted explosives such as picric acid.

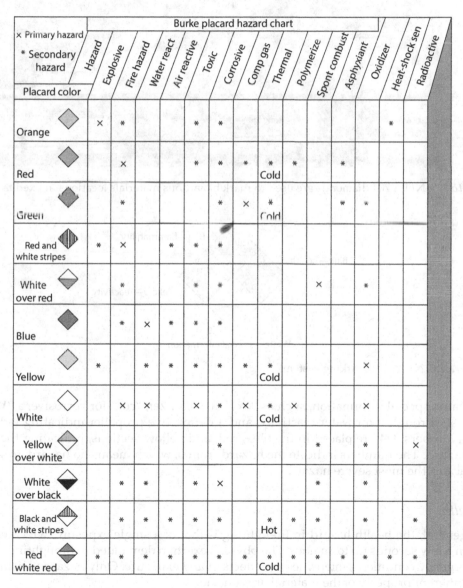

Figure 1.5 Burke placard hazard chart.

Photo 1.5 NFPA 704 diamonds are used to mark hazardous materials locations in fixed facilities.

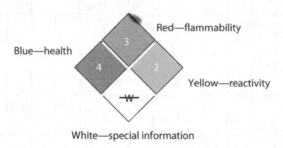

Figure 1.6 NFPA 704 marking system.

contains special information, such as "oxy" for oxidizer, "cor" for corrosive, a "W" with a slash through it for water reactivity, and a radioactive propeller indicating radioactivity. Numbers 0–4 are placed in the blue, red, and yellow sections, indicating the degree of hazard. The numbers indicate the hazard from 0, which means no hazard, to 4, which indicates the most severe hazard.

Health

In general, the health hazard in firefighting is that of a single exposure, which can vary from a few seconds up to an hour. The physical exertion demanded in firefighting, or other emergency conditions, intensifies the effects of any exposure. Only hazards arising out of an inherent property of the material are considered.

The following explanations of the four hazards are based upon protective equipment normally used by firefighters.

4. Materials too dangerous to health to expose firefighters. A few whiffs of the vapor could cause death, or the vapor or liquid could be fatal upon penetrating the firefighter's normal full-protective clothing. The normal full-protective clothing and breathing apparatus available to the average fire department will not provide adequate protection against inhalation or skin contact with these materials.
3. Materials extremely hazardous to health, but areas may be entered with extreme care. Full-protective clothing, including self-contained breathing apparatus, coat, pants, gloves, boots, and bands around legs, arms, and waist should be provided. No skin surface should be exposed.
2. Materials hazardous to health, but areas may be entered freely with full-faced mask and self-contained breathing apparatus that provides eye protection.
1. Materials only slightly hazardous to health. It may be desirable to wear self-contained breathing apparatus.
0. Materials that on exposure under fire conditions would offer no hazard beyond that of ordinary combustible material.

Flammability

Susceptibility to burning is the basis for assigning degrees within this category. The method of attacking the fire is influenced by this susceptibility factor.

4. Very flammable gases or very volatile flammable liquids. Shut off flow and keep cooling water streams on exposed tanks or containers.
3. Materials that can be ignited under almost all normal temperature conditions. Water may be ineffective because of the low flash point.
2. Materials that must be moderately heated before ignition will occur. Water spray may be used to extinguish the fire because the material can be cooled below its flash point.
1. Materials that must be preheated before ignition can occur. Water may cause frothing if it gets below the surface of the liquid and turns to steam. However, water fog gently applied to the surface will cause a frothing that will extinguish the fire.
0. Materials that will not burn.

Reactivity (stability)

The assignment of degrees in the reactivity category is based upon the susceptibility of materials to release energy either by themselves or in combination with water. Fire exposure is one of the factors considered along with conditions of shock and pressure.

4. Materials that (in themselves) are readily capable of detonation or of explosive decomposition or explosive reaction at normal temperatures and pressures. Includes materials that are sensitive to mechanical or localized thermal shock. If a chemical with this hazard rating is in an advanced or massive fire, the area should be evacuated.
3. Materials that (in themselves) are capable of detonation or of explosive decomposition or of explosive reaction, but which require a strong initiating source or which must

be heated under confinement before initiation. Includes materials that are sensitive to thermal or mechanical shock at elevated temperatures and pressures or that react explosively with water without requiring heat or confinement. Firefighting should be done from a location protected from the effects of an explosion.

2. Materials that (in themselves) are normally unstable and readily undergo violent chemical change, but do not detonate. Includes materials that can undergo chemical change with rapid release of energy at normal temperatures and pressures or that can undergo violent chemical change at elevated temperatures and pressures. Also includes those materials that may react violently with water or that may form potentially explosive mixtures with water. In advanced or massive fires, firefighting should be done from a safe distance or from a protected location.

1. Materials that (in themselves) are normally stable, but which may become unstable at elevated temperatures and pressures or which may react with water with some release of energy, but not violently. Caution must be used in approaching the fire and applying water.

0. Materials that (in themselves) are normally stable, even under fire-exposure conditions, and that are not reactive with water. Normal firefighting procedures may be used.*

NFPA 704 diamond placards are placed on the outside of buildings, storage tanks, storage sheds, and doors leading to areas where hazardous materials are present. This information is provided as a type of "stop sign" for response personnel. It says hazardous materials are present, but responders still have to obtain specific information and the identity of the chemicals. Chemicals must still be identified for specific hazards before responders can enter. NFPA 704 designations are not available for all hazardous materials. Those chemicals in this book that have been assigned NFPA 704 designations will have the numerical designation listed with the chemical characteristics. These designations are found in the *NFPA Fire Protection Guide to Hazardous Materials*.

Chemical characteristic listings and incidents

Chapters 3 through 11 will present the nine DOT hazard classes and the types of chemicals found in each class. Specific chemicals are listed, along with their physical and chemical characteristics. Examples are provided to illustrate the similarities of some chemicals in the same families and the differences of chemicals in other families. These examples are also provided because they are among the most common hazardous materials encountered by emergency responders. When reading through specific chemicals, note the multiple hazards, many of which are not placarded for the hazard class, and others that are not assigned to a hazard class at all. Along with the listings of chemicals, there are excerpts from incidents that have occurred involving many of the chemicals listed. These are provided to show the dangers of the hazardous materials and how the physical and chemical characteristics contributed to the incident outcome.

---

* Copyright 2006, National Fire Protection Association, Quincy, MA 02269. This warning system is intended to be interpreted and applied only by properly trained individuals to identify fire, health, and reactivity hazards of chemicals. The user is referred to a certain limited number of chemicals with recommended classifications in NFPA 49 and NFPA 325, which would be used as a guideline only. Whether the chemicals are classified by NFPA or not, anyone using the 704 systems to classify chemicals does so at their own risk.

Chemical and physical characteristics: Training competencies

The following discussion of the competencies required both by OSHA 1910.120 and NFPA 472 compares the level of training and the level of knowledge of chemistry that is necessary for each level. OSHA requires emergency responders to demonstrate competency at each level for which they are required to perform. OSHA identifies five levels of training for emergency responders to hazardous materials incidents. They are as follows: awareness, operations, technician, specialist, and the incident commander. OSHA regulations are not nearly as specific with competency requirements as is NFPA 472. Excerpts from the OSHA regulation are included in this section for review.

OSHA 1910.120

"Competent" means possessing the skills, knowledge, experience, and judgment to perform assigned tasks or activities satisfactorily as determined by the employer.

"Demonstration" means the showing by actual use of equipment or procedures.

"Hazardous substance" means any substance designated or listed under A through D of this definition, exposure to which results, or may result, in adverse effects on the health or safety of employees:

A. Any substance defined under section 101(14) of CERCLA.
B. Any biologic agent and other disease-causing agent, which, after release into the environment and upon exposure, ingestion, inhalation, or assimilation into any person, either directly from the environment or indirectly by ingestion through food chains, will or may reasonably be anticipated to cause, death, disease, behavioral abnormalities, cancer, genetic mutation, physiological malfunctions (including malfunctions in reproduction), or physical deformations in such persons or their offspring.
C. Any substance listed by the U.S. Department of Transportation as hazardous materials under 49 CFR 172.101 and appendices.
D. Hazardous waste as herein defined.

"Hazardous waste" means

A. A waste or combination of wastes as defined in 40 CFR 261.3
B. Those substances defined as hazardous wastes in 49 CFR 171.8

Preliminary evaluation: A preliminary evaluation of a site's characteristics shall be performed prior to site entry by a qualified person in order to aid in the selection of appropriate employee protection methods prior to site entry. Immediately, after initial site entry, a more detailed evaluation of the site's specific characteristics shall be performed by a qualified person in order to further identify existing site hazards and to further aid in the selection of the appropriate engineering controls and personal protective equipment (PPE) for the tasks to be performed.

Hazard identification: All suspected conditions that may pose inhalation or skin absorption hazards that are immediately dangerous to life or health (IDLH) or other conditions that may cause death or serious harm shall be identified during the preliminary survey and evaluated during the detailed survey. Examples of such hazards include, but are not limited to, confined space entry, potentially explosive or flammable situations, visible vapor clouds, or areas where biological indicators, such as dead animals or vegetation, are located.

Based upon the results of the preliminary site evaluation, an ensemble of PPE shall be selected and used during initial site entry, which will provide protection to a level of exposure below permissible exposure limits and published exposure levels for known or suspected hazardous substances and health hazards, and which will provide protection against other known and suspected hazards identified during the preliminary site evaluation. If there is no permissible exposure limit or published exposure level, the employer may use other published studies and information as a guide to appropriate PPE.

Risk identification: Once the presence and concentrations of specific hazardous substances and health hazards have been established, the risks associated with these substances shall be identified. Employees who will be working on the site shall be informed of any risks that have been identified. In situations covered by the Hazard Communication Standard, 29 CFR 1910.1200, training required by that standard need not be duplicated.

NOTE TO (C)(7): Risks to consider include, but are not limited to

a. Exposures exceeding the permissible exposure limits and published exposure levels
b. IDLH concentrations
c. Potential skin absorption and irritation sources
d. Potential eye irritation sources
e. Explosion sensitivity and flammability ranges
f. Oxygen deficiency

Operations level
(q)(6)(ii)(C)
Understand basic hazardous materials terms.

Technician
(q)(6)(iii)(I)
Understand basic chemical and toxicological terminology and behavior.

Specialist
(q)(6)(iv)(I)
Understand chemical, radiological, and toxicological terminology and behavior.
Recommendations from Appendix 1910.120, not mandatory.

Technical knowledge

1. Types of potential exposures to chemical, biological, and radiological hazards; types of human responses to these hazards and recognition of those responses; principles of toxicology and information about acute and chronic hazards; health and safety considerations of new technology
2. Fundamentals of chemical hazards, including but not limited to, vapor pressure, boiling points, flash points, pH, and other physical and chemical properties
3. Fire and explosion hazards of chemicals
4. General safety hazards, such as but not limited to, electrical hazards, powered equipment hazards, motor vehicle hazards, walking- and working-surface hazards, excavation hazards, and hazards associated with working in hot and cold temperature extremes

Awareness

First responders at the awareness level are not required to have a great deal of knowledge about the chemistry of hazardous materials (Photo 1.6). However, there are many shortcuts in street chemistry that could assist first responders in recognizing hazards through family characteristics. After all, first responders have the least amount of resources at their disposal when arriving on the scene of a hazardous materials release. The knowledge they bring with them through previous training and experience may be all they have to assist in the recognition of the presence of hazardous materials. Recognition of the existence of hazardous materials is the most important thing any emergency responder can do upon arrival at an incident scene. First responders at the awareness level have four basic responsibilities: recognition, notification, isolation, and protection. A basic knowledge of street chemistry can assist first responders in carrying out their responsibilities safely.

Operations

Operations-level personnel are required to have awareness training as well as an additional 8 h of operations-level training (Photo 1.7). According to OSHA and NFPA, they must be able to predict the behavior of hazardous materials and their containers. Knowledge of the physical and chemical characteristics of hazardous materials is of great benefit in making those predictions. Responders at this level are expected to

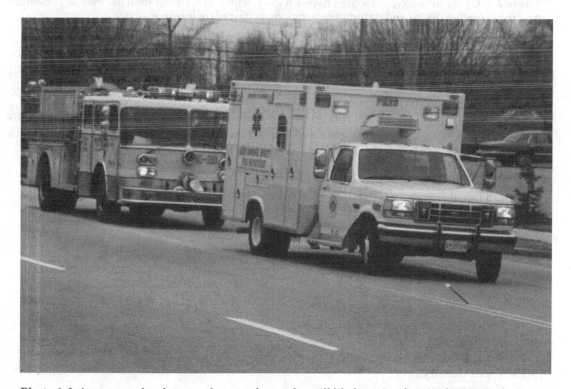

Photo 1.6 Awareness-level responders are those who will likely arrive first on the scene of a potential hazardous materials incident and must recognize the potential dangers, protect themselves and the public, and make proper notifications for assistance.

Photo 1.7 Operations-level personnel have a higher degree of training than awareness personnel and may conduct diking and damming operations as part of their response duties.

obtain information from MSDS. Much of the information on the MSDS is in the form of physical and chemical terminology. Once again, knowledge of street chemistry will benefit operations-level personnel in researching information needed to deal safely with hazardous materials.

Operations-level personnel should be able to match the following chemical and physical properties with their significance and impact on the behavior of the container and its contents. These competencies are recommended in NFPA 472.

1. Boiling point
2. Chemical reactivity
3. Corrosivity (pH)
4. Flammable range
5. Flash point
6. Ignition temperature
7. Particle size
8. Persistence
9. Physical state (solid, liquid, gas)
10. Radiation (ionizing and nonionizing)
11. Specific gravity
12. Toxic products of combustion
13. Vapor density
14. Vapor pressure
15. Water solubility

Also required of operations-level personnel is to be able to identify the health hazards of the following terms. These competencies are recommended in NFPA 472.

1. Alpha, beta, gamma, and neutron radiation
2. Asphyxiant
3. Carcinogen
4. Convulsant
5. Corrosive
6. Highly toxic
7. Irritant
8. Sensitizer, allergen
9. Target organ effects
10. Toxic

Operations-level personnel should be able to identify the UN/DOT hazard class and division of the following military/terrorist agents. These competencies are recommended in NFPA 472.

1. Blood agents
2. Biological agents and biological toxins
3. Choking agents
4. Irritants (riot control agents)
5. Nerve agents
6. Radiological materials
7. Vesicants (blister agents)

Technician

Technician-level responders are members of hazardous materials response teams. They will enter the "hot zone" and work in close proximity to hazardous materials (Photo 1.8). Technician-level personnel should have an extensive knowledge of hazardous materials chemistry. While OSHA requires only 24 h of training above the operations level, it is difficult to adequately prepare personnel with just 24 h of training. Most hazardous materials teams have developed their own training standards for team membership, and many require hundreds of hours of training. Part of that training usually includes a chemistry of hazardous materials course ranging from 3 days to 2 weeks in length. Technicians should have an in-depth knowledge of explosives; cryogenic liquids; compressed gases; liquefied gases; flammable solids, liquids, and gases; poisons; oxidizers; and corrosives. Their understanding of chemistry is critical in determining the proper type of monitoring instruments, protective clothing, decontamination, and tactical options.

Terminology required as a minimum level of knowledge of the chemistry of hazardous materials is in the following list. These competencies are recommended in NFPA 472.

1. Acid, caustic
2. Air reactivity
3. Autorefrigeration
4. Biological agents and biological toxins

Photo 1.8 Technician-level personnel are trained and equipped to wear chemical protective cloth-ing and perform rescue and mitigation operations on the incident scene.

5. Blood agents
6. Boiling point
7. Catalyst
8. Chemical change
9. Chemical interactions
10. Compound, mixture
11. Concentration
12. Critical temperature and pressure
13. Dissociation and corrosivity
14. Dose
15. Dose response
16. Expansion ratio
17. Fire point
18. Flammable range
19. Flash point
20. Half-life
21. Halogenated hydrocarbon
22. Ignition (autoignition) temperature
23. Inhibitor
24. Instability
25. Ionic and covalent compounds

26. Irritants
27. Maximum safe storage temperature (MSST)
28. Melting point and freezing point
29. Miscibility
30. Nerve agents
31. Organic and inorganic
32. Oxidation potential
33. Persistence
34. pH
35. Physical change
36. Physical state (solid, liquid, gas)
37. Polymerization
38. Radioactivity
39. Reactivity
40. Riot control agents
41. Saturated, unsaturated (straight and branched), and aromatic hydrocarbons
42. Self-accelerating decomposition temperature (SADT)
43. Solubility
44. Solution and slurry
45. Specific gravity
46. Strength
47. Sublimation
48. Temperature of product
49. Toxic products of combustion
50. Vapor density
51. Vapor pressure
52. Vesicants (blister agents)
53. Viscosity
54. Volatility

Also required of technician personnel is the knowledge of certain toxicological terms. Those include the following, which are recommended in NFPA 472:

1. Counts per minute (cpm) and kilocounts per minute (kcpm)
2. Immediately dangerous to life and health (IDLH) value
3. Incubation period
4. Infectious dose
5. Lethal concentration (LC_{50})
6. Lethal dose (LD_{50})
7. Parts per billion (ppb)
8. Parts per million (ppm)
9. Permissible exposure limit (PEL)
10. Radiation absorbed dose (rad)
11. Roentgen equivalent man (rem), millirem (mrem), microrem (μrem)
12. Threshold limit value (TLV)
13. Short term exposure limit (STEL)
14. Time weighted average (TLV-TWA)

Photo 1.9 Incident commanders need to have training that allows them to make educated decisions in order to effectively and safely manage the hazardous materials incident scene.

Incident commander

The incident commander requires training to a minimum of the operations level and competencies outlined in NFPA 472 for the incident commander (Photo 1.9). Toxicology and radiological terms the incident commander should be able to describe are the same as those required for the technician-level personnel listed previously. In addition, the incident commander should be aware of the routes of exposure for hazardous materials, acute and delayed toxicity, local and systemic effects, dose response, and synergistic effects. Detailed information about the training competencies for responders to hazardous materials incidents can be found in NFPA Standard 472. Additional information concerning hazardous materials response can be found in NFPA 471. Competencies for emergency medical personnel are outlined in NFPA 473. Many of the chemical and physical characteristics required of operations- and technician-level personnel are also required of EMS responders. However, the EMS responder is more concerned about the effect the hazardous material will have on the victims and how they can protect themselves while dealing with the treatment of the patients than mitigating the incident. This street chemistry book covers all of the chemistry requirements of NFPA 472, OSHA 1910.120, and much more to prepare emergency responders at each level to effectively deal with hazardous materials released into the atmosphere.

Review questions

1.1 Which of the following people developed the first periodic table of elements?
  A. Davy
  B. Mendeleev
  C. Plato
  D. Bush
1.2 Of the following federal agencies, which developed the term *hazardous material*?
  A. EPA
  B. OSHA
  C. DOT
  D. FEMA
1.3 Which law was passed by Congress following the Bhopal, India, incident?
  A. CERCLA
  B. Clean Water Act
  C. Occupational Safety Act
  D. EPCRA
1.4 Which emergency responders would benefit from the study of chemistry?
  A. Firefighters
  B. Police officers
  C. EMS
  D. All of the above
1.5 How many hazard classes does the DOT recognize?
  A. 7
  B. 10
  C. 9
  D. 8
1.6 DOT commonly placards which of the following?
  A. The most severe hazard
  B. The most common hazard
  C. Hazards most dangerous to responders
  D. None of the above
1.7 Hazard Class 2 chemicals present which of the following hazards?
  A. Explosive, flammable, and corrosive
  B. Toxic, oxidizer, and radioactive
  C. Flammable, poison, and oxidizer
  D. Poison, nonflammable, and flammable
1.8 Class 1 contains how many hazard subclasses?
  A. 4
  B. 2
  C. 6
  D. 1
1.9 Which of the following is not one of the three hazards of Hazard Class 4?
  A. Flammable oxidizer
  B. Spontaneously combustible
  C. Flammable solid
  D. Dangerous when wet

1.10 Which of the following is not one of the four colors of the NFPA 704 marking system?
  A. Yellow
  B. Orange
  C. Blue
  D. Red

1.11 Of the following federal regulations, which one covers hazmat training?
  A. 29 CFR
  B. 49 CFR
  C. 37 CFR
  D. 10 CFR

1.12 Hazardous materials incident commanders must be trained to which level?
  A. Awareness
  B. Operations
  C. Technician
  D. Specialist

chapter two

Basics of chemistry

Chemistry is the study of matter. It can be divided into two basic sections: inorganic and organic. *Inorganic* chemistry involves acids, bases, salts, elements except carbon-containing compounds, and the physical state of matter in which they are found. Important classes of inorganic compounds include the oxides, sulfides, sulfates, carbonates, nitrates, and halides. The study of some carbon-containing compounds such as carbon dioxide, carbonates, and cyanides is also considered part of inorganic chemistry. *Organic* chemistry involves compounds that contain carbon. Organic chemistry involves the scientific study of the structure, properties, composition, reactions, and preparation of carbon-based compounds, hydrocarbons, and their derivatives. These compounds may contain any number of other elements, including hydrogen, nitrogen, oxygen, the halogens, as well as phosphorus, silicon, and sulfur. The definition of matter is anything that occupies space and has mass. Matter can exist as a solid, liquid, or gas. Temperature and pressure can affect the physical state of a chemical, but not its properties. The hazards presented by a chemical may not be the same, depending on the physical state of the material. For example, only gases burn; solids and liquids do not burn, even though they may be listed as flammable. A solid or liquid must be heated until it produces enough vapors to burn. It is important to understand the states of matter in order to better understand the physical and chemical characteristics of hazardous materials. There can be some intermediate steps in the process of classifying solids, liquids, and gases. Some solids may have varying particle sizes, from large blocks to filings, chips, and dusts. Particle sizes of vapors may vary, from vapors that are small enough to be invisible to mists that are readily visible.

A molecule is the smallest particle of a compound that can normally exist by itself. Molecules of compounds contain different types of atoms bonded together in fixed proportions. Molecules of solids are packed together closely in an organized pattern. Because molecules are packed tightly together, they can vibrate only gently in a small space. This is why solids have a definite size and shape. When particles are this close together, they attract each other, and it takes a lot of energy to pull them apart. As a solid is heated, the molecules start to vibrate faster and eventually pull apart. Particles in liquids are farther apart; however, they are still able to attract each other. They are not arranged in a regular pattern. Liquids do not have a shape of their own, so they conform to the shape of the container in which they are placed. Particles of a gas move rapidly and are not attracted to each other. Gases have no shape of their own and conform to the space in which they exist.

Hazardous materials may undergo both chemical and physical changes. A chemical change involves a reaction that alters the composition of the substance and thereby alters its chemical identity. Chemical properties include reactivity, stability, corrosivity, toxicity, and oxidation potential. A new compound may be formed that may have different characteristics than the compounds or elements that make it up. Chlorine, for example, is a poisonous gas; sodium is a reactive metal. When they are combined, they form sodium chloride, which is neither a poison nor a reactive chemical. Physical changes involve changes in the physical state of the chemical, but do not produce a new substance, such as the physical transformation from a liquid to a gas or a liquid to a solid. Physical properties include specific gravity, vapor pressure, boiling point, vapor density, melting point, solubility, flash point, fire point, autoignition temperature, flammable range, heat content, pH, threshold limit value (TLV), and permissible exposure level (PEL).

Periodic table of elements

Basics of chemistry cannot be effectively discussed without studying the periodic table of elements (Figure 2.1). Dmitri Mendeleev, a Russian chemist, is considered the father of the periodic table based upon his work in cataloging facts about the 63 known elements of the mid-1800s. Mendeleev became convinced that groups of elements had similar "periodic" properties. He arranged the known elements according to their increasing atomic mass (weight). Blank spaces on the periodic table were left where he thought other unknown elements would fit in when discovered. British scientist Henry Moseley continued Mendeleev's work on the periodic table when he discovered that the number of protons in the nucleus of a particular type of atom was always the same. As a result of Moseley's work, the modern periodic table is based upon the atomic numbers of the elements. Elements on Moseley's periodic table are arranged in the order of increasing atomic number, which is also the number of protons in the atom of that element. The properties of the elements repeat in a regular way when the elements are arranged by increasing atomic number.

The periodic table is a method of organizing everything that is known about chemistry on one piece of paper. It shows the relationship between the elements by revealing

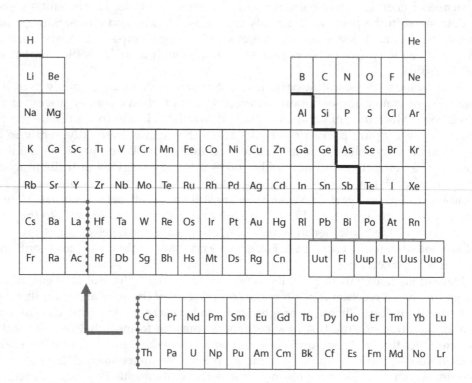

Figure 2.1 Periodic table of elements.

the tendency of their properties to repeat at regular intervals. All chemicals are derived from the elements or from combinations of elements from the periodic table. Symbols are used to represent the elements on the periodic table. The table is composed of a series of blocks representing each element. Within each block is a symbol that corresponds to the name of that particular element, which is a type of "shorthand" for the name of the element. For example, the element gold is represented by the symbol Au; chlorine is represented by the symbol Cl; and potassium is represented by the symbol K. Each symbol represents one atom of that element. Symbols may be made up of a single letter or two letters together. A single letter is always capitalized; when there are two letters, the first is capitalized and the second is lowercased. When the second letter is also capitalized, this indicates that the material described is not an element, but rather a compound. For example,

CO is the molecular formula for the compound carbon monoxide.
Co is the symbol for the element cobalt.

Symbols for each of the elements are derived from a number of sources. They may have been named for the person who discovered the element. For example, W, the symbol for tungsten, was named for Wolfram, the discoverer. Other elements are named for famous scientists, universities, cities, and states. Es is the symbol for einsteinium, named for Albert Einstein; Cm is the symbol for curium, named for Madame Curie; Bk is the symbol for berkelium, named for the city of Berkeley, CA; and Cf is the symbol for the element californium, named after the state of California. Other element names come from Latin, German, Greek, and English. In the case of sodium, Na comes from the Latin for *natrium*, which is the Latin name for sodium; Au, the symbol for gold, comes from *aurum*, meaning "shining down" in Latin; Cu (copper) is from the Latin *cuprum*, or *cyprium*, because the Roman source for copper was the island of Cyprus. Fe (iron) is from the Latin *ferrum*. Bromine means "stench" in Greek; rubidium means "red" (in color); mercury is sometimes referred to as quicksilver; sulfur is referred to as brimstone in the Bible.

Elements

On the current periodic table, there are 90 naturally occurring and 28 manmade elements. Recently, four new elements were named and added to the periodic table of elements: 111, Roentgenium (Rg); 112, Copernicium (Cn); 114, Livermorium (Lv); and 116, Flerovium (Fl). Four other elements are known to exist, but are yet to be named by the International Union of Pure and Applied Chemists (IUPAC). Not all of the elements on the periodic table are common or particularly hazardous to responders; in fact, some are so rare that they are not likely to be encountered at all. Others exist in the laboratory only for a very short period of time. There are, however, some 39 elements that we will call the *"Hazmat elements"* (Table 2.2). These elements are important to the study of the chemistry of hazardous materials. Most of the hazardous materials that response personnel will encounter include or are produced from these 39 elements. Hazardous materials personnel and students of Hazmat chemistry should be familiar with these 39 elements by symbol and name. Elements on the periodic table that contain 83 or more protons in the nucleus are radioactive; many are rare and probably will not be encountered by most response personnel. Manmade elements occur as a result of nuclear reactions and research. These elements may have existed on earth at one time, but because they are radioactive and many half-lives have passed, they no longer exist naturally.

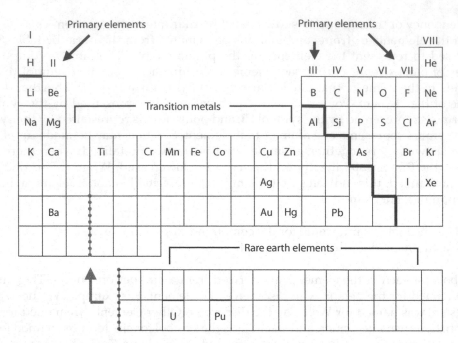

Figure 2.2 Types of elements.

Elements on the periodic table can be divided into three general groups: primary elements, transition elements, and rare earth elements (Figure 2.2). Primary elements contain a definite number of electrons in their outer shell. The number of electrons in the outer shell is also the number shown at the top of each column in the "towers" at each end of the periodic table. Transition metals may have differing numbers of electrons in their outer shells. Transition metals are located in the "valleys" between the "towers." Rare earth metals are relatively uncommon and all are radioactive. Horizontal rows on the periodic table are referred to as periods and are numbered from 1 to 7. Atomic numbers of each element increase by one as you move across the periods from left to right.

Atomic number

Atomic number is sometimes referred to as the proton number. The whole number on the periodic table for each element is a number located either above or below the symbol for the element and is known as the element's atomic number. It is sometimes represented by the letter "Z." The conventional symbol Z presumably comes from the German word *Atomzahl* (atomic number). The atomic number also equals the total number of electrons in the orbits outside the nucleus of the atom. Protons have a positive (+) charge, and electrons have a negative (–) charge. There must be an equal number of protons and electrons in the atom of the element to maintain an electrical balance. It is the number of protons that identifies a specific element. If you change the number of protons, you change the element. Protons act as a kind of "social security number" to identify a specific element.

Atomic weight

Also listed on the periodic table for each element is its atomic weight. The atomic weight (sometimes referred to as the *mass number*) is equal to the number of protons and neutrons in the nucleus and is sometimes represented by the letter "A." The letter "N" is the symbol for the neutrons in the nucleus. All of the weight of the element occurs in the nucleus of the atom. For the purposes of *"street chemistry,"* electrons do not have weight. Atomic weight is the sum of the weight of the protons and neutrons (A = Z + N). All of the weight of the element occurs in the nucleus of the atom. For the purposes of *"street chemistry,"* electrons do not have weight.

Atomic weight is located on the periodic table above or below the symbol of the element. It is the number that is *not* a whole number; location varies among periodic tables, so be sure you look for the number with the decimal point. When using atomic weight to determine molecular weight of a compound, always round it off to the nearest whole number. For example, oxygen has an atomic weight of 15.999; round the number to 16, so the atomic weight of oxygen becomes 16. Nitrogen has an atomic weight of 14.007, so you round the number to 14. The atomic weight is referred to in terms of atomic mass units (AMU) (Figure 2.3).

The periodic table is divided into two main sections by a stair-stepped line (Figure 2.4). The line starts under hydrogen, moves to the right over to boron, and then stair-steps down one element at a time to astatine or radon, depending on which periodic table is used. The 81 elements to the left and below the stair-stepped line are metals. Metals make

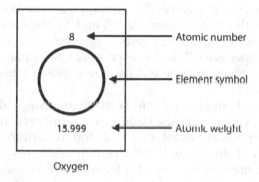

Figure 2.3 Element components.

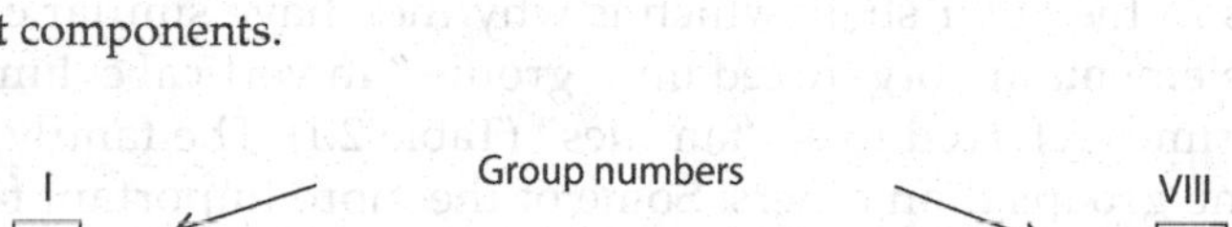

Figure 2.4 Families of elements. Group I: All have one electron in the outer shell. Group II: All have two electrons in the outer shell. Group III: All have three electrons in the outer shell. Group IV: All have four electrons in the outer shell. Group V: All have five electrons in the outer shell. Group VI: All have six electrons in the outer shell. Group VII: All have seven electrons in the outer shell. Group VIII: All have eight electrons in the outer shell.

up about 75% of all the elements. Metals lose their outer-shell electrons easily to nonmetals when forming compounds. Metals are malleable (they can be flattened), ductile (they can be drawn into a wire), and conduct heat and electricity quite well. The farther to the left of the line you are, the more metallic the properties of the element; closer to the line, properties of the element are less metallic. Metallic properties increase as you travel down a column on the periodic table. All metals are solids, except gallium, mercury, francium, and cesium, which are liquids under normal conditions.

The 17 elements to the right and above the line on the periodic table are nonmetals. Nonmetals have a strong tendency to gain electrons when forming compounds. Nonmetals may be solids, liquids, or gases; all are poor conductors of heat and electricity. Solid nonmetals are either hard and brittle or soft and crumbly.

As previously stated, numbers located at the top of the vertical columns on the periodic table indicate the number of electrons in the outer shell of the elements in that column. The exception to this statement is the transition elements located between the two *"towers"* on either end of the table. Unlike the primary elements, numbers above the transition metal columns do not indicate the number of electrons in the outer shell of those elements. Transition elements can have differing numbers of electrons in their outer shells. Atoms having the same atomic number Z but different neutron number N, and hence different atomic mass, are known as isotopes. Isotopes are variants of a particular chemical element. For example, carbon-12, carbon-13, and carbon-14 are three isotopes of the element carbon with mass numbers 12, 13, and 14 respectively. The atomic number of carbon is 6, which means that every carbon atom has six protons, so that the neutron numbers of these isotopes are 6, 7, and 8 respectively. Most naturally occurring elements exist as a mixture of isotopes, and the average atomic mass of this mixture determines the element's atomic weight.

The periodic table is organized into vertical columns, called groups or families, and horizontal rows, called periods. Groups are numbered from left to right on the table from 1 to 18. Periods are numbered from top to bottom on the table from 1 to 7. The vertical columns of the periodic table contain elements that have similar chemical characteristics in their pure elemental form. These elements have the same number of electrons in the outer shell, which is why they have similar chemical behaviors. Similar acting elements are organized into "groups" in vertical columns on the periodic table and sometimes referred to as "families" (Table 2.1). The family effect is more pronounced in some groups than others. Some of the more important families include the alkali metals in Group 1, the alkaline earth metals in Group 2, the halogens in Group 7, and the noble or inert gases in Group 8. Transition elements in the center of the periodic table are also similar in that most of them have the possibility of differing numbers of electrons in their outer shells. This is not a "family" effect because they do not have similar chemical behaviors.

Table 2.1 Element Group Names

Group number	Family name
Group 1	Alkali metals or lithium family
Group 2	Alkaline earth metals or beryllium family
Group 7	Halogens
Group 8	Noble or Inert Gases
Center of Periodic Table between "Towers"	Transition Elements (Metals)

Alkali metals in Group 1 begin with lithium and continue downward through sodium, potassium, rubidium, cesium, and francium. Hydrogen is also located in Group 1, but it is above the stair-stepped line and thus is a nonmetal. It is placed in Group 1 because it has one electron in the outer shell. Alkali metals are all solids, except for cesium and francium, which are liquids at normal temperatures. Alkali metals are all water reactive to some degree: they react violently with water, producing flammable hydrogen gas and enough heat to ignite the hydrogen gas. These elements are so reactive that they do not exist in nature in pure form, but are found as compounds of the metal, such as potassium oxide and sodium chloride. Some isotopes of cesium and all isotopes of francium are radioactive. These elements are somewhat rare, so you are not likely to see them on the street.

Alkaline earth metals located in Group 2 are less reactive than the alkali metals in Group 1. Beryllium does not react with water at all. The remaining elements in Group 2 produce varying reactions with water. Alkaline earth metals are all solids. They need to be burning or in a finer physical form (powder or shavings) before they become water reactive. Magnesium, for example, is violently water reactive when it is involved in fire. Application of water to a magnesium fire will cause violent explosions that can endanger responders. If it is necessary to fight fires involving magnesium, water should be applied from a safe distance with the use of unmanned appliances. When magnesium is found in a particulate form, it may react with water even if it is not on fire. Other elements in Group 2 are also water reactive to varying degrees. These include calcium, strontium, barium, and radium, which are radioactive.

Halogens in Group 7 are nonmetals. They may be solids, liquids, or gases. Fluorine and chlorine are gases at normal temperatures and pressures. Bromine is a liquid to 58°C and produces vapor rapidly above that temperature. Iodine is a solid. Astatine is also in Group 7. It is radioactive; however, such a small amount has ever been found that you are not likely to encounter it.

Halogens are all toxic and strong oxidizers. Fluorine is a much stronger oxidizer than oxygen; in fact, fluorine is the strongest oxidizer known to exist. In their pure elemental form, halogens do not burn; however, they will accelerate combustion much like oxygen because they are oxidizers. Some halogen compounds are components of fire-extinguishing agents that are called halons, which are being phased out because of the damage they cause to the ozone layer of the atmosphere. Elements in Group 8 of the periodic table are all gases. As a "family," they are nonflammable, nontoxic, and nonreactive. Group 8 elements are referred to as the inert or noble gases. Normally, they do not react chemically with themselves or any other chemicals. Helium is used for weather balloons and airships. Neon is used for lighting and beacons. Argon is used in lightbulbs. Krypton and xenon are used in special lightbulbs for miners and in lighthouses. Radon is radioactive and is used in tracing gas leaks and treating some forms of cancer. Radon occurs naturally in the ground and may be found in basements of homes. Levels should be monitored and may require eradication because of ill health effects created by long-term exposure.

Noble gases are not normally chemically reactive. This is because they have a complete outer shell of electrons: two in helium and eight in the rest: an "octet" is eight electrons in the outer shell; a "duet" is two electrons in the outer shell. Both the octet and duet are stable configurations of electrons in an atom. It is because of the complete outer shell of electrons that the noble gases do not react chemically. All of the other elements on the periodic table try to reach the same stable electron arrangement as the nearest noble gas, which is usually in the same period. Striving for stability is what gives elements the ability to engage in chemical reactions: the elements are trying to reach stability. Group 8 elements are all nonhazardous gases according to DOT regulations; however, they can displace oxygen in the air inside of buildings or in confined spaces and cause

Table 2.2 The Hazmat Elements

H—Hydrogen	Cu—Copper	O—Oxygen
Li—Lithium	Ag—Silver	S—Sulfur
Na—Sodium	Au—Gold	F—Fluorine
K—Potassium	Zn—Zinc	Cl—Chlorine
Be—Beryllium	Hg—Mercury	Br—Bromine
Mg—Magnesium	B—Boron	I—Iodine
Ca—Calcium	Al—Aluminum	U—Uranium
Ba—Barium	C—Carbon	He—Helium
Ti—Titanium	Si—Silicon	Ne—Neon
Cr—Chromium	N—Nitrogen	Ar—Argon
Mn—Manganese	P—Phosphorus	Kr—Krypton
Fe—Iron	As—Arsenic	Xe—Xenon
Co—Cobalt	Pu—Plutonium	Pb—Lead

asphyxiation. Inert gases are commonly shipped and stored as compressed gases or cryogenic liquids. Cryogenic liquids are quite cold and can cause thermal burns. They also have large expansion ratios. DOT does not place them in a hazard class unless they are under pressure.

Not all of the elements on the periodic table are particularly hazardous by themselves nor do they all form significant chemical compounds. For the purposes of "street chemistry," I have listed 39 elements and their characteristics, which I consider to be the important Hazmat elements. Information provided includes the symbol, physical state, family, atomic number, atomic weight, location on the periodic table, outer-shell electrons, DOT hazard class, NFPA 704 designation, and other important characteristics. Also included with each Hazmat element (Table 2.2) is some brief information on the history, properties, sources, important compounds, uses, and isotopes of those elements.

Hazmat elements

Hydrogen

Symbol: H
Metal or nonmetal: Nonmetal
Atomic number: 1
Atomic weight: 1.00794
Periodic table group: I
Outer-shell electrons: 1
Ordinary physical state: Gas
Color: Colorless
Case registry number: 1333-74-0
U.N. four digit number: 1966
DOT hazard class: 2 compressed gas
Placard: 2.1 Flammable gas
NFPA 704: 3-4-1
Cryogenic liquid: Yes
Cryogenic boiling point: −423°F

Cryogenic expansion ratio: 1–845
Boiling point: –252.9°C
Flammable range: 4%–74% in air
Autoignition temperature: 1060°F (571°C)
Vapor density: 08988 g/L
Solubility: To some extent
PEL-OSHA: Simple asphyxiant
ACGIH-TLV-TWA: Simple asphyxiant

History

The name hydrogen was derived from the Greek word, *hydro*, meaning water, and *genes*, meaning forming. Named by Lavoisier, it is the most abundant of all elements in the universe. It was recognized as an element by Henry Cavendish (1731–1810); an English chemist Cavendish described accurately hydrogen's properties but thought erroneously that the gas originated from the metal rather than from the acid. Deuterium gas ($^2H_2$, often written D_2), made up from deuterium, a heavy isotope of hydrogen, was discovered in 1931 by Harold Urey, a professor of chemistry at Chicago and California (both United States). One famous explosion and fire associated with hydrogen is the destruction of the Hindenburg in 1937 (Photo 2.1). Hydrogen was used to keep the giant airship in the air.

Photo 2.1 The Hindenburg disaster took place on Thursday, May 6, 1937, as the German passenger airship LZ129 Hindenburg caught fire and was destroyed during its attempt to dock with its mooring mast at the Lakehurst Naval Air Station. Highly flammable hydrogen fuel ignited from an unknown ignition source. (Johannes Ewers, Used with permission.)

For a number of years, it was thought that the hydrogen was to blame for the fire that broke out as the Hindenburg was coming in for a landing at the Lakehurst Naval Air Station in New Jersey. The entire ship was destroyed by flames in less than 1 min. More recent research has pointed to a different cause for the disaster, however New evidence has shown that the fabric on the outside of the ship was coated with highly flammable chemicals that were similar in composition to rocket fuel. An electrical discharge from a storm is now thought to be the true cause of ignition. The Hindenburg disaster was responsible for the loss of 35 lives. Almost all of these were caused by the people involved jumping from the burning aircraft, as opposed to the fire itself. The hydrogen on board ignited, but the flames would have burned up and away from passengers. The hydrogen fire would have burned off very quickly.

Sources

Hydrogen is estimated to make up more than 90% of all the atoms or three-quarters of the mass of the universe. H_2 gas is present in the earth's atmosphere in very small quantities, but is present to a far greater extent chemically bound as water (H_2O). Water is a constituent of many minerals. Hydrogen is the lightest element and was used for airships during the early 1900s and was the lifting agent for the ill-fated Hindenburg balloon rather than the safer helium. Hydrogen is by far the most abundant element in the universe, making up about 90% of the atoms or 75% of the mass, of the universe. Hydrogen is a major constituent of the sun and most stars. The sun burns by a number of nuclear processes but mainly through the fusion of hydrogen nuclei into helium nuclei.

Important compounds

Hydrogen is one of the key elements in hydrocarbon compounds. It is also present in the hydrocarbon derivatives and inorganic acids. Other compounds include binary compounds with halogens (known as halides) and oxygen (known as oxides). Hydrogen also bonds with oxygen to form alcohol and organic acid functional groups within the hydrocarbon derivatives. Some important hydrogen compounds are listed as follows:

Hydrogen fluoride: HF
Hydrogen chloride: HCl
Hydrogen bromide: HBr
Hydrogen iodide: HI
Water: H_2O
Hydrogen persulfide: H_2S_2
Hydrogen sulfide: H_2S
Hydrogen selenide: H_2Se
Hydrogen telluride: H_2Te
Ammonia: NH_3
Hydrazine: H_2NNH_2

Uses

Hydrogen is used for the hydrogenation of fats and oils and in the formation of hydrocarbon compounds. It is also used in large quantities in methanol production. Other uses include rocket fuel, welding, producing hydrochloric acid, reducing metallic ores, and filling balloons. The hydrogen fuel cell is a developing technology that will allow great amounts of electrical power to be obtained using a source of hydrogen gas. Located in remote regions, power plants would electrolyze seawater; the hydrogen produced would travel to distant cities by pipelines. Pollution-free hydrogen could replace natural gas, gasoline, etc., and could serve as a reducing agent in metallurgy, chemical processing, refining, etc. It could also be used to convert trash into methane and ethylene. Liquid hydrogen is important in cryogenics and in the study of superconductivity, as its melting point is only 20° above absolute zero.

Isotopes

Isotopes: H1–H2–H3

Hydrogen is the only element whose isotopes have been given different names. The ordinary or most common isotope of hydrogen, H1, is known as protium (name rarely used), the other two isotopes are H2 deuterium (a proton and a neutron) and H3 tritium (a proton and two neutrons). Deuterium is used in a variety of applications. It is used extensively in organic chemistry in order to study chemical reactions. It is also used in vitamin research. Deuterium in the form of H_2O, known as heavy water, is used as a moderator in CANDU nuclear reactors, in NMR studies, and in studies into human metabolism. Heavy water is also applied in the Sudbury Neutrino Observatory, where it is used to study the behavior of neutrinos. One atom of deuterium is found in about 6000 ordinary hydrogen atoms. H3 or tritium, is the third isotope of hydrogen and is readily produced in nuclear reactors and is used in the production of the hydrogen bomb. Tritium is readily produced in nuclear reactors and is used in the production of the hydrogen (fusion) bomb. It is also used as a radioactive agent in making luminous paints and as a tracer.

Important reactions

Hydrogen is a colorless gas, H_2, which is lighter than air. Mixtures of hydrogen gas and air do not react unless ignited with a flame or spark, in which case the result is a fire or explosion with a characteristic reddish flame whose only product is water, H_2O. Hydrogen does not react with water. Hydrogen gas, H_2, reacts with fluorine, F_2, in the dark to form hydrogen (I) fluoride. Hydrogen does not react with dilute acids and dilute bases.

Hazards to responders (Photo 2.2)

Hydrogen is a clear, odorless, and tasteless gas. It is lighter than air, which means that if a hydrogen leak occurs, it rises up into the air and dissipates, as opposed to spilling on the ground or on surfaces. Hydrogen is flammable and must be handled with care, just like other flammable substances. In order for hydrogen to ignite, it must be contained and combined with oxygen and an ignition source. If hydrogen is ignited, it burns off very quickly. It is very difficult for the naked eye to detect hydrogen burning, since it burns in the ultraviolet color range. In order for hydrogen to turn into a liquid form, it must be cooled to at least minus 423°F. If liquid hydrogen comes into contact with exposed skin, it can cause severe freeze burns. A freeze burn is similar to frostbite. In order to keep hydrogen cool enough to keep it in a liquid state, it is stored in specialized containers that are double-walled and heavily insulated. The chances of the liquid actually escaping and coming into contact with a person's skin are quite small. Hydrogen can explode, but only if it comes into contact with oxygen. Gasoline and propane, which are heavier gases than hydrogen, are more likely to explode. The fumes from each of these tend to stay close to the ground, which increases the likelihood of explosion. Hydrogen is nontoxic and is not poisonous. Using hydrogen as a fuel source does not create fumes, pollute the atmosphere,

Photo 2.2 Houston firefighters respond to a fire at the Houston Distribution Warehouse, July 24, 1995. This chemical fire went to seven alarms and mutual aid was called in to assist the Houston Fire Department. (Courtesy Houston Fire Department, Used with permission.)

or contribute to the global warming that is such a cause for concern today. If appropriate safety measures are taken, hydrogen hazards can be kept to a minimum. There are definite advantages to using hydrogen in industry as opposed to flammable substances like gasoline or propane.

Lithium

Symbol: Li
Metal or nonmetal: Metal
Atomic number: 3
Atomic weight: 6.941
Periodic table group: I
Periodic table group name: Alkali metals
Outer-shell electrons: 1
Ordinary physical state: Solid
Color: Silvery white/gray
Case registry number: 7439-93-2
U.N. four digit number: 1415
DOT hazard class: 4 flammable solid

Placard: 4.3 dangerous when wet
NFPA 704: 3-2-2 W —
Cryogenic liquid: No
Cryogenic boiling point: N/A
Cryogenic expansion ratio: N/A
Radioactive: No
Boiling point: 1347°C
Flash point: N/A
Flammable range: N/A
Autoignition temperature: N/A
Melting point: 108.54°C
Freezing point: N/A
Specific gravity: 0.534 (20°C)
Vapor density: N/A
Solubility: Elementary lithium is not very water soluble, but it does react with water. Lithium compounds such as lithium chloride, lithium carbonate, lithium phosphate, lithium fluoride, and lithium hydroxide are more or less water soluble. Lithium hydroxide, for example, has a 129 g/L solubility
PEL-OSHA: N/A
ACGIH-TLV-TWA: N/A

History

The name lithium was derived from the Greek word, meaning *lithos*, or stone and discovered by Arfvedson in 1817. It is the lightest of all metals, with a density only about half that of water. It has a fairly low melting point, but has the highest melting point among the alkali metals.

Sources

It does not occur free in nature; combined, it is found in small units in nearly all igneous rocks and in the waters of many mineral springs. Lithium is silvery in appearance, much like Na and K, other members of the alkali metal series. It reacts with water, but not as vigorously as sodium. Lithium imparts a beautiful crimson color to the flame, but when the metal burns strongly, the flame is dazzling white. Lithium is a moderately abundant element and it is present in the earth's crust in 65 ppm (parts per million). In the United States, lithium is recovered from brine pools in Nevada. Today, most commercial lithium is recovered from brine sources in Chile. World production of lithium ores and brine salts is around 40,000 tons per year and reserves are estimated to be around 7 million tons.

Important compounds

Lithium bonds with nonmetals to form salt families. Lithium takes part in a huge number of reactions, with organic reactants as well as with inorganic reactants. It reacts with oxygen to form monoxide and peroxide. It is the only alkaline metal that reacts with nitrogen at ambient temperature to produce a black nitride. It reacts easily with hydrogen at almost 500°C (930°F) to form lithium hydride.

Lithium hydride: LiH
Lithium fluoride: LiF
Lithium chloride: LiCl

Lithium chloride monohydrate: $LiCl \cdot H_2O$
Lithium iodide: LiI
Dilithium oxide: Li_2O
Dilithium peroxide: Li_2O_2
Lithium superoxide: LiO_2
Lithium sulfide: Li_2S
Dilithium selenide: Li_2Se
Dilithium telluride: $LiTe$
Trilithium nitride: Li_3N

Uses

Because the metal has the highest specific heat of any solid element, it has found use in heat-transfer applications; however, it is corrosive and requires special handling. The metal has been used as an alloying agent, is of interest in the synthesis of organic compounds, and has nuclear applications. It ranks as a leading contender as a battery anode material, as it has a high electrochemical potential. Lithium is used in special glasses and ceramics. Lithium stearate is used as an all-purpose and high-temperature lubricant. Other lithium compounds are used in dry cells and storage batteries.

Isotopes

Lithium has two isotopes that occur naturally, Li-6 and Li-7. Li-7 is used to control the pH level of the coolant in the primary water circuit of pressurized water reactors. Li-7 is also used for the production of the medical research radioisotope Be-7. Li-6 is used in thermonuclear weapons and the export, and the use of Li-6 is therefore strictly controlled.

Important reactions

Metallic lithium's reaction with water is slow compared to sodium and potassium. It also reacts with the moisture in the air. When burnt in air, it forms oxygen compounds. Lithium reacts with acids, bases, nitrogen, and halogens. Reactions with bases are exothermic.

Hazards to responders

Flammability   Many reactions may cause fire or explosion. Lithium gives off irritating or toxic fumes (or gases) in a fire. Explosion: Risk of fire and explosion on contact with combustible substances and water. On inhalation, it causes burning sensation, cough, labored breathing, shortness of breath, and sore throat. Symptoms may be delayed. Skin: Redness, skin burns, pain, and blisters. Eyes: Redness, pain, severe deep burns. Ingestion: Abdominal cramps, abdominal pain, burning sensation, nausea, shock or collapse, vomiting, and weakness.

Health effects of short-term exposure   The substance is corrosive to the eyes, the skin, and the respiratory tract. It is corrosive on ingestion. Inhalation of the substance may cause lung edema. The symptoms of lung edema often do not become manifest until a few hours have passed, and they are aggravated by physical effort. Rest and medical observation is therefore essential. Immediate administration of an appropriate spray, by a doctor or a person authorized by him or her, should be considered.

Routes of exposure The substance can be absorbed into the body by inhalation of its aerosol and by ingestion. Inhalation risk: Evaporation at 20°C is negligible; a harmful concentration of airborne particles can, however, be reached quickly when dispersed.

Chemical hazards Heating may cause violent combustion or explosion. The substance may spontaneously ignite on contact with air when finely dispersed. Upon heating, toxic fumes are formed. It reacts violently with strong oxidants, acids, and many compounds (hydrocarbons, halogens, halons, concrete, sand, and asbestos) causing fire and explosion hazard. It reacts violently with water, forming highly flammable hydrogen gas and corrosive fumes of lithium hydroxide.

Sodium

Symbol: Na
Metal or nonmetal: Metal
Atomic number: 11
Atomic weight: 22.98977
Periodic table group: I
Periodic table group name: Alkali metals
Outer-shell electrons: 1
Ordinary physical state: Solid
Color: Silvery white
Case registry number: 7440-23-5
U.N. four digit number: 1428
DOT hazard class: 4 flammable solid
Placard: 4.2 spontaneously combustible
NFPA 704: 3-3-2 W
Cryogenic liquid: No
Cryogenic boiling point: N/A
Cryogenic expansion ratio: N/A
Radioactive: No
Boiling point: 883°C
Flash point: N/A
Flammable range: N/A
Autoignition temperature: N/A
Melting point: 98°C
Freezing point: N/A
Specific gravity: 0.971
Vapor density: N/A
Solubility: Soluble in water, reacts violently with water
PEL-OSHA: N/A
ACGIH-TLV-TWA: N/A

History

Sodium is a soft, silvery-white metal. It is soft enough to cut with the edge of a coin. It was derived from English for soda and from medieval Latin, *sodanum*, a headache remedy. Long recognized in compounds, sodium was first isolated by Davy in 1807 by electrolysis of caustic soda.

Sources

Sodium is the sixth most-abundant element on earth, comprising about 2.6% of the earth's crust; it is the most abundant of the alkali group of metals. Due to its high reactivity, sodium is found in nature only as a compound and never as the free element. It is obtained commercially by electrolysis of molten sodium chloride.

Important compounds

Sodium forms binary compounds with halogens (known as halides), oxygen (known as oxides), hydrogen (known as hydrides), and some other compounds of sodium. The most common binary compound is sodium chloride, but it occurs in many other minerals. It makes up many salt compounds with nonmetal materials. Among the many compounds that are of the greatest industrial importance are common salt (NaCl), soda ash (Na_2CO_3), baking soda ($NaHCO_3$), caustic soda (NaOH), Chile saltpeter ($NaNO_3$), di- and trisodium phosphates, sodium thiosulfate (hypo, $Na_2S_2O_3$.), and borax ($Na_2B_4O_7$.).

Sodium hydride: NaH
Sodium fluoride: NaF
Sodium chloride: NaCl
Sodium iodide: NaI
Disodium oxide: Na_2O
Sodium peroxide: Na_2O_2
Sodium superoxide: NaO_2
Disodium sulfide: Na_2S
Disodium persulfide: Na_2S_2
Disodium trisulfide: Na_2S_3
Disodium hexasulfide: Na_2S_6
Disodium tetrasulfide: Na_2S_4
Disodium pentasulfide: Na_2S_5
Disodium sulfide pentahydrate: $Na_2S \cdot 5H_2O$
Disodium sulfide nonahydrate: $Na_2S \cdot 9H_2O$
Disodium selenide: Na_2Se
Disodium telluride: Na_2Te

Uses

Metallic sodium is used in the manufacture of sodamide and esters, and in the preparation of organic compounds. The metal also may be used to modify alloys such as aluminum–silicon by improving their mechanical properties and fluidity. Sodium is used to descale (smooth the surface of) metals and to purify molten metals.

Sodium vapor lamps are highly efficient in producing light from electricity and are often used for streetlighting in cities. Sodium is used as a heat transfer agent; for example, liquid sodium is used to cool nuclear reactors. Sodium chloride (table salt, NaCl) is vital for good nutrition. Sodium ions facilitate transmission of electrical signals in the nervous system and regulate the water balance between body cells and body fluids.

Isotopes

Sodium has 16 isotopes whose half-lives are known, with mass numbers 20–35. Of these, only one is stable: [23]Na.

Hazards to responders

Flammability   Severe fire risk in contact with water in any form; the reaction is exothermic and may cause hydrogen to ignite or pieces may explode during the reaction with water; ignites spontaneously in dry air when heated; to extinguish fires, use dry soda ash, salt, or lime.

Health   It forms strong caustic irritant to tissue (sodium hydroxide). Contact with skin causes irritation and possible burns, especially if the skin is wet or moist. Danger! It causes eye burns, digestive tract burns, and respiratory tract burns. It is corrosive.

Chemical hazards   Water reactive or can react with the moisture in the air.

Uses   Metallic sodium is vital in the manufacture of esters and in the preparation of organic compounds. The metal may be used to improve the structure of certain alloys, to descale metal, and to purify molten metals. Used as a coolant in nuclear reactors. Catalyst for synthetic rubber and heat transfer agent in solar-powered electric generators.

Potassium

Symbol: K
Metal or nonmetal: Metal
Atomic number: 19
Atomic weight: 39.0983
Periodic table group: I
Periodic table group name: Alkali metals
Outer-shell electrons: 1
Ordinary physical state: Solid
Color: Silvery
Case registry number: 7440-09-7
U.N. four digit number: 2257
DOT hazard class: 4 flammable solid
Placard: 4.2 spontaneously combustible
NFPA 704: 3-3-2 W
Radioactive: No
Boiling point: 759°C
Flash point: N/A
Flammable range: N/A
Autoignition temperature: N/A
Melting point: 65.5°C
Freezing point: N/A
Specific gravity: N/A
Vapor density: N/A
Solubility: Yes, but very reactive with water
PEL-OSHA: None
ACGIH-TLV-TWA: None

History

The name potassium was coined from the English, meaning potash—pot ashes; Latin, *kalium*; and Arab, *qali*, or alkali. It was discovered in 1807 by Davy, who obtained it from caustic potash (KOH); this was the first metal isolated by electrolysis. Except for lithium, it is the lightest known metal. It is soft, easily cut with a knife and is silvery in appearance immediately after a fresh surface is exposed.

Sources

The metal is the seventh most abundant and makes up about 2.4%, by weight, of the earth's crust. Most potassium minerals are insoluble, and the metal is obtained from them only with great difficulty. Potassium is never found free in nature, but is obtained by electrolysis of the hydroxide, much in the same manner as was prepared by Davy. Thermal methods also are commonly used to produce potassium.

Important compounds
Potassium hydride: KH
Potassium fluoride: KF
Potassium chloride: KCl
Potassium iodide: KI
Dipotassium oxide: K_2O
Potassium superoxide: KO_2
Dipotassium peroxide: K_2O_2
Dipotassium sulfide: K_2S
Dipotassium disulfide: K_2S_2
Dipotassium hexasulfide: K_2S_6
Dipotassium trisulfide: K_2S_3
Dipotassium pentasulfide: K_2S_5
Dipotassium tetrasulfide: K_2S_4
Dipotassium selenide: K_2Se
Dipotassium telluride: K_2Te

Uses
The greatest demand for potash has been in its use for fertilizers. Potassium is an essential constituent for plant growth and is found in most soils. An alloy of sodium and potassium (NaK) is used as a heat-transfer medium. Many potassium salts are of utmost importance, including hydroxide, nitrate, carbonate, chloride, chlorate, bromide, iodide, cyanide, sulfate, chromate, and dichromate.

Isotopes
Seventeen isotopes of potassium are known. Ordinary potassium is composed of three isotopes, one of which is $40°K$ (0.0118%), a radioactive isotope with a half-life of 1.28×10^9 years.

Important reactions
Reactions of potassium are with air, water, halogens, acids, and bases.

Hazards to responders

Flammability   As with other metals of the alkali group, it decomposes violently in water with the evolution of highly flammable hydrogen gas, which catches fire spontaneously on contact with water. When exposed to air, it can react with moisture in the air. According to the Department of Transportation (DOT) *Emergency Response Guidebook*, fires involving potassium should not be fought with water. Do not use carbon dioxide. Use graphite, soda ash, powdered sodium chloride, or suitable dry powder.

Health   Inhalation or contact with vapors, substance, or decomposition products may cause severe injury or death. It may produce corrosive solutions on contact with water. Fire will produce irritating, corrosive, and/or toxic gases. Runoff from fire control may cause pollution.

Chemical hazards   It is one of the most reactive and electropositive of the metals. It rapidly oxidizes in air and must be stored in a mineral oil, such as kerosene or in argon. Hydrogen is not compatible with acids, bases, and the halogens. Contact may result in vigorous reactions. Contact with bases may cause the potassium to burn. Potassium and its salts impart a violet color to flames.

Beryllium

Symbol: Be
Metal or nonmetal: Metal
Atomic number: 4
Atomic weight: 9.01216
Periodic table group: II
Periodic table group name: Alkaline earth metals
Outer-shell electrons: 2
Ordinary physical state: Solid
Color: Steel gray
Case registry number: 7440-41-7
U.N. four digit number: 1567 (powder)
DOT hazard class: 6 poison
Placard: Poison
NFPA 704: 1-3-0
Boiling point: 2469°C, 4476°F
Melting point: 1287°C, 2349°F
Solubility: Insoluble in water
Specific gravity: 1.848
PEL-OSHA: N/A

The current OSHA PELs for beryllium are 2 μg/m³ as an 8-h TWA, 5 μg/m³ as a ceiling not to be exceeded for more than 30 min at a time, and 25 μg/m³ as a peak exposure never to be exceeded

ACGIH-TLV-TWA: Conference of Governmental Industrial Hygienists (ACGIH) current TLV for beryllium is 0.05 μg/m³ averaged over an 8-h work shift

History
The name Beryllium comes from the Greek word meaning *beryllos* or beryl; also called glucinium or glucinum; this term is taken from the Greek word *glykys,* sweet. It was discovered by Vauquelin as the oxide in beryl and in emeralds in 1798. Wohler and Bussy independently isolated the metal in 1828 by the action of potassium on beryllium chloride. Beryllium is a very lightweight, strong, hard metal that is easy to shape.

Sources
Aquamarine and emerald are precious forms of beryl. Beryl and bertrandite are the most important commercial sources of the element and its compounds. Most of the metal is now prepared by reducing beryllium fluoride with magnesium metal. Beryllium metal did not become readily available to industry until 1957.

Important compounds
Beryllium dihydride: BeH_2
Beryllium difluoride: BeF_2

Beryllium dichloride: $BeCl_2$
Beryllium dibromide: $BeBr_2$
Beryllium diiodide: BeI_2
Beryllia: BeO
Beryllium sulfide: BeS
Beryllium selenide: $BeSe$
Beryllium telluride: $BeTe$
Triberyllium dinitride: Be_3N_2

Uses

Beryllium is used as an alloying agent in producing beryllium copper, which is extensively used for springs, electrical contacts, spot-welding electrodes, and nonsparking tools. It is applied as a structural material for high-speed aircraft, missiles, spacecraft, and communication satellites. Other uses include windshield frames, brake disks, support beams, and other structural components of the space shuttle. It is used in gyroscopes, computer parts, and instruments where lightness, stiffness, and dimensional stability are required. The oxide has a high melting point and is also used in nuclear work and ceramic applications.

Isotopes
Beryllium 7, 9, 10

Important reactions

Beryllium metal reacts with chlorine, Cl_2, or bromine, Br_2, to form the beryllium dihalides beryllium (II) chloride, $BeCl_2$, and beryllium (II) bromide, $BeBr_2$, respectively. The surface of beryllium metal is covered with a thin layer of oxide that helps protect the metal from attack by acids, but powdered beryllium metal dissolves readily in dilute acids such as sulfuric acid, H_2SO_4, hydrochloric acid, HCl, or nitric acid, HNO_3, to form solutions containing the aquated Be(II) ion together with hydrogen gas, H_2. Beryllium metal dissolves readily in dilute aqueous base solutions such as sodium hydroxide, $NaOH$, to form Be(II) complexes together with hydrogen gas, H_2. Magnesium (immediately below beryllium in the periodic table) does not do this.

Hazards to responders

Flammability   Powder may form explosive mixture with air.

Health   Beryllium and its compounds are very toxic and should be handled with great care, and special precautions must be taken when carrying out any activity that could result in the release of beryllium dust (lung cancer is a possible result of prolonged exposure to beryllium laden dust). Combustion yields beryllium oxide fume, which is toxic if inhaled. Beryllium and its compounds should not be tasted to verify the sweetish nature of beryllium. Breathing beryllium particles can lead to scarring of the lungs. This is known as chronic beryllium disease—CBD for short. It can be treated, but it cannot be cured. It is sometimes fatal.

CBD is primarily a lung disease. But it may also affect the lymph nodes, skin, spleen, liver, kidneys, and heart.

Magnesium

Symbol: Mg
Metal or nonmetal: Metal
Atomic number: 12

Atomic weight: 24.305
Periodic table group: II
Periodic table group name: Alkaline earth metals
Ordinary physical state: Solid
Color: Silvery white
Case registry number: 7439-95-4
U.N. four digit number: 1869
DOT hazard class: 4 flammable solid
Placard: 4.2 spontaneously combustible
NFPA 704: 1-0-1
Boiling point: 1090°C, 1994°F
Autoignition temperature: 650°C, 1202°F
Melting point: 650°C, 1202°F
Specific gravity: 1.738
Solubility: reacts with water if on fire
PEL-OSHA: Magnesium oxide fumes 15 mg/m^3 (total dust) TWA
ACGIH-TLV-TWA: Magnesium oxide 10 mg/m^3 (total dust) TWA

History

Named after Magnesia, a district in Thessaly, Black first recognized magnesium as an element in 1755. Davy isolated it in 1808, and Bussy prepared it in coherent form in 1831. Magnesium is the eighth most abundant element in the earth's crust. It does not occur uncombined, but is found in large deposits in the form of magnesite, dolomite, and other minerals.

Sources

The metal is now principally obtained in the United States by electrolysis of fused magnesium chloride derived from brines, wells, and seawater.

Important compounds

Compounds of magnesium are generally salts. Magnesium combines with nonmetal elements to form salt compounds.

Magnesium dihydride: MgH_2
Magnesium difluoride: MgF_2
Magnesium dichloride: $MgCl_2$
Magnesium diiodide: MgI_2
Magnesium oxide: MgO
Magnesium peroxide: MgO_2
Magnesium sulfide: MgS
Magnesium selenide: MgSe
Magnesium telluride: MgTe
Trimagnesium dinitride: Mg_3N_2

Uses

Uses include flashlight photography, flares, and pyrotechnics, including incendiary bombs. It is one-third lighter than aluminum, and in alloys, it is essential for airplane and missile construction. The hydroxide (milk of magnesia), chloride, sulfate (Epsom salts), and citrate are used in medicine.

Isotopes
Magnesium 23, 24, 25

Important reactions

Magnesium is a silvery white metal. The surface of magnesium metal is covered with a thin layer of oxide that helps protect the metal from attack by air. Once ignited, magnesium metal burns in air with a characteristic blinding bright white flame to give a mixture of white magnesium oxide, MgO, and magnesium nitride, Mg_3N_2. Magnesium oxide is more normally made by heating magnesium carbonate. Magnesium metal dissolves readily in dilute sulfuric acid to form solutions containing the aquated Mg(II) ion together with hydrogen gas, H_2. Corresponding reactions with other acids such as hydrochloric acid also give the aquated Mg(II) ion.

Hazards to responders

Flammability   Finely divided magnesium readily ignites upon heating in air and burns with a dazzling white flame. Because serious fires can occur, great care should be taken in handling magnesium metal, especially in the finely divided state. Water should not be used on burning magnesium or on magnesium fires. Flammable: Magnesium is very flammable in ribbon form, which is why it is used in fireworks, according to Hands-on Science, producing a very bright light that is hazardous to view, with temporary damage to the eyes upon direct exposure for any period of time. The metal is considered a fire risk, so caution is recommended, including the use of safety glasses. Due to the fire risk, special care has to be taken when disposing magnesium.

Explosive effects   In metal powder form, magnesium can explode when it is mixed with air. In addition, explosion can occur if exposed to certain other substances, so magnesium should be kept isolated. The Material Safety Data Sheet states that magnesium should be stored in tightly closed containers.

Health effects   According to Material Safety Data Sheet, there are some health hazards consistent with being in the presence of magnesium. These include inhaling dust or fumes from the magnesium metal, which can cause metal fume fever, resulting in chest pain, coughing, and fever; ingestion that could cause abdominal pain and diarrhea; and burns to the skin if direct contact is made. Other health hazards include irritation to the eyes and aggravation of any prior wounds or openings in the skin upon contact.

Chemical hazards   Magnesium does not react with water to any significant extent unless it is on fire. Magnesium metal does however react with steam to give magnesium oxide (MgO) (or magnesium hydroxide, $Mg(OH)_2$, with excess steam) and hydrogen gas (H_2). Magnesium is very reactive toward the halogens such as chlorine, Cl_2, or bromine, Br_2, and burns to form the dihalides magnesium(II) chloride, $MgCl_2$ and magnesium(II) bromide, $MgBr_2$, respectively.

Calcium

Symbol: Ca
Metal or nonmetal: Metal
Atomic number: 20
Atomic weight: 40.8

Periodic table group: II
Periodic table group name: Alkaline earth metals
Outer-shell electrons: 2
Ordinary physical state: Solid
Color: Silvery white
Case registry number: 7440-70-2
U.N. four digit number: 1401
DOT hazard class: Flammable solid
Placard: Dangerous when wet
NFPA 704: 1-3-2 W
Boiling point: 1484°C, 2703°F
Autoignition temperature: N/A
Melting point: 842°C, 1548°F
Specific gravity: 1.55
Solubility: Decomposes to form flammable hydrogen gas

History
The name Calcium was coined from Latin for *calx*, meaning lime: Although the Romans prepared lime in the first century under the name *calx*, the metal was not discovered until 1808.

Sources
Calcium is a metallic element, fifth in abundance in the earth's crust, of which it forms more than 3%. Calcium is an essential constituent of leaves, bones, teeth, and shells. Never found in nature uncombined, it occurs abundantly as limestone, gypsum, and fluorite.

Important compounds
Calcium dihydride: CaH_2
Calcium difluoride: CaF_2
Calcium dichloride: $CaCl_2$
Calcium diiodide: CaI_2
Calcium oxide: CaO
Calcium peroxide: CaO_2
Calcium sulfide: CaS
Calcium selenide: $CaSe$
Calcium telluride: $CaTe$
Tricalcium dinitride: Ca_3N_2

Uses
It is used as an alloying agent for aluminum, beryllium, copper, lead, and magnesium alloys. It is also used as a fertilizer ingredient.

Isotopes
Calcium 40, 42, 43, 44, 46, 48

Important reactions
Calcium is a silvery white metal. The surface of calcium metal is covered with a thin layer of oxide that helps protect the metal from attack by air, but to a lesser extent than the corresponding layer in magnesium. Once ignited, calcium metal burns in air to give a mixture of

white calcium oxide, CaO, and calcium nitride, Ca_3N_2. Calcium oxide is more normally made by heating calcium carbonate. Calcium, immediately below magnesium in the periodic table, is more reactive with air than magnesium. Calcium reacts slowly with water. This is in contrast with magnesium, immediately above calcium in the periodic table, which is virtually unreactive with cold water. The reaction forms calcium hydroxide, $Ca(OH)_2$, and hydrogen gas (H_2). The calcium metal sinks in water, and after an hour or so, bubbles of hydrogen are evident, stuck to the surface of the metal. Calcium is very reactive toward the halogens fluorine, F_2, chlorine, Cl_2, bromine, Br_2, or iodine, I_2, and burns to form the dihalides calcium(II) fluoride, CaF_2, calcium(II) chloride, $CaCl_2$, calcium(II) bromide, $CaBr_2$, and calcium(II) iodide, CaI_2, respectively. The reactions with bromine and iodine require heat to enable the formation of the products. Calcium metal dissolves readily in dilute or concentrated hydrochloric acid to form solutions containing the aquated Ca(II) ion together with hydrogen gas, H_2.

Hazards to responders

Fire hazard   Pyrophoric, ignites in air when finely divided and then burns with crimson flame. Calcium rapidly decomposes in water; the heat of reaction is sufficient that hydrolysis-released hydrogen may ignite.

Produces flammable gases on contact with water. It may ignite on contact with water or moist air. Some react vigorously or explosively on contact with water. It may be ignited by heat, sparks, or flames. May reignite after fire is extinguished. Some are transported in highly flammable liquids. Runoff may create fire or explosion hazard.

Health hazard   Contact with eyes or skin produces caustic burns.

Barium

Symbol: Ba
Metal or nonmetal: Metal
Atomic number: 56
Atomic weight: 137.33
Periodic table group: II
Periodic table group name: Alkaline earth metal
Outer-shell electrons: 2
Ordinary physical state: Solid
Color: Silvery white
Case registry number: 7440-39-3
U.N. four digit number: 1400
DOT hazard class: 4 flammable solid
Placard: 4.3 dangerous when wet
NFPA 704: Unavailable
Boiling point: 1870°C, 3398°F
Autoignition temperature: N/A
Melting point: 725°C
Density: 3.5 g cm^{-3} at 20°C
Vapor pressure: 98 Pa at 729°C
Solubility: water-soluble
PEL-OSHA: N/A

History

The name barium comes from the Greek word *barys*, meaning heavy. Baryta was distinguished from lime by Scheele in 1774; the element was discovered by Sir Humphrey Davy in 1808.

Sources

Barium is found only in combination with other elements, chiefly with sulfate and carbonate, and is prepared by electrolysis of the chloride.

Important compounds

Barium dihydride: BaH_2
Barium difluoride: BaF_2
Barium dichloride: $BaCl_2$
Barium dichloride dihydrate: $BaCl_2 \cdot 2H_2O$
Barium dibromide dihydrate: $BaBr_2 \cdot 2H_2O$
Barium diiodide: BaI_2
Barium diiodide dihydrate: $BaI_2 \cdot 2H_2O$
Barium oxide: BaO
Barium peroxide: BaO_2
Barium sulfide: BaS
Barium trisulfide: $Ba(S_3)$
Barium selenide: $BaSe$
Tribarium dinitride: Ba_3N_2

Uses

The sulfate, as permanent white, is also used in paint, in x-ray diagnostic work, and in glassmaking. Barite is extensively used as a weighting agent in oil-well drilling fluids and is used in making rubber. The carbonate has been used as a rat poison, while the nitrate and chlorate give colors in pyrotechny. All barium compounds that are water or acid soluble are poisonous.

Isotopes

Barium 130, 132, 134, 135, 136, 137, 138

Important reactions

Barium reacts readily with water, ammonia, halogens, oxygen, and most acids. It reacts with incandescence when heated with boron trifluoride. Mixtures of finely divided barium metal and a number of halogenated hydrocarbons (such as monofluorotrichloromethane, trichlorotrifluoroethane, carbon tetrachloride, trichloroethylene, or tetrachloroethylene) are explosives.

Hazards to responders

Fire hazard Barium metal has only limited use and presents an explosion hazard. It reacts with moisture in the air. It rapidly reacts with water to generate gaseous hydrogen. The heat of reaction is sufficient that the evolved hydrogen may ignite. It is pyrophoric in powdered form. Certain compounds, particularly the peroxide, nitrate, and chlorate, present fire hazards in use and storage. It produces flammable gases on contact with water. It may ignite on contact with water or moist air. Some react vigorously or explosively on contact with water. It may be ignited by heat, sparks, or flames. It may reignite after fire is extinguished. Some are transported in highly flammable liquids. Runoff may create fire or explosion hazard.

   Health hazard   Inhalation or contact with vapors, substance, or decomposition products may cause severe injury or death. It may produce corrosive solutions on contact with water. Fire will produce irritating, corrosive, and/or toxic gases. Runoff from fire control may cause pollution.
   The soluble compounds of barium (chloride, nitrate, hydroxide) are highly toxic; the inhalation of the insoluble compounds (sulfate) may give rise to pneumoconiosis. Many of the compounds, including the sulfide, oxide, and carbonate, may cause local irritation to the eyes, nose, throat, and skin.

Titanium

Symbol: Ti
Metal or nonmetal: Metal
Atomic number: 42
Atomic weight: 47.88
Periodic table group:
Periodic table group name: Transition elements
Ordinary physical state: Solid
Color: Silvery
Case registry number: 7440-32-6
U.N. four digit number: 2546 (Powder)
DOT hazard class: 4.2 spontaneously combustible
NFPA 704: Unavailable
Radioactive: No
Boiling point: 3287°C, 5949°F
Autoignition temperature: 2200°F metal, 480°F for powder
Melting point: 1668°C, 3034°F
Specific gravity: 4.506
Solubility: None, reacts with water
PEL-OSHA: None
ACGIH-TLV-TWA: None

History

The name titanium was derived from Latin word *titans*, the first sons of the earth according to Greek mythology. It was discovered by Gregor in 1791 and named by Klaproth in 1795. Impure titanium was prepared by Nilson and Pettersson in 1887; however, the pure metal (99.9%) was not made until 1910 by Hunter by heating $TiCl_4$ with *sodium* in a steel bomb. Titanium is present in meteorites and in the sun. Rocks obtained during the Apollo 17 lunar mission showed the presence of 12.1% TiO_2, and rocks obtained during earlier Apollo missions showed lower percentages. Titanium, when pure, is a lustrous white metal. It has a low density and good strength, is easily fabricated, and has excellent corrosion resistance.

Sources

Titanium is the ninth most abundant in the crust of the earth. It is almost always present in igneous rocks and in the sediments derived from them. It is present in the ash of coal, in plants, and in the human body. The metal was a laboratory curiosity until Kroll, in 1946, showed that titanium could be produced commercially by reducing titanium tetrachloride with *magnesium*. This method is largely used for producing the metal today.

Important compounds
Titanium dihydride: TiH_2
Titanium difluoride: TiF_2
Titanium trifluoride: TiF_3
Titanium tetrafluoride: TiF_4
Titanium dichloride: $TiCl_2$
Titanium trichloride: $TiCl_3$
Titanium tetrachloride: $TiCl_4$
Titanium tribromide: $TiBr_3$
Titanium tetrabromide: $TiBr_4$
Titanium diiodide: TiI_2
Titanium triiodide: TiI_3
Titanium tetraiodide: TiI_4
Titanium oxide: TiO
Titanium dioxide: TiO_2
Dititanium trioxide: Ti_2O_3
Trititanium pentoxide: Ti_3O_5
Titanium sulfide: TiS
Titanium disulfide: TiS_2
Dititanium trisulfide: Ti_2S_3

Uses
Titanium is important as an alloying agent with *aluminum, molybdenum, manganese, iron,* and other metals. Alloys of titanium are principally used for aircraft and missiles, where lightweight strength and ability to withstand extremes of temperature are important. Titanium is as strong as steel, but 45% lighter. It is 60% heavier than aluminum, but twice as strong. Titanium has potential use in desalination plants for converting seawater into freshwater. The metal has excellent resistance to seawater and is used for propeller shafts, rigging, and other parts of ships exposed to saltwater. A titanium anode coated with platinum has been used to provide cathodic protection from corrosion by saltwater. Titanium dioxide is extensively used for both house paint and artist's paint, because it is permanent and has good covering power. Titanium oxide pigment accounts for the largest use of the element. Titanium paint is an excellent reflector of infrared and is extensively used in solar observatories, where heat causes poor visibility conditions.

Isotopes
46, 47, 48, 49, 50

Important reactions
Air and water reactions Titanium is highly flammable. It is pyrophoric in dust form. It is water reactive at 371°F (700°C), releasing hydrogen, which may cause an explosion.

Hazards to responders
Fire hazard Titanium is flammable/combustible material. It may ignite on contact with moist air or moisture. It may burn rapidly with flare-burning effect. Some react

vigorously or explosively on contact with water. Some may decompose explosively when heated or involved in a fire. May reignite after fire is extinguished. Runoff may create fire or explosion hazard. Containers may explode when heated.

Health hazard   Fire will produce irritating, corrosive, and/or toxic gases. Inhalation of decomposition products may cause severe injury or death. Contact with substance may cause severe burns to skin and eyes. Runoff from fire control may cause pollution.

Chemical hazards   Titanium reacts violently with cupric oxide and lead oxide when heated. When titanium is heated with potassium chlorate, potassium nitrate, or potassium permanganate, an explosion occurs. The residue from the reaction of titanium with red fuming nitric acid exploded violently when the flask was touched. Liquid oxygen gives a detonable mixture when combined with powdered titanium. The metal, which burns in air, is the only element that burns in *nitrogen*. Titanium metal is considered to be physiologically inert.

Chromium

Symbol: Cr
Metal or nonmetal: Metal
Atomic number: 24
Atomic weight: 51.996
Periodic table group:
Periodic table group name: Transitional elements
Ordinary physical state: Solid
Color: Silvery metallic
Case registry number: 7440-47-3
U.N. four digit number: 1756
DOT hazard class: 8 corrosive
NFPA 704: Unavailable
Radioactive: No
Boiling point: 2671°C, 4840°F
Melting point: 1907°C, 3465°F
Specific gravity: 7.19
Solubility: Insoluble in water
PEL-OSHA: 5 µg/m^3 of air, calculated as an 8-h time-weighted average (TWA)
ACGIH-TLV-TWA: 0.001 mg Cr(VI)/m^3 10-h TWA

History
The name chromium was coined from the Greek word, *chroma*, meaning color. It was discovered in 1797 by Vauquelin, who prepared the metal the next year, chromium is a steel-gray, lustrous, hard metal that takes a high polish.

Sources
The principal ore is chromite, which is found in Zimbabwe, Russia, Transvaal, Turkey, Iran, Albania, Finland, Democratic Republic of Madagascar, and the Philippines.

Important compounds
All compounds of chromium are colored; the most important are the chromates of *sodium* and *potassium* and the dichromates and the *potassium* and ammonium chrome sulfates.

The dichromates are used as oxidizing agents in quantitative analysis and also in tanning leather. Other compounds are of industrial value; lead chromate is chrome-yellow, a valued pigment. Chromium compounds are used in the textile industry as mordants (substances capable of binding dyes to textile fibers) and by the aircraft and other industries for anodizing aluminum. Chromium compounds are toxic and should be handled with proper safeguards.

Chromium difluoride: CrF_2
Chromium trifluoride: CrF_3
Chromium hexafluoride: CrF_6
Chromium tetrafluoride: CrF_4
Chromium pentafluoride: CrF_5
Chromium dichloride: $CrCl_2$
Chromium trichloride: $CrCl_3$
Chromium tetrachloride: $CrCl_4$
Chromium dibromide: $CrBr_2$
Chromium tribromide: $CrBr_3$
Chromium tetrabromide: $CrBr_4$
Chromium diiodide: CrI_2
Chromium triiodide: CrI_3
Chromium tetraiodide: CrI_4
Chromium dioxide: CrO_2
Chromium trioxide: CrO_3
Dichromium trioxide: Cr_2O_3
Trichromium tetroxide: Cr_3O_4
Chromium sulfide: CrS
Dichromium trisulfide: Cr_2S_3
Chromium selenide: $CrSe$
Chromium nitride: CrN
Chromium hexacarbonyl: $Cr(CO)_6$

Uses

Chromium is used to harden steel, to manufacture stainless steel, and to form many useful alloys. It is used in plating to produce a hard, beautiful surface and to prevent corrosion. Chromium gives glass an emerald green color and is widely used as a catalyst. The refractory industry has found chromite useful for forming bricks and shapes, as it has a high melting point, moderate thermal expansion, and stability of crystalline structure.

Important reactions

Air and water reactions   It may be pyrophoric, as dust.

Hazards to responders

Flammability   Noncombustible, substance itself does not burn but may decompose upon heating to produce corrosive and/or toxic fumes. Some are oxidizers and may ignite combustibles (wood, paper, oil, clothing, etc.). Contact with metals may evolve flammable hydrogen gas. Containers may explode when heated.

Health hazard Chromium may be exposed through inhalation, ingestion, skin, and/or eye contact.

Symptoms are irritation of eyes, skin problems and lung fibrosis (histologic).

Target organs are eyes, skin, and the respiratory system.

Manganese

Symbol: Mn
Metal or nonmetal: Metal
Atomic number: 25
Atomic weight: 54.9380
Periodic table group:
Periodic table group name: Transition metals
Ordinary physical state: Solid
Color: Silvery metallic
Case registry number: 7439-96-5
U.N. four digit number: None
DOT hazard class: None
NFPA 704: Unavailable
Radioactive: No
Boiling point: 2061°C, 3742°F
Melting point: 1246°C, 2275°F
Specific gravity: 7.2–7.4
Solubility: Soluble in water
PEL-OSHA: 5 mg/m^3 ceiling
ACGIH-TLV-TWA: 0.2 mg/m^3 TWA

History

The name manganese was coined from the Latin *magnes*, meaning magnet, due to the magnetic properties of pyrolusite; also from Italian meaning manganese, a corrupt form of magnesia. It was recognized by Scheele, Bergman, and others as an element and isolated by Gahn in 1774 by reduction of the dioxide with *carbon*.

Sources

Manganese minerals are widely distributed; oxides, silicates, and carbonates are the most common. The discovery of large quantities of manganese nodules on the floor of the oceans may become a source of manganese. These nodules contain about 24% manganese, together with many other elements in lesser abundance. Most manganese today is obtained from ores found in Russia, Brazil, Australia, Republic of South Africa, Gabon, and India.

Important compounds

Manganese difluoride: MnF_2
Manganese trifluoride: MnF_3
Manganese tetrafluoride: MnF_4
Manganese dichloride: $MnCl_2$
Manganese trichloride: $MnCl_3$
Manganese dichloride dihydrate: $MnCl_2 \cdot 2H_2O$
Manganese dibromide: $MnBr_2$

Manganese diiodide: MnI_2
Manganese oxide: MnO
Manganese dioxide: MnO_2
Dimanganese trioxide: Mn_2O_3
Dimanganese heptoxide: Mn_2O_7
Trimanganese tetroxide: Mn_3O_4
Manganese sulfide: MnS
Manganese disulfide: MnS_2
Manganese selenide: $MnSe$
Manganese telluride: $MnTe$
Tetramanganese hexdecacarbonyl: $Mn_4(CO)_{16}$
Dimanganese decacarbonyl: $Mn_2(CO)_{10}$

Uses

The dioxide (pyrolusite) is used as a depolarizer in dry cells and is used to "decolorize" glass that is colored green by impurities of iron. Manganese by itself colors glass an amethyst color and is responsible for the color of true amethyst. The dioxide is also used in the preparation of oxygen and chlorine, and in drying black paints. The permanganate is a powerful oxidizing agent and is used in quantitative analysis and in medicine. Manganese is widely distributed throughout the animal kingdom. It is an important trace element and may be essential for utilization of vitamin B_1.

Isotopes

Mn 55

Important reactions

Concentrated nitric acid reacts with manganese with incandescence and a feeble explosion. Manganese or potassium ignites in nitrogen dioxide. Manganese burns with a brilliant flame when heated in sulfur dioxide vapor. Contact with concentrated hydrogen peroxide causes violent decomposition and/or ignition. Manganese burns in chlorine to form manganese(II) chloride, $MnCl_2$. It also reacts with bromine or iodine to form, respectively, manganese(II) bromide, $MnBr_2$, or manganese(II) iodide, MnI_2. The corresponding reaction between the metal and fluorine, F_2, affords the fluorides manganese(II) fluoride, MnF_2, and manganese (III) fluoride, MnF_3.

Manganese metal dissolves readily in dilute sulfuric acid to form solutions containing the aquated Mn(II) ion together with hydrogen gas, H_2. In practice, the Mn(II) is present as the virtually colorless complex ion $[Mn(OH_2)_6]^{2+}$.

Hazards to responders

Flammability Manganese dust (finely divided) has been known to be pyrophoric. Powdered manganese ignites in chlorine and burns brilliantly; with fluorine, the reaction takes place with incandescence.

Health hazard Exposure routes: Inhalation, ingestion. Symptoms: Parkinson's; asthenia, insomnia, mental confusion; metal fume fever: dry throat, cough, chest tightness, dyspnea (breathing difficulty), rales, flu-like fever; low-back pain; vomiting; malaise (vague feeling of discomfort); lassitude (weakness, exhaustion); kidney damage.

Target organs are respiratory system, central nervous system, blood, kidneys.

Iron

Symbol: Fe
Metal or nonmetal: Metal
Atomic number: 26
Atomic weight: 55.847
Periodic table group:
Periodic table group name: Transition metals
Ordinary physical state: Solid
Color: Lustrous, metallic, Grayish tinge
Case registry number: 7439-89-6
U.N. four digit number: None
DOT hazard class: None
NFPA 704: Unavailable
Radioactive: No
Boiling point: 2861°C, 5182°F
Melting point: 1538°C, 2800°F
Specific gravity: 3.6–5.4
Solubility: Soluble in water
PEL-OSHA: No limits
ACGIH-TLV-TWA: No limits

History

The name iron was coined from the Anglo-Saxon, iron; also *L. ferrum*. Iron was used pre-historically: Genesis mentions that Tubal-Cain, seven generations from Adam, was "an instructor of every artificer in brass and iron." A remarkable iron pillar, dating to about A.D. 400, remains standing today in Delhi, India. Corrosion to the pillar has been minimal, although it has been exposed to the weather since its erection. Pig iron is an alloy containing about 3% carbon, with varying amounts of *sulfur, silicon, manganese,* and *phosphorus*. Iron is hard, brittle, fairly fusible and is used to produce other alloys, including steel. Wrought iron contains only a few tenths of a percent of *carbon*; is tough, malleable, less fusible; and has usually a "fibrous" structure. Carbon steel is an alloy of iron with small amounts of Mn, S, P, and Si. Alloy steels are carbon steels with other additives, such as *nickel, chromium, vanadium.* Iron is a cheap, abundant, useful, and important metal.

Sources

Iron is a relatively abundant element in the universe. It is found in the sun and many types of stars in considerable quantity. Its nuclei are quite stable. The core of the earth, 2150 miles in radius, is thought to be largely composed of iron, with about 10% occluded hydrogen. The metal is the fourth most-abundant element, by weight, which makes up the crust of the earth. The most common ore is hematite, which is frequently seen as black sands along beaches and banks of streams.

Important compounds

Iron difluoride: FeF_2
Iron trifluoride: FeF_3
Iron dichloride: $FeCl_2$
Iron trichloride: $FeCl_3$
Iron dibromide: $FeBr_2$

Iron tribromide: $FeBr_3$
Iron diiodide: FeI_2
Iron triiodide: FeI_3
Iron oxide: FeO
Diiron trioxide: Fe_2O_3
Triiron tetraoxide: Fe_3O_4
Iron sulfide: FeS
Iron persulfide: FeS_2
Iron selenide: $FeSe$
Iron telluride: $FeTe$
Diiron nitride: Fe_2N
Iron pentacarbonyl: $Fe(CO)_5$
Diiron nonacarbonyl: $Fe_2(CO)_9$
Triiron dodecacarbonyl: $Fe_3(CO)_{12}$

Uses

Iron is a vital constituent of plant and animal life and appears in hemoglobin. The pure metal is not often encountered in commerce, but is usually alloyed with carbon or other metals.

Isotopes

Fe 54, 56, 57, 58

Important reactions

Iron metal reacts in moist air by oxidation to give a hydrated iron oxide. This does not protect the iron surface to further reaction since it flakes off, exposing more iron metal to oxidation. This process is called rusting and is familiar to any car owner. Finely divided iron powder is pyrophoric, making it a fire risk. On heating with oxygen, O_2, the result is the formation of the iron oxides Fe_2O_3 and Fe_3O_4. Air-free water has little effect upon iron metal. However, iron metal reacts in moist air by oxidation to give a hydrated iron oxide. This does not protect the iron surface from further reaction since it flakes off, exposing more iron metal to oxidation. This process is called rusting and is familiar to any car owner. Iron reacts with excess of the halogens F_2, Cl_2, and Br_2, to form ferric, that is, Fe(III), halides. This reaction is not very successful for iodine because of thermodynamic problems. The iron(III) is too oxidizing and the iodide is too reducing. The direct reaction between iron metal and iodine can be used to prepare iron (II) iodide, FeI_2. Iron metal dissolves readily in dilute sulfuric acid in the absence of oxygen to form solutions containing the aquated Fe(II) ion together with hydrogen gas, H_2. In practice, the Fe(II) is present as the complex ion $[Fe(OH_2)_6]^{2+}$. If oxygen is present, some of the Fe(II) oxidizes to Fe(III). The strongly oxidizing concentrated nitric acid, HNO_3, reacts on the surface of iron and passivates the surface.

Hazards to responders

Flammability   Iron metal powder is a fire hazard.

Health   Unless known otherwise, all iron compounds should be regarded as toxic. Iron deficiency leads to anemia. Excess iron in the body causes liver and kidney damage (hemochromatosis). Some iron compounds are suspected carcinogens.

Chemical hazards   The pure metal is highly reactive chemically and rapidly corrodes, especially in moist air or at elevated temperatures.

Cobalt

Symbol: Co
Metal or nonmetal: Metal
Atomic number: 27
Atomic weight: 58.9392
Periodic table group:
Periodic table group name: Transition metals
Ordinary physical state: Solid
Color: Lustrous, metallic, grayish tinge
Case registry number: 7440-48-4
U.N. four digit number: 3178
DOT hazard class: 4.1 flammable solid
NFPA 704: Unavailable
Radioactive: Isotopes
Boiling point: 2927°C, 5301°F
Melting point: 1495°C, 2723°F
Specific gravity: 8.92
Solubility: Insoluble in water
PEL-OSHA: 0.1 mg/m^3 of air as an 8 h TWA
ACGIH-TLV-TWA: 0.02 mg/m^3 as a TWA for a normal 8 h workday and a 40 h workweek

History

The term cobalt was coined from the German word, *Kobald*, meaning goblin or evil spirit; also from *cobalos*, the Greek word for mine. Brandt discovered cobalt in about 1735. It is a brittle, hard metal, resembling iron and nickel in appearance.

Sources

Cobalt occurs in the minerals cobaltite, smaltite, and erythrite and is often associated with *nickel, silver, lead, copper,* and *iron* ores, from which it is most frequently obtained as a by-product. It is also present in meteorites. Important ore deposits are found in Zaire, Morocco, and Canada. The U.S. Geological Survey has announced that the bottom of the north central Pacific Ocean may have cobalt-rich deposits at relatively shallow depths in water close to the Hawaiian Islands and other U.S. Pacific territories.

Important compounds

Cobalt difluoride: CoF_2
Cobalt trifluoride: CoF_3
Cobalt tetrafluoride: CoF_4
Cobalt dichloride: $CoCl_2$
Cobalt trichloride: $CoCl_3$
Cobalt dichloride dihydrate: $CoCl_2 \cdot 2H_2O$
Cobalt dibromide: $CoBr_2$
Cobalt diiodide: CoI_2
Cobalt oxide: CoO
Dicobalt trioxide: Co_2O_3
Tricobalt tetraoxide: Co_3O_4
Cobalt sulfide: CoS
Cobalt persulfide: CoS_2

Dicobalt trisulfide: Co_2S_3
Cobalt selenide: $CoSe$
Cobalt telluride: $CoTe$
Dicobalt octacarbonyl: $Co_2(CO)_8$
Tetracobalt dodecacarbonyl: $Co_4(CO)_{12}$
Hexacobalt hexadecacarbonyl: $Co_6(CO)_{16}$

Uses

It is alloyed with iron, nickel, and other metals to make alnico, an alloy of unusual magnetic strength with many important uses. Stellite alloys, containing cobalt, *chromium*, and *tungsten*, are used for high-speed, heavy-duty, high-temperature cutting tools and for dies. Cobalt is also used in other magnetic steels and stainless steels, and in alloys used in jet turbines and gas turbine generators. The metal is used in electroplating because of its appearance, hardness, and resistance to oxidation. The salts have been used for centuries to produce brilliant and permanent blue colors in porcelain, glass, pottery, tiles, and enamels. A solution of the chloride is used as a sympathetic ink. Cobalt carefully used in the form of the chloride, sulfate, acetate, or nitrate has been found effective in correcting a certain mineral-deficiency disease in animals.

Isotopes

Co 59

Radioactive:
55, 56, 57, 58, 60, 61, 62

Important reactions

Cobalt burns brilliantly when exposed to air. It is insoluble in water. Pyrophoric cobalt is a reducing agent. It decomposes acetylene in the cold as the metal becomes incandescent. It is incompatible with oxidizing agents such as ammonium nitrate, bromine pentafluoride, and nitryl fluoride.

Hazards to responders

Flammability   Literature sources indicate that the dust of this chemical is flammable.

Health hazard   Exposure routes: Inhalation, ingestion, skin and/or eye contact.

Symptoms: Cough, dyspnea (breathing difficulty), wheezing, decreased pulmonary function; weight loss; dermatitis; diffuse nodular fibrosis; respiratory hypersensitivity, asthma.

Target organs: Skin, respiratory system. Exposure to cobalt (metal fumes and dust) should be limited to 0.05 mg/m^3 (8-h TWA in a 40-h week).

Copper

Symbol: Cu
Metal or nonmetal: Metal
Atomic number: 20
Atomic weight: 63.546
Periodic table group:
Periodic table group name: Transitional metals

Ordinary physical state: Solid
Color: Copper, metallic
Case registry number: 7440-50-8
U.N. four digit number: None
DOT hazard class: None
NFPA 704: Unavailable
Radioactive: No
Boiling point: 2927°C, 5301°F
Melting point: 1084.62°C, 1984.32°F
Specific gravity: 4.1–4.3
Solubility: Copper II ions are water soluble
Density: 8.9 g cm^{-3} at 20°C
PEL-OSHA: (fume) 0.1 mg/m^3 TWA
ACGIH-TLV-TWA: (fume) 0.2 mg/m^3 TWA

History
The term copper was coined from Latin for *cuprum*, from the island of Cyprus. It is believed that copper has been mined for 5000 years. Copper is reddish and takes on a bright metallic luster. It is malleable, ductile, and a good conductor of heat and electricity (second only to *silver* in electrical conductivity).

Sources
Large copper ore deposits are found in the United States, Chile, Zambia, Zaire, Peru, and Canada. The most important copper ores are the sulfides, the oxides, and carbonates. From these, copper is obtained by smelting, leaching, and by electrolysis.

Important compounds
Copper fluoride: CuF
Copper difluoride: CuF_2
Copper chloride: CuCl
Copper dichloride: $CuCl_2$
Copper dichloride dihydrate: $CuCl_2 \cdot 2H_2O$
Copper dibromide: $CuBr_2$
Copper iodide: CuI
Copper oxide: CuO
Dicopper oxide: Cu_2O
Copper sulfide: CuS
Dicopper sulfide: Cu_2S
Copper selenide: CuSe
Dicopper selenide: Cu_2Se
Copper telluride: CuTe
Dicopper telluride: Cu_2Te

Uses
The electrical industry is one of the greatest users of copper. Copper's alloys, brass and bronze, are important: all American coins are copper alloys, and gun metals also contain copper. Copper has wide use as an agricultural poison and as an algicide in water purification. Copper compounds, such as Fehling's solution, are widely used in analytical chemistry tests for sugar.

Isotopes

Cu 63, 65

Important reactions

Copper metal is stable in air under normal conditions. At red heat, copper metal and oxygen react to form Cu_2O. The reaction between copper metal and the halogens fluorine, F_2, chlorine, Cl_2, or bromine, Br_2, affords the corresponding dihalides copper(II) fluoride, CuF_2, copper(II) chloride, $CuCl_2$, or copper(II) bromide, $CuBr_2$ respectively. Copper metal dissolves in hot concentrated sulfuric acid to form solutions containing the aquated Cu(II) ion together with hydrogen gas, H_2. In practice, the Cu(II) is present as the complex ion $[Cu(OH_2)_6]^{2+}$. Copper metal also dissolves in dilute or concentrated nitric acid, HNO_3.

Hazards to responders

Flammability   Not considered flammable.

Health   Long-term exposure to copper can cause irritation of the nose, mouth, and eyes, and it causes headaches, stomachaches, dizziness, vomiting, and diarrhea. Intentionally high uptakes of copper may cause liver and kidney damage and even death. Whether copper is carcinogenic has not been determined yet. There are scientific articles that indicate a link between long-term exposure to high concentrations of copper and a decline in intelligence with young adolescents. Whether this should be of concern is a topic for further investigation. Industrial exposure to copper fumes, dusts, or mists may result in metal fume fever with atrophic changes in nasal mucous membranes. Chronic copper poisoning results in Wilson's disease, characterized by a hepatic cirrhosis, brain damage, demyelization, renal disease, and copper deposition in the cornea.

Silver

Symbol: Ag
Metal or nonmetal: Metal
Atomic number: 47
Atomic weight: 107.8682
Periodic table group:
Periodic table group name: Transition metals
Ordinary physical state: Solid
Color: Silver
Case registry number: 7440-22-4
U.N. four digit number: None
DOT hazard class: None
NFPA 704: 0-0-1
Radioactive: No
Boiling point: 2162°C, 3924°F
Melting point: 961.78°C, 1763.2°F
Specific gravity: 10.4–10.6
Density: 10.49 g/cm³
Solubility: Insoluble in water
PEL-OSHA: 0.01 mg/m³ TWA
ACGIH-TLV-TWA: 0.1 mg/m³ TWA

History

This word has Anglo-Saxon origin, *Seolfor siolfur*, as well as Latin, *argentums*. Silver has been known since ancient times. It is mentioned in Genesis. Slag dumps in Asia Minor and on islands in the Aegean Sea indicate that man learned to separate silver from lead as early as 3000 B.C. Pure silver has a brilliant, white, metallic luster. It is a little harder than gold and is quite ductile and malleable, being exceeded only by gold and perhaps palladium.

Sources

Silver occurs both in a native form and in ores, such as argentite (Ag_2S) and horn silver (AgCl); lead, lead–zinc, copper, gold, and copper–nickel ores are principal sources. Mexico, Canada, Peru, and the United States are the primary silver producers in the Western Hemisphere. Silver is also recovered during electrolytic refining of copper. Commercial fine silver contains at least 99.9% silver. Purities of 99.999+% are available commercially.

Important compounds

Silver fluoride: AgF
Silver difluoride: AgF_2
Disilver fluoride: Ag_2F
Silver chloride: AgCl
Silver bromide: AgBr
Silver iodide: AgI
Silver oxide: AgO
Disilver oxide: Ag_2O
Disilver sulfide: Ag_2S
Disilver selenide: Ag_2Se
Disilver telluride: Ag_2Te
Silver nitrate: $AgNO_3$
Silver arsenite: Ag_3AsO_3
Silver cyanide: AgCN
Silver picrate: $C_6H_2O(NO_2)_3AgH_2O$

Uses

Sterling silver is used for jewelry, silverware, etc., where appearance is paramount. This alloy contains 92.5% silver, the remainder being copper or some other metal. Silver is of the utmost importance in photography, about 30% of the U.S. industrial consumption going into this application. It is also used for dental alloys. Silver is used in making solder and brazing alloys, electrical contacts, and high-capacity silver–zinc and silver–cadmium batteries. Silver paints are used for making printed circuits. It is used in mirror production and may be deposited on glass or metals by chemical deposition, electrode position, or by evaporation. When freshly deposited, it is the best reflector of visible light known, but tarnishes rapidly and loses much of its reflectance. Silver fulminate, a powerful explosive, is sometimes formed during the silvering process. Silver iodide is used in seeding clouds to produce rain. Silver chloride has interesting optical properties, as it can be made transparent; it also is a cement for glass. Silver nitrate, or lunar caustic, the most important silver compound, is used extensively in photography. For centuries, silver has been used traditionally for coinage by many countries of the world. In recent times, however, consumption of silver has greatly exceeded the output.

Isotopes
AG 107, 109

Important reactions
The thermally stable silver difluoride, silver(II) difluoride, AgF_2, is formed in the reaction of silver metal and fluorine, F_2.
    Silver metal dissolves in hot concentrated sulfuric acid.
    Silver metal also dissolves in dilute or concentrated nitric acid, HNO_3.

Hazards to responders
Flammability   It is not considered to be flammable.

Health   While silver itself is not considered to be toxic, most of its salts are poisonous. Exposure to silver (metal and soluble compounds, as Ag) in air should not exceed 0.01 mg/m^3 (8-h TWA in a 40-h week). Silver compounds can be absorbed in the circulatory system, and reduced silver deposited in the various tissues of the body. A condition, known as argyria, results in a grayish pigmentation of the skin and mucous membranes. Eye contact may cause severe corneal injury if liquid comes in contact with the eyes. Skin contact may cause skin irritation. Repeated and prolonged contact with skin may cause allergic dermatitis. Inhalation hazards are as follows: exposure to high concentrations of vapors may cause dizziness, breathing difficulty, headaches, or respiratory irritation. Extremely high concentrations may cause drowsiness, staggering, confusion, unconsciousness, coma, or death. Liquid or vapor may be irritating to skin, eyes, throat, or lungs. Intentional misuse by deliberately concentrating and inhaling the contents of this product can be harmful or fatal. Ingestion hazards are as follows: it is moderately toxic. It may cause stomach discomfort, nausea, vomiting, diarrhea, and narcosis. Aspiration of material into lungs if swallowed or if vomiting occurs can cause chemical pneumonitis, which can be fatal.

Gold

Symbol: Au
Metal or nonmetal: Metal
Atomic number: 79
Atomic weight: 196,9865
Periodic table group:
Periodic table group name: Transition metals
Ordinary physical state: Solid
Color: Gold
Case registry number: 7440-57-5
U.N. four digit number: None
DOT hazard class: None
NFPA 704: Unavailable
Radioactive: No
Boiling point: 2856°C, 5173°F
Melting point: 1064.18°C, 1947.52°F
Specific gravity: 19.32
Density: 19.32 g/cm^3

Solubility: Unavailable
PEL-OSHA: 0.001–0.01 mg/m^3
ACGIH-TLV-TWA: 0.001–0.01 mg/m^3

History

The term gold was coined from the Sanskrit word, *jval*; Anglo-Saxon, gold; and Latin, *aurum*. Known and highly valued from earliest times, gold is found in nature as the free metal and in tellurides; it is widely distributed and is almost always associated with quartz or pyrite. Of all the elements, gold in its pure state is undoubtedly the most beautiful. It is metallic, having a yellow color when in a mass, but when finely divided, it may be black, ruby, or purple. The Purple of Cassius is a delicate test for auric gold. It is the most malleable and ductile metal; 1 oz of gold can be beaten out to 300 ft^2. It is a soft metal and is usually alloyed to give it more strength. It is a good conductor of heat and electricity and is unaffected by air and most reagents.

Sources

It is estimated that all the gold in the world, so far refined, could be placed in a single cube 60 ft on a side. It occurs in veins and alluvial deposits and is often separated from rocks and other minerals by mining and panning operations. About two-thirds of the world's gold output comes from South Africa, and about two-thirds of the total U.S. production comes from South Dakota and Nevada. The metal is recovered from its ores by cyaniding, amalgamating, and smelting processes. Refining is also frequently done by electrolysis. Gold occurs in seawater to the extent of 0.1–2 mg/ton, depending on the location where the sample is taken. As yet, no method has been found for recovering gold from seawater profitably.

Important compounds

Gold trifluoride: AuF_3
Gold pentafluoride: AuF_5
Gold chloride: $AuCl$
Digold hexachloride: $[AuCl_3]_2$
Tetragold octachloride: Au_4Cl_8
Gold bromide: $AuBr$
Digold hexabromide: $[AuBr_3]_2$
Gold iodide: AuI
Gold triiodide: AuI_3
Digold trioxide: Au_2O_3
Digold sulfide: Au_2S
Digold trisulfide: Au_2S_3
Gold selenide: $AuSe$
Digold triselenide: Au_2Se_3
Gold ditelluride: $AuTe_2$

Uses

It is used in coinage and is a standard for monetary systems in many countries. It is also extensively used for jewelry, decoration, dental work, and plating. It is used for coating certain space satellites, as it is a good reflector of infrared and is inert.

Isotope

Au 197

Important reactions

Gold metal reacts with chlorine, Cl_2, or bromine, Br_2, to form the trihalides gold(III) chloride, $AuCl_3$, or gold(III) bromide, $AuBr_3$, respectively. On the other hand, gold metal reacts with iodine, I_2, to form the monohalide gold(I) chloride, AuI. Solutions of chlorine, Cl_2, and trimethylammonium chloride, $[NHMe_3]Cl$, in acetonitrile, MeCN, dissolve gold. Gold metal dissolves in *aqua regia*, a mixture of hydrochloric acid, HCl, and concentrated nitric acid, HNO_3, in a 3:1 ratio. The name aqua regia was coined by alchemists because of its ability to dissolve gold—the "king of metals."

Hazards to responders

Flammability   Gold is not flammable.

Health   Gold is not absorbed well by the body and its compounds are not normally particularly toxic.

Zinc

Symbol: Zn
Metal or nonmetal: Metal
Atomic number: 30
Atomic weight: 65.38
Periodic table group:
Periodic table group name: Transition metals
Ordinary physical state: Solid
Color: Bluish, pale gray
Case registry number: 7440-66-6
U.N. four digit number: 1436 (dust)
DOT hazard class: 4.1 flammable solid
NFPA 704: Unavailable
Radioactive: No
Boiling point: 907°C, 1665°F
Melting point: 419.53°C, 787.15°F
Specific gravity: 3.9–7.2
Density: 7.140 g/mL
Solubility: Insoluble in water
PEL-OSHA: None established
ACGIH-TLV-TWA: None established

History

The term zinc was coined from the German, *zink*, a word of obscure origin. Centuries before zinc was recognized as a distinct element, zinc ores were used for making brass. An alloy containing 87% zinc has been found in prehistoric ruins in Transylvania. Metallic zinc was produced in the thirteenth century A.D. in India by reducing calamine with organic substances, such as wool. The metal was rediscovered in Europe by Marggraf in 1746, who showed that it could be obtained by reducing calamine with charcoal. Zinc is a bluish-white lustrous metal. It is brittle at ordinary temperatures, but malleable at 100°C–150°C. It is a fair conductor of electricity and burns in air at high red heat, with evolution of white clouds of the oxide. It exhibits superplasticity.

Sources

The principal ores of zinc are sphalerite (sulfide), smithsonite (carbonate), calamine (silicate), and franklinite (zinc, manganese, iron oxide). One method of zinc extraction involves roasting its ores to form the oxide and reducing the oxide with coal or carbon, with subsequent distillation of the metal.

Important compounds

Zinc dihydride: ZnH_2
Zinc difluoride: ZnF_2
Zinc dichloride: $ZnCl_2$
Zinc dibromide: $ZnBr_2$
Zinc diiodide: ZnI_2
Zinc oxide: ZnO
Zinc peroxide: ZnO_2
Zinc sulfide: ZnS
Zinc selenide: $ZnSe$
Zinc telluride: $ZnTe$
Trizinc dinitride: Zn_3N_2

Uses

The metal is employed to form numerous alloys with other metals. Brass, nickel silver, typewriter metal, commercial bronze, spring bronze, German silver, soft solder, and aluminum solder are some of the more important alloys. Large quantities of zinc are used to produce die-castings, which are used extensively by the automotive, electrical, and hardware industries. An alloy called Prestal®, consisting of 78% zinc and 22% *aluminum*, is reported to be almost as strong as steel and as easy to mold as plastic. In fact, it is so moldable that it can be molded into form using inexpensive ceramics or cement die-casts. Zinc is also used extensively to galvanize other metals, such as *iron*, to prevent corrosion. Zinc oxide is a unique and useful material for modern civilization. It is widely used in the manufacture of paints, rubber products, cosmetics, pharmaceuticals, floor coverings, plastics, printing inks, soap, storage batteries, textiles, electrical equipment, and other products. Lithopone, a mixture of zinc sulfide and barium sulfate, is an important pigment. Zinc sulfide is used in making luminous dials, x-ray and TV screens, and fluorescent lights. Zinc is an essential element in the growth of human beings and animals. Tests show that zinc-deficient animals require 50% more food to gain the same weight as an animal supplied with sufficient zinc.

Isotopes

Zn 64, 66, 67, 68, 69, 70

Important reactions

Zinc dibromide, zinc(II) dibromide, $ZnBr_2$, and zinc diiodide, zinc(II) diiodide, NiI_2, are formed in the reactions of zinc metal and bromine, Br_2, or iodine, I_2. Zinc metal dissolves slowly in dilute sulfuric acid to form solutions containing the aquated Zn(II) ion together with hydrogen gas, H_2. In practice, the Zn(II) is present as the complex ion $[Zn(OH_2)_6]^{2+}$. The reactions of zinc with oxidizing acids such as nitric acid, HNO_3, are complex and depend upon precise conditions. Zinc metal dissolves in aqueous alkalies such as potassium hydroxide, KOH, to form zincates such as $[Zn(OH)_4]^{2-}$. The resulting solutions contain other species as well.

Hazards to responders

Flammability   Zinc dust is a severe fire hazard. It produces flammable gases on contact with water. It may ignite on contact with water or moist air. Some react vigorously or explosively on contact with water. It may be ignited by heat, sparks, or flames.

Health   When freshly formed ZnO is inhaled, a disorder known as the oxide shakes or zinc chills sometimes occurs. Where zinc oxide is encountered, recommendations include providing good ventilation to avoid concentration exceeding 5 mg/m^3 (time-weighted over an 8-h exposure, 40-h workweek). Zinc metal is a human skin irritant and but otherwise is nontoxic. Most common zinc compounds are not very toxic, but a few zinc salts may be carcinogens. Use of some zinc compounds is permitted around food. Pollution from industrial smoke may cause lung disease.

Mercury

Symbol: Hg
Metal or nonmetal: Metal
Atomic number: 80
Atomic weight: 200.59
Periodic table Group:
Periodic table group name: Transition metals
Ordinary physical state: Liquid
Color: Silvery white
Case registry number: 7439 97 6
U.N. four digit number: 2809
DOT hazard class: Corrosive
NFPA 704: Data unavailable
Radioactive: No
Boiling point: 356.73°C, 674.11°F
Melting point: −38.83°C, −37.89°F
Specific gravity: 13.534
Vapor density: 7.0
Vapor pressure: 0.0018 at 77°F
Solubility: Insoluble in water
PEL-OSHA: 0.1 mg/m^3 of air as a ceiling limit
ACGIH-TLV-TWA: 0.025 mg/m^3 as a TWA for a normal 8-h workday and a 40-h workweek

History

Mercury was named after the planet mercury; known to ancient Chinese and Hindus; found in Egyptian tombs of 1500 B.C. Mercury is the only common metal forming a liquid at ordinary temperatures. It only rarely occurs free in nature. The chief ore is cinnabar. The commercial unit for handling mercury is the "flask," which weighs 76 lb.

Sources

Spain and Italy produce about 50% of the world's supply of the metal. The metal is obtained by heating cinnabar in a current of air and condensing the vapor. It is a heavy, silvery-white metal; a rather poor conductor of heat, as compared with other metals; and a fair conductor of electricity.

Important compounds

It easily forms alloys with many metals, such as gold, silver, and tin, which are called amalgams. Its ease in amalgamating with gold is made use of in the recovery of gold from its ores. The most important salts are mercury chloride (corrosive sublimate—a violent poison), mercurous chloride (calomel, occasionally still used in medicine), mercury fulminate (a detonator widely used in explosives), and mercuric sulfide (vermilion, a high-grade paint pigment). Organic mercury compounds are important.

Mercury dihydride: HgH_2
Mercury difluoride: HgF_2
Dimercury difluoride: Hg_2F_2
Mercury dichloride: $HgCl_2$
Dimercury dichloride: Hg_2Cl_2
Dimercury dibromide: Hg_2Br_2
Mercury diiodide: HgI_2
Dimercury diiodide: Hg_2I_2
Mercury oxide: HgO
Dimercury oxide: Hg_2O
Mercury sulfide: HgS
Mercury selenide: $HgSe$
Mercury telluride: $HgTe$

Uses

The metal is widely used in laboratory work for making thermometers, barometers, diffusion pumps, and many other instruments. It is used in making mercury-vapor lamps and advertising signs, and is used in mercury switches and other electronic apparatus. Other uses are in making pesticides, mercury cells for caustic soda and chlorine production, dental preparations, antifouling paint, batteries, and catalysts.

Isotopes

Hg 196, 198, 199, 200, 201, 202, 204

Important reactions

Mercury forms an explosive acetylide when mixed with acetylene. Can form explosive compounds with ammonia (a residue resulting from such a reaction exploded when an attempt was made to clean it off a steel rod). Chlorine dioxide (also other oxidants, such as chlorine, bromine, nitric acid, performic acid), and mercury explode when mixed. Methyl azide in the presence of mercury is potentially explosive. Ground mixtures of sodium carbide and mercury can react vigorously. Ammonia forms explosive compounds with gold, mercury, or silver. Mercury metal reacts with fluorine, F_2, chlorine, Cl_2, bromine, Br_2, or iodine, I_2, to form the dihalides mercury(II) fluoride, HgF_2, mercury(II) chloride, $HgCl_2$, mercury(II) bromide, $HgBr_2$, or mercury(II) iodide, HgI_2, respectively.

Mercury does not react with nonoxidizing acids but does react with concentrated nitric acid, HNO_3, or concentrated sulfuric acid, H_2SO_4, to form mercury(II) compounds together with nitrogen or sulfur oxides.

Mercury dissolves slowly in dilute nitric acid to form mercury(I) nitrate, mercurous nitrate, $Hg_2(NO_3)_2$.

Hazards to responders

Flammability   It is not considered flammable.

Health Mercury is a virulent poison and is readily absorbed through the respiratory tract, the gastrointestinal tract, or unbroken skin. It acts as a cumulative poison, and dangerous levels are readily attained in air. Air saturated with mercury vapor at 20°C contains a concentration that exceeds the toxic limit many times. The danger increases at higher temperatures. It is therefore important that mercury be handled with care. Methyl mercury is a dangerous pollutant and is now widely found in groundwater and streams. It is toxic by ingestion, absorption, and inhalation of the fumes. As poisoning becomes established, slight muscular tremor, loss of appetite, nausea, and diarrhea are observed. Psychic, kidney, and cardiovascular disturbances may occur.

Boron

Symbol: B
Metal or nonmetal: Nonmetal
Atomic number: 5
Atomic weight: 10.81
Periodic table group: III
Periodic table group name: N/A
Ordinary physical state: Solid
Color: Black
Case registry number: 7440-42-8
U.N. four digit number: None
DOT hazard class: None
NFPA 704: Unavailable
Radioactive: No
Boiling point: 3927°C, 7101°F
Melting point: 2076°C, 3769°F
Specific gravity: 2.4
Density: 2340 kg/m^3 or 2.34 g/cm^3
Solubility: Soluble in water
PEL-OSHA: None identified
ACGIH-TLV-TWA: None identified

History
The term born was coined from the Argentine word, *buraq*, and the Persian, *burah*. Boron compounds have been known for thousands of years, but the element was not discovered until 1808 by Sir Humphry Davy and by Gay-Lussac and Thenard.

Sources
The element is not found free in nature, but occurs as orthoboric acid, usually found in certain volcanic springwaters, and as borates in boron and colemanite. Ulexite, another boron mineral, is interesting, as it is nature's own version of "fiber optics." Important sources of boron are ore rasorite (kernite) and tincal (borax ore). Both of these ores are found in the Mojave Desert. Tincal is the most important source of boron from the Mojave. Extensive borax deposits are also found in Turkey. High-purity crystalline boron may be prepared by the vapor-phase reduction of boron trichloride or tribromide with hydrogen on electrically heated filaments. The impure, or amorphous, boron, a brownish-black powder, can

be obtained by heating the trioxide with magnesium powder. Boron of 99.9999% purity has been produced and is available commercially.

Important compounds
Diborane(6): B_2H_6
Decaborane(14): $B_{10}H_{14}$
Hexaborane(10): B_6H_{10}
Pentaborane(9): B_5H_9
Pentaborane(11): B_5H_{11}
Tetraborane(10): B_4H_{10}
Boron trifluoride: BF_3
Diboron tetrafluoride: B_2F_4
Boron trichloride: BCl_3
Diboron tetrachloride: B_2Cl_4
Boron tribromide: BBr_3
Boron triiodide: BI_3
Diboron trioxide: B_2O_3
Diboron trisulfide: B_2S_3
Boron nitride: BN

Uses
Amorphous boron is used in pyrotechnic flares to provide a distinctive green color and in rockets as an igniter. By far, the most commercially important boron compound in terms of dollar sales is $Na_2B_4O_7 \cdot 5H_2O$. This pentahydrate is used in large quantities in the manufacture of insulation fiberglass and sodium perborate bleach. Boric acid is also an important boron compound, with major markets in textile products. Use of borax as a mild antiseptic is minor in terms of dollars and tons. Boron compounds are also extensively used in the manufacture of borosilicate glasses. Other boron compounds show promise in treating arthritis. The isotope boron-10 is used as a control for nuclear reactors, as a shield for nuclear radiation, and in instruments used for detecting neutrons. Boron nitride has remarkable properties and can be used to make a material as hard as diamond. The nitride also behaves like an electrical insulator, but conducts heat like a metal. It also has lubricating properties similar to graphite. The hydrides are easily oxidized, with considerable energy liberation, and have been studied for use as rocket fuels. Demand is increasing for boron filaments, a high-strength, lightweight material chiefly employed for advanced aerospace structures. Boron is similar to carbon in that it has a capacity to form stable, covalently bonded molecular networks. Carbonates, metalloboranes, phosphacarboranes, and other families comprise thousands of compounds.

Isotopes
B 8, 9, 10, 11, 12, 13

Important reactions
Boron reacts vigorously with the halogens fluorine, F_2, chlorine, Cl_2, bromine, Br_2, to form the trihalides boron(III) fluoride, BF_3, boron(III) chloride, BCl_3, and boron(III) bromide, BBr_3 respectively.

Crystalline boron does not react with boiling hydrochloric acid, HCl, or boiling hydrofluoric acid, HF. Powdered boron oxidizes slowly when treated with concentrated nitric acid, HNO_3.

Hazards to responders

Flammability   Dust ignites spontaneously in air; it is a severe fire and explosion hazard. It explodes with hydrogen iodide.

Health   Elemental boron and the borates are not considered toxic, and they do not require special care in handling. However, some of the more exotic boron hydrogen compounds are definitely toxic and do require care. Compounds are toxic through accumulation. Boron compounds may be carcinogenic.

Aluminum

Symbol: Al
Metal or nonmetal: Metal
Atomic number: 13
Atomic weight: 26.98154
Periodic table group: III
Periodic table group name: None
Ordinary physical state: Solid
Color: Silvery
Case registry number: 7429-90-5
U.N. four digit number: 1383
DOT hazard class: (Powder) 4.2 spontaneously combustible
NFPA 704: Unavailable
Radioactive: No
Boiling point: 2519°C, 4566°F
Flash point: N/A
Flammable range: N/A
Autoignition temperature: N/A
Melting point: 660.32°C, 1220.58°F
Specific gravity: 2.55–2.80
Vapor density: N/A
Solubility: Insoluble in water
PEL-OSHA: 15 mg/m³
ACGIH-TLV-TWA: 10 mg/m³

History

The term aluminum was coined from the Latin, *alumen*, for alum. The ancient Greeks and Romans used alum as an astringent and as a mordant in dyeing. In 1761, de Morveau proposed the name alumine for the base in alum, and Lavoisier, in 1787, thought this to be the oxide of a still-undiscovered metal. Wohler is generally credited with having isolated the metal in 1827, although an impure form was prepared by Oersted 2 years earlier. In 1807, Davy proposed the name aluminum for the metal, undiscovered at that time, and later agreed to change it to aluminum. Shortly thereafter, the name aluminum was adopted to conform to the "ium" ending of most elements, and this spelling is now in use elsewhere in the world. Aluminum was also the accepted spelling in the United States until 1925, at which time the American Chemical Society officially decided to use the name aluminum thereafter in their publications. Pure aluminum, a silvery-white metal, possesses many desirable characteristics. It is light, nonmagnetic, and nonsparking; stands second among metals in the scale of malleability, and sixth in ductility.

Sources

The method of obtaining aluminum metal by the electrolysis of alumina dissolved in cryolite was discovered in 1886 by Hall in the United States and at about the same time by Heroult in France. Cryolite, a natural ore found in Greenland, is no longer widely used in commercial production, but has been replaced by an artificial mixture of sodium, aluminum, and calcium fluorides. Aluminum can now be produced from clay, but the process is not economically feasible at present. Aluminum is the most abundant metal in the earth's crust (8.1%), but is never found free in nature. In addition to the minerals mentioned previously, it is found in granite and in many other common minerals.

Important compounds

The compounds of greatest importance are aluminum oxide, the sulfate, and the soluble sulfate with potassium (alum).

Aluminum trihydride: AlH_3
Aluminum trifluoride: AlF_3
Dialuminum hexachloride: $AlCl_3$
Dialuminum hexabromide: $[AlBr_3]_2$
Dialuminum hexaiodide: $[AlI_3]_2$
Aluminum oxide(α): Al_2O_3
Dialuminum trisulfide: Al_2S_3
Dialuminum triselenide: Al_2Se_3
Dialuminum tritelluride: Al_2Te_3
Aluminum nitride: AlN

Uses

The oxide, alumina, occurs naturally as ruby, sapphire, corundum, and emery and is used in glassmaking and refractories. It is extensively used for kitchen utensils, outside building decoration, and in thousands of industrial applications where a strong, light, easily constructed material is needed. Although its electrical conductivity is only about 60% that of copper, it is used in electrical transmission lines because of its lightweight. Pure aluminum is soft and lacks strength, but it can be alloyed with small amounts of copper, magnesium, silicon, manganese, and other elements to impart a variety of useful properties. These alloys are of vital importance in the construction of modern aircraft and rockets. Aluminum, evaporated in a vacuum, forms a highly reflective coating for both visible light and radiant heat. These coatings soon form a thin layer of the protective oxide and do not deteriorate as do silver coatings. They are used to coat telescope mirrors and to make decorative paper, packages, and toys.

Isotopes

Al 24, 25, 26, 27, 28, 29, 30

Important reactions

It reacts with metal salts, mercury and mercury compounds, nitrates, sulfates, halogens, and halogenated hydrocarbons to form compounds that are sensitive to mechanical shock. Mixtures with ammonium nitrate are used as an explosive. A mixture with powdered ammonium persulfate and water may explode. Heating with bismuth trioxide leads to an explosively violent reaction. Mixtures with finely divided bromates (also chlorates and iodates) of barium, calcium, magnesium, potassium, sodium, or zinc can explode by heat, percussion, and friction. It burns in the vapor of carbon disulfide, in sulfur dioxide, sulfur dichloride, nitrous oxide, nitric oxide, or nitrogen peroxide. A mixture with carbon tetrachloride exploded when heated to 153°C and also by impact. Mixing with chlorine trifluoride in the presence of carbon results in a violent reaction. It ignites in close contact with iodine. Three industrial explosions involving a photoflash composition containing potassium perchlorate with aluminum and magnesium powder have occurred. React with methyl chloride in the presence of small amounts of aluminum chloride to give flammable aluminum trimethyl. It gives a detonable mixture with liquid oxygen. The reaction with silver chloride, once started, proceeds with explosive violence. In an industrial accident, the accidental addition of water to a solid mixture of sodium hydrosulfite and powdered aluminum caused the generation of SO_2, heat, and more water. The aluminum powder reacted with the water and other reactants leading to an explosion that killed five workers. Aluminum metal reacts vigorously with all the halogens to form aluminum halides. So, it reacts with chlorine, Cl_2, bromine, I_2, and iodine, I_2, to form respectively aluminum(III) chloride, $AlCl_3$, aluminum(III) bromide, $AlBr_3$, and aluminum(III) iodide, AlI_3.

Aluminum metal dissolves readily in dilute sulfuric acid to form solutions containing the aquated Al(III) ion together with hydrogen gas, H_2. The corresponding reactions with dilute hydrochloric acid also give the aquated Al(III) ion. Concentrated nitric acid passivates aluminum metal. Aluminum dissolves in sodium hydroxide with the evolution of hydrogen gas, H_2, and the formation of aluminates of the type $[Al(OH)_4]^-$.

Hazards to responders

Flammability   It is highly flammable. It ignites spontaneously in air. It reacts with water to generate flammable gaseous hydrogen and heat. Aluminum powder, pyrophoric, is a reducing agent. It reacts very exothermically when mixed with metal oxides and ignited or heated (thermite process). It reacts explosively when mixed with copper oxides and heated.

Health   Fire will produce irritating, corrosive, and/or toxic gases. Inhalation of decomposition products may cause severe injury or death. Contact with substance may cause severe burns to skin and eyes.

Carbon

Symbol: C
Metal or nonmetal: Nonmetal
Atomic number: 6
Atomic weight: 12.011
Periodic table group: IV
Periodic table group name: None
Ordinary physical state: Solid
Color: Black
Case registry number: 7440-44-0
U.N. four digit number: 1361
DOT hazard class: 4.2 spontaneously combustible
NFPA 704: 0-1-0
Radioactive: No
Boiling point: 4027°C, 7281°F
Autoignition temperature: 485°C, 905°F

Melting point: 3500°C, 6400°F
Specific gravity: 2.1
Density: 2.2670 g/cm^3
Solubility: Soluble in water
PEL-OSHA: None
ACGIH-TLV-TWA: None

History

The term carbon was coined from the Latin, *carbo*, meaning charcoal. Carbon, an element of prehistoric discovery, is widely distributed in nature. It is found in abundance in the sun, stars, comets, and atmospheres of most planets. Carbon in the form of microscopic diamonds is found in some meteorites. Natural diamonds are found in kimberlite of ancient volcanic "pipes" found in South Africa, Arkansas, and elsewhere. Diamonds are now also being recovered from the ocean floor off the Cape of Good Hope. About 30% of all industrial diamonds used in the United States are now made synthetically.

Sources

Carbon is found free in nature in three allotropic forms: amorphous, graphite, and diamond. A fourth form, known as "white" carbon, is now thought to exist. Ceraphite is one of the softest known materials, while diamond is one of the hardest. "White" carbon is a transparent birefringent material. Little information is presently available about this allotrope. In combination, carbon is found as carbon dioxide in the atmosphere of the earth and dissolved in all natural waters. It is a component of great rock masses in the form of carbonates of calcium (limestone), magnesium, and iron. Coal, petroleum, and natural gas are chiefly hydrocarbons.

Important compounds

Carbon is unique among the elements in the vast number and variety of compounds it can form. With hydrogen, oxygen, nitrogen, and other elements, it forms a large number of compounds, carbon atom often linking to carbon atom. There are close to 10 million known carbon compounds, many thousands of which are vital to organic and life processes. Without carbon, the basis for life would be impossible. While it has been thought that silicon might take the place of carbon in forming a host of similar compounds, it is now not possible to form stable compounds with long chains of silicon atoms. The atmosphere of Mars contains 96.2% CO_2. Some of the most important compounds of carbon are carbon dioxide (CO_2), carbon monoxide (CO), carbon disulfide (CS_2), chloroform ($CHCl_3$), carbon tetrachloride (CCl_4), methane (CH_4), ethylene (C_2H_4), acetylene (C_2H_2), benzene (C_6H_6), acetic acid (CH_3COOH), and their derivatives.

Carbon tetrafluoride: CF_4
Carbon tetrachloride: CCl_4
Carbon tetrabromide: CBr_4
Carbon tetraiodide: CI_4
Carbon dioxide: CO_2
Carbon monoxide: CO
Carbon suboxide: C_3O_2
Carbon disulfide: CS_2
Carbon diselenide: CSe_2

Uses

Carbon Dating   This is a method that is commonly used to find the age of fossils and minerals that have been around for many centuries. A radioactivity isotope of carbon, known as carbon-14, is used for the purpose of carrying out this activity. Things that were formerly living beings can be accurately dated back to their origins using their technique.

The number of compounds that are formed by carbon atoms are around ten million. An entire branch of chemistry called organic chemistry, is devoted to the study of the properties, and uses of carbon in its many forms. The benefits of carbon for the human body, and for many other industrial purposes are unmatched, and all these properties combined make carbon a very essential element for sustaining human life.

Isotopes

Carbon has seven isotopes. In 1961, the International Union of Pure and Applied Chemistry (IUPAC) adopted the isotope carbon-12 as the basis for atomic weights. Carbon-14, an isotope with a half-life of 5715 years, has been widely used to date such materials as wood, archaeological specimens, etc. C 9, 10, 11, 12, 13, 14, 15, 16, 17.

Important reactions

Graphite reacts with fluorine, F_2, at high temperatures to make a mixture of carbon tetrafluoride, CF_4, together with some C_2F_6 and C_5F_{12}.

At room temperature, the reaction with fluorine is complex. The result is "graphite fluoride," a non-stoichiometric species with formula CF_x ($0.68 < x < 1$). This species is black when x is low, silvery at $x = 0.9$, and colorless when x is about 1.

The other halogens appear to not react with graphite.

Graphite reacts with the oxidizing acid, hot concentrated nitric acid, to form mellitic acid, $C_6(CO_2H)_6$.

Hazards to responders

Flammability   Organic gases such as methane (CH_4), ethene ($CH_2=CH_2$), and ethyne (HCCH) are dangerous as mixtures with air because of fire and explosion dangers.

Health   Elemental carbon is of very low toxicity. Health hazard data presented here is based on exposures to carbon black, not elemental carbon. Carbon compounds show the full range of toxicities. Compounds such as CO (present in car exhausts) and CN^- (cyanide, sometimes present in pollution from mining) are very toxic to mammals. Other carbon compounds are not toxic and indeed are required for life. Chronic inhalation exposure to carbon black may result in temporary or permanent damage to lungs and heart. Pneumoconiosis has been found in workers engaged in the production of carbon black. Skin conditions such as inflammation of the hair follicles and oral mucosal lesions have also been reported from skin exposure. Some simple carbon compound can be very toxic, such as carbon monoxide (CO) or cyanide (CN). It also can cross the placenta, become organically bound in developing cells, and hence endanger fetuses.

Silicon

Symbol: Si
Metal or nonmetal: Nonmetal
Atomic number: 14

Atomic weight: 28.0855
Periodic table group: IV
Periodic table group name: None
Ordinary physical state: Solid
Color: Gray-black
Case registry number: 7440-21-3
U.N. four digit number: 1346
DOT hazard class: 4.1 flammable solid
NFPA 704: Unavailable
Radioactive: No
Boiling point: 2900°C, 5252°F
Autoignition temperature: 150°C, 302°F
Melting point: 1414°C, 2577°F
Specific gravity: 2.33
Density: 2.33 g/cm^3
Solubility: Insoluble in water
PEL-OSHA: 15 mg/m^3 total particulate, 5 mg/m^3 respirable particulate

History

The word silicon was coined from the Latin words *silex* or *silicis*, meaning flint. Davy, in 1800, thought silica to be a compound and not an element; later, in 1811, Gay-Lussac and Thenard probably prepared impure amorphous silicon by heating potassium with silicon tetrafluoride. In 1824, Berzelius, generally credited with the discovery, prepared amorphous silicon by the same general method, and purified the product by removing the fluorosilicates by repeated washings. Deville, in 1854, first prepared crystalline silicon, the second allotropic form of the element.

Sources

Silicon is present in the sun and stars and is the principal component of a class of meteorites known as aerolites. It is also a component of tektites, a natural glass of uncertain origin. Silicon makes up 25.7% of the earth's crust, by weight, and is the second most abundant element, exceeded only by *oxygen*. Silicon is not found free in nature, but occurs chiefly as the oxide and as silicates. Sand, quartz, rock crystal, amethyst, agate, flint, jasper, and opal are some of the forms in which the oxide appears. Granite, hornblende, asbestos, feldspar, clay, mica, etc., are but a few of the numerous silicate minerals.

Important compounds

Silane: SiH_4
Disilicon hexahydride: Si_2H_6
Silicon tetrafluoride: SiF_4
Silicon tetrachloride: $SiCl_4$
Silicon tetraiodide: SiI_4
Silicon oxide: SiO_2
Silicon sulfide: SiS_2
Trisilicon tetranitride: Si_3N_4

Uses

Silicon is one of man's most useful elements. In the form of sand and clay, it is used to make concrete and brick; it is a useful refractory material for high-temperature work, and in the form of silicates, it is used in making enamels and pottery. Silica, as sand, is the principal ingredient of glass, one of the most inexpensive of materials with excellent mechanical, optical, thermal, and electrical properties. Glass can be made in a great variety of shapes and is used as containers, window glass, insulators, and thousands of other uses. Silicon is an important ingredient in steel; silicon carbide is one of the most important abrasives and has been used in lasers to produce coherent light of 4560 A. Silicones are important products of silicon.

Isotopes
Si 22 through 44

Important reactions
Silicon reacts vigorously with all the halogens to form silicon tetrahalides. So, it reacts with fluorine, F_2, chlorine, Cl_2, bromine, I_2, and iodine, I_2, to form respectively silicon(IV) fluoride, SiF_4, silicon(IV) chloride, $SiCl_4$, silicon(IV) bromide, $SiBr_4$, and silicon(IV) iodide, SiI_4. The reaction with fluorine takes place at room temperature, but the others requiring warming over 300°C. Silicon is attacked by bases such as aqueous sodium hydroxide to give silicates, highly complex species containing the anion $[SiO_4]^{4-}$.

Hazards to responders

Flammability   Silicon is not flammable.

Health   Silicon concentrates in no particular organ of the body but is found mainly in connective tissues and skin. Silicon is nontoxic as the element, and in all its natural forms, namely silica and silicates, which are the most abundant. Elemental silicon is an inert material. Crystalline silica (silicon dioxide) is a potent respiratory hazard. LD_{50} (oral)— 3160 mg/kg (LD_{50}: Lethal dose 50. Single dose of a substance causes the death of 50% of an animal population from exposure to the substance by any route other than inhalation. It is usually expressed as milligrams or grams of material per kilogram of animal weight). Silicon crystalline irritates the skin and eyes on contact. Inhalation will cause irritation to the lungs and mucous membrane. Irritation to the eyes will cause watering and redness. Reddening, scaling, and itching are characteristics of skin inflammation.

Nitrogen

Symbol: N
Metal or nonmetal: Nonmetal
Atomic number: 7
Atomic weight: 14.0067
Periodic table group: V
Periodic table group name: None
Ordinary physical state: Gas
Color: Colorless
Case registry number: 7727-37-9
U.N. four digit number: 1066
DOT hazard class: 2.2 non-flammable compressed gas
NFPA 704: 3-0-3 (refrigerated liquid)
Cryogenic liquid: Yes
Cryogenic boiling point: −196°C, −321°F
Cryogenic expansion ratio: 710–1

Radioactive: No
Melting point: −210.1°C, −346.18°F
Vapor density: 2.25 g/L
Solubility: Insoluble in water
PEL-OSHA: No limits (simple asphyxiant)
ACGIH-TLV-TWA: Not established (simple asphyxiant)

History

French chemist Antoine Laurent Lavoisier named nitrogen *azote*, meaning without life. From the Latin *nitrum* and Greek *nitron*, meaning native soda, and *genes*, meaning forming. Chemist and physician Daniel Rutherford discovered nitrogen in 1772. He removed oxygen and carbon dioxide from air and showed that the residual gas would not support combustion or living organisms. At the same time, there were other noted scientists working on the problem of identifying and explaining the behavior of nitrogen. They called it "burnt [or] dephlogisticated air," which means air without oxygen. The nitrogen cycle is one of the most important processes in nature for living organisms. Although nitrogen gas is relatively inert, bacteria in the soil are capable of "fixing" the nitrogen into a usable form (as a fertilizer) for plants. In other words, nature has provided a method to produce nitrogen for plants to grow. Animals eat the plant material, and the nitrogen is incorporated into their system, primarily as protein. The cycle is completed when other bacteria convert the waste nitrogen compounds back to nitrogen gas. Nitrogen has become crucial to life, being a component of all proteins.

Sources

Nitrogen gas (N_2) makes up 78.1% of the earth's air, by volume. The atmosphere of Mars, by comparison, is only 2.6% nitrogen. From an exhaustible source in our atmosphere, nitrogen gas can be obtained by liquefaction and fractional distillation. Nitrogen is found in all living systems as part of the makeup of biological compounds.

Important compounds

Nitrogen compounds are found in foods, fertilizers, poisons, and explosives. Sodium nitrate ($NaNO_3$) and potassium nitrate (KNO_3) are formed by the decomposition of organic matter, with compounds of these metals present. In certain dry areas of the world, these saltpeters are found in quantity and are used as fertilizers. Other inorganic nitrogen compounds are nitric acid (HNO_3), ammonia (NH_3), the oxides (NO, NO_2, N_2O_4, N_2O), cyanides (CN^-), etc. Nitrogen gas can be prepared by heating a water solution of ammonium nitrite (NH_4NO_3).

Ammonia   Ammonia (NH_3) is the most important commercial compound of nitrogen. It is produced by the Haber process. This process is the synthesis of ammonia by the water–gas reaction from hot coke, air, and steam. Natural gas (methane, CH_4) is reacted with steam to produce carbon dioxide and hydrogen gas (H_2) in a two-step process. Hydrogen gas and nitrogen gas are then reacted in the Haber process to produce ammonia. This colorless gas with a pungent odor is easily liquefied. In fact, the liquid is used as a nitrogen fertilizer. Ammonia is also used in the production of urea, NH_2CONH_2, which is used as a fertilizer, in the plastic industry, and in the livestock industry as a feed supplement. Ammonia is often the starting compound for many other nitrogen compounds.

Nitrogen trifluoride: NF_3
Dinitrogen tetrafluoride: N_2F_4
cis-Difluorodiazine: N_2F_2

trans-Difluorodiazine: N_2F_2
Nitrogen trichloride: NCl_3
Nitrogen triiodide: NI_3
Nitrous oxide: N_2O
Nitrogen dioxide: NO_2
Nitrogen monoxide: NO
Dinitrogen trioxide: N_2O_3
Dinitrogen tetraoxide: N_2O_4
Dinitrogen pentoxide: N_2O_5

Uses

Many industrially important compounds, such as ammonia, nitric acid, organic nitrates (propellants and explosives), and cyanides, contain nitrogen. The extremely strong bond in elemental nitrogen dominates nitrogen chemistry, causing difficulty for both organisms and industry in breaking the bond to convert the N_2 into useful compounds, but at the same time causing release of large amounts of often useful energy when the compounds burn, explode, or decay back into nitrogen gas. Nitrogen gas has a variety of applications, including serving as an inert replacement for air where oxidation is undesirable; as a modified atmosphere, pure or mixed with carbon dioxide, to preserve the freshness of packaged or bulk foods (by delaying rancidity and other forms of oxidative damage); and in ordinary incandescent lightbulbs as an inexpensive alternative to argon. Used in military aircraft fuel systems to reduce fire hazard (see inerting system), on top of liquid explosives as a safety measure; filling automotive and aircraft tires due to its inertness and lack of moisture or oxidative qualities, as opposed to air. The difference in N_2 content between air and pure N_2 is 20%. The organic and inorganic salts of nitric acid have been important historically as convenient stores of chemical energy. They include important compounds such as potassium nitrate (or saltpeter used in gunpowder) and ammonium nitrate, an important fertilizer and explosive (see ANFO). Various other nitrated organic compounds, such as nitroglycerin, trinitrotoluene, and nitrocellulose, are used as explosives and propellants for modern firearms. Nitric acid is used as an oxidizing agent in liquid-fueled rockets. Hydrazine and hydrazine derivatives find use as rocket fuels and monopropellants. When nitrates burn or explode, the formation of the powerful triple bond in the N_2 produces most of the energy of the reaction.

Isotopes

N 10 through 25

Important reactions

Inert

Hazards to responders

Flammability   Does not burn.

Health   Nitrogen as a gas is colorless, odorless, and generally considered an inert element. As a liquid (boiling point = −195.8°C), it is also colorless and odorless and is similar in appearance to water.

Asphyxiation   Liquid nitrogen in its boiling state produces colorless, odorless, flavorless nitrogen gas. Impossible to detect, this nitrogen when left uncontrolled from a spill, for example, can overtake the oxygen level in the air, causing asphyxiation and unconsciousness if inhaled. A poorly ventilated work area can also contribute to the asphyxiation risk.

Frostbite and burns   At temperatures below –100°F, liquid nitrogen is cold enough to cause frostbite on unprotected skin. If the affected area is large enough, the victim could go into shock, according to Imperial College London. Liquid nitrogen can also produce severe burns to eyes. The risk to the eyes is particularly acute as the chemical is poured, increasing the possibility of a splash to the face.

Decompression sickness   During the time spent underwater, a scuba diver's body tissue absorbs nitrogen from the compressed air that he uses to breathe. While the diver is in the water and subject to pressure at depth, the nitrogen is harmless. The problem arises if a diver ascends to the surface too quickly. The rapid decrease in pressure causes the nitrogen to form bubbles in the diver's bloodstream and tissue. On the surface, the diver will experience severe pain in his abdomen, lower back, and extremities, followed by paralysis and unconsciousness. If left untreated, decompression sickness can be fatal.

Nitrogen narcosis   When scuba divers descend to a depth of more than 30 m, they can experience a condition known as nitrogen narcosis. This is caused when the nitrogen in the compressed air that they are breathing is subject to a partial pressure that is more than three times atmospheric pressure. A diver suffering from nitrogen narcosis will experience feelings of euphoria and may entertain false beliefs in their own abilities. Manual dexterity is decreased as well as the ability for complex reasoning. The sensation is sometimes described as similar to that of alcoholic intoxication.

Phosphorus

Symbol: P
Metal or nonmetal: Nonmetal
Atomic number: 15
Atomic weight: 30.97376
Periodic table group: V
Periodic table group name: None
Ordinary physical state: Solid
Color: Colorless
Case registry number: 7723-14-0
U.N. four digit number: 1381
DOT hazard class: 4.2 spontaneously combustible
NFPA 704: 4-4-2
Radioactive: No
Boiling point: 277°C, 531°F
Flash point: 30°C, 86°F
Autoignition temperature: 300°C, 572°F
Melting point: 44.2°C, 111.6°F
Specific gravity: 2.34
Density: 1.823 g/cm^3
Solubility: Insoluble in water
PEL-OSHA: 0.1 mg/m^3
ACGIH-TLV-TWA: ACGIH TLV: TWA (8-h): 0.1 mg/m^3

History

The word phosphorus was derived from the Greek word, *phosphoros*, meaning light-bearing; also an ancient name for the planet Venus when appearing before sunrise. Brand discovered phosphorus in 1669 by preparing it from urine.

Sources

Never found free in nature, it is widely distributed in combination with minerals. Phosphate rock, which contains the mineral apatite, an impure tricalcium phosphate, is an important source of the element. Large deposits are found in Russia, Morocco, Florida, Tennessee, Utah, Idaho, and elsewhere.

Important compounds

Phosphine: PH_3
Diphosphorus tetrahydride: P_2H_4
Phosphorus trifluoride: PF_3
Phosphorus pentafluoride: PF_5
Diphosphorus tetrafluoride: P_2F_4
Phosphorus trichloride: PCl_3
Phosphorus pentachloride: PCl_5
Diphosphorus tetrachloride: P_2Cl_4
Phosphorus pentabromide: PBr_5
Diphosphorus tetrabromide: P_2Br_4
Phosphorus triiodide: PI_3
Diphosphorus tetraiodide: P_2I_4
Tetraphosphorus decaoxide: P_4O_{10}
Tetraphosphorus hexaoxide: P_4O_6
Tetraphosphorus trisulfide: P_4S_3
Tetraphosphorus decasulfide: P_4S_{10}
Tetraphosphorus hexasulfide: P_4S_6
Tetraphosphorus nonasulfide: P_4S_9
Tetraphosphorus pentasulfide(alpha): P_4S_5
Tetraphosphorus heptasulfide: P_4S_7
Tetraphosphorus tetrasulfide: P_4S_4
Tetraphosphorus triselenide: P_4Se_3

Uses

In recent years, concentrated phosphoric acids, which may contain as much as 70%–75% P_2O_5 content, have become of great importance to agriculture and farm production. Worldwide demand for fertilizers has caused record phosphate production. Phosphates are used in the production of special glasses, such as those used for sodium lamps. Bone ash, calcium phosphate, is used to create fine chinaware and to produce monocalcium phosphate, used in baking powder. Phosphorus is also important in the production of steels, phosphor bronze, and many other products. Trisodium phosphate is important as a cleaning agent, as a water softener, and for preventing boiler scale and corrosion of pipes and boiler tubes. Phosphorus is also an essential ingredient of all cell protoplasm, nervous tissue, and bones.

Isotopes

P 24 through 46

Important reactions

White phosphorus glows in the dark when exposed to damp air in a process known as chemiluminescence. White phosphorus must be handled with great care. It spontaneously ignites in air at about room temperature to form "phosphorus pentoxide." White phosphorus glows in the dark when exposed to damp air in a process known as chemiluminescence. White phosphorus, P_4, reacts vigorously with all the halogens at room temperature to form phosphorus trihalides. So it reacts with fluorine, F_2, chlorine, Cl_2, bromine, Br_2, and iodine, I_2, to form respectively phosphorus(III) fluoride, PF_3, phosphorus(III) chloride, PCl_3, phosphorus(III) bromide, PBr_3, and phosphorus(III) iodide, PI_3. White phosphorus, P_4, reacts with iodine, I_2, in carbon disulfide (CS_2) to form phosphorus(II) iodide, P_2I_4. The same compound is formed in the reaction between red phosphorus and iodine, I_2, at 180°C.

Hazards to responders

Fire hazard   Phosphorus exists in three or more allotropic forms: white (or yellow), red, and black (or violet). Ordinary phosphorus is a waxy, white solid; when pure, it is colorless and transparent. It is insoluble in water, but soluble in carbon disulfide. It takes fire spontaneously in air, burning to phosphorus pentoxide. It will reignite itself after fire is extinguished. White/Yellow: It ignites at approximately 86°F in air; ignition temperature is higher when air is dry. Black: Does not catch fire spontaneously. Red: Catches fire when heated in air to approximately 500°F and burns with the formation of the pentoxide. Burns when heated in atmosphere of chlorine. Caution: Avoid contact with potassium chlorate, potassium permanganate, peroxides, and other oxidizing agents; explosions may result on contact or friction. Upon heating, red releases toxic oxides of phosphorus and yellow emits toxic gases and vapors such as phosphoric acid fumes. Red: Avoid uncontrolled contact with oxidizing agents, or with strong alkaline hydroxides. It can react violently with oxidizing agent in presence of air and moisture, liberating phosphorus acids and toxic, spontaneously flammable phosphine gas. White/Yellow: Avoid air, all oxidizing agents including elemental sulfur, strong caustics. White/Yellow: Darkens on exposure to light. It gives off acrid fumes on exposure to air. Ignites spontaneously in air at or above 86°F. Black: stable in air. Avoid heat. Red: burning yields toxic oxides of phosphorus. White/Yellow: toxic gases and vapors such as phosphoric acid fumes are released.

Health hazard   It is classified as super toxic. The probable lethal dose is less than 5 mg/kg (a taste or less than seven drops) for 70 kg (150 lb) person. Poisonous if swallowed or if fumes are inhaled. Yellow: Fumes are irritating to the respiratory tract and cause severe ocular irritation. On contact with the skin, it may ignite and produce severe skin burns with blistering. Red: Irritates eyes.

Arsenic

Symbol: As
Metal or nonmetal: Nonmetal
Atomic number: 33
Atomic weight: 74.9216
Periodic table group: V
Periodic table group name: None
Ordinary physical state: Solid
Color: Metallic Gray

Case registry number: 7440-38-2
U.N. four digit number: 1558
DOT hazard class: 6.1 poison
NFPA 704: Unavailable
Radioactive: No
Boiling point: 614°C, 1137°F
Melting point: 817°C, 1503°F
Specific gravity: 1.97
Density: 5.727 g/m^3
Solubility: Insoluble in water
PEL-OSHA: 0.010 mg As/m^3
ACGIH-TLV-TWA: 0.01 mg/m^3

History

The word arsenic was coined from the Latin *arsenicum* and the Greek *arsenikon*, referring to yellow orpiment (arsenic trisulfide), identified with *arenikos* (Greek for male), from the belief that metals were different sexes; also from the Arabic, *az-zernikh*, referring to the orpiment from Persian, *zerni-zar*, or gold. Elemental arsenic occurs in two solid modifications: yellow and gray, or metallic, with specific gravities of 1.97 and 5.73, respectively. It is believed that Albertus Magnus obtained the element in A.D. 1250. In 1649, Schroeder published two methods of preparing the element. Mispickel, arsenopyrite (FeSAs), is the most common mineral, from which, on heating, the arsenic sublimes, leaving ferrous sulfide.

Sources

The earth's crust is an abundant natural source of arsenic. It is present in more than 200 different minerals, the most common of which is called arsenopyrite. About one-third of the arsenic in the earth's atmosphere is of natural origin. Volcanic action is the most important natural source. The next most important source is arsenic-containing vapor that is generated from solid or liquid forms of arsenic salts at low temperatures. Inorganic arsenic of geological origin is found in groundwater used as drinking water in several parts of the world, for example, Bangladesh, India, and Taiwan. Arsenic metal very rarely occurs in its pure form in nature. The most common arsenic mineral is arsenopyrite, a compound of iron, arsenic, and sulfur. Several other, less common minerals contain arsenic, including orpiment, realgar, and enargite, which are arsenic sulfides.

Important compounds

Arsine: AsH_3
Diarsenic tetrahydride: As_2H_4
Arsenic trifluoride: AsF_3
Arsenic pentafluoride: AsF_5
Arsenic trichloride: $AsCl_3$
Arsenic pentachloride: $AsCl_5$
Arsenic tribromide: $AsBr_3$
Arsenic triiodide: AsI_3
Diarsenic tetraiodide: $[AsI_2]_2$
Diarsenic trioxide: As_2O_3
Diarsenic pentoxide: As_2O_5
Diarsenic trisulfide: As_2S_3
Diarsenic pentasulfide: As_2S_5

Tetraarsenic tetrasulfide: As_4S_4
Diarsenic triselenide: As_2Se_3
Diarsenic pentaselenide: As_2Se_5
Diarsenic tritelluride: As_2Te_3

Uses

Arsenic is used in bronzing, pyrotechny, and for hardening and improving the sphericity of shot. The most important compounds are white arsenic, the sulfide, Paris green, calcium arsenate, and lead arsenate; the last three have been used as agricultural insecticides and poisons.

Isotopes

As 60 through 92

Important reactions

Arsenic reacts incandescently with bromine trifluoride, even at 10°C. Causes bromo azide to explode upon contact. Ignites if ground up together with solid potassium permanganate. It is oxidized by sodium peroxide with incandescence. A combination of finely divided arsenic with finely divided bromates (also chlorates and iodates) of barium, calcium, magnesium, potassium, sodium, or zinc can explode by heat, percussion, and friction. Bromine pentafluoride reacts readily in the cold with arsenic. Ignition usually occurs. It reacts vigorously with fluorine at ordinary temperatures. Arsenic reacts with fluorine, F_2, to form the gas pentafluoride arsenic(V) fluoride. Arsenic reacts under controlled conditions with the halogens fluorine, F_2, chlorine, Cl_2, bromine, Br_2, and iodine, I_2, to form the respective trihalides arsenic(III) fluoride, AsF_3, arsenic(III) chloride, $AsCl_3$, arsenic(III) bromide, $AsBr_3$, and arsenic(III) iodide, AsI_3.

Hazards to responders

Flammability   Behavior in fire: Burns to produce dense white fumes of highly toxic arsenic trioxide. Special Hazards of Combustion Products: Contain highly toxic arsenic trioxide and other forms of arsenic. Arsenic gas, the most dangerous form of arsenic, is produced upon contact with an acid or acid fumes.

Health hazard   Arsenic and its compounds are poisonous. It is poisonous by inhalation of dust or by ingestion. Regardless of exposure route, symptoms in most cases are characteristic of severe gastritis or gastroenteritis. All chemical forms of arsenic eventually produce similar toxic effects. Symptoms may be delayed.

Plutonium

Symbol: Pu
Metal or nonmetal: Metal
Atomic number: 94
Atomic weight: 244
Periodic table group: Transition metals
Periodic table group name: None
Ordinary physical state: Solid
Color: Silvery white
Case registry number: 7440-07-5

U.N. four digit number: N/A
DOT hazard class: N/A
NFPA 704: Unavailable
Radioactive: Yes
Boiling point: 3230°C, 5846°F
Melting point: 639.4°C, 1182.9°F
Specific gravity: 19.84
Density: 19.816 g/cm³
Solubility: Insoluble in water
PEL-OSHA: 0.2 mg/m³ as an 8-h TWA
ACGIH-TLV-TWA: 750 ppm

History

Named for the planet Pluto, plutonium was the second transuranic element of the actinide series to be discovered. The isotope ^{238}Pu was produced in 1940 by Seaborg, McMillan, Kennedy, and Wahl by deuteron bombardment of uranium in the 60-in. cyclotron at Berkeley, California. Plutonium also exists in trace quantities in naturally occurring uranium ores. It is formed in much the same manner as neptunium, by irradiation of natural uranium with the neutrons that are present.

Sources

Pu-238 was produced by Seaborg, McMillan, Kennedy, and Wahl in 1940 by deuteron bombardment of uranium. Plutonium may be found in trace amount in natural uranium ores. This plutonium is formed by irradiation of natural uranium by the neutrons that are present. Plutonium metal can be prepared by reduction of its trifluoride with alkaline earth metals.

Important compounds

Plutonium dihydride: PuH_2
Plutonium trihydride: PuH_3
Plutonium trifluoride: PuF_3
Plutonium hexafluoride: PuF_6
Plutonium tetrafluoride: PuF_4
Plutonium trichloride: $PuCl_3$
Plutonium tribromide: $PuBr_3$
Plutonium triiodide: PuI_3
Plutonium oxide: PuO
Plutonium dioxide: PuO_2
Diplutonium trioxide: Pu_2O_3
Plutonium sulfide: PuS
Plutonium disulfide: PuS_2
Diplutonium trisulfide: Pu_2S_3
Plutonium selenide: $PuSe$
Plutonium nitride: PuN

Uses

Plutonium has assumed a position of dominant importance among the transuranic elements because of its successful use as an explosive ingredient in nuclear weapons and the place it holds as a key material in the development of industrial use of nuclear power.

One kilogram is equivalent to about 22 million-kWh of heat energy. The complete detonation of a kilogram of plutonium produces an explosion equal to about 20,000 tons of TNT. Its importance depends on the nuclear property of being readily fissionable with neutrons and its availability in quantity. The world's nuclear-power reactors are now producing about 20,000 kg of plutonium a year. The various nuclear applications of plutonium are well known. ^{238}Pu has been used in the Apollo lunar missions to power seismic and other equipment on the lunar surface. As with neptunium and uranium, plutonium metal can be prepared by reduction of the trifluoride with alkaline earth metals.

Isotopes

By far of greatest importance is the isotope ^{239}Pu, with a half-life of 24,100 years, produced in extensive quantities in nuclear reactors from natural uranium: ^{238}U(n, gamma)→ ^{239}U—(beta)→ ^{239}Np—(beta)→ ^{239}Pu. Fifteen isotopes of plutonium are known. Plutonium forms binary compounds with oxygen: PuO, PuO_2, and intermediate oxides of variable composition; with the halides: PuF_3, PuF_4, $PuCl_3$, $PuBr_3$, and PuI_3; and with carbon, nitrogen, and silicon: PuC, PuN, and $PuSi_2$. Oxyhalides are also well known: PuOCl, PuOBr, and PuOI.

Important reactions

None

Hazards to responders

Flammability   Plutonium is a pyrophoric metal; it can be flammable especially as a fine powder.

Health   Because of the high rate of emission of alpha particles and the element being specifically absorbed on bone surface and collected in the liver, plutonium, as well as all of the other transuranic elements, except neptunium, are radiological poisons and must be handled with special equipment and precautions. Plutonium is a highly dangerous radiological hazard. It is now found in small quantities in some areas within the biosphere as a result of fallout from atomic bombs and from radiation leaks from nuclear facilities. It constitutes an extreme radiation hazard when even small quantities are assembled in one place. Because of the high rate of emission of α particles and the element specifically being absorbed by bone marrow, plutonium is an extreme radiological poison that must be handled only be properly trained expert personnel using very special equipment and precautions. Such personnel and equipment exist in very few locations around the world. Permitted levels of exposure to plutonium are the lowest of any element.

Chemical hazard   The metal has a silvery appearance and takes on a yellow tarnish when slightly oxidized. It is chemically reactive. A relatively large piece of plutonium is warm to the touch because of the energy given off in alpha decay. Larger pieces will produce enough heat to boil water. The metal readily dissolves in concentrated hydrochloric acid, hydroiodic acid, or perchloric acid.

Oxygen

Symbol: O
Metal or nonmetal: Nonmetal
Atomic number: 8

Atomic weight: 15.9994
Periodic table group: VI
Periodic table group name: None
Ordinary physical state: Gas
Color: Colorless (gas) pale blue (liquid)
Case registry number: 7782-44-7
U.N. four digit number: 1072
DOT hazard class: 2.2 non-flammable compressed gas
NFPA 704: 0-0-0 ox
Cryogenic liquid: Yes
Cryogenic boiling point: –187°C, –256°F
Cryogenic expansion ratio: 875–1
Radioactive: No
Boiling point: –187°C, –256°F
Melting point: –218.3°C, –360.9°F
Vapor density: 4.475 kg/m³
Solubility: Insoluble in water
PEL-OSHA: None
ACGIH-TLV-TWA: None

History

The word oxygen is coined from the Greek, *oxys*, meaning sharp or acid, and *genes*, meaning forming: acid former. For many centuries, workers occasionally realized that air was composed of more than one component. The behavior of oxygen and nitrogen as components of air led to the advancement of the phlogiston theory of combustion, which captured the minds of chemists for a century. Oxygen was prepared by several workers, including Bayen and Borch, but they did not know how to collect it, did not study its properties, and did not recognize it as an elementary substance. Priestley is generally credited with its discovery, although Scheele also discovered it independently. Its atomic weight was used as a standard of comparison for each of the other elements until 1961, when the IUPAC adopted carbon-12 as the new basis.

The gas is colorless, odorless, and tasteless. The liquid and solid forms are a pale blue color and are strongly paramagnetic.

Sources

Oxygen is the third most abundant element found in the sun, and it plays a part in the carbon–nitrogen cycle, the process once thought to give the sun and stars their energy. Oxygen under excited conditions is responsible for the bright red and yellow–green colors of the aurora borealis. A gaseous element, oxygen forms 21% of the atmosphere by volume and is obtained by liquefaction and fractional distillation. The atmosphere of Mars contains about 0.15% oxygen. The element and its compounds make up 49.2%, by weight, of the earth's crust. About 2/3rds of the human body and 9/10ths of water is oxygen.

Important compounds

Ozone (O_3), a highly active compound, is formed by the action of an electrical discharge or ultraviolet light on oxygen. Ozone's presence in the atmosphere (amounting to the equivalent of a layer 3 mm thick under ordinary pressures and temperatures) helps prevent harmful ultraviolet rays of the sun from reaching the earth's surface. Pollutants in the atmosphere may have a detrimental effect on this ozone layer. Ozone is toxic, and exposure should not exceed 0.2 mg/m^3 (8-h TWA in a 40 h workweek). Undiluted ozone has a bluish color. Liquid ozone is bluish-black, and solid ozone is violet-black. Oxygen, which is highly reactive, is a component of hundreds of thousands of organic compounds and combines with most elements.

Oxygen difluoride: OF_2

Uses

Plants and animals rely on oxygen for respiration. Hospitals frequently prescribe oxygen for patients with respiratory ailments.

Isotopes

Oxygen has nine isotopes. Natural oxygen is a mixture of three isotopes. Naturally occurring oxygen-18 is stable and available commercially, as is water (H_2O with 15% ^{18}O). Commercial oxygen consumption in the United States is estimated at 20 million short tons per year, and the demand is expected to increase substantially. Oxygen enrichment of steel blast furnaces accounts for the greatest use of the gas. Large quantities are also used in making synthesis gas for ammonia and methanol, ethylene oxide, and for oxyacetylene welding. Air separation plants produce about 99% of the gas, while electrolysis plants produce about 1%.

Important reactions

Propellant; ignites upon contact with alcohols, alkali metals, amines, ammonia, beryllium alkyls, boranes, dicyanogen, hydrazines, hydrocarbons, hydrogen, nitroalkanes, powdered metals, silanes, or thiols. Heat of water will vigorously vaporize liquid oxygen; pressures may build to dangerous levels if this occurs in a closed container. Liquid oxygen gives a detonable mixture when combined with powdered aluminum. It is a strong oxidizing agent. Forms contact explosive when in contact with asphalt pavement.

Hazards to responders

Fire hazard   Behavior in fire: Increases intensity of any fire. Mixtures of liquid oxygen and any fuel are highly explosive.

Health   Inhalation of 100% oxygen can cause nausea, dizziness, irritation of lungs, pulmonary edema, pneumonia, and collapse. Liquid may cause frostbite of eyes and skin.

Sulfur

Symbol: S
Metal or nonmetal: Nonmetal
Atomic number: 16
Atomic weight: 32.06
Periodic table group: VI
Periodic table group name: None
Ordinary physical state: Solid
Color: Lemon yellow
Case registry number: 7704-34-9
U.N. four digit number: 1350 (2448 molten)
DOT hazard class: 9 miscellaneous (molten)

NFPA 704: Dry 1-1-0, Molten 2-1-0
Boiling point: 444.72°C, 832.5°F
Melting point: 115.21°C, 239.38°F
Specific gravity: 2.07
Vapor density: 1.96 g/cm^3
Solubility: Soluble in water
PEL-OSHA: N/A
ACGIH-TLV-TWA: N/A

History

The word sulfur was coined from the Sanskrit, *sulvere*; L. sulfur. It was known to the ancients; referred to in Genesis as brimstone.

Sources

Sulfur is found in meteorites. R.W. Wood suggests that the dark area near the crater Aristarchus is a sulfur deposit. Sulfur occurs naturally in the vicinity of volcanoes and hot springs. It is widely distributed in nature as iron pyrites, galena, sphalerite, cinnabar, stibnite, gypsum, epsom salts, celestite, barite, etc. Sulfur is commercially recovered from wells sunk into the salt domes along the Gulf Coast of the United States. Using the Frasch process, heated water is forced into the wells to melt the sulfur, which is then brought to the surface. Sulfur also occurs in natural gas and petroleum crudes and must be removed from these products. Formerly, this was done chemically, which wasted the sulfur; new processes now permit recovery. Large amounts of sulfur are also recovered from Alberta gas fields. High-purity sulfur is commercially available in purities of 99.999+%. Amorphous, or "plastic," sulfur is obtained by fast cooling of the crystalline form. Crystalline sulfur seems to be made of rings, each containing eight sulfur atoms, which fit together to give a normal x-ray pattern. Organic compounds containing sulfur are important. Calcium sulfur, ammonium sulfate, carbon disulfide, sulfur dioxide, and hydrogen sulfide are but a few of the many important compounds of sulfur. In nature, sulfur can be found as the pure element and as sulfide and sulfate minerals. Elemental sulfur crystals are commonly sought after by mineral collectors for their brightly colored polyhedron shapes. Being abundant in native form, sulfur was known in ancient times, mentioned for its uses in ancient Greece, China, and Egypt. Sulfur fumes were used as fumigants, and sulfur-containing medicinal mixtures were used as balms and antiparasitics. Sulfur is referenced in the Bible as *brimstone* in English, with this name still used in several nonscientific terms [2]. It was needed to make the best quality of black gunpowder. In 1777, Antoine Lavoisier helped convince the scientific community that sulfur was a basic element, rather than a compound. Elemental sulfur was once extracted from salt domes where it sometimes occurs in nearly pure form, but this method has been obsolete since the late twentieth century. Today, almost all elemental sulfur is produced as a by-product of removing sulfur-containing contaminants from natural gas and petroleum.

Important compounds

Sulfur difluoride: SF_2
Disulfur difluoride: FSSF
Sulfur hexafluoride: SF_6
Sulfur tetrafluoride: SF_4
Disulfur decafluoride: S_2F_{10}

Sulfur difluoride sulfide: SSF_2
Sulfur dichloride: SCl_2
Disulfur dichloride: $ClSSCl$
Trisulfur dichloride: S_3Cl_2
Sulfur tetrachloride: SCl_4
Disulfur diiodide: S_2I_2
Sulfur dioxide: SO_2
Sulfur trioxide: SO_3
Disulfur oxide: S_2O
Tetrasulfur tetranitride: S_4N_4

Uses

Sulfur is a component of black gunpowder and is used in the vulcanization of natural rubber and as a fungicide. It is also used extensively in making phosphatic fertilizers. A tremendous tonnage is used to produce sulfuric acid, the most important manufactured chemical. It is used to make sulfite paper and other papers, to fumigants, and to bleach dried fruits. The element is a good insulator. Sulfur is essential to life. It is a minor constituent of fats, body fluids, and skeletal minerals.

Isotopes

(S) has 25 known isotopes with mass numbers ranging from 26 to 49, 4 of which are stable: ^{32}S (95.02%), ^{33}S (0.75%), ^{34}S (4.21%), and ^{36}S (0.02%). The preponderance of sulfur-32 is explained by its production from carbon-12 plus successive fusion capture of five helium nuclei, in the so-called alpha process of exploding type II supernovae (see silicon burning). Other than ^{35}S, the radioactive isotopes of sulfur are all comparatively short-lived. ^{35}S is formed from cosmic ray spallation of ^{40}Ar in the atmosphere. It has a half-life of 87 days. The next longest-lived radioisotope is sulfur-38, with a half-life of 170 min. The shortest-lived is ^{49}S, with a half-life shorter than 200 ns.

Important reactions

None identified

Hazards to responders

Flammability   If ignited by a spark or flame, sulfur will burn in air, yielding acrid fumes of sulfur dioxide (SO_2). The flash point of pure sulfur has been reported by various investigators at values from 370°F to 405°F (187°C–207°C). It melts at 572°F (300°C).

Health   Elemental sulfur is considered to be of low toxicity. Carbon disulfide, hydrogen sulfide, and sulfur dioxide should be handled carefully. Hydrogen sulfide in small concentrations can be metabolized, but in higher concentrations, it quickly can cause death by respiratory paralysis. It quickly deadens the sense of smell. Sulfur dioxide is a dangerous component in atmospheric air pollution. Compounds such as carbon disulfide, hydrogen sulfide, and sulfur dioxide are toxic. For example, at 0.03 ppm, we can smell hydrogen sulfide, but it is regarded as safe for 8 h of exposure. At 4 ppm, it may cause eye irritation. At 20 ppm, exposure for more than a minute causes severe injury to eye nerves. At 700 ppm, breathing stops. Death will result if there is not a quick rescue. Permanent brain damage may result.

Fluorine

Symbol: F
Metal or nonmetal: Nonmetal
Atomic number: 9
Atomic weight: 18.998403
Periodic table group: VII
Periodic table group name: Halogens
Ordinary physical state: Gas
Color: Pale yellow
Case registry number: 7782-41-4
U.N. four digit number: 9192
DOT hazard class: 2.3 poison gas
NFPA 704: 3-0-4 W
Cryogenic liquid: Yes
Cryogenic boiling point: −188.12°C, −306.62°F
Cryogenic expansion ratio: 888–1
Radioactive: No
Boiling point: −188.12°C, −306.62°F
Melting point: −219.62°C, −363.32°F
Vapor density: 1.695
Solubility: Insoluble in water
PEL-OSHA: 0.1 ppm
ACGIH-TLV-TWA: 1 ppm

History

The word fluorine was coined from Latin and French for *fluere*, meaning flow or flux. In 1529, Georgius Agricola described the use of fluorspar as a flux, and as early as 1670, Schwandhard found that glass was etched when exposed to fluorspar treated with acid. Scheele and many later investigators, including Davy, Gay-Lussac, Lavoisier, and Thenard, experimented with hydrofluoric acid, some experiments ending in tragedy. The element was finally isolated in 1866 by Moissan after nearly 74 years of continuous effort.

Sources

In stars, fluorine is rare compared to other light elements, but in earth's crust, fluorine is the 13th most abundant element. Fluorine's most important mineral, fluorite, was first formally described in 1530, in the context of smelting. The mineral's name derives from the Latin verb *fluo*, which means "flow," because fluorite was added to metal ores to lower their melting points. Suggested to be a chemical element in 1811, "fluorine" was named after the source mineral. In 1886, French chemist Henri Moissan first isolated the element. His method of electrolysis remains the industrial production method for fluorine gas. Although only a weak acid, hydrofluoric acid eats through glass and tissue and is a greater hazard than strong acids. The fluorides of lighter metals are ionic compounds (salts); those of heavier metals are volatile molecular compounds. The largest uses of inorganic fluorides are steelmaking and aluminum refining.

Important compounds

One hypothesis says that fluorine can be substituted for hydrogen wherever it occurs in organic compounds, which could lead to an astronomical number of new fluorine compounds.

Compounds of fluorine with rare gases have now been confirmed in fluorides of xenon, radon, and krypton.

Uses

The main use of elemental fluorine, uranium enrichment, was developed during the Manhattan Project. Because of the difficulty in making elemental fluorine, the vast majority of commercial fluorine is in the form of intermediate compounds. The main one being hydrofluoric acid, which is the key intermediate for the $16 billion per year global fluorochemical industry. Until World War II, there was no commercial production of elemental fluorine. The nuclear bomb project and nuclear energy applications, however, made it necessary to produce large quantities. Fluorine and its compounds are used in producing uranium (from the hexafluoride) and more than 100 commercial fluorochemicals, including many well-known high-temperature plastics. Hydrofluoric acid etches the glass of lightbulbs, etc. Fluorochlorohydrocarbons are extensively used in air-conditioning and refrigeration. The presence of fluorine as a soluble fluoride in drinking water to the extent of 2 ppm may cause mottled enamel in teeth when used by children acquiring permanent teeth; in smaller amounts, however, fluorides are added to water supplies to prevent dental cavities. Elemental fluorine has been studied as a rocket propellant, as it has an exceptionally high specific-impulse value.

Isotopes

F 14 through 31

Important reactions

It is a strong oxidizing agent, water reactive. Fluorine is the most electronegative and reactive of all elements. It is a pale yellow, corrosive gas, which reacts with most organic and inorganic substances. Finely divided metals, glass, ceramics, carbon, and even water burn in fluorine with a bright flame.

Air and water reactions   Water vapor will react combustibly with fluorine; an explosive reaction occurs between liquid fluorine and ice, after an intermediate induction period. If liquid air, which has stood for some time is treated with fluorine, a precipitate is formed, which is likely to explode. Explosive material is thought to be fluorine hydrate.

Propellant; ignites upon contact with alcohols, amines, ammonia, beryllium alkyls, boranes, dicyanogen, hydrazines, hydrocarbons, hydrogen, nitroalkanes, powdered metals, silanes, or thiols. Fluorine causes aromatic hydrocarbons and unsaturated alkanes to ignite spontaneously. Fluorine vigorously reacts with arsenic and arsenic trioxide at ordinary temperatures. Bromine mixed with fluorine at ordinary temperatures yields bromine trifluoride, with a luminous flame. Calcium silicide burns readily in fluorine. The carbonates of sodium, lithium, calcium, and lead in contact with fluorine are decomposed at ordinary temperatures with incandescence. A mixture of fluorine and carbon disulfide ignites at ordinary temperatures. The reaction between fluorine and carbon tetrachloride is violent and sometimes explosive. The uncontrolled reaction between fluorine and chlorine dioxide is explosive. Fluorine and silver cyanide react with explosive violence at ordinary temperatures. Fluorine and sodium acetate produce an explosive reaction involving the formation of diacetyl peroxide. Selenium, silicon, or sulfur ignites in fluorine gas at ordinary temperatures. Each bubble of sulfur dioxide gas led into a container of fluorine produces an explosion. Fluorine and thallous chloride react violently, melting the product.

Health   Elemental fluorine and the fluoride ion are highly toxic. The free element has a characteristic pungent odor, detectable in concentrations as low as 20 ppb, which is below the safe working level. The recommended maximum allowable concentration for a daily 8-h, time-weighted exposure in a 40-h workweek is 1 ppm. Safe handling techniques enable the transport of liquid fluorine by the ton. Elemental fluorine is highly toxic. Above a concentration of 25 ppm, fluorine causes significant irritation while attacking the eyes, respiratory tract, lungs, liver, and kidneys. At a concentration of 100 ppm, human eyes and noses are seriously damaged. Soluble fluorides are moderately toxic. For sodium fluoride, the lethal dose for adults is 5–10 g, which is equivalent to 32–64 mg of elemental fluoride per kilogram of body weight. The dose that may lead to adverse health effects is about one-fifth the lethal dose. Fluoride exposure limits are based on urine testing that has determined the human body's capacity for ridding itself of fluoride. Historically, most cases of fluoride poisoning have been caused by accidental ingestion of insecticides containing inorganic fluoride. Currently, most calls to poison control centers for possible fluoride poisoning come from the ingestion of fluoride-containing toothpaste. Malfunction of water fluoridation equipment has occurred several times, including an Alaskan incident, which affected nearly 300 people and killed 1.

Flammability   Fluorine is not flammable; however, it is a strong oxidizer and will support combustion. It may ignite other combustible materials (wood, paper, oil, etc.). Mixture with fuels may explode. Container may explode in heat of fire. Vapor explosion and poison hazard indoors, outdoors, or in sewers. Poisonous gas is produced in fire

Health   Poisonous and may be fatal if inhaled. Vapor is extremely irritating. Contact may cause burns to skin and eyes. Chronic absorption may cause osteosclerosis and calcification of ligaments.

Chemical hazard   Avoid contact with all oxidizable materials, including organic materials. This will react violently with water and most organic materials to produce heat and toxic fumes. Keep gas in tank and avoid exposure to all other materials.

Chlorine

Symbol: Cl
Metal or nonmetal: Nonmetal
Atomic number: 17
Atomic weight: 35.453
Periodic table group: VII
Periodic table group name: Halogens
Ordinary physical state: Gas
Color: Yellowish green
Case registry number: 7782-50-5
U.N. four digit number: 1017
DOT hazard class: 2.3 poison gas
NFPA 704: 3-0-0 ox
Radioactive: No

Boiling point: −34.04°C, −29.27°F
Melting point: −105.5°C, 229°F
Vapor density: 2.5
Solubility: Insoluble in water
PEL-OSHA: 3 mg/m^3
ACGIH-TLV-TWA: 0.5 ppm (1.5 mg/m^3) as a TWA for a normal 8-h workday and a 40-h workweek

History
The word chlorine was coined from the Greek, *chloros*, meaning greenish yellow. Discovered in 1774 by Scheele, who thought it contained oxygen. Chlorine was named in 1810 by Davy, who insisted it was an element.

Sources
In nature, it is found in the combined state only, chiefly with sodium as common salt (NaCl), carnallite, and sylvite. It is a member of the halogen (salt-forming) group of elements and is obtained from chlorides by the action of oxidizing agents and, more often, by electrolysis; it is a greenish-yellow gas, combining directly with nearly all elements.

Important compounds
Chlorine fluoride: ClF
Chlorine trifluoride: ClF$_3$
Chlorine pentafluoride: ClF$_5$
Chlorine dioxide: ClO$_2$
Dichlorine oxide: Cl$_2$O
Dichlorine hexoxide: Cl$_2$O$_6$
Dichlorine heptoxide: Cl$_2$O$_7$

Uses
Chlorine is widely used in making many everyday products. It is used for producing safe drinking water the world over. Even the smallest water supplies are now usually chlorinated. It is also extensively used in the production of paper products, dyestuffs, textiles, petroleum products, medicines, antiseptics, insecticides, food, solvents, paints, plastics, and many other consumer products. Most of the chlorine produced is used in the manufacture of chlorinated compounds for sanitation, pulp bleaching, disinfectants, and textile processing. Further use is in the manufacture of chlorates, chloroform, carbon tetrachloride, and in the extraction of bromine.

Isotopes
C 28 through 51

Important reactions
Strong oxidizer, chlorine reacts explosively with or supports the burning of numerous common materials. It ignites steel at 37°F (100°C) in the presence of soot, rust, carbon, or other catalysts. It ignites dry steel wool at 122°F (50°C). It reacts as either a liquid or gas with alcohols (explosion), molten aluminum (explosion), silane (explosion), bromine pentafluoride, carbon disulfide (explosion catalyzed by iron), 1-chloro-2-propyne (excess chlorine causes an explosion), dibutyl phthalate (explosion at 118°C), diethyl ether (ignition), diethyl zinc (ignition), glycerol (explosion at 70°C–80°C), methane over yellow mercury

oxide (explosion), acetylene (explosion initiated by sunlight or heating), ethylene over mercury, mercury(I) oxide, or silver(I) oxide (explosion initiated by heat or light), gasoline (exothermic reaction then detonation), naphtha–sodium hydroxide mixture (violent explosion), zinc chloride (exothermic reaction), wax (explosion), hydrogen (explosion initiated by light). It reacts as either a liquid or gas with carbides of iron, uranium, and zirconium, with hydrides of potassium sodium and copper, with tin, aluminum powder, vanadium powder, aluminum foil, brass foil, copper foil, calcium powder, iron wire, manganese powder, potassium, antimony powder, bismuth, germanium, magnesium, sodium, and zinc. It causes ignition and a mild explosion when bubbled through cold methanol. It explodes or ignites if mixed in excess with ammonia and warmed.

Hazards to responders

Flammability It may ignite other combustible materials (wood, paper, oil, etc.). Mixture with fuels may cause explosion. Container may explode in heat of fire. Vapor explosion and poison hazard indoors, outdoors, or in sewers. Hydrogen and chlorine mixtures (5%–95%) are exploded by almost any form of energy (heat, sunlight, sparks, etc.). It may combine with water or steam to produce toxic and corrosive fumes of hydrochloric acid. It emits highly toxic fumes when heated.

Health Poisonous and may be fatal if inhaled. Contact may cause burns to skin and eyes, and bronchitis or chronic lung conditions.

Bromine

Symbol. Br
Metal or nonmetal: Nonmetal
Atomic number: 35
Atomic weight: 79.904
Periodic table group: VII
Periodic table group name: Halogens
Ordinary physical state: Liquid
Color: Red brown
Case registry number: 7726-95-6
U.N. four digit number: 1744
DOT hazard class: 8 corrosive
NFPA 704: Unavailable
Radioactive: No
Boiling point: 59°C, 138°F
Melting point: –7.3°C, 19°F
Specific gravity: 3.119 at 20°C
Vapor density: 7.139
Solubility: Slightly soluble in water
PEL-OSHA: 0.1 ppm (0.7 mg/m^3)
ACGIH-TLV-TWA: 0.1 ppm, 0.66 mg/m^3 TWA

History

The word bromine was coined from the Greek, *bromos*, meaning stench. It was discovered by Balard in 1826, but not prepared in quantity until 1860.

Sources

A member of the halogen group of elements; it is obtained from natural brines from wells in Michigan and Arkansas. Little bromine is extracted today from seawater, which contains only about 85 ppm.

Important compounds

Bromine fluoride: BrF

Bromine trifluoride: BrF_3

Bromine pentafluoride: BrF_5

Bromine dioxide: BrO_2

Dibromine oxide: Br_2O

Uses

Much of the bromine output in the United States was used in the production of ethylene dibromide, a lead scavenger used in making gasoline antiknock compounds. Lead in gasoline, however, has been drastically reduced, due to environmental considerations. This will greatly affect future production of bromine. Bromine is also used in making fumigants, flameproofing agents, water purification compounds, dyes, medicinals, sanitizers, inorganic bromides for photography, etc. Organic bromides are also important.

Isotopes

Br 67 through 97

Important reactions

Elemental Bromine is a Strong oxidizing agent and bromine solution is an oxidizing agent as well. Bromine disproportionates rapidly in basic water to give bromide ion and bromate ion. The latter is also an oxidizing agent. Dissolution lowers reactivity compared to pure bromine. It reacts with reducing reagents. It reacts with hydrogen, diethylzinc, dimethylformamide, ammonia, trimethylamine, nitromethane, metal azides (silver or sodium azide). It can react with Mg, Sr, B, Al, Hg, Ti, Sn, Sb in powder or sheet form, to form bromides. It can react with methanol, ethanol, aldehydes, ketones, carboxylic acids, diethyl ether, carbonyl compounds, tetrahydrofuran, acrylonitrile, ozone, phosphorus, and natural rubber. Reactions with red phosphorus, metal azides, nitromethane, silane, and its homologues may be vigorous.

Hazards to responders

When spilled on the skin, it produces painful sores. It presents a serious health hazard, and maximum safety precautions should be taken when handling it.

Flammability   Non-combustible, substance itself does not burn but may decompose upon heating to produce corrosive and/or toxic fumes. It is a strong oxidizer and may ignite combustibles (wood, paper, oil, clothing, etc.). Contact with metals may evolve flammable hydrogen gas. Containers may explode when heated.

Health   Toxic; inhalation, ingestion, or skin contact with material may cause severe injury or death. Contact with molten substance may cause severe burns to skin and eyes. Avoid any skin contact. Effects of contact or inhalation may be delayed. Fire may produce irritating, corrosive, and/or toxic gases.

Chemical hazard   Bromine is the only liquid nonmetallic element. It is a heavy, mobile, reddish-brown liquid, volatilizing readily at room temperature to a red vapor with a strong disagreeable odor, resembling chlorine, and having an irritating effect on the eyes and throat; it is readily soluble in water or carbon disulfide, forming a red solution, and is less active than chlorine, but more so than iodine; it unites readily with many elements and has a bleaching action.

Iodine

Symbol: I
Metal or nonmetal: Nonmetal
Atomic number: 53
Atomic weight: 126.9045
Periodic table group: VII
Periodic table group name: Halogens
Ordinary physical state: Solid
Color: Violet dark gray lustrous
Case registry number: 7553-56-2
U.N. four digit number: 1759
DOT hazard class: 8 corrosive
NFPA 704: Unavailable
Radioactive: No
Boiling point: 113.7°C, 236.66°F
Melting point: 184.3°C, 363.7°F
Specific gravity: 4, 93
Vapor density: 8.8
Solubility: Insoluble in water
PEL-OSHA: 0.1 ppm (1 mg/m^3) CEILING
ACGIH-TLV-TWA: 0.1 ppm (1 mg/m^3) CEILING

History
The word iodine was coined from the Greek, *iodes*, meaning violet. Iodine was discovered by Courtois in 1811, iodine, a halogen, occurs sparingly in the form of iodides in seawater, from which it is assimilated by seaweeds, in Chilean saltpeter, and nitrate-bearing earth (known as caliche), in brines from old sea deposits, and in brackish waters from oil and salt wells.

Sources
Ultrapure iodine can be obtained from the reaction of potassium iodide with copper sulfate. Several other methods of isolating the element are known.

Important compounds
Iodine is a bluish-black, lustrous solid, volatizing at ordinary temperatures into a blue–violet gas with an irritating odor; it forms compounds with many elements, but is less active than the other halogens, which displace it from iodides. Iodine exhibits some metallic-like properties. It dissolves readily in chloroform, carbon tetrachloride, or carbon disulfide to form beautiful purple solutions. It is only slightly soluble in water.

Iodine fluoride: IF
Iodine trifluoride: IF$_3$
Iodine pentafluoride: IF$_5$

Iodine heptafluoride: IF_7
Iodine chloride: ICl
Diiodine hexachloride: $[ICl_3]_2$
Diiodine pentaoxide: I_2O_5
Diiodine tetraoxide: I_2O
Tetraiodine nonaoxide: I_4O_9

Uses

Iodine compounds are important in organic chemistry and are quite useful in medicine. Iodides and thyroxine, which contains iodine, are used internally in medicine and as a solution of KI, and iodine in alcohol is used for external wounds. Potassium iodide finds use in photography. The deep blue color in a starch solution is characteristic of the free element.

Isotopes

Thirty isotopes are recognized. Only one stable isotope, ^{127}I, is found in nature. The artificial radioisotope, ^{131}I, with a half-life of 8 days, has been used in treating the thyroid gland. The most common compounds are the iodides of sodium and potassium (KI) and the iodates (KIO_3). Lack of iodine is the cause of goiter.

Important reactions

Iodine, I_2, reacts with water to produce hypoiodite, OI^-. The position of the equilibrium depends very much upon the pH of the solution. Iodine, I_2, reacts with fluorine, F_2, at room temperature to form the pentafluoride iodine(V) fluoride. At 250°C, the same reaction affords the heptafluoride iodine(VII) fluoride. With careful control of the reaction conditions (–45°C, suspension in $CFCl_3$), it is possible to isolate the trifluoride iodine(III) fluoride. Iodine, I_2, reacts with bromine, Br_2, to form the very unstable, low melting solid, interhalogen species iodine(I) bromide. Iodine reacts with chlorine at –80°C with excess liquid chlorine to form "iodine trichloride," iodine (III) chloride, actually I_2Cl_6. Iodine reacts with chlorine in the presence of water to form iodic acid. Iodine reacts with hot concentrated nitric acid to form iodic acid. The iodic acid crystallizes out on cooling. Iodine, I_2, reacts with hot aqueous alkali to produce iodate, IO_3^-. Only one-sixth of the total iodine is converted in this reaction. Iodine is an oxidizing agent. It reacts vigorously with reducing materials. Incompatible with powdered metals in the presence of water (ignites), with gaseous or aqueous ammonia (forms explosive products), with acetylene (reacts explosively), with acetaldehyde (violent reaction), with metal azides (forms yellow explosive iodo azides), with metal hydrides (ignites), with metal carbides (ignites easily), with potassium and sodium (forms shock-sensitive explosive compounds), and with alkali-earth metals (ignites). It is incompatible with ethanol, formamide, chlorine, bromine, bromine trifluoride, and chlorine trifluoride.

Hazards to responders

Flammability   Iodine does not burn. It is a strong oxidizer and will support combustion.

Health   Care should be taken in handling and using iodine, as contact with the skin can cause lesions; iodine vapor is intensely irritating to the eyes and mucous membranes. The maximum allowable concentration of iodine in air should not exceed 1 mg/m^3

(8-h TWA in a 40-h workweek). In small doses, iodine is slightly toxic, and it is highly poisonous in large amounts. Elemental iodine is an irritant that can cause sores on the skin. Iodine vapor causes extreme eye irritation.

Exposure routes: Inhalation, ingestion, skin, and/or eye contact. Symptoms: Irritation of eyes, skin, nose; lacrimation (discharge of tears); headache; chest tightness; skin burns, rash; cutaneous hypersensitivity.

Target organs: Eyes, skin, respiratory system, central nervous system, and cardiovascular system.

Uranium

Symbol: U
Metal or nonmetal: Metal
Atomic number: 92
Atomic weight: 238.0289
Periodic table group: Transition metals
Periodic table group name: None
Ordinary physical state: Solid
Color: Metallic gray
Case registry number: 7440-61-1
U.N. four digit number: 2977
DOT hazard class: 7 radioactive, 8 corrosive
NFPA 704. Unavailable
Radioactive: Yes
Boiling point: 3900° C, 7101°F
Melting point: 1132.2°C, 2070°F
Specific gravity: 18.7
Density: 19.05 g/cm
Solubility: Insoluble in water
PEL-OSHA: 0.2 mg/m^3 of air as an 8-h TWA
ACGIH-TLV-TWA: 0.2 mg(U)/m^3

History

Uranium was named after the planet Uranus. Yellow-colored glass, containing more than 1% uranium oxide and dating back to A.D. 79 has been found near Naples, Italy. Klaproth recognized an unknown element in pitchblende (uraninite or uranium oxide) and attempted to isolate the metal in 1789. The metal was first isolated in 1841 by Peligot, who reduced the anhydrous chloride using potassium.

Sources

Uranium, not as rare as once thought, is now considered to be more plentiful than mercury, antimony, silver, or cadmium and is about as abundant as molybdenum or arsenic. It occurs in numerous minerals, such as pitchblende, uraninite, carnotite, autunite, uranophane, and tobernite. It is also found in phosphate rock, lignite, and monazite sands and can be recovered commercially from these sources. The U.S. Department of Energy purchases uranium in the form of acceptable U_3O_8 concentrates. This incentive program has greatly increased the known uranium reserves. Uranium can be prepared by reducing uranium halides with alkali or alkaline earth metals or by reducing uranium oxides by calcium, aluminum, or carbon at high temperatures. The metal can also be produced by electrolysis of KUF_5 or UF_4, dissolved in a molten mixture of $CaCl_2$ and NaCl. High-purity uranium can be prepared by the thermal decomposition of uranium halides on a hot filament.

   Important compounds
Uranium trihydride: UH_3
Uranium trifluoride: UF_3
Uranium hexafluoride: UF_6
Uranium tetrafluoride: UF_4
Uranium pentafluoride: UF_5
Tetrauranium octadecafluoride: U_4F_{18}
Tetrauranium heptadecafluoride: U_4F_{17}
Uranium trichloride: UCl_3
Uranium hexachloride: UCl_6
Uranium tetrachloride: UCl_4
Uranium pentachloride: UCl_5
Uranium tetrabromide: UBr_4
Uranium pentabromide: UBr_5
Uranium triiodide: UI_3
Uranium tetraiodide: UI_4
Uranium oxide: UO
Uranium dioxide: UO_2
Uranium trioxide: UO_3
Triuranium octaoxide: U_3O_8
Diuranium pentoxide: U_2O_5
Triuranium heptoxide: U_3O_7
Tetrauranium nonaoxide: U_4O_9
Uranium sulfide: US
Diuranium trisulfide: U_2S_3
Uranium diselenide: USe_2
Uranium triselenide: USe_3
Uranium ditelluride: UTe_2
Uranium tritelluride: UTe_3
Uranium nitride: UN
Diuranium trinitride: often written U_2N_3
Diuranium trinitride: written using HILL SYSTEM U_3N_2

   Uses
Uranium is of great importance as a nuclear fuel. Uranium-235, ^{235}U, while occurring in natural uranium to the extent of only 0.71%, is so fissionable with slow neutrons that a self-sustaining fission chain reaction can be made in a reactor constructed from natural uranium and a suitable moderator, such as heavy water or graphite, alone. One pound of completely fissioned uranium has the fuel value of over 1500 tons of coal. The uses of nuclear fuels to generate electrical power, to make isotopes for peaceful purposes, and to make explosives are well known. The estimated worldwide capacity of the 429 nuclear-power reactors in operation in January 1990 amounted to about 311,000 MW. Uranium in the United States is controlled by the U.S. Department of Energy. New uses are being found for depleted uranium, that is, uranium with the percentage of ^{235}U lowered to about 0.2%.

Uranium is used in inertial guidance devices, in gyrocompasses, as counterweights for aircraft control surfaces, as ballast for missile reentry vehicles, and as a shielding material. Uranium metal is used for x-ray targets for production of high-energy x-rays; the nitrate has been used as a photographic toner, and the acetate is used in analytical chemistry. Uranium salts have also been used for producing yellow "Vaseline" glass and glazes. Uranium and its compounds are highly toxic, both from a chemical and from a radiological standpoint.

Isotopes

Uranium has 16 isotopes, all of which are radioactive. Naturally occurring uranium nominally contains 99.28305, by weight,^{238}U, 0.7110% ^{235}U, and 0.0054% ^{234}U. Natural uranium is sufficiently radioactive to expose a photographic plate in an hour or so. Much of the internal heat of the earth is thought to be attributable to the presence of uranium and thorium. Uranuim-238, with a half-life of 4.51×10^9 years, has been used to estimate the age of igneous rocks. The origin of uranium, the highest member of the naturally occurring elements—except perhaps for traces of neptunium or plutonium—is not clearly understood. However, it may be presumed that uranium is a decay product of elements with higher atomic weight, which may have once been present on earth or elsewhere in the universe. These original elements may have been formed as a result of a primordial creation, known as the big bang, in a supernova or in some other stellar processes. U 217 through 242.

Important reactions

Uranium hexafluoride (fissile, containing more than 1% U-235) has been enriched in the fissile isotope of uranium. Naturally occurring uranium contains 0.7% U-235 (higher radioactivity) and 99.3% U-238 (lower radioactivity). It emits fumes of highly toxic metallic uranium and uranium fluorides when heated to decomposition. It reacts vigorously with aromatic hydrocarbons (benzene, toluene, xylenes) and undergoes a violent reaction with water or alcohols (methanol, ethanol). It reacts with most metals.

Hazards to responders

Fire hazard Substance does not burn. Finely divided uranium metal, being pyrophoric, presents a fire hazard. The material may react violently with fuels. Radioactivity does not change flammability or other properties of materials.

Health hazard Radiation presents minimal risk to emergency response personnel and the public during transportation accidents. Packaging durability increases as potential radiation and criticality hazards of the content increase. Chemical hazard greatly exceeds radiation hazard. Substance reacts with water and water vapor in air to form toxic and corrosive hydrogen fluoride gas and an extremely irritating and corrosive, white-colored, water-soluble residue. If inhaled, it may be fatal. Direct contact causes burns to skin, eyes, and respiratory tract. Low-level radioactive material may cause very low radiation hazard to people. Runoff from control of cargo fire may cause low-level pollution.

Chemical hazard Uranium reacts vigorously with water to form uranyl fluoride (UO_2F_2) and corrosive hydrogen fluoride (hydrofluoric acid). Uranium is a heavy, silvery-white metal, which is pyrophoric when finely divided. It is a little softer than steel and is attacked by cold water in a finely divided state. It is malleable, ductile, and slightly paramagnetic. In air, the metal becomes coated with a layer of oxide. Acids dissolve the metal, but it is unaffected by alkalies.

Helium

Symbol: He
Metal or nonmetal: Nonmetal
Atomic number: 2
Atomic weight: 4.00260
Periodic table group: VIII
Periodic table group name: Inert or noble gases
Ordinary physical state: Gas
Color: Colorless
Case registry number: 7440-59-7
U.N. four digit number: 1963
DOT hazard class: 2.2 non-flammable compressed gas
NFPA 704: Unavailable
Cryogenic liquid: Yes
Cryogenic boiling point: −269°C, −452°F
Cryogenic expansion ratio: 780–1
Radioactive: No
Boiling point: −269°C, −452°F
Melting point: −272.2°C, −458°F
Freezing point: Does not exist as a solid
Vapor density: 0.125
Solubility: Insoluble in water
PEL-OSHA: None
ACGIH-TLV-TWA: None

History

The word helium was coined from the Greek, *helios*, meaning the sun. Janssen obtained the first evidence of helium during the solar eclipse of 1868, when he detected a new line in the solar spectrum. Lockyer and Frankland suggested the name helium for the new element. In 1895, Ramsay discovered helium in the uranium mineral clevite, while it was independently discovered in cleveite by the Swedish chemists Cleve and Langlet at about the same time. Rutherford and Royds in 1907 demonstrated that alpha particles are helium nuclei.

Sources

Except for *hydrogen*, helium is the most abundant element found throughout the universe. Helium is extracted from natural gas. In fact, all natural gas contains at least trace quantities of helium. The fusion of hydrogen into helium provides the energy of the hydrogen bomb. The helium content of the atmosphere is about 1 part in 200,000. While it is present in various radioactive minerals as a decay product, the bulk of the Free World's supply is obtained from wells in Texas, Oklahoma, and Kansas.

Important compounds

While helium normally has a 0 valence, it seems to have a weak tendency to combine with certain other elements. Means of preparing helium difluoride have been studied, and species such as HeNe and the molecular ions He^+ and He^{++} have been investigated.

Uses

Helium is used as an inert gas shield in arc welding, as a protective gas in growing silicon and germanium crystals and producing titanium and zirconium, as a cooling medium for nuclear reactors, MNR and MRI imaging machines, and as a gas for supersonic wind tunnels. A mixture of helium and oxygen is used as an artificial atmosphere for divers and others working under pressure. One of the recent largest uses for helium has been for pressuring liquid-fuel rockets. A Saturn booster, like the type used on the Apollo lunar missions, required about 13 million ft^3 of helium for a firing, plus more for checkouts. Helium is also used to advertise on blimps for various companies, including Goodyear.

Isotopes
He 3 through 10

Important reactions
None identified

Hazards to responders

Helium is an inert gas   Helium has the lowest melting point of any element and is widely used in cryogenic research because its boiling point is close to absolute zero. Helium is the only liquid that cannot be solidified by lowering the temperature. It remains liquid down to absolute zero at ordinary pressures, but it can readily be solidified by increasing the pressure.

Health   Vapors may cause dizziness or asphyxiation without warning. Vapors from liquefied gas are initially heavier than air and spread along ground. Helium as a cryogenic liquid can be a simple asphyxiant and cause freeze and frostbite injury in contact with body parts.

Neon

Symbol: Ne
Metal or nonmetal: Nonmetal
Atomic number: 10
Atomic weight: 20.179
Periodic table group: VIII
Periodic table group name: Inert or noble gases
Ordinary physical state: Gas
Color: Colorless
Case registry number: 7440-01-9
U.N. four digit number: 1913
DOT hazard class: 2.2 non-flammable compressed gas
NFPA 704: Unavailable
Cryogenic liquid: Yes
Cryogenic boiling point: −246.08°C, −410.94°F
Cryogenic expansion ratio: 1438–1 (highest expansion ratio)
Radioactive: No
Boiling point: −246.08°C, −410.94°F
Vapor density: 0.70
Solubility: Insoluble in water
PEL-OSHA: None, simple asphyxiant
ACGIH-TLV-TWA: None, simple asphyxiant

History

The word neon was coined from the Greek word *neos*, meaning new. Discovered by Ramsay and Travers in 1898, neon is a rare gaseous element present in the atmosphere to the extent of 1 part in 65,000 of air.

Sources

It is obtained by liquefaction of air and separated from the other gases by fractional distillation.

Important compounds

Neon, an inert element, is, however, said to form a compound with fluorine. It is still questionable if true compounds of neon exist, but evidence is mounting in favor of their existence.

Uses

In a vacuum discharge tube, neon glows reddish orange. It has over 40 times more refrigerating capacity per unit volume than liquid helium and more than three times that of liquid hydrogen. It is compact, inert, and is less expensive than helium when it meets refrigeration requirements. Although neon advertising signs account for the bulk of its use, neon also functions in high-voltage indicators, lightning arrestors, wave meter tubes, and TV tubes. Neon and helium are used in making gas lasers.

Isotopes

Ne 16 through 34

Important reactions

None identified

Hazards to responders

Liquid neon is now commercially available and is finding important application as an economical cryogenic refrigerant. Vapors may cause dizziness or asphyxiation without warning. Vapors from liquefied gas are initially heavier than air and spread along ground. Neon can be a simple asphyxiant and cause freeze and frostbite injuries in contact with body parts.

Argon

Symbol: Ar
Metal or nonmetal: Nonmetal
Atomic number: 18
Atomic weight: 39.948
Periodic table group: VIII
Periodic table group name: Inert or noble gases
Ordinary physical state: Gas
Color: Colorless
Case registry number: 7440-37-1
U.N. four digit number: 1951
DOT hazard class: 2.2 non-flammable gas

NFPA 704: Unavailable
Cryogenic liquid: Yes
Cryogenic boiling point: −186°C, −303°F
Cryogenic expansion ratio: N/A
Radioactive: No
Boiling point: −186°C, −303°F
Melting point: −189.3°C, −308.7°F
Vapor density: 1.67
Solubility: Insoluble in water
PEL-OSHA: None, simple asphyxiant
ACGIH-TLV-TWA: None, simple asphyxiant

History
The word Argon was derived from the Greek word *argos*, meaning inactive. Its presence in air was suspected by Cavendish in 1785 and discovered by Lord Rayleigh and Sir William Ramsay in 1894.

Sources
The gas is prepared by fractionation of liquid air, because the atmosphere contains 0.94% argon. The atmosphere of Mars contains 1.6% of ^{40}Ar and 5 ppm of ^{36}Ar.

Important compounds
None identified

Uses
It is used in electric lightbulbs and in fluorescent tubes at a pressure of about 400 Pa and in filling phototubes, glow tubes, etc. Argon is also used as an inert gas shield for arc welding and cutting, as a blanket for the production of titanium and other reactive elements, and as a protective atmosphere for growing silicon and germanium crystals.

Isotopes
Ar 30 through 53

Important reactions
None identified

Hazards to responders
Neon is inert.

Health   Vapors may cause dizziness or asphyxiation without warning. Vapors from liquefied gas are initially heavier than air and spread along ground. Contact with gas or liquefied gas may cause burns, severe injury, and/or frostbite.

Chemical hazards   Argon is two- and one-half times as soluble in water as nitrogen, having about the same solubility as oxygen. Argon is colorless and odorless, both as a gas and as a liquid. Argon is considered to be an inert gas and is not known to form true chemical compounds, as do krypton, xenon, and radon.

Krypton

Symbol: Kr
Metal or nonmetal: Nonmetal
Atomic number: 36
Atomic weight: 83.80
Periodic table group: VIII
Periodic table group name: Inert or noble gases
Ordinary physical state: Gas
Color: Colorless
Case registry number: 7439-90-9
U.N. four digit number: 1056
DOT hazard class: Non-flammable compressed gas
NFPA 704: Unavailable
Cryogenic Liquid: Yes
Cryogenic boiling point: −153.22°C, −243.8°F
Cryogenic expansion ratio: 693–1
Radioactive: No
Boiling point: −153.22°C, −243.8°F
Melting point: −157.36°C, −251.25°F
Vapor density: 3.55
Solubility: Insoluble in water
PEL-OSHA: None, simple asphyxiant
ACGIH-TLV-TWA: None, simple asphyxiant

History

The word Krypton was coined from the Greek word *kryptos*, meaning hidden. It was discovered in 1898 by Ramsay and Travers in the residue left after liquid air had nearly boiled away. In 1960, it was internationally agreed that the fundamental unit of length, the meter, should be defined in terms of the orange–red spectral line of ^{86}Kr. This replaced the standard meter of Paris, which was defined in terms of a bar made of a platinum–iridium alloy. In October 1983, the meter, which originally was defined as 1/10-millionth of a quadrant of the earth's polar circumference, was again redefined by the International Bureau of Weights and Measures as the length of a path traveled by light in a vacuum during a time interval of 1/299,792,458 of a second. It is one of the noble gases. It is characterized by its brilliant green and orange spectral lines.

Sources

Krypton is present in the air to the extent of about 1 ppm. The atmosphere of Mars has been found to contain 0.3 ppm of krypton. Solid krypton is a white crystalline substance with a face-centered cubic structure, which is common to all the "rare gases."

Important compounds

None identified

Uses

Krypton clathrates (an inclusion complex in which molecules of one substance are completely enclosed within the other) have been prepared with hydroquinone and phenol. ^{85}Kr has found recent application in chemical analysis. By imbedding the isotope in various solids, kryptonates are formed. The activity of these kryptonates is sensitive to chemical reactions at the surface. Estimates of the concentration of reactants are therefore made possible. Krypton is used in certain photographic flashlamps for high-speed photography. Uses thus far have been limited because of its high cost.

Isotopes
Kr 69 through 100

Important reactions
None identified

Hazards to responders
Krypton is inert.

Health   Vapors may cause dizziness or asphyxiation without warning. Vapors from liquefied gas are initially heavier than air and spread along ground. Contact with gas or liquefied gas may cause burns, severe injury, and/or frostbite.

Xenon

Symbol: Xe
Metal or nonmetal: Nonmetal
Atomic number: 54
Atomic weight: 131.29
Periodic table group: VIII
Periodic table group name: Inert or noble gases
Ordinary physical state: Gas
Color: Colorless
Case registry number: 7440-63-3
U.N. four digit number: 2591
DOT hazard class: 2.2 non-flammable gas
NFPA 704: Unavailable
Cryogenic liquid: Yes
Cryogenic boiling point: −108°C, −162°F
Radioactive: No
Boiling point: −108°C, −162°F
Melting point: −111.7°C, −169.1°F
Vapor density: 5.761
Solubility: Insoluble in water
PEL-OSHA: None, simple asphyxiant
ACGIH-TLV-TWA: None, simple asphyxiant

History
It was coined from the Greek word *xenon*, meaning stranger. It was discovered by Ramsay and Travers in 1898 in the residue left after evaporating liquid air components; xenon is a member of the noble, or inert, gases. It is present in the atmosphere to the extent of about one part in 20 million. Xenon is present in the Martian atmosphere to the extent of 0.08 ppm. The element is also found in the gases evolved from certain mineral springs and is commercially obtained by extraction from liquid air.

Sources

Xenon is harvested from the air. It is found in the atmosphere at levels of approximately one part in 20 million. It is commercially obtained by extraction from liquid air. Xenon-133 and xenon-135 are produced by neutron irradiation in air-cooled nuclear reactors.

Important compounds

None identified

Uses

The gas is used in making electron tubes, stroboscopic lamps, bactericidal lamps, and lamps used to excite ruby lasers for generating coherent light. Xenon is used in the nuclear energy field in bubble chambers, probes, and other applications where a high molecular weight is of value. The element is also available in sealed glass containers of gas at standard pressure. Xenon is not toxic, but its compounds are highly toxic because of their strong oxidizing characteristics.

Isotopes

Xe 110 through 145

Important reactions

None identified

Hazards to responders

Xenon is inert.

Health   Vapors may cause dizziness or asphyxiation without warning. Vapors from liquefied gas are initially heavier than air and spread along ground. Contact with gas or liquefied gas may cause burns, severe injury, and/or frostbite.

Lead

Symbol: Pb
Metal or nonmetal: Metal
Atomic number: 82
Atomic weight: 207.2
Periodic table group: IV
Periodic table group name: Transition metal
Ordinary physical state: Solid
Color: Bluish white
Case registry number: 7439-92-1
U.N. four digit number: None
DOT hazard class: None
NFPA 704: Unavailable
Radioactive: No
Boiling point: 1749°C, 3180°F
Melting point: 327.46°C, 621.43°F
Specific gravity: 11.35

Solubility: Insoluble in water
PEL-OSHA: 0.050 mg/m^3
ACGIH-TLV-TWA: 0.05 mg/m^3

History

It was derived from the Anglo-Saxon, lead; Latin for *plumbum*. It was long known, mentioned in Exodus. Alchemists believed lead to be the oldest metal, and it was associated with the planet Saturn. Native lead occurs in nature, but it is rare. Its alloys include solder, type metal, and various antifriction metals. Great quantities of lead, both as the metal and as the dioxide, are used in storage batteries. Much metal also goes into cable covering, plumbing, ammunition, and in the manufacture of lead tetraethyl.

Sources

Lead is obtained chiefly from galena (PbS) by a roasting process. Anglesite, cerussite, and minim are other common lead minerals. Lead is a bluish-white metal of bright luster, is soft, highly malleable, ductile, and a poor conductor of electricity. It is resistant to corrosion; lead pipes bearing the insignia of Roman emperors used as drains from the baths are still in service. It is used in containers for corrosive liquids (such as sulfuric acid) and may be toughened by the addition of a small percentage of antimony or other metals.

Important compounds

Plumbane: PbH$_4$
Lead difluoride: PbF$_2$
Lead tetrafluoride: PbF$_4$
Lead dichloride: PbCl$_2$
Lead tetrachloride: PbCl$_4$
Lead tetrabromide: PbBr$_4$
Lead diiodide: PbI$_2$
Lead oxide: PbO
Lead dioxide: PbO$_2$
Dilead trioxide: Pb$_2$O$_3$
Trilead tetroxide: Pb$_3$O$_4$
Lead sulfide: PbS
Lead selenide: PbSe
Lead telluride: PbTe

Uses

The metal is effective as a sound absorber, is used as a radiation shield around x-ray equipment and nuclear reactors, and is used to absorb vibration. White lead, the basic carbonate, sublimed white lead, chrome yellow, and other lead compounds are used extensively in paints, although in recent years, the use of lead in paints has been drastically curtailed to eliminate or reduce health hazards. Lead oxide is used in producing fine "crystal glass" and "flint glass" of a high index of refraction for achromatic lenses. The nitrate and the acetate are soluble salts. Lead salts, such as lead arsenate, have been used as insecticides, but their use in recent years has been practically eliminated in favor of less-harmful organic compounds.

Isotopes

Pb 178 through 215

Important reactions

Finely divided lead powder is pyrophoric, however, meaning it is a fire risk. The surface of metallic lead is protected by a thin layer of lead oxide, PbO. It does not react with water under normal conditions. Lead metal reacts vigorously with fluorine, F_2, at room temperature and chlorine, Cl_2, on warming to form the poisonous dihalides lead(II) fluoride, PbF_2, and lead(II) chloride, $PbCl_2$, respectively. The surface of metallic lead is protected by a thin layer of lead oxide, PbO. This renders the lead essentially insoluble in sulfuric acid, and so, in the past, a useful container of this acid. Lead reacts slowly with hydrochloric acid and nitric acid, HNO_3. In the latter case, nitrogen oxides are formed together with lead(II) nitrate, $Pb(NO_3)_2$. Lead dissolves slowly in cold alkalies to form plumbites.

Hazards to responders

Flammability   Elemental lead is not a significant fire hazard. It does have a low melting point, and molten lead can cause serious thermal burns.

Health   Care must be used in handling lead, as it is a cumulative poison. Environmental concerns with lead poisoning have resulted in a national program to eliminate the lead in gasoline. Slightly hazardous in case of skin contact (irritant), of eye contact (irritant), of ingestion, of inhalation.

Compounds and mixtures

Two or more elements that combine chemically (through exchange or sharing of outer-shell electrons) and form chemical bonds are referred to as compounds. All compounds formed must be electrically neutral. To be successfully bonded, compounds must fulfill the octet and duet rules of bonding, with two or eight electrons in the outer shell, and must have the same number of positively charged protons in the nucleus as there are negatively charged electrons orbiting outside the nucleus. Once formed, compounds can be identified by their chemical names, chemical formulas, or in some cases chemical structures. They can also be placed in chemical families based upon the types of elements used to make up the compound. There are three types of formulae that identify chemical compounds: empirical, molecular, and structural. Structures generally apply to nonmetal compounds, including hydrocarbons and hydrocarbon-derivative families.

Chemical compounds may be combined to form mixtures. A mixture is two or more compounds blended together without any chemical bond taking place. Each of the compounds retains its own characteristic properties. Compounds when mixed form two types of mixtures: homogeneous and heterogeneous. Homogeneous means "the same kind" in Latin. In a homogeneous mixture, every part is exactly like every other part. For example, water has a molecular formula of H_2O. Pure water is homogeneous; it contains no substances other than hydrogen and oxygen. Loosely translated to include mixtures, homogeneous refers to two or more compounds or elements that are uniformly dispersed in each other. A solution is another example of a homogeneous mixture. Heterogeneous means "different kinds" in Latin. In a heterogeneous mixture, the different parts of the mixture have different properties. A heterogeneous mixture can be separated mechanically into its component parts. Some examples of heterogeneous mixtures are gasoline, the air we breathe, blood, and mayonnaise.

Solubility

Solubility is a term associated with mixing two or more compounds together. The definition of solubility from the *Condensed Chemical Dictionary* is "the ability or tendency of one substance to blend uniformly with another, e.g., solid in liquid, liquid in liquid, gas in liquid, and gas in gas." Solubility may vary from one substance to another. When researching chemicals in reference sources, relative solubility terms may include very soluble, slightly soluble, moderately soluble, and insoluble. Generally speaking, nothing is absolutely insoluble. Insoluble actually means "very sparingly soluble," that is, only trace amounts dissolve.

Compounds of the alkali metals of Group I on the periodic table are all soluble. Salts containing NH_4^+, NO_3, ClO_4, and ClO_3, and organic peroxides containing $C_2H_3O_2$ are also soluble. All chlorides (Cl), bromides (Br), and iodides (I) are soluble, except for those containing the metals Ag^+, Pb^{+2}, and Hg^{+2}. All sulfates are soluble, except for those with the metals Pb^{+2}, Ca^{+2}, Sr^{+2}, Hg^{+2}, and Ba^{+2}. All hydroxides (OH^{-1}) and metal oxides (containing O_2) are insoluble, except those of Group I on the periodic table and Ca^{+2}, Sr^{+2}, and Ba^{+2}. When metal oxides do dissolve, they give off hydroxides (their solutions do not contain O^2 ions). The following illustration is an example of a metal oxide and water reaction (note that the reaction is shown as an illustration and has not been balanced):

$$Na_2O + H_2O = NaOH$$

All compounds that contain PO_4, CO_3, SO_3, and SO_2 are insoluble, except for those of Group I on the periodic table and NH_4. Most hydrocarbon mixtures and compounds are not soluble, such as gasoline, diesel fuel, pentane, octane, etc. Compounds that are polar, such as the alcohols, ketones, aldehydes, esters, and organic acids, are soluble in water. This is because there is a rule in chemistry that says "like materials dissolve like materials." Water is also polar, so polar materials dissolve in or mix with water. Several other factors affect the solubility of a material. One is particle size: the smaller the particle, the more surface area that is exposed to the solvent; therefore, more dissolving takes place over a shorter period of time. Higher temperatures *usually* increase the rate of dissolving. The term *miscibility* is often used synonymously with the term *solubility*. Solubility is also related to polarity insofar as those materials that are polar are generally soluble in other polar materials. Miscibility, solubility, polarity, and mixtures will be discussed as they pertain to specific chemicals and families of chemicals in other chapters of this book.

Some elements are so reactive that they do not exist naturally as single atoms. They chemically bond with another atom of that same element to form "diatomic" molecules. The diatomic elements are hydrogen, oxygen, nitrogen, chlorine, bromine, iodine, and fluorine. One way to remember the diatomic elements is by using the acronym *HONClBrIF*, pronounced honk-le-brif, which includes the symbols for all of the diatomic elements. Oxygen is commonly referred to as O_2, primarily because oxygen is a diatomic element. Two oxygen atoms have covalently bonded together and act as one unit.

Much can be learned about a compound by looking at its elemental composition. Generally speaking, chemicals that contain chlorine in their formula may be toxic to some degree because chlorine is toxic. There are exceptions, such as sodium chloride, with the formula NaCl. Sodium chloride is table salt. The toxicity is low, but even if you did not know sodium chloride was table salt and treated it as a toxic material because of the chlorine, your error would be on the side of safety. If you are going to make errors when dealing with hazardous materials, always attempt to err on the side of safety. You may have "egg on your face" afterward and take some ribbing, but no one has ever died from embarrassment. On the other hand, if you are not cautious and your error is not on the side of safety, it could be fatal!

Atom

An atom is the smallest particle of an element that retains all of its elemental characteristics. The word *atom* comes from the Greek, meaning "not cut." For example, take a sheet of paper and tear it in half. Keep tearing the paper in half until you cannot tear it any more. You could then take scissors and cut the paper into smaller pieces. Eventually, you will not be able to cut the paper into a smaller piece. An atom is like that last piece of paper: you cannot have a smaller piece of an element than an atom.

Symbols of the elements on the periodic table represent one atom of each element. A single atom cannot be altered chemically. To create a smaller part of an atom of any element would require that the atom be split in a nuclear reaction. Therefore, a single atom is the smallest particle of an element that would normally be encountered. The atom comprises three major parts: electrons, protons, and neutrons (Figure 2.5). An atom is like a miniature solar system, with the nucleus in the center and electrons orbiting around it. Parts of the atom are referred to as subatomic particles. The atom is made up of positively (+) charged protons that are located in the nucleus along with neutrons. Neutrons do not have a charge, so they are electrically neutral. Orbiting in shells, or energy levels, around the nucleus are varying numbers of negatively (–) charged electrons. Electrons are important in discussing chemistry for hazardous materials responders. The shells, or orbits, about the nucleus of an atom have important characteristics:

1. Each energy level is capable of containing a specific number of electrons.
2. There is a limit on the number of electrons in the various shells.
3. The inner shell next to the nucleus *never* holds more than two electrons.
4. The last shell or outer shell *never* holds more than eight electrons.
5. The electrons in the outer shell control chemical reactions.

The only elements that occur naturally with eight electrons in the outer shell are the inert gases in family VIII of the periodic table. Actual numbers of outer-shell electrons

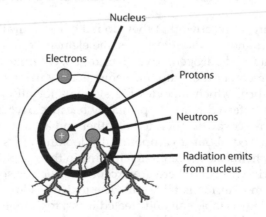

Figure 2.5 The atom.

that exist in any given element are represented by a Roman numeral at the top of each group in the vertical columns of the periodic table. The exceptions to the Roman numeral designation for the number of outer-shell electrons are the transitional element metals. The transitional metals have varying numbers of outer-shell electrons, and the process for identifying them will be discussed along with the salt compounds.

Atoms as they exist naturally are found to be electromagnetically neutral, so they must have an equal number of protons and electrons. Atoms are held together by a strong electrical attraction between the positive protons in the nucleus and the negative electrons in the shells around the nucleus. This comes from another chemistry rule that says "Opposite charges attract." Inner-shell electrons are held tightly by the nucleus. The farther away the outer shells are from the nucleus, the less control the nucleus will have over those electrons. This makes large atoms tend to lose electrons and small atoms tend to gain electrons. Electrons contained in metal elements are generally farther away from the nucleus than are nonmetal electrons; therefore, metals give up electrons and nonmetals take on electrons.

The atomic weight of an element is contained within the nucleus. While there is some weight associated with electrons, for our purposes in "street chemistry," electrons will have no weight. Protons and neutrons in the nucleus each have a weight of 1 AMU. Atoms of elements have varying numbers of protons and neutrons. The total number of neutrons and protons in each atom equals the atomic weight. The atomic number of an element equals the number of protons contained in that element. The number of neutrons is determined by subtracting the number of protons from the atomic weight. In chemistry, we are concerned more with the electrons orbiting the nucleus than the nucleus itself. We are particularly interested in the electrons in the outer shell of the atom. Chemical activity takes place between the outer-shell electrons of elements; this chemical activity forms compounds. In radioactivity, the concern is with the nucleus, where radiation is emitted. Radioactivity will be discussed further in Chapter 8.

Formulae

Just as an atom is the smallest part of an element, a single element is the smallest portion of a chemical compound that can be encountered. Chemical compounds are made up of two or more elements that have bonded together by exchanging electrons or by sharing electrons. Bonds that are shared are called covalent and bonds that exchange electrons are called ionic. Chemical compounds in addition to having names are also represented by formulae, much like elements are represented by symbols. According to the *Condensed Chemical Dictionary*, "a formula is a written representation, using symbols, of a chemical entity or relationship." There are three types of chemical formulae: empirical, molecular, and structural.

Empirical formulae indicate composition, not structure, of the relative number and kinds of atoms in a molecule of one or more compounds; for example, CH is the empirical formula for both acetylene and benzene.

The molecular formula is made up of the symbols of the elements that are in the compound and the number of atoms of each. The number of atoms of each element present is represented by a subscript number behind the symbol. The molecular formula is a kind of recipe for the compound. It lists all the ingredients and the proportions of each so that the compound can be reproduced if the formula is followed. The molecular formula also shows the actual number and kind of atoms in a chemical entity, for example, the molecular formula for sulfuric acid is H_2SO_4. Sodium chloride, table salt, has the molecular

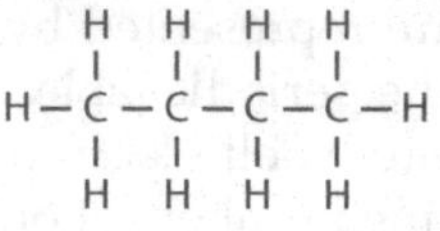

$$
\begin{array}{c}
\ \ \ H \ \ \ H \ \ \ H \ \ \ H \\
\ \ \ | \ \ \ | \ \ \ | \ \ \ | \\
H - C - C - C - C - H \\
\ \ \ | \ \ \ | \ \ \ | \ \ \ | \\
\ \ \ H \ \ \ H \ \ \ H \ \ \ H
\end{array}
$$

Figure 2.6 Structural formula.

formula NaCl. The compound has one atom of sodium and one atom of chlorine. When a molecular formula only contains one atom of each element, you do not include a subscript 1. It is understood that there is one atom of each element. Aluminum oxide has a molecular formula of Al_2O_3 (two atoms of aluminum and three atoms of oxygen).

The structural formula indicates the location of the atoms in relation to each other in a molecule, as well as the number and the location of chemical bonds. Figure 2.6 illustrates the chemical structure of the compound butane.

Electrons are shared or exchanged in the process of a chemical reaction. Once this exchange or sharing of electrons occurs, a chemical compound is formed. Opposite charges of metals and nonmetals make atoms want to come together to form salt compounds. Two basic groups of chemical compounds are formed from elements. The first group is the salts. Salts are made up of a metal and a nonmetal. For example, when combined with the nonmetal chlorine (Cl), the metal sodium (Na) forms the salt compound sodium chloride, with the molecular formula NaCl. Metals generally do not bond together. Metals that are combined are melted and mixed together to form an alloy, for example, copper and zinc are melted and mixed together to make brass. Brass is not an element. No chemical bond is involved; rather, it is a mixture of zinc and copper.

The second group of compounds is made up totally of nonmetal elements. For example, the nonmetal carbon combines with the nonmetal hydrogen to form a hydrocarbon compound. A typical hydrocarbon might be methane, with the molecular formula CH_4. Hydrocarbons will be discussed further in Chapters 4 and 5. When the outer-shell electrons of the metal lithium are given up to the nonmetal chlorine, a salt compound is formed through a chemical bond (Figures 2.7 and 2.8). The outer shell of the lithium atom is now empty, so the next shell becomes the outer shell. This shell will have two or eight electrons, which is a stable configuration; in the case of lithium, there are two electrons. Lithium is now stable and electrically satisfied. Chlorine receives the electron from the lithium and now has eight electrons in its outer shell. Chlorine is now stable and electrically satisfied. The result is that an ionically bonded compound is formed.

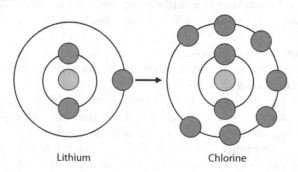

Lithium   Chlorine

Figure 2.7 Ionic bonding.

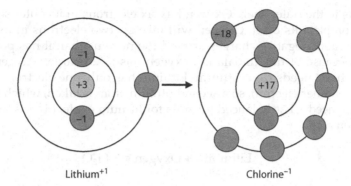

Lithium^{+1} Chlorine^{-1}

Figure 2.8 Exchange of electrons.

Ionic bonding

The process of gaining or losing electrons is called ionization or ionic bonding. When the electrons are transferred, ions are formed. Like atoms, compounds must be electrically neutral. There must be an equal number of positive protons in the nucleus and negative electrons outside in each of the atoms. When a metal gives up electrons, the metal has a positive charge because there are now more positive protons in the nucleus than negative electrons outside the nucleus. There is an electrical imbalance. An element cannot exist with an electrical imbalance.

The ion formed in the case of a metal when an atom of the metal is given up is referred to as a positive *cation*, represented by a plus (+) sign and a superscript number to the right of the element symbol. The number represents the number of electrons that were given up to the nonmetal element. When the nonmetal receives electrons from the metal, there are now more negative electrons outside the nucleus than there are positive protons in the nucleus. Again, there is an electrical imbalance. The ion of the nonmetal is referred to as a negative *anion*, represented by a minus (–) sign and a superscript number to the right of the symbol. The number represents the electrons received from the metal. Superscript numbers must balance and cancel each other out for the compound to be electrically neutral. If the superscript numbers are not equal, additional atoms of the elements involved are needed to balance the formula. If that happens, the formula will have subscript numbers indicating how many of the atoms of the specific element are present in the compound:

Calcium + chlorine = Calcium chloride

$$Ca^{+2} + Cl^{-1} = CaCl_2$$

If there are no subscripts in the formula, it is understood that there is just one atom of each of the elements shown. In the following example, each lithium atom has three electrons, two in the inner shell and one in the outer shell. Lithium will give up that one outer-shell electron to oxygen. Lithium now has one less electron than the total number of protons. There is one more positive charge in the nucleus than negative charges around the outside of the nucleus. Because of this imbalance of charges, lithium has a $^{+}1$ charge. Lithium has given up its one outer-shell electron, so the inner shell with two electrons now becomes the outer shell. Two electrons in the outer shell is a stable configuration, just like eight electrons in the outer shell. Oxygen has six electrons in its outer shell; it needs two electrons to complete its octet. Oxygen will take on two electrons in its outer shell and will have two more negative charges around the nucleus than it has positive charges in the nucleus. Because of this imbalance, oxygen has a $^-2$ charge. Oxygen can get one of the electrons that it needs from lithium. Lithium has only one electron to give; however, oxygen needs another electron, so a second lithium atom is taken, which gives oxygen the two electrons it needs. The balanced formula for lithium oxide, Li_2O, is two lithium atoms and one oxygen atom:

$$Lithium^{+1} + Oxygen^{-2} = Li_2O$$

In the next example, calcium has two electrons in its outer shell. Calcium will give up those two electrons to sulfur and will have a $^+2$ charge. Sulfur needs two electrons and will take the two electrons given up by calcium and have a $^-2$ charge. Since calcium has a $^+2$ charge and sulfur has a $^-2$ charge, the charges cancel each other out and the balanced formula is CaS, calcium sulfide. Rules for naming will be covered in "Salt" section of this chapter:

$$Calcium^{+2} + Sulfur^{-2} = CaS$$

It is important to understand how elements combine to form compounds. However, it is not the formula that will be seen most often in the real world of emergency response, but the name of the compound. When researching hazardous materials in reference books, often the formula will be listed. The formula will look more familiar if you have a basic understanding of chemistry. This will help responders to better understand the characteristics of the hazardous material.

Covalent bonding

The other family of compounds formed when elements combine is the nonmetal compound. Nonmetals comprise two or more nonmetal elements sharing electrons to form a compound. Nonmetals may be solids, liquids, or gases. The most frequently encountered groups of hazardous materials are made up of just a few nonmetal materials. They are carbon, hydrogen, oxygen, sulfur, nitrogen, phosphorus, fluorine, chlorine, bromine, and iodine. In elemental form and in compounds, these elements make up the bulk of hazardous materials found by emergency responders in most incidents. In the case of nonmetals, electrons are shared between the nonmetal elements.

Approximately 90% of covalently bonded hazardous materials are made up of carbon, hydrogen, and oxygen; the remaining 10% are composed of chlorine, nitrogen, fluorine, bromine, iodine, sulfur, and phosphorus. It is still necessary that each atom of each element has two or eight electrons in the outer shell. However, there is no exchange of those electrons. When the bonding takes place, each atom of each element brings along its electrons and shares them with the other elements. This process of sharing electrons is called covalent bonding. In the following example, carbon has four electrons in its outer shell; carbon needs to have four more to become stable. Hydrogen has one electron in its outer shell. Hydrogen can share its one electron with carbon and hydrogen will think it has a complete outer shell. Hydrogen gets the second electron it needs by sharing with carbon. This covalent bond fulfills the duet rule of bonding. Carbon still needs three more electrons. One way carbon can get three more electrons is by sharing with three more hydrogen atoms. When this happens, the carbon is complete. Carbon thinks it has eight electrons in the outer shell, and each hydrogen atom thinks it has two. The octet and duet rules of bonding are satisfied. The compound formed between carbon and hydrogen is complete, and the molecular formula is CH_4. In Figure 2.9, the shared pairs of electrons are represented by dots between the hydrogen and the carbon. This is known as the Lewis dot structure.

Electrons always share in pairs. Another way of representing pairs of shared electrons is by using a dash. The same one-carbon, four-hydrogen compound is shown with dashes representing the pairs of shared electrons (Figure 2.10).

A chemical compound becomes electrically stable through the process of sharing and exchanging electrons. The fact that compounds have become electrically stable does not mean they are no longer hazardous. Quite simply, elements combine and chemical reactions occur so that compounds can become electrically stable. These combinations of elements that form compounds create many new hazardous and nonhazardous chemicals. As elements bond together to form compounds, there is energy contained within the bonds. Breaking of bonds can release this energy, also referred to as a chemical reaction. This energy may be exothermic, endothermic, or explosive. When a chemical reaction releases heat, it is referred to as an exothermic reaction. A reaction that absorbs heat is an endothermic reaction.

Emergency responders may encounter elemental chemicals that are hazardous when released in an accident. However, most of the hazardous chemicals encountered by emergency responders will be in the form of compounds or mixtures. Compounds and mixtures present a broad range of hazards, from explosive to corrosive. It is important for emergency responders at all levels to recognize these dangers. This book will use the DOT/United Nations (UN) system of classifying hazardous materials. The remaining chapters of this book will examine the nine DOT/UN hazard classes and the types of hazardous materials they include. This hazard class system identifies only the most severe hazard presented by a material; almost every hazardous material has more than one hazard. This book will also focus on the hidden hazards of materials beyond the

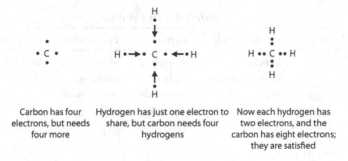

Figure 2.9 Lewis dot structure.

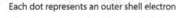

Figure 2.10 Each dash represents a pair of shared electrons.

hazard class and the corresponding placard. The physical and chemical characteristics of hazardous materials found within each of the hazard classes will be discussed in detail in the remaining chapters.

Salts

Just as the periodic table of the elements has families of elements, compounds can also be divided into families. A family of materials has particular hazards associated with it. If you can recognize to which family a material belongs from the name or the formula, you should be able to determine the hazard even if you do not know anything else about the specific chemical. If you know the hazard, you know how to handle the material properly. When dealing with salts, there are two types of naming systems used: IUPAC and common or trivial. IUPAC is a system developed by the International Union of Pure and Applied Chemistry. Their naming system was developed in order to provide common terminology in chemistry. It is primarily used in covalent bonding but will also be used to some extent in ionic bonding. In the case of salts, the IUPAC system is used when a transitional metal is in the salt compound. Transitional metals can have more than one charge when giving up electrons when forming a salt compound. Since this information is not found on the periodic table, you must be given the varying charges possible. You must also be given the varying charge type in the name of the salt or be given a correct formula. For example, copper can be found in two charges, Copper I or Copper II. Copper I indicates one electron in the outer shell and Copper II indicates two. Discussed first will be the IUPAC naming system for salts. The first family of materials we will look at is the salts (Table 2.3). If a metal reacts with a nonmetal, an ionic bond is formed, and the resulting compound is called a salt. In Figure 2.11, potassium, a metal, reacts with chlorine, a nonmetal. Potassium is located in the alkali metal family in Group I of the periodic table. This means that potassium has one electron in the outer shell; therefore, potassium will give up that one electron to the nonmetal, and there will be eight electrons in the next shell. That shell will then become the outer shell and will be stable with eight electrons.

Chlorine has seven electrons in its outer shell. Chlorine will take the electron given by potassium to give it eight electrons in its outer shell. Chlorine will become stable. The name

Table 2.3 Salt Families

Binary	Binary oxide	Peroxide	Hydroxide	Oxysalts
M+NM	M+Oxygen	M+(02)–2	M+(OH)–1	M+Oxy Rad
Not	Ends in oxide	Ends in	Ends in	
Oxygen	WR=CL, RH	peroxide	hydroxide	
Hazard		WR=CL,	WR=CL	
varies		RH	RH	
except				
NCHP				

Cyanide	Ammonium
M+CN	NH^{+1}+Nm
Ends in	Ammonium +
cyanide	Nm
Poison,	Hazard
CL	varies

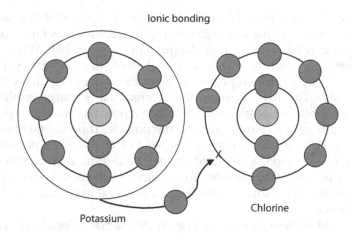

Figure 2.11 Forming a salt compound.

of the resulting compound is potassium chloride. Potassium chloride, much like sodium chloride (table salt), does not pose a serious risk to emergency responders:

$$K^{+1} + Cl^{-1} = KCl, \text{ potassium chloride, a binary salt}$$

Transition metals can also be combined with nonmetals to produce salts. Transition metals may have varying numbers of electrons in their outer shells (Table 2.4). The periodic table does not list the number of electrons in the outer shells of the transitional elements as it does for the primary elements.

Table 2.4 Transitional Element Charges

		Transition metal charges for hazmat chemistry		
Elements	Possible charges	Example compound	Naming of metal ion+ old	Naming of metal ion+ new
Copper	+1	CuCl	Copper I	Cuprous
	+2	CuCl₂	Copper II	Cupric
Iron	+2	FeCl₂	Iron II	Ferrous
	+3	FeCl₃	Iron III	Ferric
Mercury	+1	HgCl	Mercury I	Mercurous
	+2	HgCl₂	Mercury II	Mercuric
Tin	+2	SnCl₂	Tin II	Stannous
	+4	SnCl₄	Tin IV	Stannic
Cobalt	+2	CoCl₂	Cobalt II	Cobaltous
	+3	CoCl₃	Cobalt III	Cobaltic
Chromium	+2	CrCl₂	Chromium II	Chromous
	+3	CrCl₃	Chromium III	Chromic
Manganese	+2	MnCl₂	Manganese II	Manganous
	+3	MnCl₃	Manganese III	Manganic
Lead	+2	PbCl₂	Lead II	Plumbous
	+4	PbCl₄	Lead IV	Plumbic
Zinc	+2	ZnCl₂	Zine II	Zinc

The number of outer-shell electrons of the transition element in a salt compound can be determined from the name of the compound or from a correct molecular formula. For example, copper may have one or two electrons in its outer shell. When copper loses one electron during ionic bonding with a nonmetal element, it becomes Cu^{+1} because it now has one more positively charged proton in the nucleus than negatively charged electrons around the outside. This extra positive charge creates a $^{+}1$ charge on the element. If copper loses two electrons, it becomes Cu^{+2}, because it now has two more positive charges in the nucleus than negative charges around the outside. If nonmetal chlorine atoms were to pick up these electrons from copper, copper I or II chloride salts are formed. If copper loses only one electron, it is called copper I chloride, with a molecular formula of $CuCl$. When copper loses two electrons to chlorine, it becomes copper II chloride, with the molecular formula $CuCl_2$. Note that the Roman numeral in the name indicates the charge of the copper metal.

There is an older alternate naming system for the transitional metals when they ionically bond to form salt compounds. This is referred to as the trivial or common naming system. There may be chemicals encountered still using this older naming system; therefore, responders should be familiar with it. In this system, the suffixes "ic" and "ous" are used to indicate the higher and lower valence numbers (outer-shell electrons) of a transitional metal. For example, if copper I combines with chlorine, the name would be cuprous chloride. If copper I combines with oxygen, the name would be cuprous oxide. The lowest number of electrons in the outer shell of copper is one. When the metal with the lowest number of electrons is used, the suffix in the alternate naming system is "ous." When the metal with the highest number of electrons is used, the suffix is "ic." If copper II combined with phosphorus, it would create cupric phosphide. For example, copper II combined with chlorine would create cupric chloride.

The number of outer-shell electrons can also be determined from the molecular formula. First, look at the nonmetal element in a compound and determine what the charge is on that element.

The charge can be found as the group number on the periodic table at the top of the appropriate column. This group number is also the number of outer-shell electrons. The transition metal charge can then be determined (Figure 2.12) by reversing the subscript numbers in the formula to the top of the elements. For example,

The number that ends up at the top of the metal represents the number of electrons in the outer shell of that metal. Another way of viewing the same formula, Fe_1Cl_3, would be to look at the charge of the chlorine from family 7, which is −1. This −1 is placed above the Cl^{-1} in the formula. There are three atoms of chlorine in the compound. Three times one equals three. There is one atom of iron in the compound. What number times one equals three? The answer is simple, three. So the charge of the iron in the compound is three. This is represented by a +3 above the Fe^{+3} in the formula. The number of electrons in the outer shell of the iron in this compound is three (see Table 2.5).

There are three positive charges and three negative charges. The compound is in balance by using one atom of iron and three atoms of chlorine. The number above iron

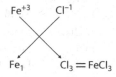

Figure 2.12 Determining transitional element charge from correct formula.

Table 2.5 Checking a
Formula for Correctness

Fe^{+3}	Cl^{-1}
X	X
Fe_1	Cl_3
+3	−3

indicates the number of electrons in the outer shell of that atom of iron. If a compound containing a transitional metal does not have subscript numbers, this indicates that the charges at the top are balanced. The compound does not require any additional atoms of those elements. In the following example, copper is combined with chlorine. From the periodic table, the charge on chlorine is ⁻1. Therefore, the charge on the copper must be ⁺1 in order to be electrically balanced, which means that the copper in this compound is copper I.

As mentioned in the older naming system, the metal that contains the lower number of outer-shell electrons ends with the suffix "ous." The metal that contains the higher number of outer-shell electrons ends with the suffix "ic." For example, iron has the possibility of two or three electrons in the outer shell. When iron II is used in a compound, the name for the iron is *ferrous*; therefore, a salt compound containing iron II and chlorine would be called *ferrous* chloride. In this case, the "ous" indicates two electrons in the outer shell of the iron. In order to use this system, you must first know the possible numbers of outer-shell electrons to determine which is the higher and which is the lower number of electrons in the outer shell. For example, mercury can have one or two, copper one or two, iron two or three, etc. If iron III were used in a compound, the name would end in "ic"; therefore, the iron III name would be *ferric*. If iron III were combined with chlorine, the name of the salt compound would be *ferric* chloride.

Salts have particular hazards, depending on which salt family they belong to. Salts generally do not burn, but can be oxidizers and support combustion. Some salts are toxic and some may be water reactive. Salt compounds can also be divided into families. There are six salt families that will be presented in this book: binary, binary oxide, peroxide, hydroxide, oxysalts, and cyanide salts.

Binary salts

Binary (meaning two) salts are made up of two elements: a metal and a nonmetal, except oxygen. They end in "ide," such as potassium *chloride*. Binary salts, as a family, have varying hazards. They may be water reactive, toxic, and, in contact with water, may form a corrosive liquid and release heat. Chemical reactions often release heat, which is referred to as an exothermic reaction. The hazard of an individual binary salt cannot be determined by the family. To determine the hazards of the binary salts, they have to be researched in reference materials. This varying hazard applies to all binary salts, except for nitrides, carbides, hydrides, and phosphides.

One helpful way to remember these four binary salts is by using the first letters of the element to form the acronym NCHP, which can be represented by "North Carolina Highway Patrol." These are compounds in which the metal has bonded with one of the nonmetals (nitrogen, carbon, phosphorus, or hydrogen). These compounds have particular hazards associated with them when they are in contact with water: nitrides give off ammonia, carbides produce acetylene, phosphides give off phosphine gas, and hydrides form hydrogen gas. In addition, a corrosive base is formed as a by-product from contact with water. The corrosive base will be the hydroxide of the metal that is attached to the nonmetal. For example, calcium carbide in contact with water will produce acetylene gas and the corrosive liquid calcium hydroxide. You have to look up the remaining binary salts to determine the hazard. In the first example, lithium metal is combined with chlorine. The resulting compound has a metal and a nonmetal other than oxygen, and the name ends in "ide"; therefore, it fits the definition of a binary salt. If lithium chloride is researched in reference sources, it is found to be soluble in water; it is not water reactive. In fact, lithium chloride does not present any significant hazard in a spill. The DOT does not list lithium chloride on its hazardous materials tables:

$$Li^{+1} + Cl^{-1} = LiCl, \text{ lithium chloride, binary salt, varying hazard}$$

The second example combines calcium metal and the nonmetal phosphorus. The compound name ends in "ide"; therefore, this is also a binary salt, which has varying hazards. Calcium phosphide, however, is one of the exceptions. Phosphides are one of the salts that have a known hazard: they give off phosphine gas and form calcium hydroxide liquid, which is a corrosive base. Even so, this is just a preliminary estimate. All materials should still be researched for additional hazards. Phosphine is a dangerous fire risk and highly toxic by inhalation:

$$Ca^{+2} + P^{-3} = Ca_3P_2, \text{ calcium phosphide, binary salt}$$

Notice the varying hazards of the following examples: lithium fluoride is a strong irritant to the eyes and skin; potassium bromide is toxic by ingestion and inhalation; sodium chloride is table salt, a medical concern when ingested in excess, but certainly of no significant hazard to emergency responders. However, if sodium chloride is washed into a farmer's field as a result of an incident, the farmer may not be able to grow crops in that field for many years.

Binary oxides (metal oxides)

The next group of salts is known as the binary, or metal, oxides. They are also made up of two elements: a metal and a nonmetal, but in this case, the nonmetal can only be oxygen. They end in "oxide," such as aluminum oxide. As a group, they are water reactive and, when in contact with water, almost always produce heat and form a corrosive liquid. However, they do not give off oxygen, because there is not an excess of oxygen. There is only one oxygen atom, and it is held tightly by the oxide salt and is not released as free-oxygen gas. In the following example, potassium metal is combined with one oxygen atom. The name of the salt compound is potassium oxide. This compound meets the definition of a binary oxide salt.

$K^{+1} + O^{-2} = K_2O$, potassium oxide, binary oxide salt, when in contact with water, releases heat and forms potassium hydroxide, a corrosive liquid.

Peroxide salts

Peroxides are composed of a metal and a nonmetal peroxide radical, O_2^{-2}. The prefix "per" in front of a compound or element name means that the material is "loaded" with atoms of a particular element. In the case of the peroxides, they are loaded with oxygen. A radical in the salt families is two or more nonmetals covalently bonded together, acting as a single unit with a particular negative (–) charge on the radical. The charge will not be found on the periodic table. The charges must be found on a listing of radical charges, as shown in Table 2.4. In the case of the peroxide salt, two oxygen atoms have bonded together and are acting as one unit with a -2 charge. When a peroxide comes in contact with water, heat is produced, a corrosive liquid is formed, and oxygen is released. This makes peroxides particularly dangerous in the presence of fire. Peroxides release oxygen because, unlike the oxide salts, there is an excess of oxygen present. In the following example, sodium metal combines with the peroxide radical to form the compound sodium peroxide. The name ends in "peroxide," so this is a peroxide salt. When in contact with water, sodium peroxide is a dangerous fire and explosion risk, and a strong oxidizing agent.

$Na^{+1} + O_2^{-2} = Na_2O_2$, sodium peroxide, a peroxide salt gives off heat and a corrosive liquid, sodium hydroxide, and releases oxygen when in contact with water.

Hydroxide salts

Hydroxide salts are made up of a metal and the nonmetal hydroxide radical $-OH^{-1}$. The name always ends with the word "hydroxide." They are water reactive and, when in contact with water, release heat and form a corrosive liquid. In the following example, calcium metal is combined with the hydroxide radical; the resulting compound is calcium hydroxide, a hydroxide salt. $Ca^{+2} + OH^{-1} = Ca(OH)^2$, calcium hydroxide, a hydroxide salt, releases heat and forms a corrosive liquid, calcium hydroxide, in contact with water.

Complex ions (polyatomic)

With the salts discussed so far, we have bonded a metal with a nonmetal to form an ionic bond and a salt. Nonmetals can be covalently bonded together to form a radical that represents a salt family and can also be bonded to the single metal. This radial replaces the normal single nonmetal we have used so far. It acts like a singular atom and has its own charge as a radical that is used when bonded to the metal. The complex ion takes on a metal ion(s) to form a balanced compound. Examples of complex ions that will be used in the upcoming salt compounds include oxy-radicals like ClO_3, NO_3, and MnO_3. Cyanide salts sometimes called metal cyanides use the complex ion CN. Hydroxide salts use the complex ion OH. Peroxide salts use two oxygen atoms bonded together and represented by O_2, which is also a complex ion. Finally, ammonium salts use the complex ion NH_4.

Oxysalts

Oxysalts are made up of a metal and an oxy-radical. The names end in "ate" or "ite" and may have the prefixes "per" or "hypo." Generally, as a group, they do not react with water; they dissolve in water. Some of the "hypo-ites" technically do react with water to release chlorine, but the reaction is mild. Oxysalts are oxidizers as a family; they will release oxygen, which accelerates combustion if fire is present. Another hazard occurs when oxysalts dissolve in water and the water is soaked into another material, such as packaging or firefighter turnouts. Water will evaporate, and the oxysalt will be left in the material. If the material is then exposed to heat or fire, the material will burn rapidly because the oxysalt in the material accelerates the combustion. Nine oxysalt radicals will be presented with this group. There are other oxy-radicals, but the ones chosen are considered most important to emergency response personnel.

Table 2.6 Copper I Chloride

+1	−1
Cu	Cl

The first six oxy-radicals all have ⁻1 charges: FO_3 (fluorate), ClO_3 (chlorate), BrO_3 (bromate), IO_3 (iodate), NO_3 (nitrate), and MnO_3 (manganate). The next three have ⁻2 charges: CO_3 (carbonate), CrO_4 (chromate), and SO_4 (sulfate). The last three oxy-radicals are PO_4 (phosphate), BO_3 (borate), and AsO_4 (arsenate) all of which have a ⁻3 charge (Table 2.6). All of the radicals listed are considered to be in their base state.

The base state is the normal number of oxygen atoms present in that oxy-radical. When a metal is added to any oxy-radical in the base state, the compound suffix is "ate," such as sodium *phosphate:* $Na^{+1} + PO_4{}^{-3} = Na_3PO_4$.

In the following example, the metal potassium is combined with the oxy-radical carbonate; the resulting compound is potassium carbonate. The compound does not have a prefix on the oxy-radical and the suffix is "ate"; therefore, it is the base state of the compound:

$$K^{+1} + CO_3{}^{-2} = K_2CO_3, \text{ potassium carbonate}$$

Potassium carbonate is an oxysalt in the base state, an oxidizer. Oxy-radicals may be found with varying numbers of oxygen atoms than the base state. Regardless of the number of oxygen atoms in the oxy-radical, the charge of the radical does not change (Table 2.7).

When naming compounds with an additional oxygen atom, the prefix *per* is used to indicate excess oxygen over the base state (Table 2.8); the suffix is still "ate." An example is sodium *persulfate:* $Na^{-1} + SO_5{}^{-3} = Na_3SO_5$. In the following example, the metal potassium is combined with the oxy-radical perchlorate: the resulting compound is potassium perchlorate. The level of oxygen is one above the base state. Notice that the charge on the oxy-radical is still −1, even though the number of oxygen atoms has changed. Potassium perchlorate is a fire risk in contact with organic materials, a strong oxidizer, and a strong irritant.

Table 2.7 Oxysalt Radicals

Oxy radicals		
−1	−2	−3
FO_3	CO_3	PO_4
ClO_3	CrO_4	BO_3
BrO_3	SO_4	AsO_4
IO_3		
NO_3		
MnO_3		

Table 2.8 Naming Oxysalts

+1 Oxygen prefix	Per-_____ate
Base State ending	_____ate
−1 Oxygen ending	_____ite
−2 Oxygen prefix	Hypo-_____ite

- $K^{+1} + ClO_4^{-1} = KClO_4$, potassium perchlorate.
- When the number of oxygen atoms is one less than the base state of an oxy-radical, the suffix of the oxy-radical name will be "ite"; an example is magnesium *sulfite*: $Mg^{+2} + SO_2^{-2} = MgSO_2$. In the following example, the metal sodium is combined with the oxy-radical phosphite. There is now one less oxygen than the base state. The charge on the oxy-radical has not changed. In addition to being an oxidizer, sodium phosphite is also used as an antidote in mercuric chloride poisoning.
- $Na^{+1} + PO_3^{-3} = Na_3PO_3$, sodium phosphite.

Finally, an oxy-radical can have two less oxygen atoms than the base state. The oxy-radical name will now have a *hypo* prefix and the suffix will be "ite." An example would be aluminum *hypo*phosph*ite*. $Al^{+3} + PO_2^{-3} = AlPO_2$. In the following example, calcium is combined with the oxy-radical hypochlorite; the resulting compound is calcium hypochlorite, a common swimming pool chlorinator. Calcium hypochlorite is an oxidizer and a fire risk when in contact with organic materials:

$$Ca^{+2} + ClO^{-1} = Ca(ClO)_2, \text{ calcium hypochlorite}$$

Cyanide salts

The next salt family to be discussed is the cyanide salts. They are made up of a metal and the cyanide radical, CN, which has a −1 charge. The name of the resulting compound ends in the word *cyanide*; an example is potassium cyanide: $K^{+1} + CN^{-1} = KCN$. Cyanide salts are toxic materials that dissolve in water to form a hydrogen cyanide solution. Hydroxide ions are produced that will make the solution basic. Cyanide salts react with acids to produce hydrogen cyanide gas, which has an almond-like odor. Hydrogen cyanide gas is used in gas chambers. Cyanides are deadly poisons and can be found as salts or in solution and may produce a toxic gas when heated. When the cyanide ion enters the body, it forms a complex ion with the copper ion located in the cells. These copper ions are essential to the enzyme that allows the cell to use oxygen from the blood. This enzyme is deactivated by the binding of the cyanide and copper ions. In the following example, sodium metal is combined with the cyanide radical. The resulting compound is sodium cyanide. Sodium cyanide is toxic by ingestion and inhalation:

$$Na^{+1} + CN^{-1} = NaCN, \text{ sodium cyanide}$$

Ammonium salts

- The last salt family to be discussed will be the ammonium salts. While there are no metals in the ammonium salts, the ammonium ion has similar properties to the heavier alkali metals and is often considered a very close relative. They are made up of the ammonium radical NH_4^{-1}, which has a −1 electrical charge and a nonmetal. Ammonium compounds are named by using the word *ammonium* followed by the name of the attached anion. Attached anions are very similar to oxysalts in appearance and generally follow the naming process of the oxysalts. Hazards of ammonium salts vary depending on the chemical attachment. Examples of ammonium salts are given later. The first is ammonium chloride. An ammonium ion is attached to a chlorine, which is a nonmetal.

- $NH_4 + Cl = NH_4Cl$, ammonium chloride.
- Ammonium chloride fumes are toxic with a TLV of 10 mg/m^3 of air. It is soluble in water and strongly endothermic. Ammonium bromide is a combination of the ammonium ion and bromine.
- $NH_4 + Br = NH_4Br$, ammonium bromide.

Ammonium bromide is soluble in water and alcohol and is not particularly hazardous under normal conditions. Ammonium sulfite is a compound of the ammonium radical in combination with the radical SO_3:

$$NH_4 + SO_3 = (NH_4)2SO_3, \text{ ammonium sulfite}$$

Ammonium sulfite is water soluble and not particularly hazardous. The last example of an ammonium salt presented here is ammonium nitrate, also known as Norway saltpeter. This compound is made up of the ammonium ion and the radical NO_3:

$$NH_3 + NO_3 = NH_3NO_3, \text{ ammonium nitrate}$$

Ammonium nitrate is an oxidizer, which may explode under confinement and high temperatures. When mixed with fuel oil, a deflagrating explosive material is created. Ammonium nitrate and fuel oil were used as the explosive in the Alfred P. Murrah Federal Building terrorist attack in Oklahoma City and the first terror attack on the World Trade Center in New York City in the mid-1990s.

Inorganic nonsalts

Inorganic nonsalt compounds bond together covalently, but are considered inorganic because of a general lack of carbon. They are not salts because no metals are present and the bonding is not ionic. Inorganic nonsalts may contain carbon, but they are not hydrocarbon compounds or derivatives of hydrocarbon compounds. Inorganic nonsalts are organized into families much like the salt families. These include binary nonsalts, nonmetal oxides, binary acids (inorganic acids), oxyacids (inorganic acids), and inorganic cyanides.

Binary nonsalts

Binary nonsalts are composed of a nonmetal + a nonmetal, NM + NM, except oxygen, much like the binary salts that are made up of a metal and a nonmetal, except oxygen. When naming the compounds, name the first nonmetal in its elemental form. Name the second nonmetal starting with the Greek number that identifies the number of atoms of that element in the compound and end in "ide." For example, combine bromine and fluorine using one atom of bromine and five atoms of fluorine. The first nonsalt is bromine, and the second nonsalt is fluorine. The ending for fluorine is dropped and an "ide" is added for the ending. There are five atoms of fluorine. The Greek number for five is penta (Table 2.9), resulting in bromine pentafluoride:

$$Br + F_5 = BrF_5, \text{ bromine pentafluoride}$$

Table 2.9 Greek Prefixes

One	Mono
Two	Di (bis)
Three	Tri (tris)
Four	Tetra
Five	Penta
Six	Hexa
Seven	Hepta
Eight	Octa
Nine	Nona
Ten	Deca

Bromine pentafluoride is explosive, water reactive, and has corrosive vapors with a TLV of 0.1 ppm.

Additional examples:

Phosphorus trichloride, PCl_3, reacts with water to create hydrogen chloride and phosphoric acid.
Iodine pentafluoride, IF_5, reacts with water, is a dangerous fire risk, and is corrosive.
Boron trifluoride, BF_5, is toxic by inhalation and corrosive to tissues.

Binary nonsalts have a variety of hazards, so you will have to look them up to determine hazards of specific compounds.

Nonmetal oxides

Nonmetal oxides are composed of a nonmetal plus oxygen, NM + Ox, x represents the number of oxygen atoms in the compound. Naming starts with the nonmetal followed by the number of oxygen atoms, represented by the Greek number and ending in "oxide." For example, combine sulfur and oxygen. Sulfur is the nonmetal, and if combined with three atoms of oxygen would be named sulfur, with the Greek name for three, which is "tri" and ending in oxide. Thus sulfur trioxide. Sulfur trioxide is a strong oxidizer that reacts violently with water to produce sulfuric acid:

$$S + O_3 = SO_3, \text{ sulfur trioxide}$$

Additional examples

Sulfur dioxide, SO_2, is toxic by inhalation, irritant, TLV 2 ppm in air.
Carbon monoxide, CO, is highly flammable, flammable limits are 12%–75%, it is toxic by inhalation, TLV 50 ppm. Carbon monoxide has an affinity for blood hemoglobin over 200 times that of oxygen.
Carbon dioxide, CO_2, is simple asphyxiant.
Nitrogen dioxide, NO_2, is toxic by inhalation, strong irritant.

Most binary oxides are toxic gases.

Binary acids (inorganic acids)

Binary acids are composed of the nonmetal hydrogen and a halogen from Group VII on the periodic table. As a group, they are sometimes referred to as acid gases or hydrogen halides. They are named by using hydro for the hydrogen and adding the halogen plus the "ic" ending, followed by the word *acid*. For example, hydrogen, hydro, plus fluorine, drop the "ine" and add ic, fluoric becomes hydrofluoric and add the word acid, hydrofluoric acid:

$$H + F = HF, \text{ hydrofluoric acid}$$

Additional examples

Hydrochloric acid, HCl, is toxic by inhalation and ingestion, strong irritant.
Hydrobromic acid, HBr, is a strong irritant to eyes and skin.
Hydroiodic acid, HI, is a strong irritant to eyes and skin.

These acids are made by dissolving a gas in water. As a result, they will off-gas when containers are open. This is not the same as a fuming acid, which is overloaded with the gas to the point that when opened, large amount of gas is released. Binary acids are corrosive, do not burn, and off-gas to some extent.

Oxyacids (inorganic acids)

Oxyacids are ionically bonded to a complex anion (polyatomic nonmetal). They are named much like the oxysalts based upon the number of oxygens in the polyatomic nonmetal and a valance charge based upon the oxy-radical attached (Table 2.10). Name ends in "ic" or "ous" depending on the state. The normal base state suffix is "ic." One oxygen above normal, add the prefix "per" and "ic" as the suffix. One oxygen below normal has the suffix "ous." Two oxygens below have the prefix "hypo" and the suffix "ous." The entire name ends in acid. For example, hydrogen and the oxy-radical ClO_3 combined becomes chloric acid (Table 2.11):

$$H + ClO_3 + acid = \text{chloric acid}$$

Table 2.10 Oxyacids

−1	−2	−3
ClO_3	CO_3	PO_4
BrO_3	CrO_4	
IO_3	SO_4	
NO_3		
MnO_3		

Table 2.11 Naming Oxyacids

+1 Oxygen prefix	Per-_____-ic
Normal state	_____-ic
−1 Oxygen ending	_____-ous
−2 Oxygen prefix	Hypo_____-ous

Additional examples

> *Perchloric acid*, $HClO_4$, strong oxidizing agent, will ignite vigorously in contact with organic materials, or detonate by shock or heat. It is toxic by ingestion and inhalation. It is a strong irritant.
>
> *Chlorous acid*, $HClO_2$, is a powerful oxidizing agent.
>
> *Hypochlorous acid*, $HClO$, is an irritant to skin and eyes.
>
> *Phosphoric acid*, H_3PO_4, is toxic by ingestion and inhalation, irritant to skin and eyes, TLV: 1 mg/m^3 of air.
>
> *Carbonic acid*, H_2CO_3, is a weak acid.
>
> *Hypochlorous acid*, $HClO$, is an irritant to skin and eyes.
>
> *Nitric acid*, HNO_3, is dangerous fire risk in contact with organic materials. It is highly toxic by inhalation, corrosive to skin and mucous membranes, and a strong oxidizing agent. TLV: 2 ppm in air.
>
> *Nitrous acid*, HNO_2, is a weak acid.
>
> *Persulfuric acid (Caro's acid)*, H_2SO_5, is a strong irritant to eyes, skin, and mucous membranes. It is a strong oxidizer and may explode in contact with organic materials.
>
> *Sulfuric acid*, H_2SO_4, is toxic irritant to tissue. TLV: 1 mg/cm^3 in air.
>
> *Sulfurous acid*, H_2SO_3, is toxic by ingestion and inhalation, strong irritant to tissue.
>
> *Hyposulfurous acid*, H_2SO_2 (does not exist)

Inorganic cyanides

It is bonded covalently with a nonmetal (except carbon) plus the cyanide anion. It is named with the nonmetal first with the cyanide ending or cyanogen plus the nonmetal. Cyanide compounds are extremely toxic by interrupting aerobic metabolism:

$$H + CN = HCN, \text{ hydrogen cyanide}$$

Also known as hydrocyanic acid or prussic acid, hydrogen cyanide is used as a military warfare agent with the designation of CN.

Examples

> *Cyanogen chloride*, $ClCN$, which has a military designation of CK.
>
> *Cyanogen bromide*, $BrCN$, is also called as bromine cyanide.

Most cyanide compounds are gases ($BrCN$ is a crystalline) and are extremely toxic by inhalation. They act by interrupting the cellular ability to use oxygen (cellular asphyxiant), sometimes referred to as chemical asphyxiation. They are primarily used as chemical warfare agents by the military.

Nonmetal compounds

Nonmetals and their compounds may be solids, liquids, or gases. Some may burn; some are toxic; and they can also be reactive, corrosive, and oxidizers. The largest quantities of hazardous materials encountered are made up of nonmetal (nonsalt) materials. These materials can also be divided into families. Hydrocarbon fuel is the most commonly encountered hazardous material. Products such as gasoline, diesel fuel, and fuel oil are all used to power our vehicles or heat our homes and businesses. Propane, butane, and natural gas are also fuels, but are not mixtures; they are pure compounds. They are transported frequently and stored in large quantities and can present frequent problems to responders. They form a family called the hydrocarbons because they are made up primarily of carbon and hydrogen. Hydrocarbons are flammable and may be toxic or cause asphyxiation by displacing oxygen in the air.

Nonmetal compounds are combinations of nonmetallic elements that combine in a covalent bonding process. Since there are no electrons exchanged, there are no electrical charges or superscripts above the elements in a formula. Subscripts are an indication of the actual number of atoms of that element in the compound. When elements bond covalently, the bonding electrons are shared between the elements: $C + S_2 = CS_2$, carbon disulfide. Since the bond is covalent and electrons are shared, in addition to the molecular formula, a structural formula showing the pairs of shared electrons can be drawn. In the following example, carbon needs four electrons and each sulfur needs two in order to satisfy the octet rule of bonding. The carbon is sharing two electrons with each of the two sulfur atoms present. The result is that the carbon thinks it has eight electrons in its outer shell and each sulfur thinks it has eight electrons in its outer shell. The carbon and the sulfur atoms are satisfied. The compound formed is carbon disulfide, which is a poison by absorption; it is a highly flammable, dangerous fire and explosion risk, has a wide flammable range from 1% to 50%, and can be ignited by friction. Carbon disulfide also has a low ignition temperature and can be ignited by a steam pipe or a lightbulb. The structural formula for carbon disulfide is shown in the following:

$$S = C = S$$

Carbon may bond with itself to satisfy its need for electrons. This happens frequently in the hydrocarbon families (to be discussed in detail in Chapters 4 and 5). There are four subfamilies of hydrocarbons, known as alkanes, alkenes, alkynes, and aromatics. (These families will be discussed in detail in Chapters 4 and 5.) The alkane and aromatic families of hydrocarbons occur naturally; the alkenes and alkynes are manmade. The naming of the hydrocarbon and hydrocarbon-derivative families in this book will cover both the IUPAC and the trivial naming system, which uses prefixes indicating the number of carbons in the compound (Table 2.12).

The suffix of the hydrocarbon name reflects the family (Table 2.13) and type of bond between the carbons in the compound. Hydrocarbons are named with a prefix indicating the number of carbons in the compound. The suffix for the name indicates the hydrocarbon family. Alkane hydrocarbons have single bonds between the carbons and end in "ane." They are considered saturated. Saturated compounds have all single bonds and

Table 2.12 Hydrocarbon Prefixes

Meth/form—1 Carbon
Eth/acet—2 Carbons
Prop—3 Carbons
But—4 Carbons
Pent—5 Carbons
Hex—6 Carbons
Hept—7 Carbons
Oct—8 Carbons
Non—9 Carbons
Dec—10 Carbons

Table 2.13 Hydrocarbon Families

Alkanes	Alkenes	Alkynes	Aromatics	BTX
Single bond	Double bond	Triple bond	Resonant bond	Benzene
Saturated	Unsaturated	Unsaturated	Acts saturated	Toluene
Ends in -ane	Ends in -ene	Ends in yne		Xylene

cannot have anything else added without removing a hydrogen. For example, the single carbon alkane begins with "meth" and ends in ane:

$$\text{Meth} + \text{ane} = \text{Methane, } CH_4$$

Alkene hydrocarbons use the same prefixes as the alkane family and end in "ene." Alkene hydrocarbons have one or more double bonds. They are considered unsaturated because double bonds can be broken and other compounds created. For example, two carbons connected together with a double bond have the prefix "eth" and ends in "ene":

$$\text{Eth} + \text{ene} = \text{Ethene, } C_2H_6$$

Alkyne hydrocarbons also use the same prefixes and end in "yne." A two-carbon double-bonded compound begins with "eth" and ends in "yne." It is named ethyne but has a common trade name acetylene:

$$\text{Eth} + \text{yne} = \text{Ethyne (acetylene) } C_2H_2$$

Aromatic hydrocarbons have a unique structure and bonding system. The primary aromatic compounds are benzene, toluene, xylene, and styrene. They are neither single, double, nor triple bonded. Aromatics have a resonant bond, which is cyclic in structure. Electrons in the benzene ring are not attached to any one carbon. They are sometimes referred to as delocalized electrons. Electrons of this type are not associated with a single atom or one covalent bond. Delocalized electrons are contained within an orbital that extends over several adjacent atoms. Generally, the resonance of benzene leads to a fairly stable compound. The parent group is benzene, characterized by a six-carbon cyclic structure. It has a molecular formula C_6H_6. Toluene has the benzene ring with a methyl radical attached to a carbon where the hydrogen has been removed. The formula is $C_6H_5CH_3$. Xylene has two methyl radicals attached on the ring where two hydrogen atoms have been removed. It has a formula of $C_6H_4(CH_3)_2$. Styrene has the benzene ring with one hydrogen atom removed. It has a two-carbon radical attached with a double bond. The formula is $C_6H_5CHCH_2$.

The other naming system for the hydrocarbon families is called the IUPAC system and was developed by the IUPAC. (An outline of the IUPAC system for naming organic compounds is located in the Appendix of this book.)

Hydrocarbons are used to make other families of chemicals, known as hydrocarbon derivatives. Radicals of the hydrocarbon families are made by removing at least one hydrogen from the hydrocarbon and replacing it with a nonmetal other than carbon or hydrogen. Fourteen of these hydrocarbon derivatives will be discussed in detail in the appropriate chapters associated with their major hazards: alkyl halides, nitros, nitriles, isocyanates, sulfides, thiols, carbamates, organophosphates, amines, ethers, peroxides, alcohols, ketones, aldehydes, esters, and organic acids.

Table 2.14 Hydrocarbon Derivative Families

Family	General formula	Hazard
Alkyl halide	R–X X=F,Cl,Br,I	Toxic/flammable
Nitro	R–NO$_2$	Explosive
Amine	R–NH$_2$	Toxic/flammable
Nitrile	R–CN	Toxic/flammable
Isocyanate	R–NCO	Toxic/flammable
Sulfide	R–S–R	Toxic/flammable
Thiol	R–SH	Toxic/flammable
Ether	R–O–R	Flammable, WFR
Peroxide	R–O–O–R	Oxidizer, explosive
Alcohol	R–OH	Flammable/toxic, WFR
Ketone	R–CO–R	Flammable/toxic
Aldehyde	R–CHO	Flammable/toxic, WFR
Ester	R–COOH	Corrosive, flammable

Just a few elements in addition to carbon and hydrogen are combined to make the 14 hydrocarbon derivative families discussed in this book (Table 2.14). These materials can be toxic, flammable, corrosive, explosive, or reactive; some may be oxidizers and some may polymerize.

Physical and chemical terms

Physical and chemical terms, such as boiling point, flash point, ignition temperature, pH, sublimation, water reactivity, spontaneous combustion, toxicity, and others will be discussed in the remaining chapters of the book where appropriate. Many of the reference books used by emergency responders contain physical and chemical terms, and it is important that responders understand the significance of each of them. There are numerous hazards that chemicals can present to emergency responders; very few hazardous materials have only one hazard. In addition to the nine DOT/UN hazard classes, some materials have hidden hazards. Not all placards will tell that a material is water reactive, air reactive, may be explosive, or may polymerize. These hidden hazards will be discussed throughout the remaining chapters of this book. Many of the hazardous materials and their hazards that will be discussed in this book can be placed in families with similar characteristics. Potential hazardous materials can also be identified by endings and names in compounds. Table 2.15

Table 2.15 Hints to Hazardous Materials Compounds

Names ending in							
-al	-ate	-ane	-azo	-ene	-ine	-ite	-ol
-one	-oyl	-yde	-ane	-yl	-yne		
Names that include							
acet	acid	alkali	amyl	azide	bis	caustic	cis
hepyl	hydride	iso	mono	naptha	oxy	penta	
per	tetra	trans	tris	vinyl	nitrile	cyan	

identifies some of the hints to hazardous materials. These are just hints to the possible presence of hazardous materials; the chemicals still need to be researched before any tactical operations are undertaken.

Review questions

(Answers located in the Appendix)

2.1   Chemistry is the study of
   A. Isomers
   B. Matter
   C. Isotopes
   D. Reactions

2.2   Elements on the periodic table are represented by
   A. The atomic number
   B. The atomic weight
   C. Symbols
   D. Molecules

2.3   Match the following family names with the column in which they are located on the periodic table.
   A. Noble gases   Group I
   B. Bromine family   Group II
   C. Alkaline earth metals   Group VII
   D. Alkali metals   Group VIII
   E. Halogens   Not a Group

2.4   When metals and nonmetals combine, the bond that is formed is called
   A. Combination
   B. Ionic
   C. Covalent
   D. Atomic

2.5   When nonmetals combine, the bond that is formed is called
   A. Covalent
   B. Non-covalent
   C. Ionic
   D. Isomeric

2.6   Atoms are made up of three subatomic particles; which of the following is not one of the particles?
   A. Neutrons
   B. Protons
   C. Neurons
   D. Electrons

2.7   An atom of an element is said to be electrically stable when it has how many electrons in its outer shell?
   A. Six
   B. Eight
   C. Two
   D. Both B and C

2.8 There are two rules of bonding; which of the following is not one of the names for the
bonding rules?
A. Octet
B. Isomeric
C. Duet
D. None of the above

2.9 Balance the formulas, if needed; give the name, the salt family, and the hazards for
the following salts (salts with transitional metals are already balanced):

NaCl	$CaPO_4$	AlO_2	$CuBr_2$	KOH
LiO	MgClO	HgO_2	NaF	$FeCO_3$

2.10 Provide a balanced formula, family name, and hazard(s) for the following salts:

Calcium hypochlorite	Aluminum chloride	Lithium hydroxide
Copper II peroxide	Sodium oxide	Potassium iodide
Magnesium phosphide	Mercury I perchlorate	Iron III fluorate

chapter three

Explosives

Since the first modern armies met on the battlefield in the Byzantine Empire, leaders have attempted to gain an advantage using superior firepower over their enemies. Explosives could provide that advantage. Dynamite was one of the first high explosive materials developed, not for the battlefield, but for construction to clear rock and large parcels of land. Dynamite did, however, lead to more effective explosives such as Semtex and C-4, which are used for military and terrorism purposes. Explosives in the United States are regulated by the Bureau of Alcohol, Tobacco and Firearms (ATF) in fixed storage and the Department of Transportation when in transit. The first Department of Transportation/United Nations (DOT/UN) hazard class deals with explosives. The DOT defines an *explosive* in 49 CFR 173.50 as "any substance or article, including a device, which is designed to function by explosion or which, by chemical reaction within itself, is able to function in a similar manner even if not designed to function by explosion." This definition applies only to chemicals designed to create explosions. Another definition, taken from the *National Fire Academy Tactical Considerations Student Manual*, is "a substance or a mixture of substances, which, when subjected to heat, impact, friction, or other suitable initial impulse, undergoes a very rapid chemical transformation, forming other, more stable, products entirely or largely gaseous, whose combined volume is much greater than the original substance."

Other chemicals have explosive potential under certain conditions, but may not be placarded or recognized as explosives. These materials will be found in fixed facilities as well as in transportation, and may not have any markings or warnings that they have explosive properties. Some examples are ethers, potassium metal, formaldehyde, organic peroxides, and perchloric acid. Ethers are Class 3 Flammable Liquids that have a single oxygen atom in their composition; as ether ages, oxygen from the air can combine with this oxygen atom to form a peroxide molecule. The oxygen-to-oxygen single bond that is formed is a highly reactive and unstable bond: –O–O–.

This same type of bond is present in nitro compounds that are explosive and will be presented later in this chapter. Peroxides formed in compounds in this manner are shock- and heat-sensitive; they may explode simply by being moved or shaken. Formaldehyde solutions are Class 3 Flammable Liquids, and potassium metal is a Class 4.3 Flammable Solid, Dangerous When Wet. These materials form explosive peroxides just like ether and are equally dangerous as they age. Organic peroxides are Class 5.2 Oxidizers. Many organic peroxides are temperature-sensitive; they are usually stored under refrigeration. If the temperature is elevated, organic peroxides may decompose explosively. Perchloric acid is a corrosive by hazard class. However, perchloric acid is also a strong oxidizer and will explode when shocked or heated. Chemical oxidizers are one of the necessary components for a chemical explosive to function. These chemical oxidizers should be treated with the same respect as the explosives in Class 1.

Definition of explosion

An *explosion* is defined in the National Fire Protection Association (NFPA) *Fire Protection Handbook* as "a rapid release of high-pressure gas into the environment" (Photo 3.1). This release of high-pressure gas occurs regardless of the type of explosion that has produced it. The high-pressure energy is dissipated by a shock wave that radiates from the blast center. This shock wave creates an overpressure in the surrounding area that can affect personnel, equipment, and structures (Table 3.1). An overpressure of just 0.5–1 psi can break windows and knock down personnel. At 5 psi, eardrums can rupture, and wooden utility poles can be snapped in two. Ninety-nine percent of people exposed to overpressures of 65 psi or more would die.

Categories of explosions

According to the NFPA *Fire Protection Handbook*, there are two general categories of explosions: *physical* and *chemical*. In a physical explosion, the high-pressure gas is produced by mechanical means, i.e., even if chemicals are present in the container, they are not affected chemically by the explosion. In a chemical explosion, the high-pressure gas is generated by the chemical reaction that takes place.

Phases of explosions

There are two phases of an explosion, the positive and the negative. The positive phase occurs first as the blast wave travels outward, releasing its energy to objects it comes in

Photo 3.1 Explosions may be followed by fireballs, shock waves, and flying shrapnel.

Table 3.1 Overpressure Damage

Examples of overpressure damage to property	
0.5–1 psi	Window glass breakage
1–2 psi	Buckling of corrugated steel and aluminum wood siding and framing blown in
2–3 psi	Shattering of concrete or cinder block walls
3–4 psi	Steel panel building collapse
	Oil storage tank rupture
5 psi	Wooden utility poles snapping failure
7 psi	Overturning loaded rail cars
7–8 psi	Brick walls shearing and flexure failures
Examples of injuries and death to personnel	
1 psi	Knockdown of personnel
5 psi	Eardrum rupture
15 psi	Lung damage
35 psi	Threshold for fatalities
50 psi	50% fatalities
65 psi	99% fatalities

contact with. This is also known as the blast pressure, generally, the most destructive element of an explosion. If the explosion is a detonation, the waves travel equally in all directions away from the center of the explosion. The negative phase occurs right after the positive phase stops. A partial vacuum is created near the center of the explosion by all of the outward movement of air from the blast pressure. During the negative phase, the debris, smoke, and gases produced by the blast are drawn back toward the center of the explosion origin, then rise in a thermal column vertically into the air, and are eventually carried downwind by the air currents. The negative phase may last up to three times as long as the positive phase. If the explosion is a result of an exothermic (heat-producing) chemical reaction, the shock wave may be preceded by a high-temperature thermal wave that can ignite combustible materials. Some explosives are designed to disperse projectiles when the explosion occurs, e.g., antipersonnel munitions and hand grenades. Other explosives may be in metal or plastic containers to provide the necessary confinement for an explosion to occur. These containers may become projectiles when the explosion occurs. Projectiles from an explosion travel out equally in all directions from the blast center just as the blast pressure does. In order for the waves or projectiles coming off the explosion to travel equally in all directions, the velocity (speed) of the release of the high-pressure gases must be supersonic or faster than the speed of sound. This occurs only in a detonation, not in a deflagration.

Mechanical overpressure explosions

The two major explosion types can be divided further into four subgroups that result in the release of high-pressure gas. The first occurs as the result of the physical overpressurization of a container, causing the container to burst, as in the case of a child's balloon bursting when too much air is placed in it. The container fails because it can no longer hold the pressure that has built up inside. Overpressure can occur in containers that may not have pressure-relief valves or if the pressure-relief valve fails to operate. Overpressure does not have to occur as a result of filling a container. As heat is applied to a container from ambient temperature increases, or from radiant heat, the pressure increases inside the container. If this increase in pressure is not relieved, the container may fail. Failure will occur at the weakest point in the container. If a container has been damaged as a result of an accident, the damaged point of the container may become the weakest point; even if a relief is present on the container, the container can fail.

Mechanical/chemical explosions

The second subgroup of explosion occurs via physical or chemical means, as in the case of a hot water heater or boiler explosion. The water inside the heater turns to steam when overheated, which results in a pressure increase inside the container, and the container fails at its weakest point. Hot water heaters and boilers have pressure-relief valves, but they may become corroded and plugged by materials in the water. Relief valves should be tested periodically to make sure they are operational. Hot water heater/boiler explosions are by far the most common type of accidental explosion.

Container failure can also occur in containers that hold liquefied compressed gases. As the temperature of the liquid in the container increases, so does the vapor content in the container; as the vapor content increases, so does the vapor pressure. This increase in temperature may be the result of increases in ambient temperature, radiant heat from a fire, or other heat source, such as direct flame impingement on the pressure container. If the container has been damaged in an accident, it may fail at the point of damage, before the relief valve can function to relieve the pressure buildup. The tank wall can also be weakened by direct flame impingement on the vapor space of the container. Flame impingement weakens the metal; the tank can no longer hold the buildup of pressure, and tank failure occurs. When the heat from flame impingement is at the liquid level, the heat is absorbed by the liquid so the tank is not weakened. However, as the liquid absorbs the heat, more liquid is turned to vapor. This increase in vapor increases the pressure in the container. Pressure-relief valves are designed to relieve increased pressure caused by increases in ambient temperature or from radiant heat sources. The increase in pressure caused by direct flame impingement may overpower the relief valve, and again the container may fail.

Chemical explosions

The third subgroup of explosion involves a chemical reaction: the combustion of a gas mixture. Many of the explosive materials that the DOT regulates and allows for transport can cause a chemical explosion. Therefore, most explosions and explosive materials in this hazard class involve materials that explode by chemical means, with the high pressure created by a chemical reaction. Chemical explosions are really nothing more than a rapidly burning fire. The components that allow this fire to burn rapidly enough to produce an explosion are the presence of a chemical oxidizer and confinement of the material. There are other chemicals offered for transportation and found in fixed facilities that are not classified as explosives but can explode through chemical reaction. Responders must be aware of the fact that most chemicals have multiple hazards that may not be depicted by placards or information on shipping papers.

Dust explosions

This third subgroup of potentially explosive materials also involves a chemical-reaction type of explosion, sometimes referred to as combustible dusts Table 3.2. These are, in many cases, ordinary combustible materials or other chemicals that, because of their physical

Table 3.2 Explosive Grain Dusts

Coals	Crude rubber	Peanut hulls
Soy protein	Sugar	Aluminum
Cork	Cornstarch	Flour
Magnesium	Pea flour	Titanium
Zirconium	Walnut shells	Silicon

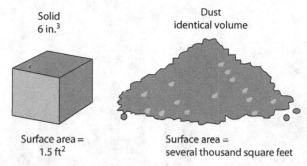

Figure 3.1 Surface area comparison.

size, have an increased surface area Figure 3.1. This increased surface area exposes more of the particles to oxygen when they are suspended in air. When these materials are suspended in air, they can become explosive if an ignition source is present.

One of the major facilities where dust explosions occur is grain elevators; explosions occur when grain dust is suspended in air in the presence of an ignition source (Photo 3.2). The primary danger area where the explosion is likely to occur within most elevators is the "leg," or the inclined conveyor, the mechanism within the elevator that moves the grain from the entry point to the storage point. For a dust explosion to occur, five factors must be present: an ignition source, a fuel (the dust), oxygen, a mixing of the dust and the oxygen, and confinement. The explosion will not occur unless the dust is suspended in air within an enclosure at a concentration that is above its lower explosive limit (Table 3.3).

There are three phases in a dust explosion (Figure 3.2): initiation, primary explosion, and secondary explosion. Initiation occurs when an ignition source contacts a combustible dust that has been suspended in air. This creates the primary explosion, which shakes more dust loose from the confined area and suspends it in air. The secondary explosion then occurs, which is usually the larger of the two explosions because there is more fuel present. Combustible dusts may be present in many different types of facilities. Common places for combustible dusts to be found are in grain elevators, flour mills, woodworking shops, and dry-bulk transport trucks. Dusts in facilities have caused many explosions over the years that have killed and injured employees. An explosion occurred in a facility on the East Coast that had many hazardous materials on site. At first, it was thought that one of the chemicals had exploded. The fire department and the hazmat team were called to the scene. Investigation revealed that the explosion occurred in a dust-collection system; it was a combustible-dust explosion. Dust explosions can be prevented by proper housekeeping and maintenance practices at facilities where these types of dusts are present.

Photo 3.2 Grain elevators are a primary source of dust explosions.

Table 3.3 Explosive Properties of Grain Dust

Type of dust	Ign temp (°F)	LEL
Wheat/corn/oats	806	55
Wheat flour	380	50
Cornstarch	734	40
Rice	824	50
Soy flour	1004	60

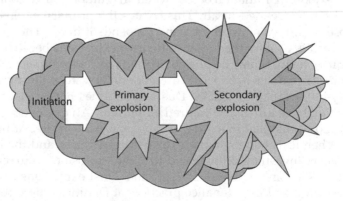

Figure 3.2 Three phases of grain dust explosion.

Nuclear explosions

The fourth subgroup of explosion is nuclear explosion, which is the result of a tactical decision, a weapon that malfunctions, or an act of terrorism (Photo 3.3). There are two types of nuclear explosions: air burst and ground burst. Both are engaged as a result of a tactical objective. An *air burst* is designed to knock out all electronic equipment, disrupting communications and computer usage. This type of explosion does not create fallout, because it does not reach the ground and does not suck up debris in the negative phase of the explosion. A *ground burst* is designed for mass destruction of everything it contacts. Initially, during the positive phase of the ground blast, there is a thermal wave that is released first, followed by a shock wave. During the negative phase of the explosion, the debris from the explosion is drawn into the cloud, travels downwind, and then falls back to the ground. The debris, while in the cloud, is contaminated with radioactive particles, and radioactive fallout is created. With the onset of terrorism that has occurred over the past several years, another type of nuclear bomb has become a concern for emergency responders, the dirty bomb. While the dirty bomb is not technically a nuclear explosion, it does create radioactive contamination similar to a ground burst. A dirty bomb uses conventional explosive materials to disseminate radioactive materials to cause contamination of areas where they are released. This topic will be discussed further in Chapter 9.

Photo 3.3 Nuclear explosions can be air bursts or ground bursts. Air bursts are designed to knock out electronics but do not contain fallout. Ground bursts produce nuclear fallout.

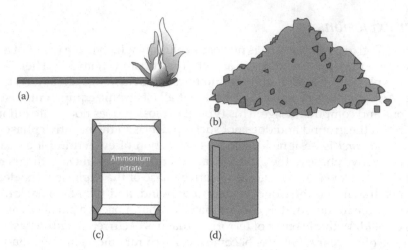

Figure 3.3 Four components of chemical explosion: (a) heat, (b) fuel, (c) chemical oxidizer, and (d) confinement.

Components of a chemical explosion

In simple terms, an explosive that functions via chemical reaction creates a rapidly burning fire that is made possible by the presence of a chemical oxidizer. Atmospheric oxygen does not provide enough oxygen for a chemical explosion to take place. Four components must be present for a chemical explosion to occur: fuel, heat (initiator, source of ignition), a chemical oxidizer, and confinement of the materials (Figure 3.3). The materials themselves can provide the confinement. Note the similarity between the requirements for an explosion and the fire triangle. There are other types of explosions that produce high-pressure gases, some of which will be discussed later.

Types of chemical explosives

Explosives can be divided into two primary groups, high explosives and low explosives, based upon the speed at which the chemical transformation takes place, usually expressed in feet per second. Low explosives change physical state from a solid to a gas rather slowly. The low explosive burns gradually over a somewhat sustained period of time. This action is typically used as a pushing and shoving action on the object against which it is placed. The primary uses of low explosives are as propelling charges and for powder trains, such as in time fuses. Examples of low explosives include black powder and smokeless powder.

High explosives change from a solid to a gas almost immediately, an action referred to as detonation. A high explosive is detonated by heat or shock, which sets up a detonating wave. This wave passes through the entire mass of explosive material instantly. This sudden creation of gases and the extremely rapid extension produce a shattering effect that can overcome great obstructions. Examples of high explosives are trinitrotoluene (TNT) and dynamite.

Explosives can be further subdivided into families: (Figure 3.4) inorganic, organic, aromatic, and aliphatic. Inorganic compounds include fulminates, ammonium nitrate, and azides. Organic explosives include aromatic, with a benzene ring, and aliphatic, with straight chains of hydrocarbons. Aromatics include trinitrobenzene (TNB), single substitute TNB, multiple substitute TNB, and multiple TNB rings. Aliphatic explosives include aliphatic nitrate esters, aliphatic nitramines, and nitro aliphatics. Fulminates and azides are usually too unstable to be transported, so will likely be encountered in fixed facilities. Examples of fulminates

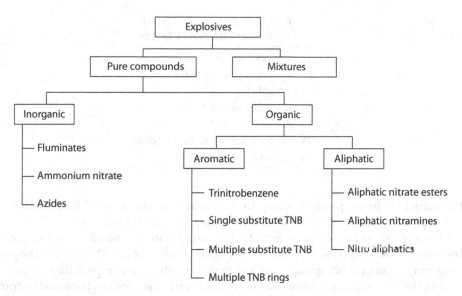

Figure 3.4 Families of explosives.

are mercury fulminate $Hg(ONC)_2$, silver fulminate AgONC, and silver acetylide (C_2Ag_2). Examples of azide explosives include lead azide $Pb(N_3)_2$ and silver azide AgN_3. Organic explosives are the most common type of explosives in use and more likely to be transported than inorganic explosives. Examples of organic aromatic explosives are TNB $C_6H_3(NO_2)_3$ and TNT $C_6H_2CH_3(NO_2)_3$. An example of aliphatic explosives is nitroglycerine $C_2H_5(ONO_2)_3$ (Figure 3.4).

The DOT explosives hazard Class 1 is divided into six subclasses: 1.1–1.6 (49 CFR 173.20) (Table 3.4). Because of their potential danger, subclasses 1.1–1.3 require placarding of the highway transportation vehicle regardless of the quantity of explosives carried (49 CFR 172.504) (Table 3.5). Railroad shipments must always be placarded in all hazard classes, regardless of the quantity shipped. Subclasses 1.4–1.6 (49 CFR 172.504) (Table 3.6) fall under the 1001-lb rule (49 CFR 172.504(1)), which requires 1001 lb or more of explosives

Table 3.4 DOT Explosives Subclasses

Explosive 1.1	Mass explosion hazard
Explosive 1.2	Projection hazard
Explosive 1.3	Fire, minor blast, minor projection hazards
Explosive 1.4	Device with minor explosion hazard
Explosive 1.5	Very insensitive explosives
Explosive 1.6	Extremely insensitive explosives

Table 3.5 Placard of Table I
Hazardous Materials Classes

Explosives 1.1–1.3
Poison gas 2.3
Poison 6.1
Dangerous when wet 4.3
Radioactive yellow III

Table 3.6 Placard of Table II
Hazardous Materials Classes

Explosives 1.4–1.6
Compressed gases 2.1–2.2
Flammable liquids 3
Flammable solids 4.1–4.2
Oxidizers 5.1–5.2
Poison 6.1 (Noninhalation hazard)
Corrosive 8

---

on the vehicle before a placard is required. This means there could be 1000 lb or less of a 1.4–1.6 explosive, and no placard would be required at all (Photo 3.4)!

In the explosives hazard class, next to the hazard subclass number on the placard, there will be a letter, known as the compatibility group letter. 49 CFR 173.52, identifies the procedures for assigning compatibility group numbers to shipments. The compatibility groups are used to prevent the increase in explosive hazard should certain explosives be stored or transported together. This information is designed for the shippers of the explosive materials and has little, if any, emergency response value. All explosives should be treated as if they were Class 1.1, because all explosives will explode under certain conditions. Responders are not likely to recognize when an explosive material is in a condition to cause an explosion. Therefore, responders should treat all explosives as if they were the worst type of explosive they might encounter until such point when they can positively identify the explosive material.

Photo 3.4 Dry-bulk tanker transporting explosives Class 1.5 blasting agent on the highway.

Forbidden explosives for transportation

In addition to the explosives that the DOT has approved for transportation, there are forbidden explosives that are too unstable or dangerous to be transported. Some may be transported only if wetted. While the forbidden explosives may not be transported, they can be encountered in fixed facilities. The following are some examples of explosives that are forbidden to be transported.

Nitrogen triiodide (black unstable crystals) explodes at the slightest touch when dry. When handled, it is kept wet with ether. It is too sensitive to be used as an explosive, because it cannot be stored, handled, or transported. Azides, such as lead azide and hydrazoic azide, are highly unstable. *Lead azide* is a severe explosion risk and should be handled under water; it is also a primary detonating compound. *Hydrazoic acid* or *hydrogen azide* is a dangerous explosion risk when shocked or heated. It is the gas-forming agent in many air bag systems in automobiles and escape chutes in airplanes. Sodium acetylide is an unstable powder, a salt of acetylene. It is a severe explosion risk when shocked or heated. It is used in the manufacture of detonators. Metal fulminates, such as *mercury fulminate*, explode readily when dry. They are used in the manufacture of caps and detonators for producing explosions.

Explosives that contain a chlorate along with an ammonium salt or an acidic substance, including a salt of a weak base and a strong acid, are forbidden in transportation; e.g., *ammonium chlorate*, which is shock-sensitive, can detonate when exposed to heat or vibration. It is used in the production of explosives. *Ammonium perchlorate* is also shock-sensitive and may explode when exposed to heat or by spontaneous chemical reaction. This is the material that was involved in the explosion at the Pepcon plant in Henderson, Nevada. It is also used in the production of explosives, pyrotechnics, etching and engraving, and jet and rocket propellants.

Packages of explosives that are leaking, damaged, unstable, condemned, or contain deteriorated propellants are forbidden in transportation. Nitroglycerine, diethylene glycol dinitrate, or other liquid explosives are not authorized. *Diethylene glycol dinitrate* is a severe explosion hazard when shocked or heated. It is used as a plasticizer in solid rocket propellants. Other forbidden explosives include fireworks that combine an explosive and a detonator or fireworks that contain yellow or white phosphorus. Toy torpedoes exceeding 0.906-in. outside dimension or containing a mixture of potassium chlorate, black antimony (antimony sulfide), and sulfur are prohibited if the weight of the explosive material in the device exceeds 0.01 oz. The Hazardous Materials Table in CFR 49 Part 172.101 lists all specific restricted explosives in various modes of transportation and those forbidden from shipment. Even though these materials are not transported, they may be found in fixed facilities of various types, such as research facilities, defense contractors, solid rocket fuel plants, explosives suppliers, and others.

Types of chemical explosions

Detonation is an instantaneous decomposition of the explosive material in which all of the solid material changes to a gas instantaneously with the release of high heat and pressure shock waves. Detonation is the only type of chemical explosion that will produce a true shock wave. A material that detonates is considered a high-yield explosive (Table 3.7). Blast pressures can be as much as 700 tons/in.^{2} Pressure and heat waves travel away from the center of the blast equally in all directions. The reaction occurs at supersonic speed or, in

Table 3.7 Characteristics of High and Low Explosives

	High	Low
Initiation method	Primary by ignition Secondary by detonation	By ignition
Conversion to gas	Microseconds	Milliseconds
Consumption velocity	1–6 mile/s	Few inches to feet per second
Velocity of flame front	1–6 mile/s	1/3–1 mile/s
Pressure of explosion	50,000–4,000,000 psi	Up to 50,000 psi

other words, faster than the speed of sound, which is 1250 ft/s. Many of the reactions occur between 3,300 and 29,900 ft/s or over 20,300 miles/h!

Often the terms "explosion" and "detonation" are used interchangeably, which is not accurate. An explosion may be a detonation; however, an explosion can occur that is not a detonation. In either case, each occurs rapidly, and the difference cannot be distinguished easily by the human senses. The only way you can distinguish a detonation from a deflagration is by hearing the sound of the explosion. In a detonation, the explosion will be visualized, and the shock wave sent off before the explosion is actually heard. In a deflagration, the explosion will be heard almost immediately. The two terms only apply to the speeds of the explosions; they do not infer that one is any less dangerous than the other.

Deflagration is a rapid autocombustion that occurs at a subsonic speed, less than 1250 ft/s. The solid material changes to a gas relatively slowly. A material that deflagrates is considered a low-yield explosive. A material that is designed to deflagrate may, however, under the right conditions produce a detonation. The explosion in Kansas City that killed six firefighters involved ammonium nitrate mixed with fuel oil. Ammonium nitrate is listed as a Class 1.5 Insensitive Explosive, designed to produce a deflagration. The ammonium nitrate and fuel oil (ANFO) mixture involved in Kansas City was on fire and, as a result, produced a detonation that may have been caused by the application of water by the firefighters.

Explosive effects

There are primary and secondary effects of explosions. The primary effects are blast pressure, thermal wave, and fragmentation. The blast pressure has two phases, the positive and negative. In the *positive phase*, the high-pressure gas, heat wave, and any projectiles travel outward. During the *negative phase*, a partial vacuum is produced, sucking materials back toward the area of origin.

The next effect is fragmentation. Fragmentation may come from the container that held the explosive material and from materials in close proximity. Objects in the path of the explosion are broken into small parts by the force of the blast pressure, creating fragments. These fragments may have jagged or sharp edges. The fragments will travel away from the blast center at high speeds, greater than 2700 ft/s, faster than a speeding bullet!

The final effect of an explosion is thermal, the generation of heat by the explosion. The amount of heat produced will depend on the type of material that is involved. There is a flash and a fireball associated with almost any chemical explosion. The more rapid the explosion, the greater the effects that will be produced by the heat. Although the total heat produced may be similar in each explosion, a detonation will produce the most heat over a larger area, because of the speed of the explosion.

Secondary effects of an explosion are shock-wave modification and fire and shock-wave transfer. There are three ways that a shock wave can be modified: it may be reflected, focused, or shielded. Reflection refers to the shock wave striking a solid surface and bouncing off. When a shock wave strikes a concave (curved) surface, the force of the shock wave is focused, or concentrated, on an object or small area once it bounces off the concave surface. This effect is similar to the principle behind satellite dishes. When a signal reaches a satellite dish from the satellite in space, the signal is focused on the electronic sensor protruding out of the front of the satellite dish. Shielding simply means that the shock wave encounters an object too substantial to be damaged by the wave, so the shock wave goes around the object or is absorbed by it. The area immediately behind the object provides a place of shelter from the shock wave. Fire and shock-wave transfer involve the transfer of the shock-wave energy and fire to other objects, causing fires and destruction.

Yield vs. order

The yield of an explosive is associated with the rate or speed at which the explosion occurs. This is an indication of whether an explosive will detonate or deflagrate. A high-yield explosive detonates, and the blast pressure shatters materials that it contacts. Examples of high-yield explosives are dynamite, TNT, nitroglycerine, detcord (Photo 3.5), C-3, C-4, and explosive bombs.

Photo 3.5 Blasting caps used to detonate explosive materials.

Table 3.8 Commercial Explosives

Primary	Secondary
Mercury fulminate	Nitroglycerine
Lead azide	Ammonium nitrate
Diazodinitrophenol	Trinitrotoluene
Lead styphnate	Dinitrotoluene
Nitromannite	Nitrostarch

A low-yield explosive deflagrates and is used to push and shove materials. Examples of low-yield explosives are black powder and commercial ammonium nitrate. Deflagrating materials are often used to move rocks in road construction, in quarries, and in mining (Table 3.8).

Order has to do with the extent and the rate of a detonation. A high-order detonation is one in which all of the explosive material is consumed in the explosion, and the explosion occurs at the proper rate. The proper rate in this case would be supersonic. Thus, a low-order explosion would occur as an incomplete detonation or at less than the desired rate. Yield involves the specific explosive material that is used, and order indicates the way in which the explosive detonated. The hazards to emergency responders are obvious. If an explosion is of low order, not all of the explosive material has been consumed, and therefore the remaining material presents a hazard. Whether high or low yield, high or low order, all explosives should be treated as high-yield, high-order Class 1.1 explosives.

Division 1.1–1.3 explosives

Division 1.1 explosives present a mass explosion hazard. They are sensitive to heat and shock, and they may either detonate or deflagrate when they explode. Division 1.2 explosives have a projection hazard, but not a mass explosion hazard. Division 1.3 explosives have a fire hazard and a minor blast hazard or a minor projection hazard or both, but not a mass explosion hazard.

Explosive families of compounds

Inorganic explosive compounds

Fulminates, ammonium nitrate, and azides are inorganic explosive compounds. Fulminate ions (CNO) are unstable; thus the salts of fulminates are friction-sensitive explosives. Metals are attached to the CNO ion forming an explosive metal fulminate. Fulminates are among the oldest explosive compounds and have been around since the 1800s. Mercury fulminate was the first inorganic explosive discovered. Mercury II fulminate, $Hg(ONC)_2$, is the most common fulminate and is used as the primary explosive in detonators. A fulminate primary explosive can also be made from silver. Silver fulminate, AgONC, has very little practical value due to its extreme sensitivity to impact, heat, pressure, and electricity. The compound becomes progressively sensitive as it is aggregated, even in small amounts; the touch of a falling feather, the impact of a single water droplet, or a small static discharge are all capable of explosively detonating an unconfined pile of silver fulminate no larger than a dime and no heavier than a few milligrams. Aggregation of larger quantities is impossible due to the compound's tendency to self-detonate under its own weight.

Silver acetylide (silver carbide) (C_2Ag_2) is also an inorganic explosive that is highly sensitive and cannot be used in detonators. Silver acetylide is a primary explosive. It is a white powder that is sensitive to light. Generally, a chemical explosive must be confined for an explosion to take place. However, silver acetylide maintains a high energy density and will detonate without confinement. Dry silver acetylide poses an explosion hazard when exposed to heat, shock, or friction. When dry, it should not be stored indoors. Because it is light sensitive, it should be stored in a dark room. It should also be stored in an amber bottle.

Ammonium nitrate, NH_4NO_3, is an inorganic explosive material (oxidizer) that can be an explosive all by itself under certain conditions (Texas City). It does not work well under normal conditions and is hard to initiate, but response personnel should still deal with ammonium nitrate incidents with a great deal of caution. Commonly, ammonium nitrate is mixed with other explosives or fuels that make it a more effective explosive. Ammonium nitrate is a strong oxidizer and the primary component of ANFO, a commercially available explosive material. *Ammonium nitrate, NH_4NO_3,* is classified as an oxidizer. It is a colorless or white-to-gray crystal that is soluble in water. It decomposes at 210°C, releasing nitrous oxide gas. The four-digit UN identification number is 1942 with an organic coating and 2067 as the fertilizer grade. There are a number of other mixtures of ammonium nitrate that have four-digit numbers; they can be found in the Hazardous Materials Tables and in the DOT's *Emergency Response Guide.*

Metal azides

Metal azides are inorganic explosive compounds composed of the N_3 ion attached to a metal. Heavy metal azides are very explosive when heated or shaken. Those include silver azide (AgN_3) and lead azide (PbN_3). Lead azide is more explosive than other azides and is used in detonators that initiate secondary explosives. Sodium azide, NaN_3, decomposes explosively upon heating above 275°C. Sodium azide releases diatomic nitrogen and is used in air bag and airline escape chute deployment. Sodium azide is highly toxic and behaves like cyanide inside the body (chemical asphyxiation). Response personnel should be very careful around automobile accidents where air bags have deployed. The white powder residue is likely to contain sodium azide. Most inorganic and organic azides are prepared directly or indirectly from sodium azide. Sodium azide is used in the production of metal azide explosive compounds and as a detonator.

Aliphatic explosive compounds (nitro hydrocarbon derivatives)

Explosives may also be organized into families; in this case, aliphatic nitro compounds are a hydrocarbon-derivative family. Nitro is the one hydrocarbon-derivative family that is classified with an explosive as its primary hazard. However, there are some nitro compounds that have other primary hazards, such as nitrobenzene, which is a poison. This is an exception to the general hazard and, for safety purposes, consider nitros explosive as a group. The nitro group is represented by a nitrogen covalently bonded to two oxygen atoms. Nitrogen must have three connections to complete the octet rule of bonding (Figure 3.5). The oxygens have a single bond between themselves. This oxygen-to-oxygen single bond is highly unstable and can come apart explosively.

The other bonding spot on the nitrogen is attached to a hydrocarbon or hydrocarbon-derivative backbone of some type. These backbones may include methane and others. A nitro compound is a hydrocarbon with one or more hydrogen atoms removed and

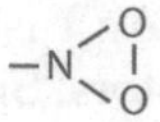

$$-N\begin{smallmatrix} \nearrow O \\ \\ \searrow O \end{smallmatrix}$$

Figure 3.5 Nitro radical structure.

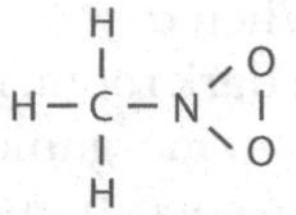

$$H-\underset{\underset{H}{|}}{\overset{\overset{H}{|}}{C}}-N\begin{smallmatrix} \nearrow O \\ \\ \searrow O \end{smallmatrix}$$

CH_3NO_2

Figure 3.6 Nitromethane.

replaced by the nitro functional group NO_2. If more than one nitro radical is used, they are represented by the Greek prefix indicating the number; "di-" for two, "tri-" for three, and "tetra-" indicating four. When naming compounds from the nitro group, the word "nitro" is used first and the end is the hydrocarbon to which the nitro is attached. In Figure 3.6, one nitro functional group is attached to the hydrocarbon radical for one carbon, "meth-," and all hydrocarbon bonds are single so the ending is "ane," which forms the compound methane. Methane has had one hydrogen atom removed, which becomes the methyl radical CH_3 to create a place to attach the nitrogen on the nitro functional group NO_2.

$$CH_3 + NO_2 = CH_3NO_2, \text{ nitromethane}$$

Nitromethane, CH_3NO_2, is a colorless liquid that is soluble in water. The specific gravity is 1.13, which is heavier than water. Nitromethane is a dangerous fire and explosion risk and is shock- and heat-sensitive. It may detonate from nearby explosions. The boiling point is 213°F, and the flash point is 95°F. The flammable range lists only a lower explosive limit, which is 7.3% in air; an upper limit has not been established. The ignition temperature is 785°F. Nitromethane may decompose explosively above 599°F, if confined, and is a dangerous fire and explosion risk, as well as toxic by ingestion and inhalation. The TLV is 100 ppm in air. The four-digit UN identification number is 1261; the NFPA 704 designation is health 1, flammability 3, and reactivity 4. Nitromethane is used in drag racing to give the fuel in the engine an extra kick to increase speed. It is also used in polymers and rocket fuel.

Another example of a nitro compound is nitroglycerine, which utilizes resonance bonding to allow three oxygen atoms to be attached to one nitrogen while the nitrogen is also attached to a carbon. This resonance bonding is similar to the bonding used for aromatic hydrocarbons. The electrons necessary for bonding oxygen to nitrogen are considered to be in a state of resonance or mesomerism and not associated with a single atom or covalent bond (delocalized electrons):

$$CH_2 + NO_3 + CH + NO_3 + CH_2 + NO_3 = C_3H_5N_3O_9, \text{ nitroglycerine}$$

Nitroglycerine, $C_3H_5N_3O_9$: the IUPAC name for the compound is 1,2,3-trinitroxypropane. Nitroglycerine is a pale yellow, viscous liquid. It is slightly soluble in water, with a specific gravity of 1.6, which is heavier than water. It is a severe explosion risk and will explode spontaneously at 424°F. It is much less sensitive to shock when it is frozen. Nitroglycerine freezes at about 55°F. It is highly sensitive to shock and heat, and is toxic by ingestion, inhalation, and skin absorption. The TLV is 0.05 ppm in air. Nitroglycerine is forbidden in transportation unless sensitized.

When in solution with alcohol at not more than 1% nitroglycerine, the four-digit UN identification number is 1204. When in solution with alcohol and more than 1%, but not more than 5% nitroglycerine, the four-digit UN identification number is 3064. The NFPA 704 designation for nitroglycerine is health 2, flammability 2, and reactivity 4. The primary uses are in explosives and dynamite manufacture (Photo 3.6), in medicine as a vasodilator, in combating oil-well fires, and as a rocket propellant. The structure for nitroglycerine is shown in Figure 3.7. Notice that there are three nitro functional groups. These nitro groups were attached to the alcohol glycerol after three hydrogens were removed.

Photo 3.6 Dynamite.

$$CH_2NO_3CHNO_3CH_2NO_3$$

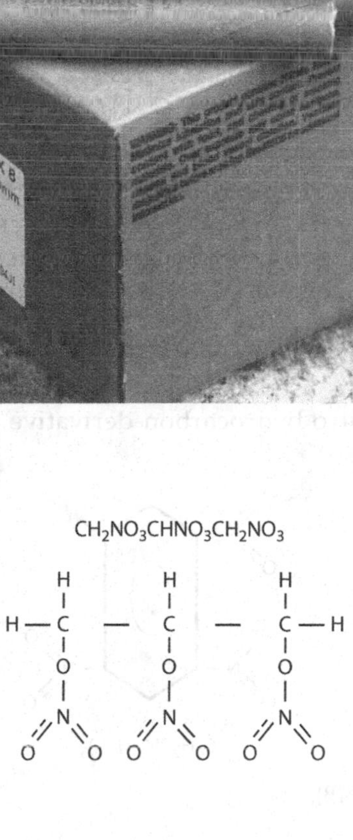

Figure 3.7 Nitroglycerine.

Aromatic explosive compounds

Nitro compounds are also formed when nitro functional groups are added to the benzene and toluene rings and radicals that are formed for hydrocarbon derivatives. The first such compound is the nitrated benzene ring, which is called TNB, $C_6H_3(NO_2)_3$. Three nitro functional groups are added to the benzene ring, thus the prefix "tri" is used in the naming followed by the functional group name "nitro" and ending in benzene, which is the aromatic before the nitro functional groups were added. Other aromatic explosive compounds are created from the first:

$$C_6H_3 + NO_2 + NO_2 + NO_2 = C_6H_3(NO_2)_3, TNB$$

Trinitrobenzene (TNB), $C_6H_3(NO_2)_3$, *IUPAC name 1,3,5-trinitrobenzene*, is an aromatic explosive material. TNB is dangerous and explodes by heat or shock. It is yellow crystal in color. The four-digit UN identification number is 3367 when wetted with not less than 10% water and 1354 when wetted with not less than 30% water. TNB is a member of the nitro hydrocarbon-derivative family (Figure 3.8).

The common explosive TNT is created by adding a methyl radical CH_3 to nitrobenzene. This compound is formed when three nitro functional groups are attached to the created toluene ring by adding the methyl radical. Because the prefix "tri" is located in front of nitro, there are three nitro groups attached. It is named by the nitro functional group with the prefix "tri" and ending in the name toluene because that is the hydrocarbon the nitro functional groups were added to:

$$C_6H_2 + CH_3 + NO_2 + NO_2 + NO_2 = C_6H_2CH_3(NO_2)_3, TNT$$

Trinitrotoluene (TNT), $CH_3C_6H_2(NO_2)_3$, *IUPAC name 2-methyl-1,3,5-trinitrobenzene*, is flammable, a dangerous fire risk, and a moderate explosion risk. It is light cream-to-rust in color and is usually found in 0.5- or 1-lb blocks. It is fairly stable in storage. TNT will detonate only if vigorously shocked or heated to 450°F; it is toxic by inhalation, ingestion, and skin absorption. The four-digit UN identification number is 1356, when wetted with not less than 30% water. Other mixtures are listed in the hazardous materials tables with several ID numbers. TNT is one of the common ingredients used in military explosives and is used as a blast-effect measurement for other explosives. TNT is a member of the nitro hydrocarbon-derivative family. In the structure shown

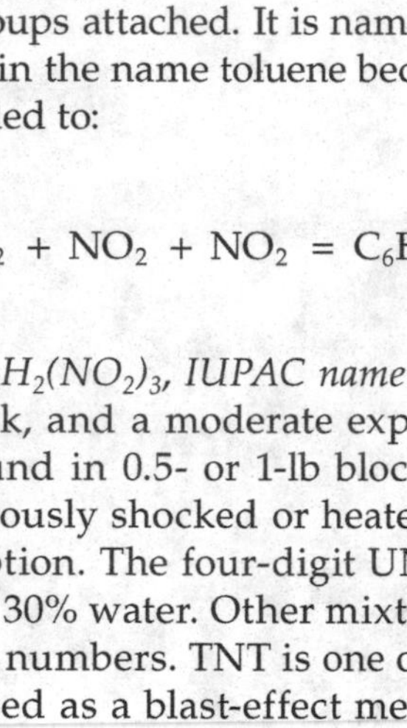

$C_6H_2CH_3(NO_2)_3$

Figure 3.8 Trinitrobenzene (TNB).

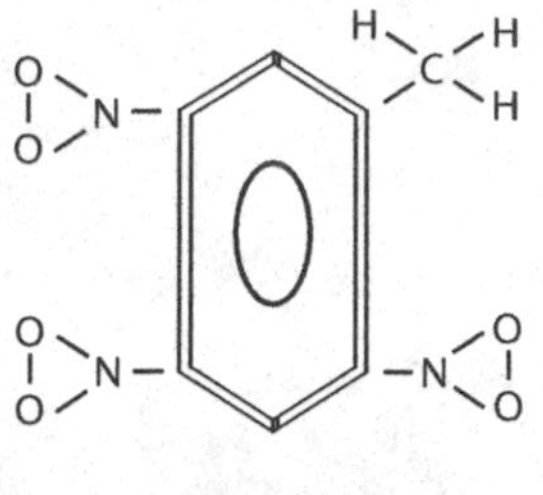

$$CH_3C_6H_2(NO_2)_3$$

Figure 3.9 Trinitrotoluene (TNT).

in Figure 3.9, toluene is the backbone for TNT. Three hydrogens are removed from the toluene ring and three nitro functional groups are attached.

Another compound that uses the benzene ring is the explosive trinitrophenol (TNP), also known as picric acid. This compound is formed when three nitro functional groups are attached to the benzene ring. Because the prefix "tri" is located in front of nitro, there are three nitro groups attached. It is named by the nitro functional group with the prefix "tri" and ending in the name phenol because that is the hydrocarbon functional group that the nitro functional groups were added to. Remember from Chapter 2, the radical of benzene is phenol. The ending "ol" is also a clue that the compound is an alcohol. However, though technically an alcohol, phenol has different characteristics from the common alcohol family. Thus, TNP is made from phenol by adding nitro radicals to the structure:

$$C_6H_2 + OH + NO_2 + NO_2 + NO_2 = C_6H_2OH(NO_2)_3, TNP$$

Trinitrophenol (TNP; picric acid), $C_6H_2OH(NO_2)_3$, IUPAC name 2,4,6-trinitrophenol, is composed of yellow crystals that are soluble in water. It is a high explosive, is shock- and heat-sensitive, and will explode spontaneously at 572°F. Trinitrophenol is reactive with metals or metallic salts and is toxic by skin absorption. The TLV is 0.1 mg/m^3 of air. When shipped in 10%–30% water, it is stable unless the water content drops below 10% or it dries out completely (Photo 3.7). The four-digit UN identification number is 1344 when shipped with not less than 10% water. The NFPA 704 designation is health 3, flammability 4, and reactivity 4 (Figure 3.10).

The primary uses are in explosives, matches, electric batteries, etching copper, and textile dyeing. Picric acid is often found in chemical labs in high schools and colleges, and can be a severe explosion hazard if the moisture content of the container is gone. Picric acid was used by the Japanese during World War II as a main charge explosive filler. When in contact with metal, picric acid will form other picrates, which are extremely sensitive to heat, shock, and friction. Great care should be taken when handling World War II souvenirs, because of the possible presence of these picrates.

When the structure of picric acid is compared with the structure of TNT, the only difference is the fuel that the nitro functional groups were placed on; the number of nitro groups is exactly the same. The explosive power of picric acid is similar to that of TNT. There are other Class 1.1–1.3 materials that are nitro compounds and some Class 1.1–1.3 materials that are made up of other chemicals. *Black powder* is a low-order explosive made up of a mixture of potassium or sodium nitrate, charcoal, and sulfur in 75%, 15%, and 10%

Photo 3.7 Dry, shock-sensitive picric acid on the shelf of a high school chemistry classroom.

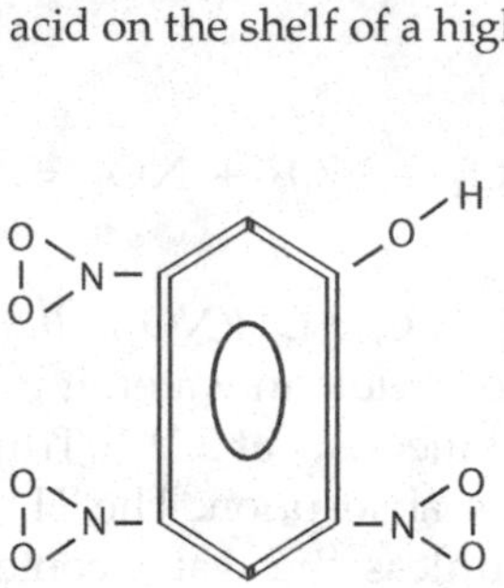

$$C_6H_2OH(NO_2)_3$$

Figure 3.10 Trinitrophenol (picric acid).

proportions, respectively. It has an appearance of a fine powder to dense pellets, which may be black or have a grayish-black color. It is a dangerous fire and explosion risk, is sensitive to heat, and will deflagrate rapidly.

Incidents

A number of accidents have occurred over the years involving 1.1–1.3 (former Class A and B explosive materials). In Waco, Georgia, an automobile collided with a truck carrying 25,414 lb of explosives, resulting in a fire and the explosion of the dynamite cargo. The fire started as a result of gasoline and diesel fuels spilling from the vehicles and then igniting. Heat transfer from the fires caused the nitroglycerine-based dynamite to detonate, killing two firefighters, a tow truck driver, and two bystanders; injuring another 33 persons; and

Photo 3.8 Leaking natural gas from a transmission line(s) under a building housing a firearms store was ignited by gun fired in the firing range at the firearms store in Richmond, Indiana. (Courtesy of Richmond Fire Department, Richmond, IN.)

causing over $1 million in property damage. The National Transportation Safety Board (NTSB) investigated the incident and, as part of their report, listed the following contributing factors that led to the injuries and deaths:

1. The lack of a workable system to warn everyone within the danger zone of an explosion
2. The failure to notify emergency service personnel promptly and accurately of the hazards
3. The decision of the firefighters to try and contain the hazardous fire
4. Bystanders disregard or lack of understanding of the truck driver's warnings

Richmond, Indiana, April 6, 1968, 1:47 p.m., leaking natural gas from a transmission line(s) under a building housing a firearms store was ignited by gun fired in the firing range at the firearms store. A secondary explosion of up to a ton of smokeless powder, along with some black powder that was stored in the basement, killed 41 persons, injured another 150, and caused over $2 million in property damage. Three buildings were destroyed by the explosion and five others damaged (Photo 3.8).

Explosives subclasses 1.4–1.6

Division 1.4 is made up of explosives that present a minor explosion hazard. The explosive effects are largely confined to the package, and no projection of fragments of appreciable size or range is to be expected. An external fire must not cause virtually instantaneous explosion of almost the entire contents of the package.

The DOT considers Division 1.5 (blasting agents) insensitive explosives. This division comprises substances that have a mass explosion hazard, but are so insensitive that there is little probability of initiation or of transition from burning to detonation under normal conditions of transport. (The probability of transition from burning to detonation is greater when large quantities are transported in a vessel or in storage facilities.) Division 1.6 explosives are considered extremely insensitive and do not have a mass explosion hazard. This division comprises articles that contain only extremely insensitive detonating substances and that demonstrate a negligible probability of accidental initiation or propagation. (The risk from articles of Division 1.6 is limited to the explosion of a single article.)

Most of the listings in 49 CFR 172.101 Hazardous Materials Tables for 1.4–1.6 explosives are for small premanufactured explosive devices: small arms ammunition, signal cartridges, tear gas cartridges, detonating cord, detonators for blasting, explosive pest control devices, fireworks, aerial flares, practice grenades, signals, and railway track explosive smoke signals. Recent changes in DOT rules now allow certain Class 1.1 and 1.2 explosives to be shipped in small quantities in special packages as Class 1.4. Normally, Class 1.1 through 1.3 explosives must be placarded regardless of the quantity. The problem was that bomb squads and other explosive experts ordering small amounts of C-4, sheet explosives, DEXS caulk, and slip boosters had to pay high shipping costs, as much as $1000 in some cases, to common carriers. Special packaging was designed and underwent thorough testing by the U.S. Department of Mines. The new DOT rules allow small amounts of explosives to be shipped by UPS, Federal Express, and other small-package shipping companies.

Class 1.4 explosives do not require placarding until 1001 lb or more are shipped. One of the main chemical explosives listed from groups 1.4 to 1.6 is 1.5 ammonium nitrate (Photo 3.9). When mixed with fuel oil, it becomes a blasting agent. Subjected to confinement or high heat, it may explode but does not readily detonate. Fertilizer-grade ammonium nitrate, which is a strong oxidizer above 33.5%, may also explode if it becomes contaminated. Fertilizer-grade ammonium nitrate was used in the bombings of the World Trade Center in New York City in 1993 and the Federal Building in Oklahoma City in 1995. The bombs were made up of ammonium nitrate or urea nitrate (the oxidizer) and fuel oil, which provides the fuel. It is estimated that the World Trade Center bomb contained 1200 lb of urea nitrate, and the Oklahoma City bomb over 4800 lb of ammonium nitrate. The bomb that destroyed the 8th Marine in Beirut, Lebanon, contained explosives with the power of 12,000 pounds of TNT.

Explosive chemicals

There are a number of other chemicals that are not explosives, but have explosive potential under certain conditions. Oxygen is an oxidizer that causes organic materials to burn explosively. Chlorine is also an oxidizer and when in contact with organic materials the mixture becomes explosive.

Ether is an organic compound that forms explosive peroxides when in contact with air. When a container of ether is opened, oxygen from the air bonds with the single oxygen in each ether molecule and forms an organic peroxide. These peroxides are very unstable and become sensitive to shock, heat, and friction. Moving or shaking a container can cause an explosion. Ethers are also very flammable, with wide flammable ranges. Fire is likely to follow an explosion of an ether container. Some examples include ethyl ether, isopropyl ether, methyl tert-butyl ether (MTBE), and propylene oxide.

Photo 3.9 Mixer truck used to transport ammonium nitrate and fuel oil and to mix into a blasting agent on site.

Ethyl ether, (diethyl ether) (*ethoxyethane, IUPAC*) is a colorless, volatile, mobile liquid. It is slightly soluble in water, with a specific gravity of 0.7, which is lighter than water. It is a severe fire and explosion risk when exposed to heat or flame. The compound forms explosive peroxides from the oxygen in the air as it ages. The flammable range is wide, from 1.85% to 48% in air. Boiling point is 95°F (35°C), flash point is –49°F (–45°C), and ignition temperature is 356°F (180°C). Vapor density is 2.6, which is heavier than air.

Isopropyl ether (diisopropyl ether) is highly flammable, with a wide flammable range of 1.4%–21% in air. Boiling point is 156°F (68°C), flash point is –18°F (–27°C), and ignition temperature is 830°F (443°C). Vapor density is 3.5, which is heavier than air. In addition to flammability, isopropyl ether is toxic by inhalation and a strong irritant, with a TLV of 250 ppm in air.

Methyl tert-butyl ether, (MTBE), with molecular formula $(CH_3)_3COCH_3$, is a volatile, flammable, and colorless liquid that is sparingly soluble in water. MTBE is primarily used as a gasoline additive. Although the main danger is explosion, MTBE can also form explosive peroxides when in contact with air. It is not as prone to peroxide formation as other ethers, but it can happen after long periods.

Butadiene may also form explosive peroxides when exposed to air.

Potassium metal (K) is a metallic element from family one on the periodic table of elements. Potassium metal can form peroxides and superoxides at room temperature and may explode violently when handled. Simply cutting a piece of potassium metal with a knife to conduct an experiment could cause an explosion. These chemicals will be discussed in detail in Chapters 2 and 5.

Incidents

Several major incidents have occurred with ammonium nitrate:

Texas City, Texas, April 16, 1947, started out like any other spring day along the Gulf Coast in the town of Texas City, population 16,000 people. The SS *Grandcamp* was at the port taking on a load of ammonium nitrate fertilizer to be shipped to Europe as part of the rebuilding process following World War II. The 32.5% ammonium nitrate fertilizer was placed into the hold of the ship along with a cargo of small arms ammunition. Approximately 17,000,000 lb (7,700 tons) of ammonium nitrate was loaded onto the ship. Also in the harbor that fateful day was the SS *High Flyer* located approximately 600 ft from the *Grandcamp* on the same dock and loaded with 2,000,000 lb (900 tons) of ammonium nitrate and 4,000,000 lb (1,800 tons) of sulfur.

By comparison, the bomb used in the bombing of the Oklahoma City Federal Building contained 5000 lb of ammonium nitrate and the bomb used at the World Trade Center in New York City contained 1500 lb of ammonium nitrate. Ammonium nitrate at a concentration of 32.5% is classified by the U.S. DOT as an oxidizer and used primarily as a fertilizer for agricultural and home use. Ammonium nitrate is also used as an oxidizer in the manufacture of explosives. Ammonium nitrate that was loaded on the two ships at the Port of Texas City had been coated with paraffin (a hydrocarbon product) and other chemicals to prevent caking of the material. By adding paraffin to the ammonium nitrate oxidizer, it is like combining fuel and oxygen fulfilling two requirements of the fire triangle. You now have two of the three materials necessary for a fire to occur. All that is missing is heat. If you add confinement, you would have all the ingredients for a massive explosion. Because ammonium nitrate is also an oxidizer under certain conditions, it can become explosive. For an explosion to occur involving ammonium nitrate, confinement is necessary for the fuel and oxidizer before the heat is applied. As longshoremen were loading the ammonium nitrate on the ship, they reported that the bags were warm. This could have been caused by an exothermic (heat releasing) chemical reaction going on within the bags of ammonium nitrate. A chemical process resulting in spontaneous combustion can occur with chemical oxidizers at elevated temperatures.

Around 8:10 a.m., fire was reported deep in cargo hold number 4 of the *Grandcamp*. Hold number 4 was located toward the rear of the ship at the lower level (Photo 3.10). Two fire extinguishers and a gallon jug of drinking water were applied to the fire area by the ship's crew with little effect. The captain of the ship ordered the hold sealed and steam injected into the burning hold (the heat and confinement necessary for an explosion!). A common method of shipboard firefighting at that time was to seal the cargo hold where the fire was occurring and inject steam into the hold to extinguish the fire. This confinement and injection of steam is likely to have resulted in the elevation of the temperature of the ammonium nitrate and the explosion that occurred. Texas City's volunteer fire department responded to the initial report of fire with their four fire engines led by Chief Henry J. Baumgartner (Photo 3.11). It is unknown if the fire department had any knowledge of the dangers of fires involving ammonium nitrate or potential explosives. A fire boat from Galveston was requested at about the same time by the ship's captain. Twenty-seven of the department's 28 volunteers answered the call along with the Republic Oil Refining Company firefighting team. They set up their hose streams along the dock and applied water into the burning hold of the *Grandcamp*. Smoke poured from the hold followed by flames at around 9:00 a.m. The unusual color of the smoke caught many people's eye. Some called it a peach color, others called it reddish orange. At approximately 9:12 a.m., an explosion occurred within the hold of the *Grandcamp*.

Photo 3.10 Firefighters spraying water on the *Grandcamp* deck. (Courtesy of City of Texas City, Moore Memorial Public Library.)

Instantly, all 27 members of the Texas City Volunteer Fire Department at the scene were killed, and some bodies were disintegrated by the heat and blast pressure of the explosion. All that remained of their fire engines were piles of twisted metal (Photo 3.12). Texas City lost all but one of their firefighters and all of their apparatus in the explosion. The force of the explosion knocked two small airplanes out of the sky and was heard over 150 miles away. Debris fell on homes and businesses, setting many buildings on fire. People on the streets of Galveston, 11 miles away, were knocked to the ground by the force of the blast. Chemical plants and petroleum storage facilities along the shores of the bay were set afire as well. The explosion set fire to the USS *High Flyer*, which was at the dock near the *Grandcamp*, also loaded with ammonium nitrate. An anchor from the *Grandcamp* weighing over 3000 lb was propelled two miles away and landed in a 10 ft crater. A seismologist in Denver, Colorado, recorded the shock waves from the explosion and thought an atomic bomb had been detonated in Texas. In Omaha, Nebraska, the Strategic Air Command briefly elevated the U.S. defense condition (Defcon), believing there was a nuclear attack taking place. When it was all over, more than 405 identified people were dead, more than 3500 injured. There were also 63 people who died and could not be identified, 19 volunteer firefighters were never identified and may have been among the 63 people who could not be identified. More than 100 people were presumed

Exh. 9d

Photo 3.11 One of the last photographs taken of the Texas City Volunteer Fire Department crew before the explosion. Marion "Jack" Westmorland holding the hose (bottom right corner). Fire Chief Buamgartner (center) with hat on. Both were killed in the explosion. (Courtesy of City of Texas City, Moore Memorial Public Library.)

dead, as their bodies were never found. Some believed that hundreds more were killed but unaccounted for including visiting seamen, noncensus laborers and their families, and untold numbers of travelers. There were people located as close as 70 ft from the ship when it exploded actually survived. Refinery infrastructure and pipelines, including about 50 oil storage tanks, were extensively damaged or destroyed from the blast pressure and resulting fires that burned for days. Monsanto Chemical Companies plant was heavily damaged, and 143 of the civilian deaths occurred there among employees. Over 500 homes were destroyed and hundreds of others damaged, leaving over 2000 people homeless. Property loss was listed at over 100 million dollars. That is close to one-half billion dollars by today's standards. Bulk cargo-handling operations never again resumed at the Port of Texas City. When the Texas City Disaster occurred in 1947, there were no fire department hazmat teams or hazardous materials training in existence in the United States, self-contained breathing apparatuses were generally unheard of, no placarding and labeling system existed, and there was little government regulation of hazardous materials.

Photo 3.12 Damaged fire engine sits amid debris near a dock warehouse that appears to have only metal framework remaining following the explosion. (Courtesy of City of Texas City, Moore Memorial Public Library.)

Texas City firefighters who lost their lives on April 16, 1947 (Photo 3.13)
Henry J. Baumgartner, Fire Chief
Joseph Milton Braddy, Assistant Chief
Sebastian B. Nunez, Captain
William Carl Johnson, Captain
Marshall B. Stafford, Lieutenant
William D. Pentycuff, Lieutenant
Privates:
Zolan Davis
William C. O'Sullivan
Roy Louis Durio
Marcel Pentycuff
Archie Boyce Emsoff
Harvey Alonzo Menge
Henry John Findeisen
Jimmy Reddicks
Virgil D. Fereday
Robert Dee Smith

Photo 3.13 Memorial to the Texas City firefighters who died in the explosions located on the south lawn of the Texas State Capitol Building in Austin, Texas.

Edward Henry Henricksen
Joel Clifton Stafford
William Fred Hughes
Maurice R. Neely
Lloyd George Cain
Marion D. Westmoreland
Frank P. Jolly
Clarence J. Wood
William Louis Kaiser
Clarence Rome Vestal
Jacob Otto Meadows

On November 29, 1988, at approximately 03:40 h, the Kansas City, Missouri, Fire Department received a call for a fire at a highway construction site. The fire was reported by a security guard at the site to be in a small pickup truck; however, a woman in the background, another security guard, could be heard saying "the explosives are on fire." Pumper 41 was dispatched to the site with a captain and two firefighters (in Kansas City, fire apparatus with a pump is called a pumper). Dispatch cautioned Pumper 41 that there may be explosives at the site. Pumper 41 arrived on scene at 03:46 and found there were two separate fires burning, and a second pumper company was requested. Pumper 41 also requested that dispatch warn Pumper 30 of the potential for explosives at the site. Pumper 30 was dispatched and arrived on scene at 03:52. At 04:08, 22 min after Pumper 41 arrived and approximately 16 min after Pumper 30 arrived, the magazine exploded killing all six firefighters assigned to Pumper 41 and Pumper 30.

Battalion chief 107 and his driver were just arriving on scene and stopped about ¼ mile from the explosion. They received minor injuries when the windshield of their vehicle was blown in. Following the first explosion, the battalion chief ordered firefighters to withdraw from the area and a command post was set up at a safe distance from the site (Photo 3.14). Approximately 40 min after the first explosion, a second blast occurred, followed by several smaller explosions. It is likely that the actions of the battalion chief prevented additional deaths and injuries. It was reported by the Kansas City Fire Department that the first explosion involved a split load of materials in the trailer/magazine. One compartment had approximately 3500 lb of ANFO mixture. The rest of the contents were approximately 17,000 lb of ANFO mixture with 5% aluminum pellets. In the second trailer/magazine, there was approximately 1000 30 lb "socks" of ANFO mixture with 5% aluminum pellets. In comparison, the World Trade Center bombing was 1500 lb of ammonium nitrate with additives, and the Oklahoma City bombing was 5000 lb of ammonium nitrate. Pumper 41 was damaged beyond recognition as a piece of fire apparatus by the explosion (Photo 3.15). Pumper 30 received significant damage as well but could still be identified as a fire department vehicle.

As a direct result of the 1988 explosion, the Kansas City, Missouri Fire Department's Hazardous Materials Team was placed in service in 1989. Pumper Companies 30 and 41 and their personnel were lost in the explosion. The numbers 30 and 41 were added together to form the number for Hazmat 71 in honor of the firefighters killed in the explosion. Because of this incident and others, OSHA now requires the use of the DOT placarding and labeling system for fixed-facility storage of hazardous materials (Photo 3.16). When they are transported, and until the materials are used up or the containers purged, the placards and labels must remain on the containers.

Photo 3.14 Aerial view of the blast site where 6 Kansas City, Missouri, firefighters lost their lives in an ammonium nitrate explosion. (Courtesy of Kanas City Fire Department, Kanas City, MO.)

Firefighters killed in Kansas City, Missouri explosion (Photo 3.17)
Captain Gerald C. Halloran 57
Thomas M. Fry 41
Luther E. Hurd 31
Captain James H. Kilventon Jr. 54
Robert D. McKarnin 42
Michael R. Oldham 32

Marshall's Creek, Pennsylvania, June 26, 1964, a fire in a cargo truck, thought to be caused by spontaneous heating following the blowout of a tire, quickly spread to the cargo compartment of a 28-ft tractor-trailer truck. The driver had disconnected the trailer off the roadway and had driven the tractor several miles to a service station. While the driver was gone, the tires ignited. The truck was carrying 4000 lb of 60% standard gelatin dynamite in boxes and 26,000 lb of nitrocarbonitrate blasting agent in 50 lb bags. The compound is a mixture of 850 lb of ammonium nitrate fertilizer to 7 gal of No. 2 diesel fuel. Another passing tractor-trailer driver reported the fire to the Marshalls Creek Fire Department. The driver reported that there were no markings on the trailer. Three fire engines responded, and an attack line was pulled to fight the blaze. As the firefighters approached the trailer, a detonation occurred. The fire had reached the explosive cargo, and the resulting explosion

Photo 3.15 Engine 41 was damaged in the explosion in Kansas City, Missouri, almost beyond recognition as a piece of fire apparatus. (Courtesy of Kanas City Fire Department, Kanas City, Mo.)

killed six people, including three firefighters, the truck driver who reported the fire, and two bystanders. Property damage was over $600,000, including all three of the fire engines.

Firefighters killed in Marshalls Creek, Pennsylvania
F. Earl Miller
Leonard R. Mosier
Edward F. Hines

Homemade explosives/terrorist explosives

Several terrorist incidents have occurred in United States over the past 20 years that involved bombs and the use of explosives and devices that were homemade. When terrorist bombings are discussed within the United States, it immediately brings to mind the Unabomber; bombings at the World Trade Center in New York City, New York; the Oklahoma City, Oklahoma Murrah Federal Building; Olympic Park in Atlanta, Georgia; the abortion clinic and night club in Atlanta, Georgia; and the abortion clinic in Birmingham, Alabama. Secondary explosive devices were set up to injure or kill response personnel in two of the Atlanta incidents. During the bombing of the abortion clinic in Atlanta, the secondary device went off. A secondary device was found at the nightclub in Atlanta and was disposed off before it went off. Responders were aware of the possibility of secondary devices and took proper precautions. Explosive devices used in the previously mentioned incidents included ammonium nitrate vehicle bombs at the World Trade Center and Oklahoma City. Homemade pipe bombs and as many as 10–20 sticks of dynamite were

Photo 3.16 Fixed-storage bunkers for explosive materials are heavily regulated by state and federal laws.

Photo 3.17 Memorial dedicated to the memory of the six Kansas City, Missouri, firefighters from Pumpers 31 and 40 who lost their lives in an ammonium nitrate explosion.

used in the family planning clinic and bar in Atlanta bombings and the family planning clinic bombing in Birmingham. Responders should not only be aware of potential secondary devices, but also have a basic understanding of the characteristics and effects of explosives and explosive devices. For the first time in nearly 30 years, emergency responders were subjected to a secondary explosive device in Atlanta, Georgia. It was 10:35 a.m. and the Fulton County, Georgia, Fire Department received a call for a transformer explosion at 275 Carpenter Drive in Sandy Springs. One engine company was dispatched, and as firefighters arrived they found extensive damage. It was clear that the type of damage found by firefighters could not have been caused by an electrical transformer explosion. There was little doubt that a bomb had exploded at the Sandy Springs Professional Building. Targeted was the Atlanta North Side Family Planning Services Center. This same family planning clinic had been bombed once before, 13 years prior. Firefighters and law enforcement personnel had been on the scene about an hour when a second blast occurred, destroying the fire chief's car. Six people were injured: two bystanders, one firefighter, one FBI agent, and one ATF agent. Dynamite had been placed in an old ammunition box near the west entrance of the building's parking lot. Two parked cars in the building parking lot provided shielding from the brunt of the blast force for other response personnel, or there may have been many more injuries or deaths.

Following the family clinic bombing in Sandy Springs, two more bombings occurred: one in Atlanta, and the other in Birmingham, Alabama. The Atlanta incident also involved the planting of a secondary device. At approximately 10:00 p.m., on February 22, 1997, an explosion occurred at the Otherside Lounge, a gay bar in Atlanta. Five patrons were injured when the bomb scattered large nails into the crowd of approximately 150 patrons. A secondary device was found in a backpack outside of the building. Police were able to destroy the device using a robot before it went off. The Olympic Park bombing had also involved a backpack that held three pipe bombs containing nails. The New Women All Women Health Center in Birmingham was the target of a January 29, 1998, blast. This explosion resulted in the death of an off-duty police officer working as a security guard and the serious injury of a nurse. A secondary device was located and rendered safe before an explosion could occur.

Bombings at the World Trade Center and Oklahoma City Federal Building involved a chemical blasting agent. The bombs were made up of ammonium nitrate or urea nitrate (the oxidizer) and fuel oil, which provides the fuel. It is estimated that the World Trade Center bomb contained 1200 lb of urea nitrate, and the Oklahoma City bomb over 4800 lb of ammonium nitrate. Bombs used in the New York City and Oklahoma City explosions were vehicle bombs. These types are characteristically large and powerful. Vehicles are fitted with quantities of explosive materials, coupled with timed or remote triggering devices. In both of the aforementioned cases, the vehicle used was a rental moving van. Ammonium nitrate and fuel oil mixtures were used in both the Oklahoma City and World Trade Center bombings. Explosions that occurred at the World Trade Center and the Oklahoma City Federal Building were caused by what are considered to be low-yield fillers that deflagrate. However, the material that was used in those bombings detonated, producing a great deal of physical damage to the structures, more so in Oklahoma City than at the World Trade Center. The bomb used in the Oklahoma City bombing was four times larger than the World Trade Center bomb.

On February 26, 1993, at 12:18 p.m., international terrorism struck the United States. The World Trade Center in New York City was bombed by Islamic fundamentalists under suspected mastermind Ramzi Yousef (Photo 3.18). The World Trade Center boasts two high-rise towers. Each tower was 110 stories tall, with several other buildings completing the complex. When fully occupied, there are over 150,000 people in the buildings. Noontime is

Photo 3.18 World Trade Center bombing scene. (From The New York City Fire Department, Firefighter John Strandberg, FDNY Photo Unit. Used with permission.)

the busiest period of day at the complex and would be the ideal time to launch a terrorist attack to produce a large number of deaths and injuries. Components of the World Trade Center bomb included pellets and bottled hydrogen. The van that Yousef used had four 20 ft (6 m) long fuses, all covered in surgical tubing. He calculated that the fuse would trigger the bomb in 12 min after he would use a cheap cigarette lighter to light the fuse. The materials to build the bomb cost some 300 explosive effects from the 1300-lb bomb caused a crater 180 ft deep, 100 ft long, and 200 ft wide in the underground parking garage. The crater was six levels deep. A rented truck was used to transport the bomb into the parking garage. In that attack, 6 people died and 1042 were injured.

Oklahoma City, Oklahoma, April 19, 1995, was the second anniversary of the end of the siege of the Branch Davidian compound in Waco, Texas. At approximately 9:02 a.m., an explosion ripped through the Alfred P. Murrah Federal Building in Oklahoma City, Oklahoma, killing 168 people and injuring another 600 (Photo 3.19). Over 800 buildings sustained some type of damage from ground shock and blast pressure. Of the buildings damaged, 50 would have to be demolished. Windows were broken as far as 2 miles from the blast site and the blast was heard 50 miles away. It registered 3.5 on the open-ended Richter scale in Denver, Colorado. Across the street in a parking lot, 60 cars were completely destroyed (Photo 3.20). Property loss was estimated to be over $250 million. A 4000-lb bomb made up of an ANFO mixture detonated inside a Ryder rental truck that was parked in front of the Alfred P.

Photo 3.19 Oklahoma City Federal Building, following explosion. (From City of Oklahoma City, Public Information and Marketing. Used with permission.)

Photo 3.20 Firefighters responding to reported explosion in downtown Oklahoma City first found automobiles ablaze obscuring view of the federal building. (From City of Oklahoma City, Public Information and Marketing. Used with permission.)

Murrah Federal Building. This was the same type of explosive material used in the bombing of the World Trade Center in New York City, but a much larger amount. The Alfred P. Murrah Federal Building was a nine-story high-rise constructed of reinforced concrete. The front of the building was ripped off by the blast, leaving it open on all nine floors. A pile of debris three stories high was left in front of the building as a result of the explosion, which ripped away 80% of the building front.

It was reported to me by one of the first arriving battalion chiefs from Oklahoma City Fire Department that the Murrah Federal Building was not the first choice of targets for Timothy McVeigh, the convicted and executed bomber. McVeigh had planned on blowing up the federal courthouse across the street to the South of the Federal Building. Road construction narrowing the street and blocking direct access to the courthouse the day of the bombing caused a change of plan and the destruction of the Murrah Federal Building.

Homemade bombs can take on any form and are limited only by the imagination of the person(s) making the bomb. Step-by-step instructions and diagrams are available from the Internet, in books, and through videotapes for making many different types of bombs. These include exploding lightbulbs, computer diskette bombs, tennis ball bombs, fertilizer bombs, napalm, mailbox bombs, car bombs, paint bombs, contact bombs, plastic explosives from bleach, and smoke bombs, to name a few! Instructions are also available to make your own potassium nitrate, a common ingredient of black powder.

Acetone peroxide, $C_9H_{18}O_6$, also *TATP triacetone triperoxide* is an organic peroxide and a primary high explosive. It takes the form of a white crystalline powder with a distinctive bleach-like odor. It is susceptible to heat, friction, and shock. The instability is greatly altered by impurities. It is not easily soluble in water. It is more stable and less sensitive when wet. TATP is an explosive that can be made with liquids that are commonly available from pharmacies and hardware stores. Primary ingredients include hydrogen peroxide (hair bleach), sulfuric acid, and acetone. After mixing, the liquid mixture is evaporated, which leaves crystals. TATP can be used as an explosive or as a booster to set off larger charges. It is extremely sensitive and is stored under refrigeration to help maintain stability. Acetone peroxide is also known as "Mother of Satan" because it is highly unstable and liable to detonate with the slightest shock or rise in temperature. Response personnel should be very cautious if any of the precursor chemicals or empty containers is found in or around an occupancy (Figure 3.11).

Pentaerythritol tetranitrate, PETN, $C_5H_8O_{12}N_4$, a nitrate ester of pentaerythritol, is a white crystalline substance that feels powdery to the touch. In its pure form, PETN melts at 285°F (141.3°C). PETN is most well known as an explosive. Mixed with a plasticizer, it forms a plastic explosive. It's the high explosive of choice because it is stable and safe to handle, but it requires a primary explosive to detonate it. This compound was used by the 2001 shoe Bomber, in the 2009 Christmas Day bomb plot and in the 2010 cargo plane bomb plot. It is a shock-sensitive explosive used for demolition, blasting caps, and detonating compositions (Figure 3.12) ("Primacord").

Military explosives

Military explosives are noted for their high shattering power accompanied by rapid detonation velocities. They must be stable, because they are often kept in storage for long periods of time. Because of their intended use, they must detonate dependably after being stored and do so under a variety of conditions. The military explosives used most commonly are TNT, C-3, C-4, and RDX cyclonite. These explosives release large quantities of toxic gases when they explode (Table 3.9).

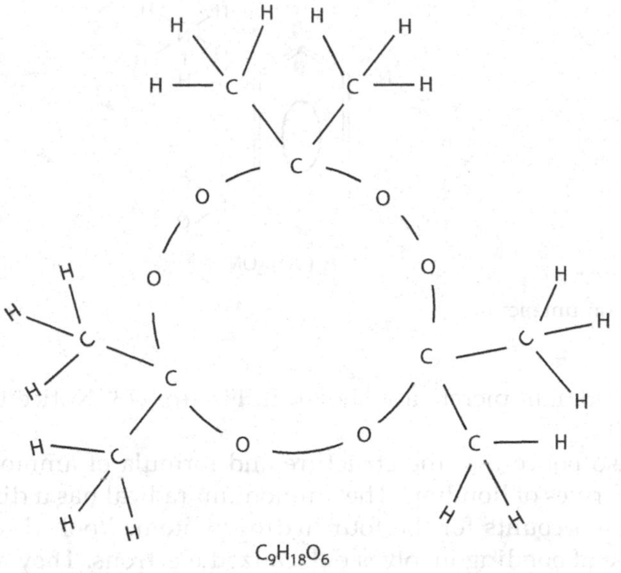

$C_9H_{18}O_6$

Figure 3.11 Acetone peroxide (TATP).

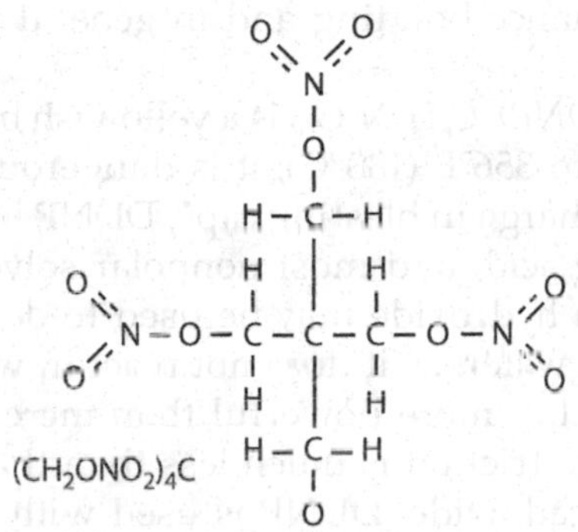

$(CH_2ONO_2)_4C$

Figure 3.12 Pentaerythritol tetranitrate (PETN).

Table 3.9 Military Explosives

Primary	Secondary
Mercury fulminate	Nitroglycerine
Lead azide	Ammonium picrate
Diazodinitrophenol	Trinitrotoluene
Lead styphnate	RDX
Nitromannite	Picric acid

Ammonium picrate, Dunnite, Explosive D, $C_6H_2(NO_2)_3ONH_4$, IUPAC name ammonium 2,4,6-trinitrophenolate, is a high explosive when dry and flammable when wet. It is composed of yellow crystals that are slightly soluble in water. The four-digit UN identification number is 1310 for ammonium picrate wetted with not less than 10% water. It is used in pyrotechnics and other explosive compounds. The structure and molecular

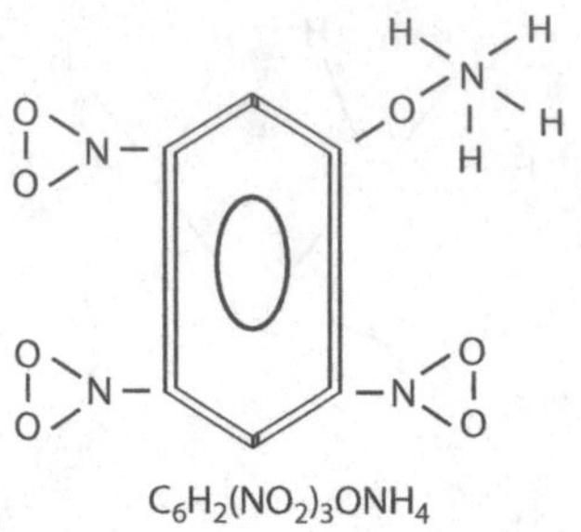

$C_6H_2(NO_2)_3ONH_4$

Figure 3.13 Ammonium picrate.

formula for ammonium picrate are shown in Figure 3.13. Notice the similarity to picric acid and TNT.

You may also notice that the structure and formula of ammonium picrate do not follow the usual rules of bonding. The ammonium radical has a different bonding configuration, which accounts for the four hydrogen atoms hooked to the nitrogen atom. This hybrid type of bonding involves delocalized electrons. They are electrons that are not associated with any one atom or one covalent bond. Delocalized electrons are contained within an orbital that extends over several adjacent atoms. This is also sometimes referred to as resonance bonding and in general is beyond the scope of "street chemistry."

Diazodinitrophenol, (DDNP), $C_6H_2N_4O_5$, is a yellowish brown powder. DDNP explodes when shocked or heated to 356°F (180°C); it is dangerous and is used as an initiating explosive. It is a primary charge in blasting caps. DDNP is soluble in acetic acid, acetone, concentrated hydrochloric acid, and most nonpolar solvents but is insoluble in water. A solution of cold sodium hydroxide may be used to destroy it. DDNP may be desensitized by immersing it in water, as it does not react in water at normal temperature. It is less sensitive to impact but more powerful than mercury fulminate and lead azide. The sensitivity of DDNP to friction is much less than that of mercury fulminate, but it is approximately that of lead azide. DDNP is used with other materials to form priming mixtures, particularly where a high sensitivity to flame or heat is desired. DDNP is often used as an initiating explosive in propellant primer devices and is a substitute for lead styphnate in what are termed "nontoxic" (lead-free) priming explosive compositions (Figure 3.14).

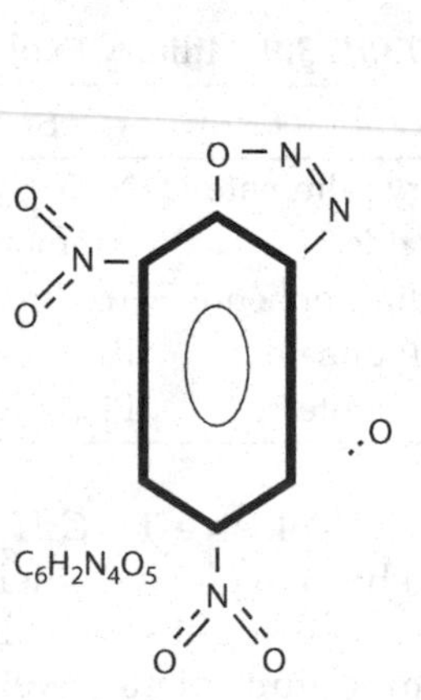

$C_6H_2N_4O_5$

Figure 3.14 Diazodinitrophenol (DDNP).

Lead Styphnate, (lead 2,4,6-trinitroresorcinate, $C_6HN_3O_8Pb$), whose name is derived from styphnic acid, is an explosive used as a component in primer and detonator mixtures for less sensitive secondary explosives; is a slurry or wet mass of orange-yellow crystals; must be shipped wet with at least 20% water or water and denatured ethyl alcohol mixture; may explode due to shock, heat, flame, or friction if dried; and is used as an initiating explosive. The primary hazard is the blast of an instantaneous explosion and not flying projectiles and fragments. There are two forms of lead styphnate: six-sided monohydrate crystals and small rectangular crystals. Lead styphnate varies in color from yellow to brown. Lead styphnate is particularly sensitive to fire and the discharge of static electricity. When dry, it can be readily detonated by static discharges from the human body. The longer and narrower the crystals, the more susceptible lead styphnate is to static electricity. Lead styphnate does not react with metals and is less sensitive to shock and friction than mercury fulminate or lead azide. Lead styphnate is only slightly soluble in water and methyl alcohol and may be neutralized by a sodium carbonate solution. It is stable in storage, even at elevated temperatures. As with other lead-containing compounds, lead styphnate is inherently toxic to humans if ingested, i.e., can cause heavy metal poisoning (Figure 3.15).

Nitromannite, Mannitol hexanitrate, $C_6H_8(ONO_2)_6$, is a powerful explosive. Physically, it is a powdery solid at normal temperature ranges, with density 1.6 g/cm³. The chemical name is hexanitromannitol, and it is also known by the names nitromannite, MHN, nitromannitol, nitranitol, or mannitrin. It is less stable than nitroglycerin, and it is used in detonators. Mannitol hexanitrate is a secondary explosive formed by the nitration of mannitol, a sugar alcohol. The product is used in medicine as a vasodilator and as an explosive in blasting caps. Its sensitivity is considerably high, particularly at high temperatures >167°F (>75°C) where it is more sensitive than nitroglycerine. It has the highest brisance of any known conventional explosive, even more than nitroglycerine (Figure 3.16).

RDX, Cyclonite, $N(NO_2)CH_2N(NO_2)CH_2N(NO_2)CH_2$, is a high explosive, easily initiated by mercury fulminate, toxic by inhalation and skin contact, TLV: 1.5 g/m⁴ of air, explosive 1.5 times as powerful as TNT, and an explosive nitroamine widely used in military and industrial applications. It was developed as an explosive that was more powerful than TNT, and it saw wide use in World War II. RDX is also known as cyclonite. Its chemical name is cyclotrimethylenetrinitramine; name variants include cyclotrimethylene-trinitramine and cyclotrimethylene trinitramine.

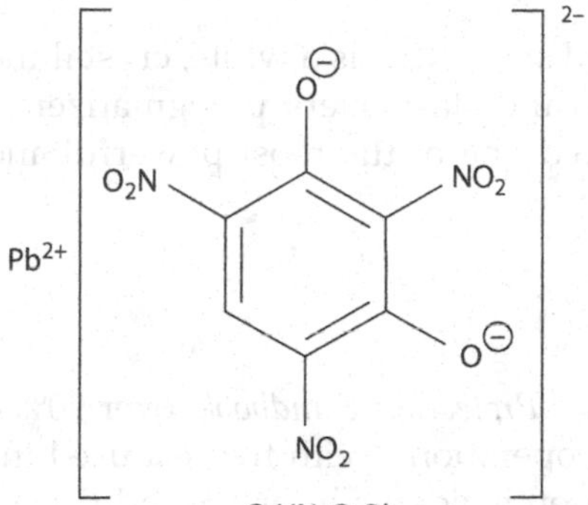

$C_6HN_3O_8Pb$

Figure 3.15 Lead styphnate.

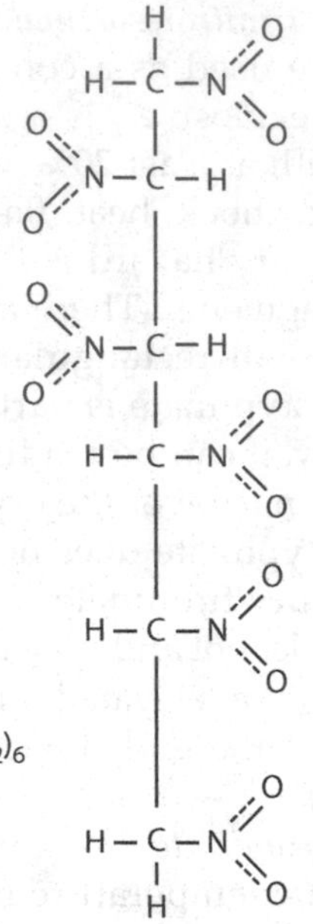

$C_6H_8(ONO_2)_6$

Figure 3.16 Nitromannite.

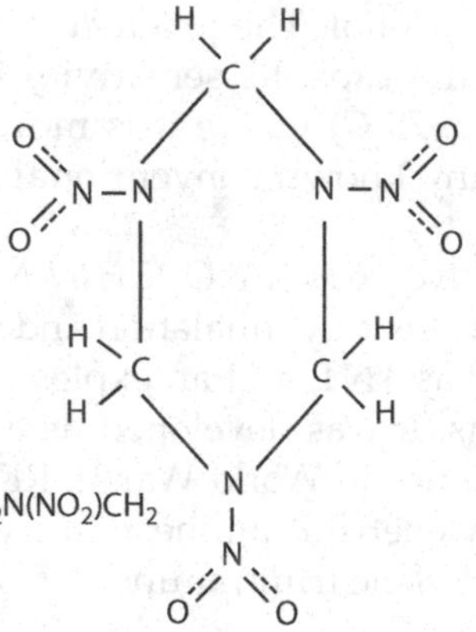

$N(NO_2)CH_2N(NO_2)CH_2N(NO_2)CH_2$

Figure 3.17 RDX, cyclonite.

In its pure, synthesized state, RDX is a white, crystalline solid. It is often used in mixtures with other explosives and plasticizers, phlegmatizers, or desensitizers. RDX is stable in storage and is considered one of the most powerful and brisant of the military high explosives (Figure 3.17).

Summary

According to the NFPA *Fire Protection Handbook*, over 90% of explosives in the industrial world are used in mining operations, with the rest used in construction. When responding to fixed or transportation incidents in or around these types of operations, be on the lookout for explosives. When responding to transportation incidents, always consider the possibility of explosives being present. Fire is the principal cause of accidents involving explosive materials. Look for explosive signs, such as placards and labels. Evacuate the area according to the distances listed in the *Emergency Response Guidebook* orange section. If no other evacuation information is available, a 2000-ft minimum distance should be observed, according to the NFPA *Fire Protection Handbook*. There is one rule of thumb in responding to incidents where explosives are involved: *DO NOT FIGHT FIRES IF THE FIRE HAS REACHED THE EXPLOSIVE CARGO!*

Review questions

3.1 An explosion has two phases, which of the following is not one of those phases?
   A. Negative
   B. Positive
   C. Thermal
   D. All of the above
3.2 List the five types of explosions.
3.3 Name the two types of chemical explosions.
3.4 Name and draw structures and provide molecular formulas as appropriate for the following nitro compounds.

   CH_3NO_2 Nitro propane $C_6H_2(NO_2)_3CH_3$ Tri-nitro phenol

3.5 Name the four components necessary for a chemical explosion to occur.
3.6 A chemical explosion is really nothing more than a rapidly burning fire. Which of the following makes the rapid burning possible?
   A. High-yield explosive material
   B. A chemical oxidizer
   C. The additive "rapid burn"
   D. High-octane fuel
3.7 A detonation occurs at which of the following speeds?
   A. Supersonic
   B. Faster than light
   C. Subsonic
   D. 55 miles per hour
3.8 A deflagration occurs at which of the following speeds?
   A. Supersonic
   B. Faster than light
   C. Subsonic
   D. 85 miles per hour
3.9 Which of the following terms refers to the efficiency of an explosion?
   A. Order
   B. Confinement
   C. Yield
   D. None of the above
3.10 Which of the following terms refers to the speed of an explosion?
   A. Chemical oxidizer
   B. Order
   C. Yield
   D. All of the above

3.11 List the six DOT subclasses of explosive materials permitted in transportation.

3.12 Which of the following is not a Class I Explosive Material?

    A. Dynamite

    B. Picric acid (20% Water)

    C. Ammonium nitrate

    D. Blasting caps

3.13 Of the following blast waves, which is released first during an explosion?

    A. Shock

    B. Fragmentation

    C. Thermal

    D. None of the above

3.14 Blast waves travel away from the center of an explosion in which direction?

    A. Downwind

    B. Only in opposite directions

    C. Equally in all directions

    D. Only upwind

3.15 Which of the following nuclear explosions creates radioactive fallout?

    A. Ground burst

    B. Neutron explosion

    C. Air burst

    D. Electron ionization

chapter four

Compressed gases

DOT Hazard Class 2 is composed of gases that are under pressure. They may be entirely in the gaseous state in the case of compressed gases or they may be liquefied gases such as propane and other liquefied petroleum gases (LPGs). The primary hazard is the pressure, which can cause violent container failure when an accident occurs. If the gas is flammable, poisonous, or causes asphyxiation, the hazard is compounded. Liquefied gas pressure containers can be dangerous under accident conditions; ambient temperature changes, flame impingement, and damage to containers can cause boiling liquid expanding vapor explosions (BLEVEs).

Compressed gas containers may become missiles propelled by high pressure, creating an impact hazard when valves are knocked off (Photo 4.1). Compressed gas containers exposed to fire or radiant heat may rupture and become airborne from rapid pressure release. Class 2 is divided into three subclasses: 2.1 gases are flammable, 2.2 gases are non-flammable, and 2.3 gases are poisonous. Each compressed gas category presents its own special hazards, in addition to the hazard of being under pressure, in a specially designed and regulated pressure container. Container pressures range from 5 psi to as much as 6000 psi (Table 4.1). The higher the pressure, the more substantial the container must be constructed to contain the pressure; the higher the pressure, the greater the danger when the pressure is released or the container fails.

Gases may also have other hazards, such as flammability, toxicity, and reactivity. Gases may be heavier or lighter than air; however, most gases are heavier. Seven lighter-than-air gases along with some common compressed gases are listed in Table 4.2.

Gases may also be liquefied. Liquefaction takes advantage of a gas's ability to be liquefied by pressure or cooling, or a combination of both. Liquefied gases have a large liquid-to-gas expansion ratio. This allows a larger quantity of gas to be shipped as a liquid compared to a gas. It is much more economical to ship a compressed gas as a liquid.

Three terms are important in understanding the process of the liquefaction of gases by pressure: critical point, critical temperature, and critical pressure. Critical point is the point at which a gas will exist as a gas or as a liquid. When a gas is heated to its critical temperature at its critical pressure, it becomes a liquid (Table 4.3). Critical temperature is the maximum temperature at which a liquid (in this case, a liquefied gas) can be heated and still remains a liquid. For example, for butane, the critical temperature is 305°F (151°C). As more heat is added, more of the liquid vaporizes. At the critical temperature, no amount of pressure can keep the liquid from turning into a gas. Critical pressure is the maximum pressure required to liquefy a gas that has been cooled to a temperature below its critical temperature. The critical pressure for butane is 525 psi. In order to liquefy any gas, it must be cooled to below its critical temperature. For example, the critical temperature of butane is 305°F (151°C); at 305°F (151°C), it must be pressurized to 525 psi to become a liquefied compressed gas.

Because liquefied gases have large liquid-to-gas expansion ratios, this also increases the hazard of the gas during an emergency (Figure 4.1). A small amount of a liquid leaking from a container can form a large gas cloud. Larger leaks will produce larger vapor clouds.

Photo 4.1 Compressed gas cylinders contain very high pressure. If a valve is knocked off or they are involved in fire, they can become rockets and injure or kill anyone who gets in their way.

Table 4.1 Container Pressure

Atmospheric pressure	0–5 psi
Low pressure	5–100 psi
High pressure	100–3000 psi
Ultra high pressure	3000–6000 psi

Table 4.2 Common DOT Class II Compressed Gases

Common class 2 gases			
Lighter than air		Heavier than air	
Helium	He	Argon	Ar
Acetylene*	C_2H_2	Propane*	C_3H_8
Hydrogen*	H_2	Butane*	C_4H_{10}
Ammonia*	NH_3	Chlorine	Cl_2
Methane*	CH_4	Phosgene	$COCl_2$
Nitrogen	N_2	Hydrogen sulfide*	H_2S
Ethylene*	C_2H_4	Butadiene*	C_4H_6

*Flammable gases.

Table 4.3 Critical Temperature and Pressure

Gas	Boiling point	Critical temp °F	Critical pressure psi
Ammonia	−28	266	1691
Butane	31	306	555
Carbon dioxide	−110	88	1073
Hydrogen	−422	−390	294
Nitrogen	−320	−231	485
Oxygen	−297	−180	735
Propane	−44	206	617

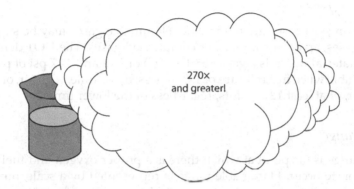

Figure 4.1 Liquid-to-gas expansion ratio.

This increases the danger of flammability if an ignition source is present and of asphyxiation or toxicity when vapor clouds form. Some liquefied gases, such as propane and butane, are ambient temperature liquids. Except for railroad shipments, they are shipped and stored in uninsulated tanks, and the temperature of the liquid inside is close to the ambient air temperature. Railroad tanks are insulated to prevent fire and radiant heat from reaching containers during an accident.

Cryogenic liquefied gases, such as hydrogen, nitrogen, oxygen, and argon are extremely cold liquids. They are not in a DOT hazard class unless they are under pressure. The definition of a cryogenic liquid is any liquid with a boiling point below −130°F. They will be shipped and stored in insulated containers. Sometimes, they are stored under pressure. The liquids inside containers will have temperatures ranging from −130°F (54°C) to −452°F (−268° below zero). Cryogenics are discussed further in "Nonflammable gas compounds" section at the end of this chapter.

When ambient temperatures are cold, the liquefied compressed gases will also be cold. Water from the booster tanks on fire apparatus can be 70°F (21°C) or higher. Thus, water, if applied to a tank surface to cool it during a fire, could actually be heating the liquid in the tank rather than cooling it. Cryogenic liquids are already colder than the water at any temperature, and the water will act as a superheated material, causing the cryogenic to heat up and vaporize faster. This difference in temperatures can cause problems for responders unless the containers are handled properly. Care should be taken when applying water to "cool" containers.

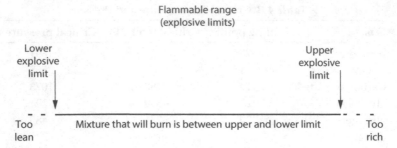

Figure 4.2 Flammable range.

Flammable gases

Class 2.1 compressed gases are flammable. Flammable gases may be shipped and stored as liquefied gases, cryogenic liquids, or compressed gases. The DOT defines a flammable gas as "any material which is a gas at 68°F (20°C) or less and 14.7 psi of pressure or above, which is ignitable when in a mixture of 13% or less, by volume, with air, or has a flammable range with air, of at least 12% points, regardless of the lower limit."

Flammable range

Flammable range is the point at which there is a proper oxygen and fuel mixture present for combustion to occur. Flammable range is represented by a scale, numbered from 0% to 100% (Figure 4.2). Two common terms used to express where the mixture is located within the range are the upper explosive limit (UEL) and lower explosive limit (LEL). When a flammable gas/air mixture is above the UEL, it is considered too rich to burn, which means there is plenty of fuel for combustion to occur, but not enough oxygen. When a flammable gas/air mixture is below its LEL, there is plenty of oxygen for combustion to occur, but not enough fuel. For combustion to take place, a flammable gas and air mixture must be between its UEL and LEL.

Different flammable gases have different flammable ranges (Table 4.4). Most common hydrocarbon fuel gases have ranges from 1% to 13%. Some flammable materials have wide flammable ranges, such as hydrogen and acetylene. The only way to determine if a flammable gas is within its flammable range is to use monitoring instruments. These instruments check for a percentage of the LEL. The rule of thumb, according to the EPA, is when you reach 10% of the LEL, it becomes too dangerous for personnel to proceed any further.

Table 4.4 Flammable Range
of Common Materials

Acetylene	2.5%–80%
Ethyl ether	1.85%–48%
Methyl alcohol	6%–36.5%
Formaldehyde	7%–73%
Propane	2.4%–9.5%
Ammonia	16%–25%
Methane	5%–15%

Vapor density

Another important physical characteristic of gases is vapor density. Vapor density is the relationship between the molecular weight of a gas and that of the air. If you separate component gases in air into nitrogen, oxygen, and argon, and the rare gases helium, neon, krypton, and xenon, you can measure the weight of a molecule of air (Table 4.5). The molecular weight of an average air molecule is 29 atomic mass units (AMUs). Therefore, gases that have a molecular weight of less than 29 AMUs will be lighter than air, and those with a weight greater than 29 will be heavier. For example, the gas methane has one carbon atom, which weighs 12, and 4 hydrogen atoms, which weigh 1 each. Adding them together gives you a molecular weight for methane of 16, which is less than 29, so methane is lighter than air.

When determining vapor pressure from reference books, air is given a value weight of 1 when compared to other gases; any gas that has a vapor density greater than 1 will be heavier than air. Propane has a vapor density of 1.56; therefore, propane is heavier than air. If a gas has a vapor density less than 1, it will be lighter than air. Natural gas (methane) has a vapor density of 0.554; therefore, natural gas is lighter than air. An explosion in a structure that causes damage near the foundation usually indicates a heavier-than-air gas, such as propane. Because propane is heavier than air, it stays low to the ground and goes into basements and confined spaces. A natural gas explosion will cause damage in the upper part of the structure because natural gas is lighter than air and will rise to the upper parts of the structure.

DOT has established criteria for a material to be classified as a flammable gas in transportation. First of all, its LEL must be below 13%. Some materials have wide flammable ranges that make them much more dangerous than materials with narrow ranges. The DOT states that if a material has a flammable range greater than 12% points, regardless of the lower limit, the gas is also classified as flammable. Anhydrous ammonia that is classified by DOT as a nonflammable gas burns under certain conditions (Photo 4.2). It is not considered flammable by DOT because it does not meet their definition.

Propane and butane are two common flammable liquefied compressed gases. Propane has a flammable range of 2.4%–9.5% in air, and butane 1.9%–8.5% in air. Propane and butane have boiling points of –44°F (–42°C) and 31°F (0°C), respectively. Both materials are above their boiling points under ambient temperature conditions in many parts of

Table 4.5 Gas Contents of Air

Contents of air	
Substance	% by volume
Nitrogen, N_2	78.08
Oxygen, O_2	20.95
Argon, Ar	0.93
Carbon dioxide, CO_2	0.033
Neon, Ne	0.0018
Helium, He	0.00052
Methane, CH_4	0.0002
Krypton, Kr	0.00011
Nitrogen (I) oxide, N_2O	0.00005
Hydrogen, H_2	0.00005

Photo 4.2 December 11, 1983, ammonia explosion heavily damaged the Borden's Ice Cream plant at Milam and Calhoun Streets in Houston. (Courtesy Houston, Texas Fire Department.)

the country year-round. This makes the materials highly dangerous when a leak or fire occurs, especially if there is a fire with flame impingement on the container. The vapor density of propane is 1.56 and that of butane is 2.07. Thus, both propane and butane vapors are heavier than air.

Propane and butane are shipped on the highway in MC/DOT 331 uninsulated containers. On the rail, they are shipped in insulated, pressurized tank cars. The railcars are insulated to keep flame impingement away from the tank surface to extend the time it would take for tank failure to occur. Propane and butane are stored in uninsulated bulk pressure containers shaped much like the MC/DOT 331. They may also be found in varying sizes of portable pressure containers. Because the gases are already above their boiling points, flame impingement, radiant heat transfer, or increases in ambient temperature can cause the materials to boil faster. Faster boiling causes an increase in pressure within the container. Even though the containers are specially designed to withstand pressure and have relief valves provided to release excess pressure, there are limits to the pressure they can tolerate. If the pressure buildup in the container exceeds the ability of the tank to hold the pressure or of the relief valve to relieve the pressure, the container will fail, resulting in a BLEVE.

Flammable gas elements

Hydrogen, H_2, is a nonmetallic, diatomic, elemental flammable gas. Hydrogen is one of the most flammable materials known. It burns clean, without smoke or visible flame. The only way to detect a hydrogen fire is from the radiant heat. Hydrogen may be found in transportation or storage as a cryogenic liquid (refrigerated liquid), a compressed gas, or a liquefied compressed gas. The flammable range of hydrogen is wide, at 4%–75%; its boiling point is −423°F (−252°C); and its ignition temperature is 1075°F (579°C). Hydrogen gas is slightly soluble in water and is noncorrosive. It is an asphyxiant gas and can displace the oxygen in the air or in a confined space. There are traces of hydrogen in the atmosphere, and it is abundant in the sun and stars. In fact, it is the most abundant element in the universe. Hydrogen has a vapor density of 0.069; therefore, it is lighter than air. As a cryogenic liquid, hydrogen has a four-digit UN identification number of 1066; as a compressed gas, its four-digit UN identification number is 1049. The NFPA 704 classification for hydrogen is health 3, flammability 4, and reactivity 0. It is usually shipped in insulated containers, insulated MC 338 tank trucks, tank cars, and tube trailers. Hydrogen is used in the production of ammonia, hydrogenation of vegetable oils, fuel for nuclear engines, hydrofining of petroleum, and cryogenic research.

Hydrocarbon families

Three of the hydrocarbon families of hazardous materials contain compounds, which are gases: alkanes, alkenes, and alkynes. Hydrocarbon families are compounds made up of carbon and hydrogen (Photos 4.3–4.5). Carbon has a unique ability to bond with itself almost indefinitely. It forms long-chained compounds, with varying numbers of carbon atoms and hydrogen atoms filling the remaining bonds. Names of the compounds are based upon the number of carbons bonded together in the compound and the type of bond between the carbons. The number of carbons in a compound is indicated in the name of the compound by a specific prefix (Table 4.6).

Photo 4.3 Fixed-facility tube bank with compressed hydrogen gas.

Photo 4.4 Fixed-facility propane tank. Note: condensation on side, indicating liquid level.

Photo 4.5 Highway transportation tube trailer with compressed hydrogen gas.

Table 4.6 Hydrocarbon Prefixes

Meth/form—1 carbon
Eth/acet—2 carbons
Prop—3 carbons
But—4 carbons
Pent—5 carbons
Hex—6 carbons
Hept—7 carbons
Oct—8 carbons
Non—9 carbons
Dec—10 carbons

All three of the families use the same prefixes to indicate the number of carbons. For example, a one-carbon prefix is "meth," two carbons "eth," three carbons "prop," and four carbons "but." The hypothetical number of carbons and prefixes is endless. For the purposes of this book, only the prefixes from 1- to 10-carbon chains will be considered. Those compounds that have more than four carbons are usually liquids and will be discussed in Chapter 5, Flammable Liquids. A system for naming compounds with more than 10 carbons is located in the Appendix, page under the IUPAC System.

Alkanes

Alkanes are the first hydrocarbon family to be presented. The alkane family has only single bonds between the carbons. The smallest alkane contains one carbon. The prefix for a one-carbon compound is "meth." The ending for the alkane family indicating all single bonds is "ane." Therefore, a one-carbon alkane is called methane, with a molecular formula of CH_4. A two-carbon compound has the prefix "eth" and the ending "ane," thus, the compound is called ethane, with a molecular formula of C_2H_6. A three-carbon compound has the prefix "prop" and the ending "ane." The compound is called propane, with a molecular formula of C_3H_8 (Photos 4.4–4.6).

LPGs, which include propane, butane, isobutane propylene, and mixtures, are consistently one of the highest-volume hazardous materials shipped by rail. A four-carbon compound is named butane, with the molecular formula of C_4H_{10}. Structures, molecular formulas, and some physical characteristics of these alkane-compressed gas compounds are shown in Figure 4.3. Notice the differences in boiling points and flammable ranges. As the carbon content increases, the boiling point of the compounds increases. In addition to carbon content, polarity and branching of compounds will affect physical characteristics. This concept will be discussed in more detail in Chapter 5. For now, just be aware that there are relationships between the physical characteristics of flammable gases, which include ignition temperature, heat output, vapor content, and vapor pressure.

Alkane hydrocarbons occur naturally, and their major hazard is flammability. Gases may also act as asphyxiants by displacing oxygen in the air. Alkanes are considered

Photo 4.6 MC331 used to transport liquefied compressed gases.

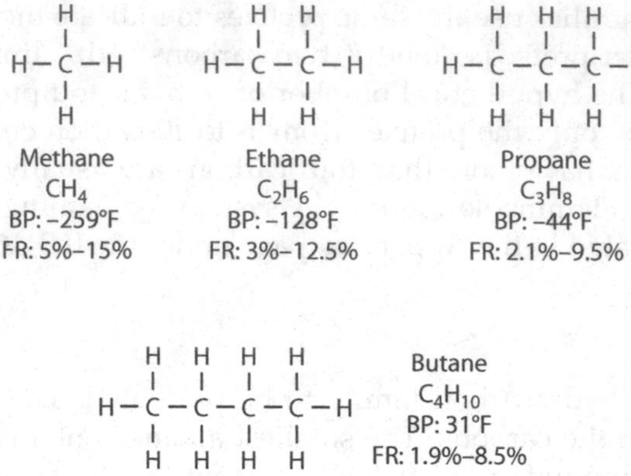

$$H-\overset{\overset{\displaystyle H}{|}}{\underset{\underset{\displaystyle H}{|}}{C}}-H$$

Methane
CH$_4$
BP: −259°F
FR: 5%–15%

Ethane
C$_2$H$_6$
BP: −128°F
FR: 3%–12.5%

Propane
C$_3$H$_8$
BP: −44°F
FR: 2.1%–9.5%

Butane
C$_4$H$_{10}$
BP: 31°F
FR: 1.9%–8.5%

Figure 4.3 Alkane gases.

saturated hydrocarbons because all of the single bonds are full. No other elements can be added to the compound without physically removing one or more of the hydrogen atoms attached to the carbons. When hydrogen atoms are removed, radicals are formed, and these radicals are used to make hydrocarbon derivatives. The type of hydrocarbon family can be estimated from the molecular formula of the compound by looking at the ratio of carbons to hydrogen atoms. In the case of alkanes, there are twice as many hydrogen atoms as carbons, as well as two hydrogen atoms. For example, ethane has two carbons: $2 \times 2 = 4 + 2 = 6$, so there are two carbons and six hydrogen atoms in ethane. Butane has four carbons: $4 \times 2 = 8 + 2 = 10$, so there are four carbons and 10 hydrogen atoms in butane. The molecular formula for butane is C$_4$H$_{10}$.

Isomers

Hydrocarbon structures may be altered so that their physical characteristics make them more economically valuable. One such alteration is called branching or isomers. An *isomer* is a hydrocarbon that has the same molecular formula, i.e., the same number of hydrogen atoms and carbons, but a different structural form. The molecular formula for butane is C$_4$H$_{10}$; the molecular formula for the isomer of butane would be *i*-C$_4$H$_{10}$.

The formula stays the same, but the structure is different. In order to determine the difference, a prefix has to be added to the molecular formula and the name. When a structure is without a branch, it is sometimes referred to as normal, or straight-chained, i.e., all the carbons are connected together, end-to-end, in a chain. In Figure 4.4, in both

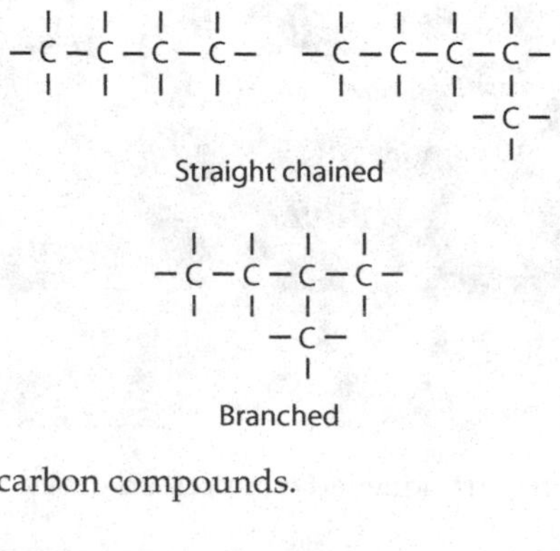

Straight chained

Branched

Figure 4.4 Branching of hydrocarbon compounds.

examples, all of the carbons are connected together end-to-end. In the second example, the end carbon has been placed below the chain, but is still connected end-to-end with the rest of the carbons. So the second example is not a branch or an isomer, even though it might appear to have a "branch." The fact is it is just the same as the first example, except for the arrangement of the carbons in the chain. To have a branched hydrocarbon, the branch cannot be on the end carbons, but must be on the carbons between the ends (see the third example).

The straight-chained hydrocarbon is sometimes referred to as the "normal" configuration. When a material is listed in a reference book, you may see a small "n-" in front of the name or molecular formula; this indicates that it is the "normal," or straight-chained, form of the compound. If the material is the isomer, or branched form, a small "i" will be placed in front of the molecular formula and the prefix "iso" in front of the name. For example, butane can be normal, which would be written as normal butane, or *n*-butane. The branched form of butane would be written isobutane. The molecular formula would be C_4H_{10} or n-C_4H_{10} for the normal, and i-C_4H_{10} for the branched compound (Figure 4.5). Normal is not used all of the time with the normal form of the compound; it is found mostly in reference books and on laboratory containers. If no prefixes appear in the name, it is understood to be the normal form. However, "iso" or the small "i" must be used to designate the branched form of the compound. Without this designation, there is no way to determine the branching from the name or the molecular formula.

The physical effect that branching has on a hydrocarbon is the lowering of the boiling point of that material. For example, butane is a liquefied compressed gas and has a boiling point of 31°F. One of the primary uses of butane is as a fuel for household and industrial purposes. Propane and butane tanks are usually located outdoors, and the liquefied gas takes on whatever the ambient temperature happens to be. In many parts of the country, the ambient temperature is below 31°F (0°C) much of the winter. Because of the low ambient temperatures, butane would not be above its boiling point, and therefore would not be producing enough vapor to be used as a fuel. However, by changing the structure of butane and making it the branched isobutane, the boiling point becomes 10°F (–12°C). Thus, with the lower boiling point, isobutane can be used as a fuel at lower ambient temperatures.

One way of determining if the structure is branched or straight-chained is to try to draw a line through all of the carbons connected together in a chain without lifting the pencil or having to backtrack to reach another carbon (Figure 4.6). In the following examples, the structure on the left is normal, or straight-chained, because the line can be drawn through all of the carbons without backtracking. The structure on the right,

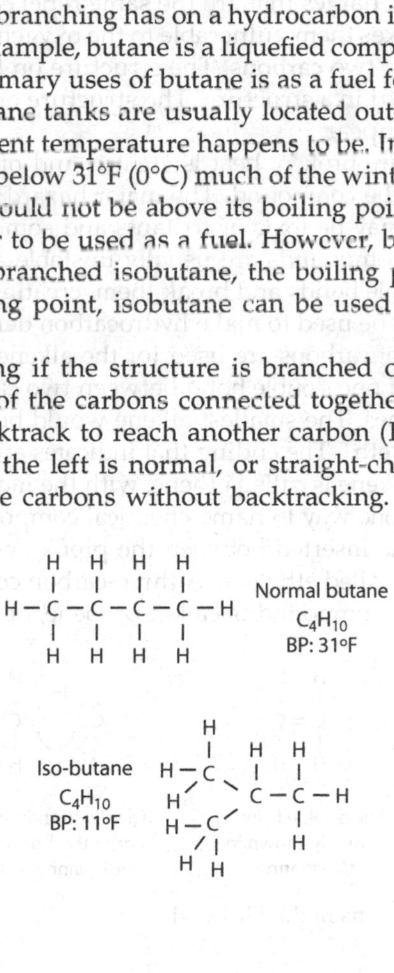

Figure 4.5 Branching butane.

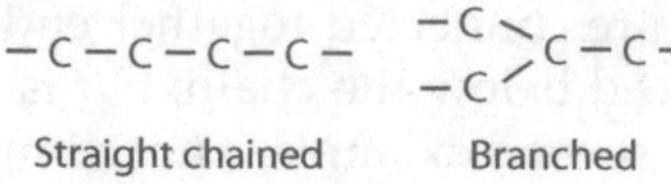

$$-C-C-C-C- \qquad \begin{array}{c} -C \\ {\diagup}{\diagdown} \\ -C \end{array} C-C-$$

Straight chained   Branched

Figure 4.6 Determining a branched structure.

however, requires lifting the pencil or backtracking to draw the line to the branched carbon. Therefore, the compound on the right is the isomer, or branched, compound.

Branching can also occur in the alkene and alkyne families. In order for branching to occur in hydrocarbons, there must be at least a four-carbon compound. Propane cannot be branched until the hydrocarbon derivatives, i.e., elements other than carbon and hydrogen, are added to the structure of the compound. Other types of branching will be discussed in "Hydrocarbon derivatives" section of Chapter 5.

Alkenes

The alkene family has one or more double bonds between the carbons in the chain. Alkene compounds do not occur naturally; they are manmade. They are considered unsaturated because the double bond can be broken by heat or oxygen. The double bond is actually out-of-plane electrons between the carbons. The charges on the four electrons between the carbons are negative. Charges that are the same repel each other. This forces the electrons out of plane and makes them vulnerable to the oxygen from the air. Figure 4.7 shows the double bonds between two carbons. The structure on the left shows how the double bond is usually represented in a structure. The structure on the right shows how the out-of-plane electrons really appear.

When double bonds are broken, heat is created and other elements, including atmospheric oxygen, attach to the compound. The major hazard of the alkenes is flammability. Some of the compounds may be toxic or irritants and some are suspected of being carcinogenic. Double-bonded compounds are usually unstable and reactive. Oxygen from the air can react with the double bonds and break them, creating heat and forming other compounds. Alkenes may also be used to make hydrocarbon derivatives. The same prefixes for determining the number of carbons are used for the alkenes as for the alkanes. However, since there must be at least one double bond between two carbons to have an alkene, there are no single-carbon alkenes. The smallest alkene would be two carbons. The prefix for a two-carbon compound is "eth." The ending that indicates at least one double bond is "ene." Therefore, a two-carbon alkene is called ethene, with the molecular formula of C_2H_4. Many times, there is more than one way to name chemical compounds. In the case of the alkene family, sometimes a "yl" is inserted between the prefix and the ending "ene." Therefore, ethene may sometimes be called ethylene. A three-carbon compound has the prefix "prop" and the ending "ene." The compound is called propene, or propylene, with the molecular

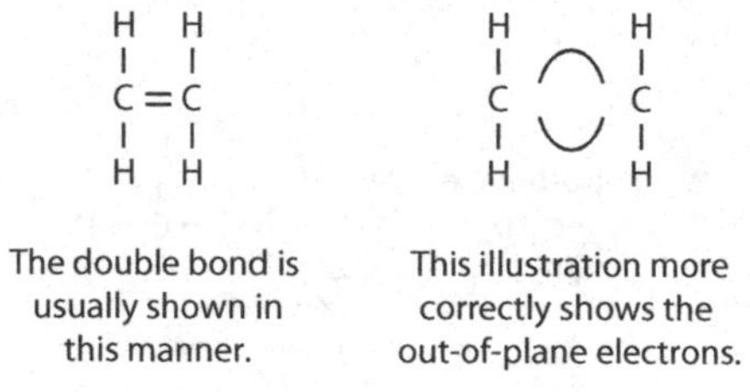

$$\begin{array}{ccc} H & & H \\ | & & | \\ C & = & C \\ | & & | \\ H & & H \end{array} \qquad\qquad \begin{array}{ccc} H & & H \\ | & & | \\ C & & C \\ | & & | \\ H & & H \end{array}$$

The double bond is   This illustration more
usually shown in   correctly shows the
this manner.   out-of-plane electrons.

Figure 4.7 Out-of-plane electrons in double bond.

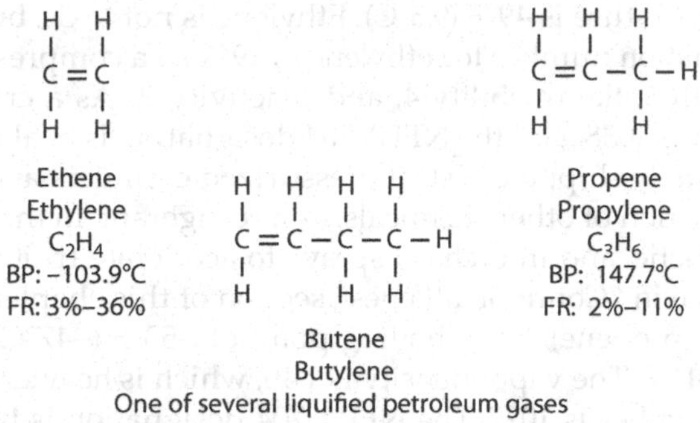

```
    H   H                           H   H   H
    |   |                           |   |   |
    C = C                           C = C - C - H
    |   |                           |       |
    H   H                           H       H
```

Ethene	Propene
Ethylene	Propylene
C_2H_4	C_3H_6
BP: −103.9°C	BP: 147.7°C
FR: 3%–36%	FR: 2%–11%

```
            H   H   H   H
            |   |   |   |
            C = C - C - C - H
            |       |   |
            H       H   H
```

Butene

Butylene

One of several liquified petroleum gases

Figure 4.8 Alkene LPGs.

formula C_3H_6. A four-carbon compound would have the prefix "but" and the ending "ene." The compound is named butene, or butylene, with the molecular formula of C_4H_8. The structures for these alkene compounds are shown in some compounds of alkenes have more than one double bond in the structure (Figure 4.8). Naming of the prefix for the number of carbons is the same as with the other alkenes. The ending "ene" is still used to indicate double bonds in the alkene family. There is, however, a prefix used to indicate the number of double bonds in the compound. The prefix "di" is inserted before the "ene" in the name to indicate two double bonds. For example, butene with two double bonds is called butadiene. The prefix "tri" is inserted to indicate three double bonds. Hexene, with three double bonds, is called hexatriene. There must be at least four carbons before two double bonds can be present.

Two double bonds next to each other are highly unstable and will not hold together long. Examples of two and three double bonds in compounds are shown in Figure 4.9.

When trying to estimate the hydrocarbon family from the molecular formula, there is a ratio that indicates the alkenes. With the alkene family, there are twice as many hydrogen atoms as carbons. Propene has three carbons: $3 \times 2 - 6$, so there are three-carbon atoms and six hydrogen atoms in propene (C_3H_6). In the case of alkenes that have more than one double bond, the ratio does not work. It is better to look at a molecular formula and draw out the structure rather than to try to guess from the ratio of the molecular formula. While the ratios work most of the time, there are exceptions, such as the two and three double-bonded compounds.

Common alkenes

Ethylene, C_2H_4 (ethene), has a boiling point of −155°F (−103°C) and is a dangerous fire and explosion risk. The flammable range is fairly wide, with an LEL of 3% and a UEL of 36%. The vapor density is 0.975, which is slightly lighter than air. The critical pressure is 744 psi,

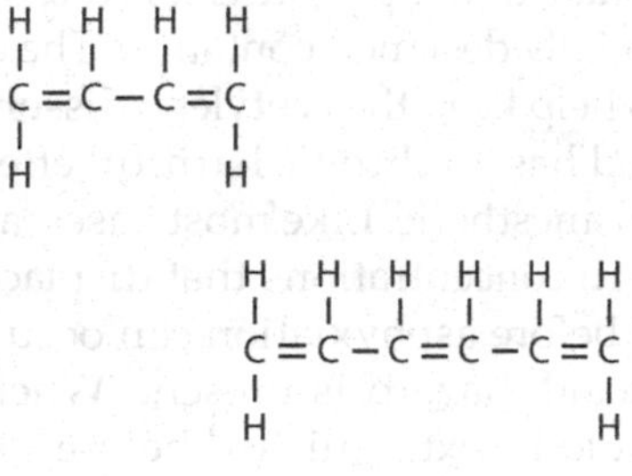

```
        H   H   H   H
        |   |   |   |
        C = C - C = C
        |           |
        H           H
```

```
        H   H   H   H   H   H
        |   |   |   |   |   |
        C = C - C = C - C = C
        |                   |
        H                   H
```

Figure 4.9 Butadiene and hexatriene.

and the critical temperature is 49°F (9.5°C). Ethylene is not toxic, but can be an asphyxiant gas. The UN designation number for ethylene is 1962 as a compressed gas. The NFPA 704 designation is health 3, flammability 4, and reactivity 2. As a cryogenic liquid, the UN designation number is 1038, and the NFPA 704 designation is health 1, flammability 4, and reactivity 2. It is usually shipped in steel pressurized cylinders and tank barges. Ethylene is used in the production of other chemicals, as a refrigerant, in the welding and cutting of metals, as an anesthetic, and in orchard sprays to accelerate fruit ripening. The structure for ethylene is shown in "Common alkenes" section of this chapter.

Propylene, C_3H_6 (propene), has a boiling point of −53°F (−47°C). The flammable range of propylene is 2%–11%. The vapor density is 1.46, which is heavier than air. The four-digit UN identification number is 1077. The NFPA 704 designation is health 1, flammability 4, and reactivity 1. It is not toxic, but can be an asphyxiant gas by displacing the oxygen in the air. It is usually shipped as a pressurized liquid in cylinders, tank cars, and tank barges. The structure for propylene is shown in "Common alkenes" section of this chapter.

Butadiene, C_4H_6, has a boiling point of 24°F (−4°C) and a flammable range of 2%–11%. The vapor density is 1.93 psi, which means it is much heavier than air. It is highly flammable and may polymerize. Butadiene may form explosive peroxides in contact with air. It has a four-digit UN identification number of 1010. The NFPA 704 designation is health 2, flammability 4, and reactivity 2. Butadiene must be inhibited during transportation and storage. It is usually shipped in steel pressurized cylinders, tank cars, and tank barges. The structure is shown in "Common alkenes" section of this chapter.

Alkynes

The alkyne hydrocarbons have at least one triple bond between the carbons in the chain. The ending for the alkyne family is "yne." Alkynes are unsaturated, with the triple bonds being reactive to heat and the oxygen in the air. As with the alkene family, there are no one-carbon alkynes. The most commercially valuable alkyne is the two-carbon compound. The prefix for two carbons is "eth," so the chemical name for the two-carbon, triple-bond compound is ethyne, with the molecular formula of C_2H_2. This is probably the only alkyne that emergency responders will ever encounter. Ethyne, however, is known by the commercial name *acetylene*. This is a trade name, so it is not derived from any of the naming rules of the hydrocarbon families. Acetylene is a highly flammable colorless gas with a flammable range of 2.5%–80%. Pure acetylene is odorless; however, the ordinary commercial purity has a distinct garlic-like odor. Acetylene can be liquefied and solidified; however, both forms are highly unstable. The vapor density is 0.91, so it is slightly lighter than air. Acetylene is produced when water is reacted with calcium carbide and other binary carbide salts. When these salts come in contact with water, acetylene gas is released. Acetylene is unstable and burns rich, with a smoky flame. The material may burn within its container. In fact, acetylene is so unstable that it can detonate under pressure. It is dissolved in a solvent, such as acetone, to keep it stable within a specially designed container. The container has a honeycomb mesh of ceramic material inside to help keep the acetylene dissolved in the acetone.

Acetylene is nontoxic and has no chronic harmful effects, even in high concentrations. In fact, it has been used as an anesthetic. Like most gases, acetylene can be a simple asphyxiant if present in high-enough concentrations that displace the oxygen in the air. The LEL of acetylene is reached well before asphyxiation can occur, and the danger of explosion is reached before any other health hazard is present. When fighting fires involving acetylene containers, the fire should be extinguished before closing the valve to the container. This is because the acetylene has such a wide flammable range that it can burn inside

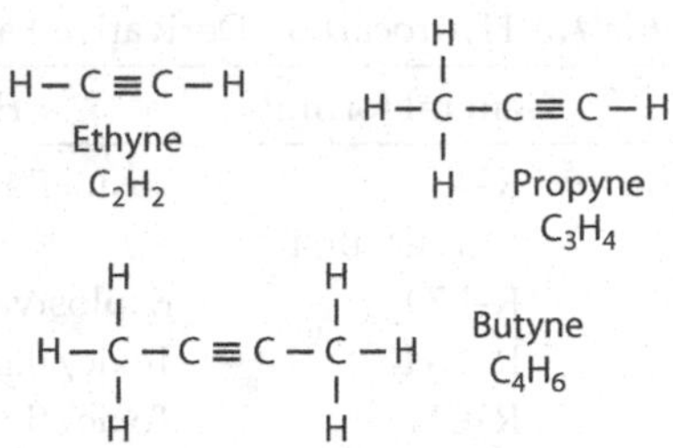

Figure 4.10 Alkyne gases.

the container. Acetylene is incompatible with bromine, chlorine, fluorine, copper, silver, mercury, and their compounds. Acetylene has a four-digit UN identification number of 1001. The NFPA 704 designation is health 1, flammability 4, and reactivity 3. Reactivity is reduced to 2 when the acetylene is dissolved in acetone.

There are other alkyne compounds, but they do not have much commercial value and will not be commonly encountered. A three-carbon compound with one triple bond has "prop" as a prefix for three carbons and is called propyne, with a molecular formula of C_3H_4. It is listed in the *Condensed Chemical Dictionary* as propyne, but you are referred to *methylacetylene* for information. It is listed as a dangerous fire risk, and it is toxic by inhalation. Propyne is used as a specialty fuel and as a chemical intermediate. A four-carbon alkyne has the prefix "but," and the compound is called butyne, with the molecular formula of C_4H_6. The chemical listing is under the name *ethylacetylene*, and it is designated as a dangerous fire risk. It is also used as a specialty fuel and as a chemical intermediate. The structures for ethyne, propyne, and butyne are shown in Figure 4.10.

While there are no commercially valuable two or three triple-bonded compounds, the same rules for naming them would apply as in the alkenes: the prefixes "di" for two and "tri" for three would be used. There is a ratio of carbon atoms to hydrogen atoms that can be used to identify the compound from the molecular formula. With the alkyne family, there are twice as many hydrogen atoms as carbons −2. Ethyne has two carbons: $2 \times 2 = 4 - 2 = 2$. So there are two carbon atoms and two hydrogen atoms in the compound ethyne, with a molecular formula of C_2H_2.

Hydrocarbon derivatives

Hydrocarbon-derivative compounds do not occur naturally (Table 4.7). They are manmade from hydrocarbon compounds, as discussed earlier, with some additional elements added. Hydrocarbon derivatives belong to families just as the hydrocarbons. In order to make hydrocarbon derivatives, hydrogen needs to be removed from the alkane family. Alkene hydrocarbons have one or more double bonds that can be broken and other elements added. They may or may not have hydrogen removed. Elements commonly added to hydrocarbon compounds to create hydrocarbon derivatives include oxygen, nitrogen, fluorine, chlorine, bromine, and iodine. Together with the hydrocarbons, these elements make up over 50% of all hazardous materials.

There are some hydrocarbon-derivative functional groups that have flammable gas compounds in their families. Alkyl halides are listed with toxicity as a primary hazard. However, there are some flammable alkyl halides. Vinyl chloride and methyl chloride are alkyl halides. Vinyl chloride and methyl chloride are extremely flammable gases. The amines are also primarily toxic as a group; there are, however, some flammable amine

Table 4.7 Hydrocarbon Derivative Families

Family	General formula	Hazard
Alkyl halide	R–X	Toxic/flammable
	X=F, Cl, Br, I	
Nitro	R–NO$_2$	Explosive
Amine	R–NH$_2$	Toxic/flammable
Nitrile	R–CN	Toxic/flammable
Isocyanate	R–NCO	Toxic/flammable
Sulfide	R–S–R	Toxic/flammable
Thiol	R–SH	Toxic/flammable
Ether	R–O–R	Flammable, WFR
Peroxide	R–O–O–R	Oxidizer, explosive
Alcohol	R–OH	Flammable/toxic, WFR
Ketone	R–CO–R	Flammable/toxic
Aldehyde	R–CHO	Flammable/toxic, WFR
Ester	R–COOH	Corrosive, flammable

gases. Methylamine, dimethylamine, ethylamine, and propylamine are all flammable gases. There are a few ether flammable gases, most of which do not use the trivial naming system and may not be recognized as ethers. For example, propylene oxide is an ether compound. Methyl ether may be found as a compressed gas or a liquid. In the aldehyde family, most compounds are liquids, except for the one-carbon aldehyde, formaldehyde. Hydrocarbon-derivative functional groups will be discussed in detail in Chapter 5, Flammable Liquids.

Methyl chloride, CH$_3$Cl, is an alkyl-halide hydrocarbon derivative. It is a colorless compressed gas or liquid with a faintly sweet ether-like odor. It is a dangerous fire risk, with a flammable range of 10.7%–17% in air. The critical temperature is approximately 225°F (107°C), and the critical pressure is 970 psi. It is slightly soluble in water. The vapor density is 1.8, which is heavier than air. The boiling point is −11°F (−23°C), and the flash point is 32°F (0°C). The ignition temperature is 1170°F (632°C). It is a narcotic, producing psychogenic effects. The TLV is 50 ppm in air. The four-digit UN identification number is 1063. The NFPA 704 designation is health 1, flammability 4, and reactivity 0. The primary uses are as a catalyst in low-temperature polymerization, as a refrigerant, as a low-temperature solvent, as an herbicide, and as a topical anesthetic. The structure for methyl chloride is shown in Figure 4.11.

Dimethylamine, (CH$_3$)$_2$NH, an amine hydrocarbon derivative, is a gas with an ammonia-like odor. It is a dangerous fire risk, with a flammable range of 2.8%–14% in air. It is insoluble in water. The vapor density is 1.55, which is heavier than air. The boiling point is 44°F (6°C), and the ignition temperature is 806°F (430°C). Dimethylamine is an irritant, with a TLV of 10 ppm in air. The four-digit UN identification number is 1032. The NFPA 704 designation is health 3, flammability 4, and reactivity 0. The primary uses are in electroplating

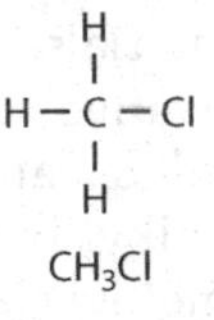

$$H-\overset{\overset{\displaystyle H}{|}}{\underset{\underset{\displaystyle H}{|}}{C}}-Cl$$

CH$_3$Cl

Figure 4.11 Methyl chloride.

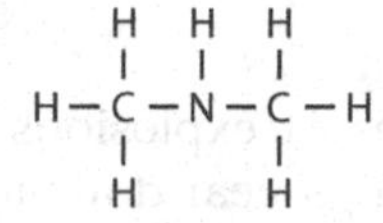

$$
\begin{array}{ccccc}
\text{H} & & \text{H} & & \text{H} \\
| & & | & & | \\
\text{H}-\text{C} & - & \text{N} & - & \text{C}-\text{H} \\
| & & | & & | \\
\text{H} & & & & \text{H} \\
\end{array}
$$

(CH₃)₂NH

Figure 4.12 Dimethylamine.

and as gasoline stabilizers, pharmaceuticals, missile fuels, pesticides, and rocket propellants. The structure for dimethylamine is shown in Figure 4.12.

Common hydrocarbon derivatives

Vinyl chloride, C_2H_3Cl, is a compressed gas that is easily liquefied. Vinyl chloride is the most important vinyl monomer and may polymerize if exposed to heat. It has an ether-like odor, and phenol is added as an inhibitor during shipment and storage. It is highly flammable, with a flash point of –108°F (–77°C) and a boiling point of 7°F (–13°C). The flammable range is 3.6%–33% in air, with an ignition temperature of 882°F (472°C). Vinyl chloride is insoluble in water and has a specific gravity of 0.91, which is lighter than water. The vapor density is 2.16, which is heavier than air. It is toxic by inhalation, ingestion, and skin absorption. Vinyl chloride is a known human carcinogen. The TLV is 5 ppm in air. The four-digit UN identification number is 1086. The NFPA 704 designation is health 2, flammability 4, and reactivity 2; uninhibited, the values would be higher for reactivity. The primary uses are in making polyvinyl chloride and as an additive in plastics. The structure for vinyl chloride is shown in Figure 4.13.

Formaldehyde, HCHO, is an aldehyde hydrocarbon derivative. Formaldehyde is a gas with a strong, pungent odor, and it readily polymerizes. Commercially, it is offered as a 37%–50% solution, which may contain up to 15% methanol to inhibit polymerization. Boiling points for solutions range from 206° (96°C) to 212°F (100°C). Commercial solutions have the trade name *formalin.* Formaldehyde is flammable, with a wide flammable range of 7%–73% in air. The boiling point is –3°F (–12°C) and the flash point is 185°F (85°C). The ignition temperature for the gas is 572°F (300°C). It is water-soluble. The vapor density is 1, which is the same as air. Formaldehyde is also toxic by inhalation, a strong irritant, and a carcinogen. The TLV is 1 ppm in air. Nonflammable solutions are Class 9 Miscellaneous Hazardous Materials, with a four-digit UN identification number of 2209. The NFPA 704 designation is health 3, flammability 2, and reactivity 0. Flammable solutions are Class 3 Flammable Liquids, with a four-digit UN identification number of 1198. The NFPA 704 designation is health 3, flammability 4, and reactivity 0. The structure for formaldehyde is shown in Figure 4.14.

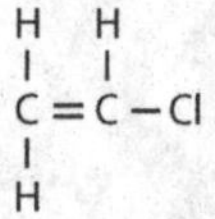

$$
\begin{array}{ccc}
\text{H} & & \text{H} \\
| & & | \\
\text{C} & = & \text{C}-\text{Cl} \\
| & & \\
\text{H} & & \\
\end{array}
$$

C₂H₃Cl

Figure 4.13 Vinyl chloride.

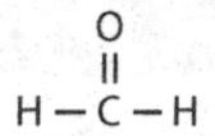

$$
\begin{array}{c}
\text{O} \\
|| \\
\text{H}-\text{C}-\text{H} \\
\end{array}
$$

HCHO

Figure 4.14 Formaldehyde.

Incidents

Dallas, Texas, July 25, 2007, a series of explosions at a gas facility sent flaming debris raining onto highways and buildings near downtown and injured at least three people (Photo 4.7). Authorities evacuated a half-mile area surrounding the Southwest Industrial Gases, Inc. facility and shut down parts nearby Interstates 30 and 35 as the explosions continued for more than half an hour. Because of the very high danger factor involved with this fire, cylinders of acetylene gas were exploding and sending projectiles as far as 1/4 mile away. The fire started as an industrial accident at around 9:15 a.m. when some cylinders were being filled and a pigtail failed that led to the massive fire, which shot fireballs as high as 200 ft and sent debris flying over 500 ft in the air and up to 1/4 mile away with large pieces being blown across two nearby Interstate Highways, which were soon closed for over 11 h. This industrial gas supplier had close to 1000 filled canisters in their work yard, which made it very dangerous to enter since many of them had safety valves that had opened, which started venting the flammable gases that ignited; the valves opened because of the extreme heat that was enveloping the canisters. There were just three injuries that required hospitalization and no deaths from this accident. The incident began around 9:30 a.m. when a malfunctioning connector was used to join acetylene tanks during the filling process. He said the three people injured included the manager and a worker at the facility, and a truck driver. Environmental Protection Agency emergency responders were on the scene, and the U.S. Chemical Safety Board was sending a team of investigators.

There have been numerous incidents over the years involving pressurized containers and flammable gases. Two of the most recent incidents have resulted in the deaths of four firefighters, two in Carthage, Illinois, and two in Albert City, Iowa (Photo 4.8). Both incidents were similar in that they involved storage tanks on farms, one a commercial farm, and the other a private farm. Because the Albert City explosion involved a commercial facility, the incident was investigated by the U.S. Chemical Safety and Hazard

Photo 4.7 A massive fire that shot fireballs as high as 200 ft and sent debris flying over 500 ft in the air and up to 1/4 mile away with large pieces being blown across two nearby Interstate Highways, which were soon closed for over 11 h. (From Dallas news photographer Michael Mulvey.)

Photo 4.8 Albert City, Iowa, propane explosion killed two firefighters; tank was blown into a chicken house.

Investigation Board. During the Albert City incident, the tank was engulfed in flames due to leaking propane underneath the tank. It was this flame impingement that resulted in the BLEVE that killed the firefighters. According to the investigative report, the firefighters were positioned too close to the burning propane tank when it exploded. They were under the impression they would be protected during an explosion if they stayed away from the ends of the tank. Three major factors resulting in the deaths were noted by the safety board's investigation:

1. Protection of aboveground piping was inadequate.
2. The diameter of the pipe downstream from an excess-flow valve was too narrow, which prevented the valve from functioning properly.
3. Firefighter training for responding to BLEVEs was inadequate.

In *Kingman, Arizona, on July 5, 1973, at 1:30 p.m.*, a 33,500-gal railroad tank car containing LPG was being off-loaded at a gas distribution plant. As the liquid lines were attached to the tank, a leak was detected. During attempts by workers to tighten the fittings to stop the leak, a fire occurred. Both workers were severely burned, and one later died from injuries. The Kingman Fire Department received the alarm at 1:57 p.m. In July 1973, the Kingman Fire Department was a combination force of six career firefighters and 36 volunteers operating out of two stations. One career member was on duty in each station at all times. Kingman's equipment in service at the time of the explosion included four engines and a rescue vehicle. Station 2 was located just a half mile west of the Doxol Gas Distribution Plant, the site of the explosion.

Shortly after the fire department arrived, the liquid line failed and flame impingement began on the vapor space of the tank car (Photo 4.9). Nineteen minutes after the flames began contacting the surface of the tank car, a BLEVE occurred at 2:10 p.m. Eleven Kingman firefighters, two career and nine volunteers, died as a result of burns from the explosion. Three were killed instantly and eight more passed away over the next week. The fireball and radiant heat set five buildings on fire, including a tire company, restaurant, truck stop, and the gas company office building and started several brushfires. A large section of the 20-ton tank was propelled 1200 ft by the explosion. More than $1 million in property damage was reported (Photo 4.10). Railroad conductor Hank Graham, who took many of

Photo 4.9 Kingman, Arizona, railcar fire just prior to explosion. Note: flame impingement on vapor space of tank. (Courtesy of Hank Graham.)

Photo 4.10 Kingman, Arizona, propane explosion. Note: pieces of tank car propelled by the explosion. (Courtesy of Hank Graham.)

the now-famous Kingman photographs, had the hair burned off his arms as he took pictures. Eleven Kingman firefighters were killed in the explosion. Only one firefighter from Kingman, who was close to the scene, survived. Ninety-five spectators on nearby Highway 66 were injured by the blast. The explosion set fire to lumber stored nearby, a tire company, and other businesses within 900 ft of the blast. In spite of the terrible loss of life, *no other hazmat incident has occurred in the United States that has had more of a positive impact on the fire service than the Kingman incident.* Many changes in procedures and regulations occurred across the fire service as a result of this explosion. We owe uncounted saved lives to those brave men who gave theirs in Kingman that fateful day in July 1973.

Firefighters who died in the Kingman propane explosion (Photos 4.11 and 4.12)
Donald G. Webb, 30
Arthur A. Stringer, 25
Frank (Butch) Henry, 28
Christopher G. Sanders, 38
Alan H. Hansen, 34
John O. Campbell, 42

Photo 4.11 Memorial in Firefighters Park in Kingman, Arizona, dedicated to the 11 firefighters who lost their lives fighting the propane tank fire. (Courtesy of Kingman City Fire Department, Kingman, AZ.)

Photo 4.12 Bricks are being sold to raise funds for the memorial. This brick was provided to the author as a gift and the message is in sincere appreciation of those firefighters who lost their lives and all of the assistance given to the author by the Kingman Fire Department. (Courtesy of Kingman City Fire Department, Kingman, AZ.)

Joseph M. Chambers, 37
M.B. (Jimmy) Cox, 55
William L. Casson, 52
Roger A. Hubka, 27
Richard Lee Williams, 47

A quite different type of incident occurred in *Waverly, Tennessee*, February 24, 1978, involving two 28,000-gal propane tank cars. At approximately 10:30 p.m., a Louisville and Northern (L&N) train heading from Nashville to Memphis derailed in this community of 6000 residents. Investigators determined that a wheel on a gondola car, overheated from a hand brake left in the applied position, broke apart East of Waverly.

A wheel truck damaged by the breaking wheel managed to remain with the train for 7 miles before it finally came loose from the car causing the derailment. Twenty-four of the 92 cars of the train left the tracks in the center of downtown Waverly. Two of the derailed tank cars, which contained liquefied propane gas, played a major role in the incident that unfolded over several days. The derailment created a mass of piled railcars; a portion of tank #83013 had been damaged and weakened by the derailment. Initially, 50–100 people were evacuated within one-quarter mile of the accident site as a precaution, although no leaks, fires, or explosions occurred when the train derailed. There were approximately 1/2 in. of snow on the ground; it was cloudy, and the temperature had been in the 1920s for several days.

The task of cleaning up the crash site began about mid-day on Wednesday (Photo 4.13). By Friday, the situation was felt to be well under control, security was relaxed, and people were allowed to return to their homes. Spectators and nonessential personnel were allowed into the immediate area of the derailment. Workmen smoked and used acetylene cutting torches at will in the process of cleaning up the site. A tank truck was brought in to off-load the damaged propane tank cars. Friday afternoon, the temperature began to rise, causing an increase in pressure inside the damaged tank. Normally, the pressure-relief valve would function to release pressure created by increases in ambient temperature.

Photo 4.13 Waverly, Tennessee, propane tank car UTLX 83013 before BLEVE. (Courtesy of Waverly Fire Department.)

Photo 4.14 Waverly, Tennessee, propane tank car UTLX 83013 following BLEVE. (Courtesy of Waverly Fire Department.)

However, in this instance, the damaged portion of the tank was weaker than the relief-valve pressure setting. At approximately 3:00 p.m., the tank could no longer withstand the increased pressure and opened up, releasing into the environment liquid propane, which instantly vaporized. The vapors quickly found an ignition source, and flames shot 1000 ft into the air.

The giant fireball was visible for 30 miles around Waverly. The resulting explosion and fire killed 16 people, including the Waverly fire and police chiefs and 5 firefighters (Photo 4.14). Fifty-four people were injured, many severely burned. The incident in Waverly was the high-water mark of hazardous materials incidents in the United States. In terms of loss of life to emergency responders and citizens resulting from train derailments, this incident was the last in which so many people were to die. The Waverly incident also resulted in many changes in both tactics for dealing with LPG fires in containers and safety equipment on rail tank cars. It is easy to look at incidents like Waverly and play "Monday morning quarterback." But the fact remains that hazardous materials including LPGs are dangerous if not handled properly. We need to study past incidents and identify the lessons that can be learned. Incident commanders (ICs), fire officers, and firefighters need to be aware of the dangers of propane and other hazardous materials and make sure they conduct a risk benefit analysis before determining tactical options that commit response personnel to harm's way. Let us take steps to help ensure that those that lost their lives in Waverly and other hazardous materials incidents did not do so in vane. Let us not be so slow to respond to the lessons learned from previous incidents.

Firefighters who were killed in the propane explosion (Photo 4.15)
Wilber J. York, 65 Waverly Fire Chief
James E. Ham, 58
Melvin B. Holcombe, 43
Tommy Hornburger, 19
Terry L. Hamm, 20
Guy O. Barnett, 45 Chief of Police

Photo 4.15 Memorial to those who lost their lives in the propane explosion in Waverly, Tennessee, including four firefighters, Waverly's fire chief and police chief.

Crescent City, Illinois, June 21, 1970, was the site of yet another accident involving propane in rail transportation. Sixteen cars of an eastbound Toledo, Peoria, and Western Railroad Company's Train No. 20 derailed in the center of town at approximately 6:40 a.m., including 10 cars each containing 34,000 gal of liquid propane. Two additional propane tanks remained on the tracks. During the derailment, one of the propane tank cars was punctured by the coupler of another car, causing a leak that ignited almost immediately. Flames reached several hundred feet into the air. A nearby house and business were set on fire from the radiant heat, injuring several residents. Relief valves on the other tank cars began to open as the pressure built up from the surrounding fires. Chief Carlson and the 20-man Crescent City volunteer fire department (two were out of town) responded quickly with their two pieces of fire apparatus 1956 and 1961 International Harvester front mounted 500 gpm pumpers (still in service today as brushfire units). As Crescent City firefighters arrived, they tried to contain the fire that was burning intensely around the railcars. Initial firefighting efforts were hampered by a lack of electricity that was knocked out by the derailment, which prevented the city's water pumps from functioning. Firefighters were able to take water directly from the city water tower to fight the fire until help arrived from other communities. Water was hauled by privately owned tractor trailers from surrounding towns, and other fire departments also responded with tankers. Initial calls went out to several area fire departments for assistance, and many other departments responded to the scene on their own. Ultimately, fire companies from 33 surrounding towns (some from as far away as Indiana) appeared with 58 pieces of equipment and 250 firefighters. An Illinois State Police Sergeant located at Watseka, Illinois, about 6.3 miles East of Crescent City was notified of the derailment shortly after it happened and proceeded immediately to the scene. He arrived at approximately 6:45 a.m. and sized up the situation. When he determined that a tank car was being heated by the fire and contained propane, he notified police officers in the area to evacuate the town and warned firefighters to move back to a safer location to fight the fires. His actions may very well have prevented serious injury and loss of life of firefighters, police officers, and residents of the community when the propane tank cars started exploding.

Photo 4.16 Fire companies from 33 surrounding towns (some from as far away as Indiana) appeared with 58 pieces of equipment and 250 firefighters responded to calls for help in Crescent City, Illinois, as burning tank cars of propane exploded following a derailment in the center of town. (Courtesy of Irma Hill.)

The first explosion (BLEVE) occurred around 7:33 a.m., almost 1 h after the derailment. In that first blast, several firefighters and bystanders were injured, and some fire equipment was damaged (Photo 4.16). Additional explosions occurred at 9:20, 9:30, 9:45, 9:55, and 10:10 a.m. Parts of tank cars were propelled all over town, setting fires and damaging structures. However, there were no injuries to civilians from the explosions, because of a quick evacuation after the derailment; 66 firefighters, police officers, and press personnel were injured by the explosions, 11 required hospitalization, but there were no fatalities. Most of the responding firefighters had little training in dealing with propane fires. Some of the injured firefighters were not wearing their personal protective clothing. Others sustained burns to hands and heads when helmets were blown off by the force of the explosions. Firefighters at the time had no hand protection, and their plastic helmets did not have ear protection or chin straps. Twenty-four living quarters were destroyed by fire and three homes destroyed by "flying" tank cars; numerous other homes received damage. Eighteen businesses were destroyed. Because of the dangers of the burning propane and tanks being heated by the fires, the town of Crescent City was evacuated and people remained out of their homes and businesses for 36 h. The remaining propane tanks were allowed to burn, which took some 56 h after the derailment. According to the National Transportation Safety board, overheating caused the breaking of the L-4 Journal of the 20th car in the train. The exact cause of the overheating was not determined, but a motorist spotted smoke coming from one of the train cars as the train was approximately ten miles west of Crescent City.

Every 5 years on the anniversary of the 1970 disaster, the community of Crescent City holds a "Fireball Festival" to commemorate and remember the good fortune that no lives were lost and there were no serious injuries during the train derailment and exploding propane tank cars in 1970. While there were numerous injuries, the town was fortunate that no one lost their life in the disaster. During the festival, numerous events are staged including games, races, variety show, basketball tournament, softball tournament, craft show and flea market, and a grand parade. Additional information is available at the "Fireball Festival" on Facebook at http://www.Facebook.com/Pages/Crescent-City-fireball-Festival/257325402764.

The city of Weyauwega, Wisconsin, population 1700, was evacuated for 20 days following a train derailment in the middle of town on March 4, 1996, at 5:55 a.m. (Photo 4.17). Thirty-four cars were derailed, including seven containing LPG, seven containing propane, and two containing sodium hydroxide. The resulting fires from leaking propane damaged a feed mill and a storage building. There were no explosions as a result of the derailment and resulting fires. The IC, Jim Baehnman, assistant chief of the Weyauwega Fire Department, said, "The tone of the response from the beginning was not time-driven, but rather safety-driven." He also stated that the incident in Waverly, Tennessee, was considered in planning tactics to deal with this incident. These factors may have very well accounted for the fact that not a single death or injury occurred as a direct result of this incident.

When dealing with emergencies involving pressurized containers and flammable gases, great caution should be taken. Flame impingement on the vapor space of a container is a "no-win" situation (Figure 4.15). If a BLEVE is going to occur, it is just a matter of time. To try to fight a fire under those conditions is to play Russian roulette. The NFPA *Fire Protection Guide* says that BLEVE times range from 8 to 30 min, with the average time being 15 min. There is usually no way to know how long the flame impingement has been going on prior to the fire department arrival and no way to know exactly when the BLEVE will occur. If the only threat is to the emergency responders, there is little reason to risk their lives needlessly. If the impingement is on the liquid space, the liquid will absorb the heat for a period of time and will boil faster as it does. There will be an increase in pressure within the tank as the liquid boils faster. This can still be a dangerous situation if not handled properly. Conditions involving the tank must be monitored constantly for changes, including liquid level, pressure increases, and signs of tank failure. Precautions should be taken

Photo 4.17 Weyauwega, Wisconsin, train derailment burning feed mill from propane tank car fires. (Photograph by Robert Ehrenberg of the Weyauwega Fire Department. Used with permission.)

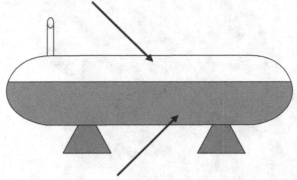

Flame impingement on vapor space will cause metal failure

The heat from flame impingement on the liquid level is absorbed
by the liquid causing increased vapor production but no metal failure

Figure 4.15 Flame impingement propane tanks.

to ensure personnel safety when fighting fires involving flammable gases. Flammable gas fires should not be extinguished until the source of the gas has been shut off.

It is much safer to have the gas on fire and know where it is than to have the gas leaking and going where it wants to go. Flammable gases are more dangerous than flammable liquids. Emergency responders should do a risk/benefit analysis before determining tactics for dealing with flammable gas emergencies. This analysis should be based upon the physical and chemical characteristics of the gas, the container characteristics, ambient conditions, and the life-safety factor of the public and the firefighters. If the risk/benefit analysis determines that the only lives in danger are those of the firefighters, then they should initiate unmanned hose streams and withdraw to a safe location and let the incident take its course. Property can be replaced; firefighter lives cannot. Lessons learned from the previously mentioned incidents should be used to develop standard operating procedures.

Houston experienced a tank car incident of their own on Mykawa Road, following a derailment on October 19, 1971, at approximately 1:15 p.m. (Photo 4.18). Sixteen cars of an 82 car Missouri Pacific train derailed. Hazardous materials on the train included six cars total of vinyl chloride, one each of acetone, caustic soda, formaldehyde, plasticine, and butadiene. Two of the derailed cars were involved in fire, a vinyl chloride car and the butadiene car. The butadiene tank car BLEVEed approximately 40 min following the derailment and 20 min after the arrival of the Houston Fire Department, which resulted in one firefighter fatality and 37 firefighter injuries. The primary reason so many firefighters were hurt was that fire officials did not have information on the chemicals in the railcars. They did not fully understand the potential danger of the burning chemicals and tank cars. Firefighter Truxton Hathaway was assigned to the Fire Department Training Academy. When the explosion occurred, Hathaway took a camera to the scene to record the incident for future training purposes. He arrived on the scene and began recording the incident when the second tank car exploded, sending a wall of fire over Truxton and killing him instantly.

Water supply was an initial problem with the nearest hydrant ¼ mile away from the derailment site. Two alarms of fire equipment responded with seven additional engines requested to supply water to the scene pumping in tandem from the nearest hydrant. The butadiene tank was 100 ft long and overturned during the derailment. This placed the tank car pressure-relief valve in the ground and covered it with liquid that prevented the relief valve from functioning properly. As a result of the explosion, parts of the tank car traveled

Photo 4.18 Burning butadiene tank car on Mykawa Road in Houston, Texas, October 19, 1971, prior to BLEVE of the tank car. Firefighter Andy Nelson can be seen climbing the aerial ladder to flow water onto the fire from the ladder pipe. (Courtesy of Houston, Texas fire Department.)

400 ft from its original location. Video of this incident has been circulating through the fire service for years showing a firefighter on an aerial ladder engulfed when the explosion takes place. That firefighter, Andy Nelson, was burned, but survived. District Chief V.E. Rogers, who was today's equivalent of the IC, was burned over 50% of his body and spent 3 months in the hospital. It was thought that the Mykawa incident was what motivated Chief V.E. Rogers to form a hazardous materials team in Houston. However, the team was formed as a result of Chief Rogers attending a chief's conference where a presentation was made by Ron Gore about the new Jacksonville, Florida Hazardous Materials Response Team. Chief Rogers returned home and directed District Chief Max H. McRae to organize Houston's team.

Houston firefighter killed at Mykawa Road
Firefighter Truxton Hathaway

Nonflammable compressed gases

Class 2.2 includes gases that are nonflammable and nonpoisonous. Gases in this subclass can be compressed, liquefied, cryogenic, or gases in solution. Although these gases are nonflammable, the containers can still BLEVE, under flame impingement conditions on the tank or from a fire involving other materials. BLEVE can also occur from a damaged or weakened container or from an overpressure of the container caused by overfilling. Increases in pressure caused by increases in ambient temperature can also cause container failure.

The DOT definition of a nonflammable gas is "a material that exerts in the packaging an absolute pressure of 41 psi or greater at 20°C, and does not meet the definitions of Division 2.1 or 2.3." If the pressure in a container is less than 41 psi, a gas does not belong in this category. Many cryogenic materials are shipped at atmospheric pressure and so are not considered compressed gases. Cryogenics are not considered a DOT hazard class as a group except when under pressure or regulated by some other hazard class. If cryogenic gases are shipped above 41 psi, they are considered a compressed gas. Cryogenics may be required to be placarded if they have other hazards, such as flammable, poison, or oxidizer. If cryogenics do not have another hazard, they are not required to be placarded under DOT regulations. Compressed gases that are shipped as liquefied gases, such as cryogenics, exhibit other hazards not indicated by the placard. Liquefied refrigerated gases, such as cryogenics, are extremely cold materials; boiling points are −130°F (−90°C) or greater (Table 4.8). Liquid helium has a boiling point of −452°F (−268°C); it is the coldest material known. It is also the only material on earth that never exists as a solid under normal temperatures and pressures; it exists only as a cryogenic liquid or as a gas.

Gases are processed into cryogenic liquids by a combination of pressurization, cooling, and ultimate release of pressure. Therefore, cryogenics do not require pressure to keep them in the liquid state, unless they will be in the container for a long period of time; then they are pressurized. Cryogenics are kept cold by the temperature of the liquid and the insulated containers. The cryogenic liquefaction process begins when gases are placed into a large processing container. They are pressurized to 1500 psi. The process of pressurizing a gas causes the molecules to move faster, causing more collisions with each other and the walls of the container. This causes heat to be generated. An example of this heating process occurs while filling a self-contained breathing apparatus (SCBA) bottle—the top feels hot. Continuing with the liquefaction process, once the pressure of 1500 psi is reached, the material is cooled to 32°F (0°C) using ice water. When the gas is cooled, the pressure is once again increased, this time up to 2000 psi, with an accompanying increase in temperature. The gas is then cooled to −40°F (−40°C) using liquid ammonia. As the gas is cooled, all of the pressure is released. The resulting temperature decrease turns the cryogenic gas into a liquid.

Several gases found on the periodic table are extracted from the air and turned into cryogenic liquids. These include neon, argon, krypton, xenon, oxygen, and nitrogen. All but oxygen are considered inert, i.e., they are nontoxic, nonflammable, and nonreactive. To extract these gases from the air, the air is first turned into a cryogenic liquid. Then the liquid is run through a type of distillation process, where each component gas is drawn

Table 4.8 Boiling Points of Cryogenic Liquids

Helium	−452°F	Air	−318°F
Neon	−411°F	Fluorine	−307°F
Nitrogen	−321°F	Hydrogen	−423°F
Argon	−303°F	Methane	−257°F
Oxygen	−297°F	Nitric oxide	−241°F
Krypton	−244°F	CO	−312°F
Xenon	−162°F	NF_3	−200°F

off as it reaches its boiling point. Gases are then liquefied by the same process mentioned previously. Some common cryogenic gases and their characteristics are listed in the following paragraphs.

Helium, He, is a gaseous nonmetallic element from the noble gas family, family eight on the periodic table. Helium is colorless, odorless, and tasteless. It is nonflammable, nontoxic, and nonreactive. Helium has a boiling point of −452°F (−268°C) and is slightly soluble in water. Even though helium is an inert gas, it can still displace oxygen and cause asphyxiation. Helium has a vapor density of 0.1785, which is lighter than air. It is derived from natural gas by liquefaction of all other components. Helium has a four-digit UN identification number of 1046 as a compressed gas and 1963 as a cryogenic liquid. Helium is used to pressurize rocket fuels, in welding, in inflation of weather and research balloons, in luminous signs, geological dating, lasers, and as a coolant for nuclear-fusion power plants.

Neon, Ne, is a gaseous nonmetallic element from the noble gas family. It is colorless, odorless, and tasteless and is present in the Earth's atmosphere at 0.0012% of normal air. It is nonflammable, nontoxic, and nonreactive and does not form chemical compounds with any other chemicals. Neon is, however, an asphyxiant gas and will displace oxygen in the air. The boiling point of neon is −410°F (−245°C), and it is slightly soluble in water. Neon has a vapor density of 0.6964, which is lighter than air. The four-digit UN identification number is 1065, when compressed, and 1913 as a cryogenic liquid. Its primary uses are in luminescent electric tubes and photoelectric bulbs. It is also used in high-voltage indicators, lasers (liquid), and cryogenic research.

Argon, Ar, is a gaseous nonmetallic element of family eight. It is present in the Earth's atmosphere to 0.94%, by volume. It is colorless, odorless, and tasteless. Argon does not combine with any other chemicals to form compounds. It has a boiling point of −302°F (−185°C), and it is slightly soluble in water. Argon has a vapor density of 1.38, which makes it heavier than air. The four-digit UN identification number is 1006, as a compressed gas, and 1951 as a cryogenic liquid. Argon is used as an inert shield in arc welding, in electric and specialized lightbulbs (neon, fluorescent, and sodium vapor), in Geiger-counter tubes, and in lasers.

Krypton, Kr, is a gaseous nonmetallic element of family eight. It is present in the Earth's atmosphere to 0.000108%, by volume. It is a colorless and odorless gas. Krypton is nonflammable, nontoxic, and nonreactive. It is, however, an asphyxiant gas and can displace oxygen in the air. At cryogenic temperatures, krypton exists as a white crystalline substance with a melting point of 250°F (121°C). The boiling point of krypton is −243°F (−152°C). Krypton is known to combine with fluorine at liquid nitrogen temperature by means of electric discharges or ionizing radiation to form KrF_2 or KrF_4. These materials decompose at room temperature. Krypton is slightly water-soluble. The vapor density is 2.818, which is heavier than air. The four-digit UN identification number is 1056, for the compressed gas, and 1970 for the cryogenic liquid. Krypton is used in incandescent bulbs, fluorescent light tubes, lasers, and high-speed photography.

Xenon, Xe, is a gaseous nonmetallic element from family eight. It is a colorless and odorless gas or liquid. It is nonflammable and nontoxic at standard temperatures and pressures, but is an asphyxiant and will displace oxygen in the air. The boiling point is −162°F (−107°C), and the vapor density is 05.987, which is heavier than air. Xenon is chemically unreactive; however, it is not completely inert. The four-digit UN identification number is 2036 for the compressed gas and 2591 for the cryogenic liquid. Xenon is used in luminescent tubes, flashlamps in photography, lasers, and as an anesthesia.

Xenon compounds: Xenon combines with fluorine through a process of mixing the gases, heating in a nickel vessel to 752°F (400°C), and cooling. The resulting compound is xenon tetrafluoride, XeF_4, composed of large, colorless crystals. Compounds of xenon difluoride, XeF_2, and hexafluoride, XeF_6, can also be formed in a similar manner. The hexafluoride compound melts to a yellow liquid at 122°F (50°C) and boils at 168°F (75°C). Xenon and fluorine compounds will also combine with oxygen to form oxytetrafluoride, $XeOF_4$, which is a volatile liquid at room temperature. Compounds formed with fluorine must be protected from moisture to prevent the formation of xenon trioxide, XeO_3, which is a dangerous explosive when dried out. The solution of xenon trioxide is a stable weak acid, which is a strong oxidizing agent.

Cryogenic liquids have large expansion ratios, some as much as 900 or more to 1 (Table 4.9). Because of this expansion ratio, if the cryogenic liquid is flammable or toxic, these hazards are intensified because of the potential for large gas cloud production from a small amount of liquid. As the size of a leak increases, so does the size of the vapor cloud.

Even though some cryogenic liquids do not require placards, these materials can still pose a serious danger to responders. Gases formed by warming of cryogenic liquids can displace oxygen in the air, which can harm responders by asphyxiation. Normal atmospheric oxygen content is about 21%. When the oxygen in the lungs and ultimately the blood is reduced, unoxygenated blood reaches the brain, and the brain shuts down. It may only be a few seconds between the first breath and collapse. Being quite cold, cryogenic liquids can cause frostbite and solidification of body parts. When the parts thaw out, the tissue is irreparably damaged.

Cryogenic liquids are shipped and stored in special containers (Photo 4.19). On the highway, the MC 338 tanker is used to transport cryogenic liquids. The tank is usually not pressurized, but is heavily insulated to keep the liquids cold. A heat exchanger is located underneath the belly of the tank truck to facilitate the off-loading of product as a gas. Railcars are also specially designed to keep the cryogenic liquids cold inside the containers to minimize the boiling off of the gas. Fixed-storage containers of cryogenic liquids are usually tall, small-diameter tanks (Photo 4.20). These are insulated and resemble large vacuum bottles that keep the liquid cold. Cryogenic storage containers are also under pressure to keep the material liquefied. A heat exchanger is used to turn the liquid back into a gas for use (Photo 4.21). It is composed of a series of metal tubes with fins around the outside. As the liquid runs through the tubes, it is warmed and turns into a gas. Other gases, such as hydrogen, are liquefied, sometimes made into cryogenics, and placarded as flammable gases. Liquid oxygen is placarded as an oxidizer or nonflammable compressed gas.

Table 4.9 Expansion Ratios of Cryogenic Liquids

Argon	841/1
Ethane	487/1
Fluorine	981/1
Helium	754/1
Hydrogen	840/1
LNG[a]	637/1
Nitrogen	697/1
Oxygen	862/1

[a]Liquefied natural gas.

Photo 4.19 MC 338 used for the transportation of cryogenic liquids.

Photo 4.20 Fixed-facility cryogenic liquid oxygen container with heat exchanger outside of hospital.

Photo 4.21 Heat exchanger underneath MC 338 allows the cryogenic liquid to be off-loaded as a gas.

MRI and NMR facilities

Magnetic resonance imaging (MRI) and nuclear magnetic resonance (NMR) scanners are found in hospitals, nuclear medicine clinics, radiation departments, and research facilities both private and at universities across the country (Photo 4.22). MRI is a test that uses a magnetic field and pulses of radio wave energy to make pictures of organs and structures inside the body. In many cases, MRI gives different information about structures in the body than that can be seen with an x-ray, ultrasound, or computed tomography scan. MRI also may show problems that cannot be seen with other imaging methods. NMR utilizes magnetic fields for research purposes involving radiation and

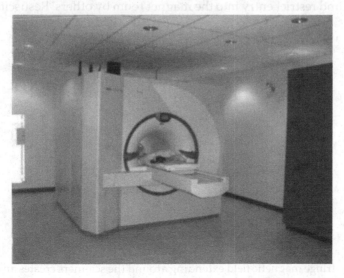

Photo 4.22 Primary hazards for emergency response personnel at MRI and NMR facilities are the magnetic fields and cryogenic liquids used to cool the magnet.

atoms and the identification of structures of organic compounds. While the MRI is used for medical imaging and the NMR is used for research, both have one thing in common; "the invisible force." This invisible force is a powerful magnetic field. These invisible forces cannot be detected by the human senses. Response personnel need to rely on education, signage, on-site visits, and knowledgeable facility personnel to become familiar with and understand the hazards posed by MRI and NMR facilities during emergency responses. Approved signage is located throughout the facilities noting dangerous locations. Pay close attention to the signs.

Primary hazards for emergency response personnel at MRI and NMR facilities are the magnetic fields and cryogenic liquids used to cool the magnet. Magnets can range from 30,000 to 60,000 times greater than the magnetic field of the earth.

Metal implants, prostheses, and foreign metallic bodies (even those not ferromagnetic) can move or dislodge causing severe injury. Firefighter equipment including but not limited to SCBA, axes and other tools, radios, flashlights, stretchers and defibrillators, and any other objects that are ferrous (iron-containing) are not safe to be used around the magnet. Also metal belt buckles and steel-tipped shoes and any steel protection in firefighter boots. This includes portable fire extinguishers as well. Most facilities have nonferromagnetic fire extinguishers on-site. Accidents have occurred where tools, mop buckets, floor polishers, chairs, acetylene and oxygen tanks, and other equipment have been drawn into the magnets during daily operations and during maintenance and repairs in a facility. Cell phones, ID cards, and credit cards may not work after exposure to a magnetic field. In 2002, a 6 year old boy was killed during an MRI scan when a metal oxygen tank was drawn from across the room into the scanner by the magnetic force and fractured his skull and caused brain damage (Photo 4.23). During 2003, a repairman broke his arm when a piece of metal pinned him to an MRI. He carried the metal into the room, and the force of the magnet attracted the metal and the repairman carrying the metal against the MRI machine. Firefighters were called to release him from the machine and treat injuries.

When responding to an accident where someone has been pinned by a metal object, determine whether the object pinning the victim can be removed without causing further injury. If removal is successful, immediately evacuate the victim to an area outside the magnet room and restrict entry into the magnet room by others. Resuscitation or treatment

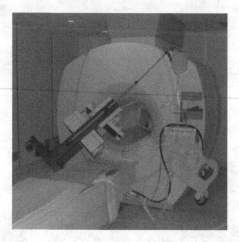

Photo 4.23 The fringe magnetic field extending around the scanners creates an environment where ferromagnetic objects become high-speed, potentially fatal projectiles when drawn toward the magnet.

aided by ferromagnetic devices can be administered once the victim is outside the magnet room. If a life-threatening emergency exists and there is no other way to free the victim without eliminating the magnetic field, then it will be necessary to initiate a magnet quench (bring down the magnetic field). The magnet quench procedure will create a dangerous environment. Expect a loud noise from the escape of cryogens and a release of a dense white fog. There is a high risk of asphyxiation and potential for frostbite. As the magnetic field decreases, the object pinning the victim may fall and could cause further damage. Do not perform this procedure unless you are prepared to immediately evacuate yourselves and the victim if oxygen is displaced from the room. In 2006, two people were injured when an MRI machine they were moving exploded. It is reported that nitrogen and helium still in the machine mixed causing the explosion. One of the injured had shrapnel in his head and the other an injured arm. Firefighters were called to the facility following the explosion to investigate and treat the injured. The power of these magnets should not be underestimated, and great care should be exercised by emergency response personnel working near them. All ferromagnetic metals should be removed without exception before entering the facility magnet area. If metal cannot be removed, personnel should not enter.

Magnets at MRI and NMR facilities are always on, and there is always a magnetic field present, even if power to the facility or the magnet is disconnected. The only way magnets can be shut down is by quenching. This can occur by instigation through a quench switch located in facility or spontaneous quenching caused by a malfunction in the system. Instigation of a quench using the switch can discharge the magnet in about 20 s. Even a partially quenched magnet can still pose a hazard and can still attract ferrous objects such as gas cylinders. Instigated quenching is the removal of the nitrogen and helium from the magnet and exhausting to the outside of the building. Spontaneous quenching results from a breakdown in the system that may result in some vapors of nitrogen and helium displacing the oxygen in the room and causing simple asphyxiation to anyone not wearing SCBA. When entering area where oxygen deficiency is suspected or possible, portable oxygen monitoring equipment should be used. During instigated quenching, it has also been reported that exhausting equipment has failed resulting in vapors in the room as well. Quenching can not only be a dangerous operation, but also be very expensive. It can cost over \$50,000 to reenergize the magnet and result in a 1–2 month downtime. If the magnet needs to be repaired as a result of the quenching, it can cost over \$100,000 with several months of downtime. If the machine cannot be repaired, replacement can cost over 2 million dollars. Quenching should only be done under the supervision of facility personnel when there is a life-threatening situation, like a person pinned to the magnet by a large object that cannot be removed by hand or as the result of a large fire so that firefighters can enter the room safely. If an emergency quench is necessary, pressure generated by the quench may prevent doors from opening, so prop open the magnet room door. Allow no others to enter the room through the open door. All personnel must know to leave the room and not return until the cryogenic gases have dissipated and the room is safe to reoccupy. In addition to the asphyxiation hazards of cryogenic liquid nitrogen and helium, they are very cold materials and can cause thermal burns and solidification of body parts. Contact with the liquids can solidify personal protective equipment normally worn by emergency responders. Avoid contact with the materials. Special gloves are used within the facilities for handling cold or frosted surfaces. These gloves will not provide any protection if submerged in the liquids themselves. Solidification of hands and fingers can occur quickly. Cryogenic liquids also have large liquid-to-vapor expansion ratios, and a small amount of liquid can cause a significant amount of vapor. Helium has a boiling point of −452°F (−268°C). Helium is the only substance on Earth that does not exist as a solid. There is nothing colder to solidify helium. Helium's expansion ratio is 754–1. One gallon of liquid helium would expand to 754 gal of helium vapor. Nitrogen has a boiling point of −321°F (−196°C). Nitrogen's expansion ratio is 697–1. One gallon of nitrogen liquid would expand to 697 gallons of nitrogen vapor. Extreme caution should be exercised when working in areas where cryogenic liquid or vapor has been released. Both helium and nitrogen gases are odorless and colorless. However, cryogenic gases are so cold that they can freeze water vapor in the air causing a vapor cloud appearance. Facilities where MRI and NMR are used will be equipped with oxygen monitors and alarms to indicate whether the air is safe to breathe without respiratory protection or if respiratory protection is necessary. If the alarm is sounding, it is likely that helium or nitrogen vapors are present and have displaced oxygen in the area. Helium and nitrogen vapors are lighter than air so crawling on the floor would provide an atmosphere of less helium and nitrogen and more oxygen. Material safety data sheets (MSDSs) on helium and nitrogen should be available at the facility along with any other chemicals in use.

During firefighting or other emergency operations at an NMR facility, care should be taken to not overturn magnets; they are very top-heavy and not very stable. Cryogenic containers called Dewars are also top-heavy and easily overturned. In addition to other hazards mentioned earlier, there are radiation hazards present in NMR facilities, which may present unique hazards to emergency response personnel. Become familiar with any MRI or NMR facilities that may be present in your response areas. Consult knowledgeable magnetic resonance personnel if they are on-site. They are the best source of information when it comes to dealing with the magnetic fields and associated cryogenic hazards during any emergency at a facility.

There are some "foolers" in the nonflammable compressed gas subclass. Anhydrous ammonia, for example, is regulated by the DOT as a nonflammable compressed gas. The United States is the only country in the world that placards anhydrous ammonia in this manner. Such placarding of anhydrous ammonia as a nonflammable gas is a result of lobbying efforts by the agricultural fertilizer industry. In most other parts of the world, anhydrous ammonia is placarded as a poison gas, not to mention that it is also flammable under certain conditions. If anhydrous ammonia leaks inside a building or in a confined space, it may very well be found within its flammable range and burn if an ignition source is present. If anhydrous ammonia met the DOT definition of a flammable gas, which is a LEL of less than 13% or a flammable range of greater than 12% points, it would be placarded as such. It does not, however, meet the DOT definition; it has an LEL of 16% and a flammable range of 16%–25%. Therefore, it is placarded as a nonflammable gas even though, in reality, it should be placarded by its most severe hazard, which is a poison gas.

Hydrocarbon derivatives

A few hydrocarbon derivatives from the alkyl-halide family are 2.2 nonflammable compressed gases. This illustrates the wide range of hazards of the alkyl halides as a group. Some are flammable, some are toxic, and some are nonflammable and nontoxic. Some are fire-extinguishing agents! They can still act as asphyxiants and displace the oxygen in the air. It is important to remember that the primary hazard of the alkyl halides is toxicity. Some of them are also flammable; therefore, all must be assumed to be toxic and flammable until the individual chemical is researched and the actual hazards are determined. It is interesting to note that while the DOT lists tetrafluoromethane as a nonflammable, nonpoisonous gas, the *Condensed Chemical Dictionary* lists the compound as toxic by inhalation. The *NIOSH Pocket Guide to Chemical Hazards* does not list the compound. The best

```
        F
        |
   F — C — F
        |
        F
```

Tetrafluoromethane
CF_4

Figure 4.16 Tetrafluoromethane.

source of information about this compound and others may be the MSDS. Examples of nonflammable Class 2.2 alkyl halides are tetrafluoromethane and trifluoromethane.

Tetrafluoromethane, CF_4, also known as carbon tetrafluoride and fluorocarbon 14, is a colorless gas that is slightly soluble in water. It is nonflammable, but is listed as toxic by inhalation in the *Condensed Chemical Dictionary*. The four-digit UN identification number is 3159. The primary uses are as a refrigerant and gaseous insulator. The structure is illustrated in Figure 4.16.

Trifluoromethane, CHF_3, also known as fluoroform, propellant 23, and refrigerant 23, is a colorless gas that is nonflammable. There are no hazards listed for trifluoromethane. It may be an asphyxiant gas and displace oxygen in the air and in confined spaces. The four-digit UN identification number is 2035 for the compressed gas and 3136 for the cryogenic liquid. The primary uses for trifluoromethane are as a refrigerant, as a direct coolant for infrared detector cells, and as a blowing agent for urethane foams. The structure is shown in Figure 4.17.

Nonflammable gas compounds

Nitrogen, N_2, is colorless, odorless, and tasteless and makes up 78% of the air in the Earth's atmosphere. Nitrogen has a boiling point of $-321°F$ ($-196°C$) and is slightly soluble in water. Nitrogen is nonflammable and nontoxic. It may, however, displace oxygen and be an asphyxiant gas. The vapor density of nitrogen is 0.96737, which makes it slightly lighter than air. The four-digit UN identification number is 1066 as a compressed gas and 1977 as a cryogenic liquid. As a cryogenic liquid, the NFPA 704 designation is health 3, flammability 0, and reactivity 0. Nitrogen is used in the production of ammonia, cyanides, and explosives; as an inert purging agent; and as a component in fertilizers. It is usually shipped in insulated containers, insulated MC 338 tank trucks, and tank cars.

Oxygen, O_2, like nitrogen, is a nonmetallic elemental gas. Oxygen, although nontoxic, is highly reactive with hydrocarbon-based materials and is an oxidizer. Oxygen makes up approximately 21% of the air breathed. The boiling point of oxygen is $-297°F$ ($-182°C$). It is nonflammable, but supports combustion. Oxygen can explode when exposed to heat or organic materials. The vapor density of oxygen is 1.105, which makes it slightly heavier than air. Oxygen is incompatible with oils, grease, hydrogen, flammable liquids, solids, and gases. The four-digit UN identification number for oxygen is 1072 as a compressed gas and 1073 as a cryogenic liquid. The NFPA 704 designation for liquid oxygen is health 3,

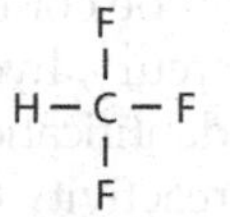

```
        F
        |
   H — C — F
        |
        F
```

Trifluoromethane
CHF_3

Figure 4.17 Trifluoromethane.

Photo 4.24 Dewar container of cryogenic nitrogen.

flammability 0, and reactivity 0. Liquid oxygen is shipped in Dewar flasks (Photo 4.24) and MC 338 tank trucks. It may also be encountered in cryogenic railcars.

Liquid oxygen in contact with an asphalt surface, such as a parking lot or highway, can create a contact explosive; dropping an object, driving, or even walking on the area can cause an explosion to occur (Photo 4.25).

Anhydrous ammonia, NH_3, seeks water, "anhydrous" meaning "without water." This can be particularly dangerous to responders because ammonia can seek water in the eyes, lungs, and other moist parts of the body (Photo 4.26). Ammonia is toxic, with a TLV of 25 ppm; the inhalation of concentrated fumes can be fatal above 2000 ppm. Ammonia odor is detectable at 1–0 ppm. It has a boiling point of –28°F (–2°C) and a flammable range of 16%–25% in air, although it does not meet the DOT definition of a flammable gas. The ignition temperature is 1204°F (651°C). The vapor density of ammonia is 0.6819, which makes ammonia gas lighter than air. It is also water-soluble. Hose streams can be used to control vapor clouds of ammonia gas. The runoff created, however, is ammonium hydroxide, which is corrosive, so the runoff should be contained.

Ammonia is incompatible with mercury, hydrogen fluoride, calcium hypochlorite, chlorine, and bromine. The four-digit UN identification number is 1005, and the NFPA 704 designation is health 3, flammability 1, and reactivity 0. Ammonia may be shipped as a cryogenic liquid or a liquefied compressed gas. Ammonia is used as an agricultural fertilizer (Photo 4.27) and as a coolant in cold-storage buildings and food lockers. It is usually shipped in MC 331 tank trucks, railcars (Photo 4.28), barges, and steel cylinders. Anhydrous ammonia is the second

Photo 4.25 Liquid oxygen in contact with blacktop or asphalt surfaces can form a contact explosive. Driving over it, dropping a tool on it, or even walking on the surface can cause the explosion to occur.

Photo 4.26 Farm tanks are required by law to have a minimum 5 gal water tank on top of the farm tank to use in an emergency if ammonia gets on your skin or eyes.

most-released chemical from fixed facilities on the EPA listing, with 3586 accidents resulting in the release of over 19 million pounds of ammonia. Anhydrous ammonia is consistently among the seven highest-volume hazardous materials shipped on the railroads.

Mild exposure to anhydrous ammonia can cause irritation to eye, nose, and lung tissues. When NH_3 is mixed with moisture in the lungs, it causes severe irritation. Ammonium hydroxide is actually produced in the lungs. Prolonged breathing can cause suffocation. The human eye is a complex organ made up of nerves, veins, and cells. The front of the human eye is covered by membranes, which resist exposure to dust and dirt. However, these cannot keep out anhydrous ammonia, because the entire eye is about 80% water. A shot of ammonia under pressure can cause extensive, almost immediate, damage to the eye. The ammonia extracts the fluid and destroys eye cells and tissue in minutes. If you get a shot of anhydrous ammonia in your eye, the first few seconds are crucial. Immediately flush the eyes with copious amounts of water.

Photo 4.27 Trailer used to haul anhydrous ammonia to be used as a fertilizer in farming.

Photo 4.28 Anhydrous ammonia is shipped in pressure tank cars throughout the country in the early spring in time for the agricultural planting season.

If wearing contact lenses, remove them. Your eyes will fight to stay closed because of the extreme pain, but they must be held open so the water can flush out the ammonia. Continue to flush the eyes for at least 15 min. Get professional medical help as soon as possible to prevent permanent damage. If water is not available, fruit juice or cool coffee can be used to flush the eyes. Remove contaminated clothing and thoroughly wash the skin.

Clothing frozen to skin by liquid ammonia can be loosened with liberal application of water. Wet clothing and body thoroughly, and then remove the clothing. Leave burns exposed to the air and do not cover with clothing or dressings. Immediately after first-aid treatment with water, get the burn victim to a physician. Do not apply salves, ointments, or oils, as these cause ammonia to burn deeper. Let a physician determine the proper medical treatment. Remove the victim to an area free from fumes if an accident occurs. If the patient is overcome by ammonia fumes and stops breathing, get him or her to fresh air and give artificial respiration. The patient should be placed in a reclining position with head and shoulders elevated. Basic life support should be administered if needed. Oxygen has been found useful in treating victims who have inhaled ammonia fumes. Administer 100% oxygen at atmospheric pressure. Any person who has been burned or overcome by ammonia should be placed under a physician's care as soon as possible. Begin irrigation with water immediately. The rescuer should use freshwater, if possible. If the incident is a farm accident, there is a requirement for water tanks for irrigation of the eyes and the body on the anhydrous ammonia tank.

Open water in the vicinity of an anhydrous ammonia leak may have picked up enough NH_3 to be a caustic aqua-ammonia solution. This could aggravate the damage if used in the eyes or for washing burns. The victim should be kept warm, especially to minimize shock. If the nose and throat are affected, irrigate them with water continuously for at least 15 min. Take care not to cause the victim to choke. If the patient can swallow, encourage drinking lots of some type of citrus drink, such as lemonade or fruit juice. The acidity will counteract some of the effect of the anhydrous ammonia.

Ammonia and propane are often shipped in the same type of container. In the late winter and early spring, the containers are purged of the propane and used for ammonia. In the late summer and early fall, the containers are purged of ammonia and used for propane. One of the primary uses of ammonia is as a fertilizer (Photo 4.29). The tanks used for farm application of ammonia can also be used for propane in the winter for heating and grain-drying purposes. Ammonia will attack copper, zinc, and their alloys of brass and bronze. Propane valves, fittings, and piping are often made of these materials. If the valves and other fixtures are not changed to steel before being used for ammonia, or if the ammonia is not completely purged from the containers before propane use, serious accidents can occur. Ammonia can damage fittings and cause leaks of the highly flammable propane, which may result in fires. There have also

Photo 4.29 Leaking ammonia can cause vegetation to turn brown. This cornfield was exposed to leaking ammonia. Notice the foliage: the corn is too short to be turning brown naturally.

been situations in which propane has been transported in ammonia containers and the placards for ammonia have been left on the containers. This can present a serious hazard for emergency responders.

Carbon dioxide, CO_2, is colorless and odorless. It can also be a solid (dry ice), which will undergo sublimation and turn back into carbon dioxide gas. Carbon dioxide may also be encountered as a cryogenic liquid. CO_2 is miscible with water and is nonflammable and nontoxic, but can be an asphyxiant and displace oxygen. In 1993, two workers were killed aboard a cargo ship when a carbon dioxide fire-extinguishing system discharged. The oxygen in the area was displaced by the carbon dioxide, and the men were asphyxiated. Carbon dioxide has a vapor density of 1.53, which is heavier than air. It has a four-digit UN number of 2187 as a cryogenic and 1013 as a compressed gas. The NFPA 704 designation is health 3, flammability 0, and reactivity 0. It is used primarily in carbonated beverages and fire-extinguishing systems.

Incidents

According to NFPA studies on ammonia incidents between 1929 and 1969, there were 36 incidents in which released ammonia gas was ignited; 28 resulted in a combustion explosion. All of the explosions occurred indoors.

On January 18, 2002, a Canadian Pacific freight train derailed outside Minot, North Dakota. Five of the cars carried anhydrous ammonia. Leaking ammonia killed one person and sent dozens of others to hospitals for treatment. Ten of those seeking treatment were admitted to the hospital. Some local residents were evacuated, while others were asked to shelter in place. Civil defense sirens and local radio and television stations alerted residents.

In Shreveport, Louisiana, one firefighter was killed and one badly burned in a fire involving anhydrous ammonia. A leak developed inside a cold storage plant. The firefighters donned Level A chemical protective clothing and went inside to try to stop the leak. Something caused a spark, and the anhydrous ammonia caught fire.

In Verdigris, Oklahoma, a tank car was being filled with anhydrous ammonia. It was unknown that there was a weakened place on the tank car, which gave way from the pressure of the ammonia and resulted in a BLEVE. There was no fire, just a vapor cloud that traveled downwind, defoliating trees and turning other vegetation brown. One worker who was filling the tank car was killed in the incident.

In Delaware County, Pennsylvania, ammonia was being removed from an abandoned cold-storage facility when a leak occurred. Firefighters were exposed to ammonia and complained of irritation and burning of the face and other exposed skin surfaces. Several were transported to local hospitals for treatment after going through decontamination at the scene (Photo 4.30).

At the Knouse Foods facility in Orrtanna, Pennsylvania, two workers were killed while performing routine maintenance on an ammonia system in a cold-storage building used for storing fresh fruit (Photo 4.31). Both of the workers were members of the local volunteer fire company, and one was the assistant chief. Local firefighters were called to the scene for rescue when they discovered they two of their own were in need of rescue. Several firefighters were injured by the ammonia vapors while trying to rescue the workers. Ammonia is a nonflammable compressed gas according to the DOT. However, not only will it burn, especially inside of a building, it is considered a poison gas everywhere else in the world but the United States. It is also corrosive.

Photo 4.30 Firefighters responded to a reported ammonia leak at an abandoned locker plant. It was a summer day with temperatures in the upper 90s. The entire first alarm assignment was exposed to the ammonia gas and complained of burning skin.

Photo 4.31 Two off duty volunteer firefighters were killed at the Knouse Foods cold-storage facility while performing routine maintenance.

Anhydrous ammonia is a common, but dangerous, material when not properly handled. Firefighters should wear full Level A chemical protective clothing when exposed to ammonia vapors. Be aware, however, that ammonia is also flammable, and Level A protective clothing provides no thermal protection. When responding to incidents involving ammonia, use caution. Firefighter turnouts do not provide adequate protection from ammonia vapors. If there are victims exposed to ammonia for any length of time, the chance of rescue is slim. Do not expose unprotected rescue personnel to ammonia vapor. Victims can be decontaminated using emergency decon procedures by first responders. This will reduce the impact of NH_3 on victims.

Poison gases

Subclass 2.3 materials are an inhalation hazard, and some may also be absorbed through the skin. Firefighters are exposed to certain types of toxic materials whenever they fight a fire. These toxic materials are by-products of the combustion process. Toxic fire gases include carbon monoxide (CO), hydrogen chloride (HCl), hydrogen cyanide (HCN), sulfur dioxide (SO_2), nitrogen dioxide (NO_2), ammonia (NH_3), hydrogen sulfide (H_2S), and phosgene ($COCl_2$). These toxic materials kill thousands of persons each year. There are, however, poisons in every community that are potentially more dangerous toxic materials capable of killing tens of thousands of people in a matter of minutes.

The DOT definition of poison gas is "a material that is a gas at 68°F or less at 14.7 psi and is so toxic to humans as to pose a hazard to health during transportation, or in the absence of adequate data on human toxicity, is presumed to be toxic to humans because when tested on laboratory animals it has an LC_{50} value of not more than 5000 ml/m³" (Photo 4.32). These materials are considered so toxic that, when transported, the vehicle must be placarded regardless of the quantity. The potential exists for 2.3 materials to affect large populations by creating toxic gas clouds. In order to understand the toxic effects of poisons, it is necessary to know some toxicological terminology. One thing to remember about toxicological data is that little human data are available. The data available are the result of tests on laboratory animals. Toxicity for humans is really nothing more than an educated guess. Most of the terms mentioned here are applied to workplace exposures. Acceptable exposures in many cases are 8 h a day, 40 h a week. Concentrations encountered on the scene of an incident will be much higher than any ordinary workplace exposure, but for a shorter period of time.

TLVTWA is the threshold limit value–time-weighted average concentration for a normal 8-h workday and a 40-h workweek, to which nearly all workers may be repeatedly exposed, day after day, without adverse effect.

IDLH is immediately dangerous to life and health. IDLH determines the highest concentration that a person can be exposed to for a maximum of 30 min and still escape without any irreversible health effects.

LC_{50} is the lethal concentration by inhalation for 50% of the laboratory animals tested.

Photo 4.32 Level A chemical protection is necessary when dealing with poison gases.

STEL is the short-term exposure limit, defined as a 15-min TWA exposure, which should not be exceeded at any time during a workday, even if the 8-h TWA is within the TLV-TWA. Exposures above the TLV-TWA up to the STEL should not be longer than 15 min and should not occur more than four times a day. There should be at least 60 min between successive exposures in this range.

Concentration is the amount of one substance found in a given volume of another substance. Depending on the materials involved, there are many different ways of expressing concentration. Two of the most common ways are ppm (parts per million) and milligrams per kilogram (mg/kg). Toxicology will be discussed further in Chapter 8. Examples of some 2.3 poison gases are fluorine, chlorine, carbon monoxide, hydrogen sulfide, phosgene, phosphine, and chloropicrin–methyl bromide mixture.

Carbon monoxide, CO, is an odorless, colorless, tasteless gas that is toxic by inhalation, with a TLV of 50 ppm. A 1% concentration is lethal to adults if inhaled for 1 min. Carbon monoxide binds to the blood hemoglobin 220 times tighter than oxygen. The more carbon monoxide that binds to the blood, the less oxygen can be carried. Carbon monoxide will prevent oxygen from being taken into the blood, thus causing a type of chemical asphyxiation. In addition to its primary hazard of toxicity, it is also highly flammable and is a dangerous fire and explosion risk. The boiling point is −313°F (156°C) and the ignition temperature is 1292°F (700°C). Carbon monoxide has a flammable range from 12% to 75%. Its vapor density is 0.967, which is slightly less than that of air. It is slightly soluble in water. The four-digit UN identification number is 1016 as a compressed gas and 9202 as a cryogenic liquid. The NFPA 704 designation for carbon monoxide is health 3, flammability 4, and reactivity 0. The primary uses are in the synthesis of organic compounds, such as aldehydes, acrylates, alcohols, and in metallurgy.

Hydrogen sulfide, H_2S, is a colorless gas with an odor like rotten egg. It is the only common material that can halt respiration. It is toxic by inhalation with a TLV of 10 ppm. The minimal perceptible odor is found at concentrations of 0.13 ppm; at 4.60 ppm, the odor is moderate; at 10 ppm, tearing begins; at 27 ppm, there is a strong, unpleasant, but not intolerable, odor. When TLV reaches 100 ppm, coughing begins, eye irritation occurs, and loss of sense of smell begins after 2–15 min. Marked eye irritation occurs at 200–300 ppm, and respiratory irritation after 1 h of exposure. Loss of consciousness takes place at 500–700 ppm, with the possibility of death in 30 min to 1 h. Concentrations of 700–1000 ppm cause unconsciousness, cessation of respiration, and death. Instant unconsciousness occurs at 1000–2000 ppm, with cessation of respiration and death in a few minutes. Death may occur even if the victim is removed to fresh air at once. Hydrogen sulfide is highly flammable and a dangerous fire and explosion risk. The boiling point is −76°F. The flammable range is 4.3%–46%. The ignition temperature is 500°F. Its vapor density is 1.189, which is heavier than air. Hydrogen sulfide is soluble in water. It is incompatible with oxidizing gases and fuming nitric acid. The four-digit UN identification number is 1053. The NFPA 704 designation for hydrogen sulfide is health 4, flammability 4, and reactivity 0. It is used in the purification of hydrochloric and sulfuric acids and is a source of hydrogen and sulfur. Hydrogen sulfide is usually shipped in steel pressure cylinders.

Fluorine, F_2, is a nonmetallic elemental gas from the halogens, which is family seven on the periodic table. It is the most electronegative and powerful oxidizing agent known. It reacts vigorously with most oxidizable substances at room temperature, frequently causing combustion. Fluoride compounds form with all elements except helium, neon, and argon. Fluorine is a pale yellow gas with a pungent odor. It is nonflammable, but will support combustion because it is an oxidizer. The boiling point is −307°F (152°C). The vapor density is 1.31, which is heavier than air. Fluorine is water-reactive. The primary hazard is toxicity; fluorine is toxic by inhalation and extremely irritating to tissues. The TLV is 1 ppm, and the IDLH is 25 ppm in air. Fluorine is incompatible with and should be isolated from everything! The four-digit UN identification number is 1045. The NFPA 704 designation is health 4, flammability 0, and reactivity 4. The white section at the bottom of the diamond contains a W with a slash through it, indicating water reactivity. Because of the strong reactivity with other materials, it is shipped in special steel containers. The primary uses are in the production of metallic and other fluorides, fluorocarbons, fluoridation of drinking water, and in toothpaste.

Boron trifluoride, BF₃, is a colorless gas with a vapor density of 2.34, which is heavier than air. It is water-soluble and does not support combustion. It is also water-reactive, toxic by inhalation, and corrosive to skin and tissue. The TLV is 1 ppm, and the IDLH is 100 ppm in air. The boiling point is −148°F (64°C). The four-digit UN identification number is 1008. The NFPA 704 designation is health 4, flammability 0, and reactivity 1. The primary uses are as a catalyst in organic synthesis, in instruments for measuring neutron intensity, in soldering fluxes, and in gas brazing.

Dichlorosilane, H₂SiCl₂, is a pyrophoric, water-reactive gas. It is flammable, with a wide flammable range of 4.1%–99% in air. The boiling point is 47°F (8°C), and the flash point is −35°F (−37°C). The ignition temperature is 136°F (57°C). The vapor density is 3.48, which is heavier than air. It is immiscible in water and highly water-reactive. Contact with water releases hydrogen chloride gas. It is toxic by inhalation and skin absorption. Hydrogen chloride causes severe eye and skin burns and is irritating to the skin, eyes, and respiratory system. The four-digit UN identification number is 2189. The NFPA 704 designation is health 4, flammability 4, and reactivity 2. The white area at the bottom of the diamond contains a W with a slash through it, indicating water reactivity. It is shipped in carbon steel cylinders.

Phosgene, COCl₂, is a clear to colorless gas or fuming liquid, with a strong stifling or musty hay-type odor. It is slightly soluble in water. The vapor density is 3.41, which is heavier than air. Phosgene is a strong irritant to the eyes, is highly toxic by inhalation, and may be fatal if inhaled. The TLV is 0.1 ppm, and the IDLH is 2 ppm in air. The boiling point is 46°F, and it is noncombustible. When carbon tetrachloride comes in contact with a hot surface, phosgene gas is evolved, which is one of the main reasons that carbon tetrachloride fire extinguishers are no longer approved. The four-digit UN identification number is 1076. The NFPA 704 designation is health 4, flammability 0, and reactivity 1. It is shipped in steel cylinders, special tank cars, and tank trucks. The primary uses are in organic synthesis, including isocyanates, polyurethane, and polycarbonate resins; in carbamate, organic carbonates, and chloroformate pesticides; and in herbicides. The structure for phosgene is shown in Figure 4.18.

Phosphine, PH₃, a nonmetallic compound, is the gas evolved when binary phosphide salts come in contact with water. It is colorless, with a disagreeable, garlic-like, or decaying fish odor. It is toxic by inhalation and is a strong irritant. It has a TLV of 0.3 ppm and

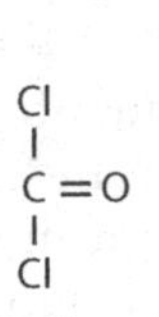

$$\begin{array}{c} Cl \\ | \\ C = O \\ | \\ Cl \end{array}$$

COCl₂

Figure 4.18 Phosgene.

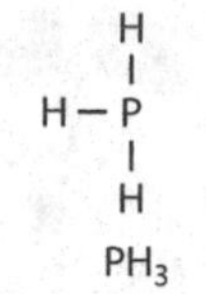

$$H$$
$$|$$
$$H-P$$
$$|$$
$$H$$
$$PH_3$$

Figure 4.19 Phosphine.

an IDLH of 200 ppm in air. It is also highly flammable (pyrophoric) and will spontaneously ignite in air. The flammable range is extremely wide, at 1.6%–98% in air. It is slightly soluble in cold water. The vapor density is 1.17, which is heavier than air. The four-digit UN identification number is 2199. The NFPA 704 designation is health 4, flammability 4, and reactivity 2. It is shipped in steel cylinders. The primary uses are in organic compounds, as a polymerization initiator, and as a synthetic dye. The structure is shown in Figure 4.19.

Diborane, B_2H_6, is a colorless gas with a nauseating sweet odor. It decomposes in water and is highly reactive with oxidizing materials, including chlorine. It is toxic by inhalation and a strong irritant, with a TLV of 0.1 ppm in air. The IDLH is 40 ppm. In addition to being toxic, diborane is also a dangerous fire risk. It is pyrophoric and will ignite upon exposure to air. The boiling point is –135°F and the flammable range is 0.8%–88% in air. The ignition temperature is 100° (37°C) to 140°F (60°C), and the flash point is 130°F (54°C). Diborane will react violently with halogenated fire-extinguishing agents, such as the halons. The four-digit UN identification number is 1911. The NFPA 704 designation is health 4, flammability 4, and reactivity 3. The white section of the diamond has a W with a slash through it, indicating water reactivity. The primary uses are as a polymerization catalyst, as fuel for air-breathing engines and rockets, as a reducing agent, and as a doping agent for p-type semiconductors. The structure is shown in Figure 4.20.

Poison gases may be encountered as gases, liquefied gases, or cryogenics. The placard will indicate poison gas; it will not tell you the material has been liquefied or turned into a cryogenic liquid. The container type will help determine the physical state of the materials. If there is a four-digit identification number on the placard, it may provide physical state information when looked up in the *Emergency Response Guidebook*.

Chlorine, Cl_2, an elemental gas, is one of the most common poison gases transported and stored. Chlorine does not occur freely in nature. Chlorine is derived from the minerals halite (rock salt), sylvite, and carnallite and is found as a chloride ion in seawater. Chlorine is nonflammable. Its vapor density is about 2.45, which makes it heavier than air. Chlorine is toxic, with a TLV of 0.5 ppm in air, and is also a strong oxidizer. Chlorine will behave much the same way as oxygen in accelerating combustion during a fire. It is also corrosive. The specific gravity of chlorine is 1.56, which makes it heavier than water, and it is only slightly soluble in cold water. Chlorine is incompatible with ammonia, petroleum

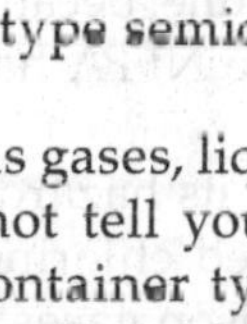

$$H \quad H$$
$$| \quad |$$
$$H-B-B-H$$
$$| \quad |$$
$$H \quad H$$
$$B_2H_6$$

Figure 4.20 Diborane.

Photo 4.33 Ton containers of chlorine.

gases, acetylene, butane, butadiene, hydrogen, sodium, benzene, and finely divided metals. The four-digit UN identification number is 1017. The NFPA 704 designation for chlorine is health 3, flammability 0, and reactivity 0. It may be encountered in 150-lb cylinders, 1-ton containers, and tank car quantities. Chlorine is nonflammable, but a BLEVE is possible if the container is exposed to flame, because it is a liquefied gas. However, the likelihood of container failure is low. The NFPA has never recorded an incident of a BLEVE involving chlorine.

Because chlorine is so common, its hazards are sometimes taken for granted. There was a time when firefighters handled chlorine leaks with turnouts and SCBAs; that is no longer an acceptable practice. Poison gases pose a large threat not only to the public, but also to emergency responders. To properly protect responders, full Level A chemical protective clothing and SCBAs must be worn for protection. Chlorine is used as a swimming pool chlorinator, as a water-treatment chemical, and for many other industrial uses (Photo 4.33). Chlorine is the fourth most-released chemical from fixed facilities on the EPA listing, with 2099 accidents resulting in the release of over 84 million pounds of chlorine. Chlorine is not often involved in transportation incidents, but when it is, it causes serious injuries (Photo 4.34).

Ethylene oxide, CH_2OCH_2, is a colorless gas at room temperature. It is miscible with water and has a specific gravity of 0.9, which is lighter than water. It is an irritant to the skin and eyes, with a TLV of 1 ppm in air. Ethylene oxide is a suspected human carcinogen. In addition to toxicity, it is highly flammable, with a wide flammable range of 3%–100% in air. The flash point is −20°F (−28°C), and the boiling point is 120°F (48°C). The ignition temperature is 1058°F (570°C). The vapor density is 1.5, which is heavier than air. The four-digit UN identification number is 1040. The NFPA 704 designation is health 3, flammability 4, and reactivity 3. The primary uses are in the manufacture of ethylene glycol and acrylonitrile, as a fumigant, and as a rocket propellant. The structure is shown in Figure 4.21.

Photo 4.34 Pressure railcar of chlorine.

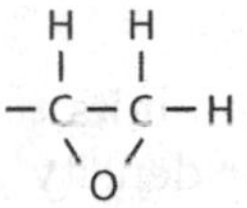

$$H-\overset{\overset{\displaystyle H}{|}}{C}-\overset{\overset{\displaystyle H}{|}}{C}-H$$

(H₂O)C H₂

Figure 4.21 Ethylene oxide.

Incidents

In Atlanta, Georgia, a small pressurized cylinder fell from a truck in the garage of the Hilton Hotel. The resulting leak of chlorine, a 2.3 poison gas, sent 33 people to the hospital, including 6 firefighters and 4 police officers. Eight people were killed and 88 others were injured as a result of leaking chlorine from a railroad tank car during a derailment in Youngstown, Florida. The liquid chlorine car ruptured, releasing a toxic cloud of chlorine. Chlorine is 2.5 times heavier than air, stays close to the ground, and has an expansion ratio of 460:1, which means that 1 gal of liquid chlorine will vaporize into 460 gal of chlorine gas. The chlorine gas settled into a low area on a nearby highway. As cars passed through the chlorine cloud, they stalled. Drivers were overcome by the chlorine, and eight of them died. A few breaths of chlorine at 1000 ppm concentration can be fatal. The concentrations at the accident scene were estimated by environmental personnel to be 10,000–100,000 ppm.

Summary

Particular attention must be paid to compressed gases in emergency response situations. Compressed gases present responders with multiple hazards, including poisons, flammables, oxidizers, cryogenics, and the hazard of the pressure in the container. If the container fails or opens up, it can become a projectile or throw pieces of the container over a mile from the incident scene. Learn to recognize pressure containers and be cautious when there is flame impingement on a pressure container.

Review questions

4.1 Which of the following is true of cryogenic liquids?
   A. They have wide flammable ranges
   B. They are extremely cold materials
   C. They have large expansion ratios
   D. All of the above

4.2 Hazard Class 2 is composed of compressed gases that may have which of the following hazards?
   A. Flammability
   B. Elevated temperature
   C. Sublimation
   D. Volatility

4.3 The flammable range of a 2.1 compressed gas occurs at which of the following locations?
   A. Below the lower limit
   B. Within the first 12%
   C. Above the upper limit
   D. Between the upper and lower limits

4.4 A compressed gas with a vapor density greater than 1 will have what relationship with air?
   A. Heavier
   B. Equal weight
   C. Lighter
   D. Will mix with air

4.5 Which of the following hydrocarbon compounds has at least one double bond?
   A. Aromatic
   B. Alkyne
   C. Alkane
   D. Alkene

4.6 Which of the following hydrocarbon compounds is considered saturated?
   A. Alkene
   B. Alkane
   C. Alkyne
   D. Aromatic

4.7 Acetylene belongs to which of the following hydrocarbon families?
   A. Alkyne
   B. Aromatic
   C. Alkene
   D. Alkane

4.8 Provide names and formulae for the compounds represented by the following structures.

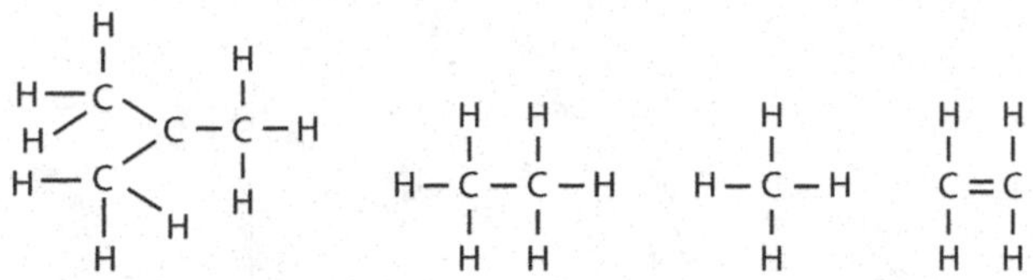

4.9 Indicate whether the following compounds are alkanes, alkenes, or alkynes.

$$CH_4 \quad C_2H_2 \quad C_3H_6 \quad C_2H_6 \quad C_3H_4$$

4.10 Match the following with the appropriate hydrocarbon family.
   A. Alkane Saturated
   B. Alkene Unsaturated
   C. Alkyne

chapter five

Flammable liquids

Class 3 materials are liquids that are flammable. Flammable liquids are involved in more fires than flammable gases because they are more abundant. Vapors of many flammable liquids are heavier than air. Most flammable liquids have a specific gravity of less than 1, so they float on water. They may also be incompatible with ammonium nitrate, chromic acid, hydrogen peroxide, sodium peroxide, nitric acid, and the halogens. According to the DOT, flammable liquids "have a flash point of not more than 141°F (60°C), or [are] any material in a liquid phase with a flash point at or above 100°F (37°C), that is intentionally heated and offered for transportation or transported at or above its flash point in bulk packaging." There is an exception to this definition that involves flammable liquids with a flash point between 100°F and 140°F (37°C and 60°C). Those liquids may be reclassified as combustible liquids and, at the option of the shipper, may be placarded flammable, combustible, or fuel oil. Even though the DOT wanted all liquids up to 140°F (60°C) to be placarded flammable, this exception was made because of public comments, particularly from the fuel oil industry. Combustible liquids are defined as "materials that do not meet the definition of any other hazard class specified in the DOT flammable liquid regulations and have flash points above 141°F (60.5°C) and below 200°F (93°C)."

The NFPA uses a classification system for flammable and combustible liquids in fixed storage facilities (Table 5.1). This system is part of the consensus standard NFPA 30, the Flammable and Combustible Liquids Code. NFPA's system further divides the flammable and combustible liquid categories into subdivisions based upon the flash points and boiling points of the liquids. NFPA is an optional fixed storage classification system that is adopted by local jurisdictions and does not apply to transportation of hazardous materials. DOT regulations are promulgated as a result of laws passed by congress and supersede NFPA 30. Examples of liquids in the various classification categories are listed in Table 5.2.

As previously mentioned, all of the DOT hazard classes identify only the most severe hazard of materials in the hazard class. All classes have hidden hazards that are both chemical and physical in nature. Flammability is not the only hazard associated with Class 3 flammable liquids. They may also be poisonous or corrosive. For general purposes, there are no UN/DOT subclasses of flammable liquids. Emergency responders should realize that all materials with red placards will burn under certain conditions. Appropriate precautions should be taken when dealing with flammable liquids. The dividing line for flammable and combustible liquids is 140°F (60°C) in the DOT regulations and 100°F (37°C) in the NFPA standard. Those liquids with flash points below 100°F and 140°F (37°C and 60°C), respectively, are considered flammable; those above are considered combustible. The problem with classifying flammable and combustible liquids in emergency response situations is ambient temperature and radiant heat.

Ambient temperatures near or above 100°F (37°C) are common in many parts of the country. The radiant heat from the sun or an exposure fire can reach temperatures well

Table 5.1 NFPA 30 Flammable and
Combustible Liquids Classifications

Flammable

Class IA = FP < 73°F—BP < 100°F

Class IB = FP < 73°F—BP > 100°F

Class IC = FP > 73°F—BP < 100°F

Combustible

Class II = FP > 100°F < 140°F

Class IIIA = FP > 140°F < 200°F

Class IIIB = FP > 200°F

Table 5.2 NFPA Flammable and Combustible
Liquids Examples

Class I	Class II	Class III
Acetone	Acetic acid	Benzyl chloride
Benzene	Fuel oil	Corn oil
Carbon disulfide	Kerosene	Linseed oil
Gasoline	Decane	Nitro benzene
Methanol	Pentanol	Parathion

above 100°F (37°C). Many of the liquids classified as combustible have flash points at or near 100°F (37°C). Surfaces such as roadways may have temperatures well above the flash points of combustible liquids. When a combustible liquid is spilled on a surface, it may be heated above its flash point and the combustible liquid may act like a flammable liquid.

Because of the uncertainty of the potential flammability of combustible liquids, they will be referred to as flammable liquids throughout this book. It is highly recommended that they be treated the same way on the incident scene for the purpose of responder and public safety. Flash point will be discussed in detail later in this chapter. It is important, however, to note at this point that the flash point temperature is the most critical factor in determining if a flammable liquid will burn. Important precautions to prevent ignition from occurring are to control ignition sources at the incident scene and keep personnel from contacting the liquid without proper protective clothing.

According to DOT and EPA statistics, flammable liquids are involved in over 64% of all hazardous materials incidents (Photo 5.1). This should not be surprising since flammable liquids are used as motor fuels for highway vehicles, railroad locomotives, marine vessels, and aircraft. Additionally, many flammable liquids are used to heat homes and businesses. Effective handling of flammable liquids at an incident scene requires that emergency responders have a basic understanding of the physical characteristics of flammable liquids.

Effects of temperature on flammable liquids

Many of the physical characteristics of flammable liquids involve temperature. It is important to understand that there are different temperature scales listed in reference books; make sure which scale is used when materials are researched. There is a big difference

Photo 5.1 Flammable liquids must be at their flash point before combustion can occur if an ignition source is present and the material is within its flammable range

between the temperatures of the Fahrenheit and centigrade (Celsius) scales. The Fahrenheit scale is familiar to most emergency responders because it is the temperature used most commonly in the United States. The centigrade scale is used predominantly throughout the rest of the world. It is also used within the scientific and technical community of the United States. Reference books used by emergency responders to obtain information on hazardous materials may also use the centigrade scale. Temperature conversion formulas used to convert centigrade to Fahrenheit and Fahrenheit to centigrade are shown in Table 5.3.

Table 5.3 Temperature Conversion Formulas

From °C to °F: $°F = \dfrac{9 \times °C}{5} + 32$

Example:

$C\ Temp = 40° = \dfrac{9 \times 40}{5} + 32 = \dfrac{360}{5} = 72 + 32 = 104°F$

From °F to °C: $°C = \dfrac{5(°F - 32)}{9}$

Example:

$F\ Temp = 104° = \dfrac{5(104 - 32)}{9} = \dfrac{5 \times 72}{9} = \dfrac{360}{9} = 40°C$

Another temperature scale that may be encountered on a less frequent basis is the Kelvin scale, also known as absolute temperature. This scale is used principally in theoretical physics and chemistry and in some engineering calculations. Absolute temperatures are expressed either in degrees Kelvin or in degrees Rankine, corresponding, respectively, to the centigrade and Fahrenheit scales.

Temperatures in Kelvin are obtained by adding 273° to the centigrade temperature (if above 0°C) or subtracting the centigrade temperature from 273 (if below 0°C). Degrees Rankine are obtained by subtracting 460 from the Fahrenheit temperature. Absolute zero is the temperature at which the volume of a perfect gas theoretically becomes zero and all thermal motion ceases, which occurs at −273.13°C or −459.4°F.

Boiling point

The boiling point of a flammable liquid is the first physical characteristic that will be discussed. It is a physical characteristic that is affected by the temperature of the liquid and atmospheric pressure. Boiling point is defined as "the temperature at which the vapor pressure of a liquid equals the atmospheric pressure of the air." Atmospheric pressure is 14.7 psi at sea level. Liquids naturally want to become gases; it is atmospheric pressure that keeps a liquid from becoming a gas at normal temperatures and pressures. Atmospheric pressure decreases as altitude increases. The higher the altitude, the lower the boiling point of any given material (Table 5.4). For example, water boils at 212°F (100°C) at sea level. In Denver, Colorado, the altitude is 1 mile (5280 ft) above sea level; water boils at approximately 203°F (95°C). At Pikes Peak, Colorado, the altitude is more than 14,000 ft above sea level, and water boils at approximately 186°F (85.5°C). Atmospheric pressure decreases as altitude increases because the air is thinner at higher altitudes.

Atmospheric pressure is always pushing down on the surface of a liquid in an open tank or in a spill. With atmospheric pressure pushing down on a liquid, there is not much vapor moving away from the surface of the liquid. There is a direct relationship between the boiling point of a liquid and the amount of vapor present in a spill or in a container. Inside a container, increased vapor will increase the pressure in the tank. In an open spill, the vapor content will be greater. If a flammable liquid is above its boiling point, more vapor will be produced. The more heat that is applied to the container or spill, the more vapor that is produced, the higher the pressure in the container or the further the vapor will travel away from a spill.

If a flammable liquid is in a closed container, the vapor pressure will increase inside the container as the temperature of the liquid increases. This increase in temperature can come from many different sources. Increases in ambient temperature, radiant heat from the sun, or a nearby fire can increase the vapor pressure in a container. With containers that have pressure-relief valve(s) installed, as the pressure increases in a container, it will

Table 5.4 Altitude Effect on Water Boiling Point

Altitude	Atmospheric pressure (psi)	Boiling point (°F)
Sea level	14.7	212
3,300 ft	13.03	203
5,280 (1 mile)	12.26	201
10,000 ft	10.17	194
13,000 ft	9.00	188

Values are approximate.

reach the setting on the pressure-relief valve and the relief valve will function. If this pressure increase occurs in a container that does not have a relief valve, the container may rupture. Rupture may also occur in a container with a relief valve if the pressure rises too fast for the relief valve to vent the material into the air or if the relief valve is not working properly. In either case, the rupture may be violent, with a fireball and flying pieces of tank that can travel over a mile from the blast site.

Factors affecting boiling point

Molecular weight

A number of outside factors can help determine the relative boiling point of a flammable liquid. They include molecular weight, polarity, and branching. The first consideration in determining boiling point is the molecular weight of a compound. Each element in a compound has an atomic weight, listed on the periodic table. Most flammable liquids are made up of hydrogen, carbon, and a few other elements. They can be organized into families called hydrocarbons and hydrocarbon derivatives.

Hydrocarbons are made up of just carbon and hydrogen. Carbon has an atomic weight of 12, so each carbon atom in a compound weighs 12 AMU. Hydrogen has an atomic weight of 1, so each hydrogen atom in a compound will weigh 1 AMU. Methane is the smallest hydrocarbon, made up of one carbon atom and four hydrogen atoms. The hydrocarbon compound methane has a weight of 16 AMU. In the compound butane, there are four carbon atoms, each weighing 12 AMU (which equals 48) and 10 hydrogen atoms, each weighing 1 AMU (which equals 10). Butane therefore weighs 58 AMU. Molecular weight of a compound is the sum of all the weights of all the atoms. As more carbons are added to a hydrocarbon compound, the heavier it will be. The heavier it is, the more energy it will take to get the liquid to boil and overcome atmospheric pressure. Therefore, the heavier a compound is, the higher the boiling point it will have. In Figure 5.1, there are three hydrocarbon compounds. Compare the molecular weights of each and look at their boiling points. You can readily see the relationship between heavy compounds and high boiling points.

Polarity

Hydrocarbon derivatives are compounds with other elements in addition to hydrogen and carbon. Weight will still determine boiling points when comparing hydrocarbon derivatives within the same family. However, when comparing different families, the concept of polarity has to be considered with some of the compounds. Polarity is the second factor that affects boiling point. There is a rule in chemistry that says, "Like materials dissolve like materials." Materials that are polar will generally mix together either totally or partially. Another term used when materials mix together is miscibility. If materials are miscible, they will mix with some other material(s). If a material is immiscible, it will not mix with another.

All polar compounds are alike in terms of being polar. Therefore, polar compounds are soluble in polar compounds. One of the main reasons polarity is discussed in terms of

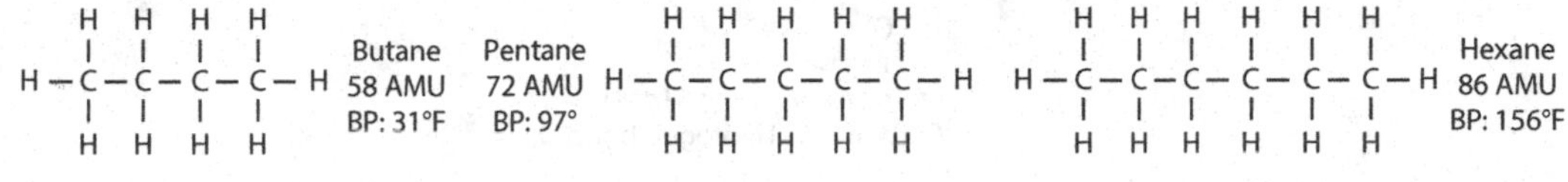

Figure 5.1 Effect of molecular weight on boiling point.

emergency response is because of the foam used for fighting flammable liquid fires. Two general types of foam are used to extinguish flammable liquid fires: hydrocarbon foam and polar solvent foam, which is sometimes referred to as alcohol-type foam. The reason that different types of foam are necessary for flammable liquid fires is polarity.

Water is a polar compound. Because "like dissolves like," water is miscible with most polar solvents. The main ingredient of firefighting foam is water, along with foam concentrate and air. If regular foam is put on a polar solvent liquid, such as alcohol, the water will be removed from the foam by the polar solvent and the foam blanket will break down. To effectively extinguish fires with polar solvents, it is necessary to use polar solvent or alcohol-type foam. This information becomes even more important today as ethanol, a highly polar alcohol, is being added to hydrocarbon fuels and being used in higher percentages. The higher the percentage of ethanol used, the more likely polar solvent foam will be needed. In some fuels, ethanol makes up 85% of the mixture.

Polarity is presented here to discuss the types of flammable liquids that are polar solvents and the effect that polarity has on them. Polar solvents, such as alcohols, aldehydes, organic acids, and ketones, require special foam to extinguish fires. Choosing the right type of foam is part of the process of effectively managing a flammable liquid incident. Polar liquids tend to have higher boiling points than nonpolar liquids. Polarity is said to have the effect of raising the boiling point of a liquid. There are two types of structure that represent polarity in hydrocarbon-derivative compounds: the carbonyl structure and hydrogen bonding (Figure 5.2). Organic acids that are considered super polar have both types.

A carbonyl contains a carbon-to-oxygen double bond. Most double bonds are reactive; however, with the carbonyl family, except for aldehyde, the double bond is protected by hydrocarbon radicals on either side of the carbonyl. This prevents oxygen from getting to the bond and breaking it. Therefore, not all carbonyl compounds are going to polymerize. Ester compounds have the double bond, and their hazard is polymerization. Some individual compounds may have this hazard in other families, but in general they do not polymerize.

Double bonds are stable within the carbonyl families, except for aldehydes and esters. The carbonyl structure is shown in Figure 5.3, first by itself, and then in the hydrocarbon functional groups ketone, aldehyde, organic acid, and ester. All four compounds are polar because of the carbonyl, and the organic acid also has a hydrogen bond. The amount of polarity is generally the same between the ketone, ester, and aldehyde. The polarity of an acid is higher because of the hydrogen bond and the carbonyl in the same compound, like a double dose of polarity.

When oxygen and hydrogen covalently bond, the bond is polar. Water has a hydrogen–oxygen bond that gives water polarity. Water is a liquid between 32°F and 212°F and has a molecular weight of 18 AMU. The molecular weight of an average air molecule is 29 AMU. Therefore, the water molecule is lighter than air! Water should be a gas at normal temperatures and pressures, but it is a liquid due to polarity. Even though water has a molecular weight of 18 AMU, it has a boiling point of 212°F (100°C)! If water was not polar, there would be no life on earth as we know it.

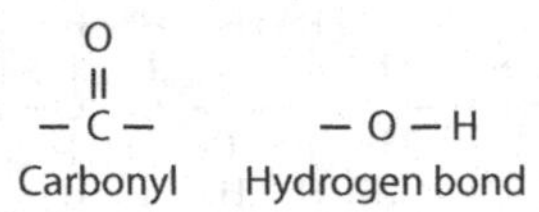

$$
\begin{array}{cc}
\text{O} & \\
\| & \\
-\text{C}- & \quad -\text{O}-\text{H} \\
\text{Carbonyl} & \text{Hydrogen bond}
\end{array}
$$

Figure 5.2 Carbonyl and hydrogen bonds.

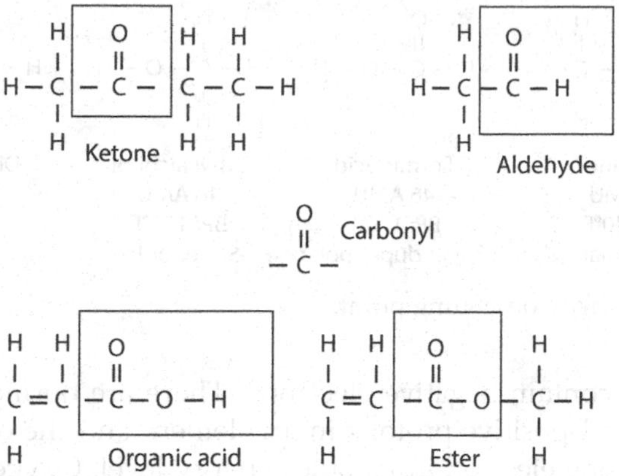

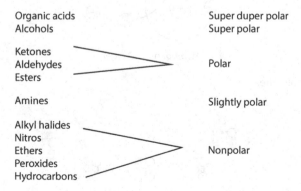

Figure 5.3 Carbonyl compounds.

The hydrogen bond is also a polar bond. There are two hydrocarbon-derivative families presented in this book that have hydrogen bonding: alcohols and organic acids. The organic acid also has a carbonyl bond. This double polarity makes organic acid the most polar material among the hydrocarbon derivatives. Many materials exhibit some degree of polarity. However, only the polarity of carbonyl and hydroxyl groups of hydrocarbon derivatives will be discussed here. All other materials mentioned will be considered nonpolar. The hydrocarbon-derivative compounds are listed in Figure 5.4. They are ranked in the order of descending degree of polarity. Organic acids are the most polar of all the materials listed and will be referred to as "super-duper polar." Alcohols are the second most polar compounds, and they will be referred to as "super polar." The remaining carbonyls are polar, with the exception of amines, which are "slightly polar." All other compounds discussed will be considered nonpolar. In the structures in Figure 5.5, compare the polarity of the compounds listed and notice the effect that polarity has on boiling point.

Formic acid has the highest boiling point even though it weighs less than dimethyl ketone. Methyl alcohol has the next highest boiling point; it also weighs less than dimethyl ketone. The weights of methyl alcohol and ethane are about the same, yet the difference in boiling point is 275°F (135°C).

To fully understand the concept of polarity, we must revisit the structure of the atom presented in Chapter 2. The nucleus contains positive protons, and the energy levels

Organic acids	Super duper polar
Alcohols	Super polar
Ketones	
Aldehydes	Polar
Esters	
Amines	Slightly polar
Alkyl halides	
Nitros	
Ethers	Nonpolar
Peroxides	
Hydrocarbons	

Figure 5.4 Polarity of hydrocarbon derivatives.

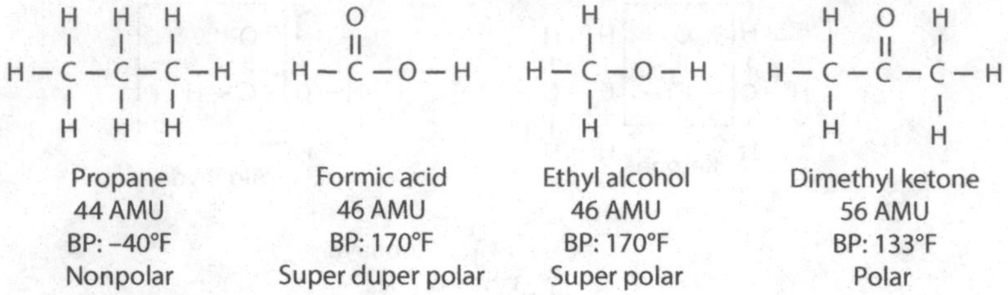

H H H	O	H	H O H
| | |	||	|	| || |
H—C—C—C—H	H—C—O—H	H—C—O—H	H—C—C—C—H
| | |		|	| |
H H H		H	H H

Propane	Formic acid	Ethyl alcohol	Dimethyl ketone
44 AMU	46 AMU	46 AMU	56 AMU
BP: −40°F	BP: 170°F	BP: 170°F	BP: 133°F
Nonpolar	Super duper polar	Super polar	Polar

Figure 5.5 Effect of polarity on boiling point.

outside the nucleus contain negative electrons. There are normally an equal number of negative electrons and positive protons in an element and the elements of a compound. Hydrogen has only one electron in its outer energy level. Oxygen has 16 electrons in its outer energy levels. As hydrogen and oxygen bond together, oxygen has a tendency to draw the one electron from hydrogen toward the oxygen side of the covalent bond. This exposes the positive nucleus of hydrogen, creating a slightly positive side to the hydrogen end of the molecule. Oxygen has drawn the electron from hydrogen toward its nucleus. In doing so, oxygen has more negative electrons near the nucleus than positive protons inside. This creates a slightly negative field around the oxygen side of the molecule. This concept is represented in Figure 5.6.

Hydrogen–oxygen bonded molecules are then attracted to other hydrogen–oxygen molecules because of the rule in chemistry that says, "Opposite charges attract." Positive ends of the hydrogen–oxygen molecules are attracted to the negative ends of other hydrogen–oxygen molecules. This attraction holds the molecules together, so that it takes more energy to break them apart and cause the compound to boil. This attraction of molecules is illustrated in Figure 5.7. To illustrate this concept, consider two bags of groceries weighing the same amount. If you place one hand on each, they can be lifted fairly easily. However, if you place glue under one and allow it to set, it will be difficult to lift that bag of groceries. It

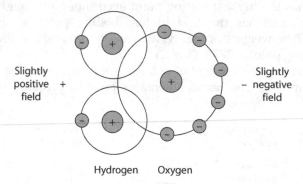

Slightly positive + field   Slightly − negative field

Hydrogen   Oxygen

Figure 5.6 Water molecule.

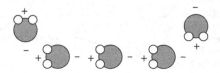

Figure 5.7 Polar attraction of water molecule.

has a kind of "stickum" on the bottom. Polarity acts like the stickum placed on the grocery bag and makes polar compounds regardless of weight harder to boil.

Branching

Branching is the last factor that affects the boiling point of a flammable liquid. (Branching was discussed in Chapter 4 under Isomers in "Hydrocarbon families" section.) Branched compounds are all manmade. Since branching does not occur naturally, it is done for a particular purpose, usually because there is an economic value.

When a compound is branched, it has a lower boiling point than the unbranched version of the liquid. In Figure 5.8, the structures for pentane and isopentane are shown with the corresponding boiling points. The effect that branching has on the boiling point of a liquid is clearly visible.

Another type of structure that has an effect on the boiling point of a liquid is called cyclic. Cyclic compounds with five, six, or seven carbon atoms are highly stable and materials that are cyclic tend to have higher boiling points than straight-chained compounds with the same number of carbons. Those with less than five carbons are reactive and come apart easily. Cyclic compounds with more than seven carbon atoms tend to fragment. Cyclic compounds will be discussed later in this chapter under Hydrocarbons and the Aromatic Compounds. In Figure 5.9, hexane is compared to cyclohexane. Notice the difference that the cyclic structure has on the boiling point.

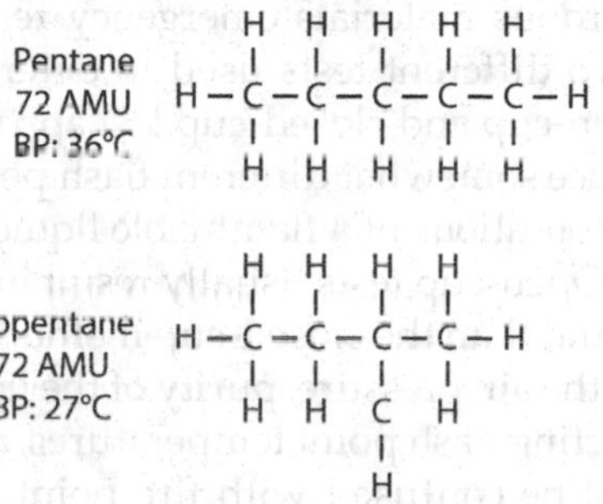

Figure 5.8 Effect of branching on boiling point.

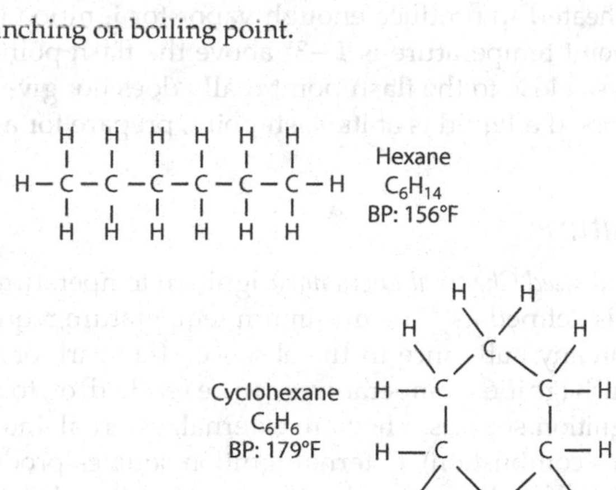

Figure 5.9 Cyclic structure effect on boiling point.

Flash point

Next and probably most important of the physical characteristics of a flammable liquid is its flash point. Flash point is the most important information for emergency responders to obtain about a flammable liquid. Flash point, more than any other characteristic, helps to define the flammability hazard of a liquid in any given situation. If a flammable liquid is not at its flash point temperature, *it will not burn*. Flash point is defined as "the minimum temperature to which a liquid must be heated to produce enough vapor to allow a vapor flash to occur (if an ignition source is present)." After all, it is the vapor that burns, not the liquid, so the amount of vapor present is critical in determining whether the vapor will burn.

Flash point is a measurement of the temperature of the liquid. Therefore, even if the ambient temperature is not at the flash point temperature, the liquid may have been heated to its flash point by some external heat source. For example, the radiant heat of the sun, heat from a fire, or heat from a chemical process may heat the liquid to its flash point. If an ignition source that produces a temperature at or higher than the flash point of the liquid is present, ignition can and probably will occur. While flash point is the first and most important physical characteristic to consider, it is not the only one. Flash point facilitates combustion, but several other factors must come into play once a liquid is at or above its flash point temperature. In order for ignition to occur, if a flammable liquid is at its flash point temperature, there has to be an ignition source present that has a temperature at or above the ignition temperature of the vapor.

Reference books, chemical data bases, and the Internet that are used to research chemical characteristics in hazardous materials emergency response may show different flash point values. There are two different tests used to determine the flash point of a liquid. They are known as the open-cup and closed-cup test apparatus. The differences in the testing procedures often produce somewhat different flash point temperatures. Open-cup flash point tests try to simulate conditions of a flammable liquid in the open, such as a spill from a container to the ground. Open-cup tests usually result in a higher flash point temperature for the same flammable liquid than the closed-cup method. The flash point of a liquid varies with the oxygen content of the air, pressure, purity of the liquid, and the method of testing. If reference books give conflicting flash point temperatures, *use the lowest flash point value given*.

Flash point should not be confused with fire point. Fire point is the temperature at which the liquid is heated to produce enough vapor for ignition and sustained combustion to occur. Fire point temperature is 1°–3° above the flash point temperature. The fact that the fire point is so close to the flash point really does not give it much significance to emergency responders: if a liquid is at its flash point, prepare for a fire.

Ignition temperature

According to the *Condensed Chemical Dictionary*, ignition temperature, also known as autoignition temperature, is defined as "The minimum temperature required to initiate self-sustained combustion in any substance in the absence of a spark or flame." There are three basic methods in which ignition temperatures can be reached, or, to put it another way, there are three types of ignition sources. They are external, external–internal (autoignition), and internal (spontaneous combustion). External ignition sources produce heat that enters the vaporspace or liquid itself and transfers their heat energy directly to the flammable material. Examples of external ignition sources are open flames, sparks (electrical, static, or frictional), and heated objects. Sparks are capable of developing temperatures ranging from 2000°F to 6000°F (1093°C–3315°C). External–internal (autoignition) sources heat the vapor or liquid through an indirect method. Three types of indirect methods are radiant-heat transfer, convection-heat transfer, and combustion-heat transfer. From any of these sources, heat is transferred until ignition temperature is reached and ignition occurs, without the presence of any open flame or spark. Spontaneous ignition is the final type of ignition source. In this case, the material itself produces heat sufficient to reach its own ignition temperature. This can occur in two ways: as a result of the biological processes of some microorganisms and slow oxidation. Biological processes are usually not associated with flammable liquids, but rather with organic materials, such as hay and straw. Activity of biological organisms within the material generates heat, and heat is confined by the materials until ignition temperature is reached and ignition occurs. Slow oxidation is a chemical reaction. Chemical reactions may produce heat. If the heat is insulated from dissipating to the outside of the material, it will continue to build up. As the heat builds, the material is heated from within. The process continues until the ignition temperature of the material is reached and ignition occurs.

As an example of autoignition, consider a pan of cooking oil on an electric stove. Cooking oils are animal or vegetable oils. They are combustible liquids with high boiling points and flash points. If the heat is turned up too high on a stove with a pan of cooking oil, the oil may catch fire. Many kitchen fires occur because of cooking oils or grease being overheated. The reason for this is ignition temperature. Liquids that have high boiling and flash points conversely have low ignition temperatures. Corn oil, commonly used as cooking oil, has an ignition temperature of 460°F (237°C). If corn oil is heated on a stove to 460°F (237°C), it will autoignite. Gasoline on the other hand has a low boiling point and low flash point and has a high ignition temperature of approximately 800°F, depending on the blend. If gasoline were placed on the same stove without an ignition source, the stove would not produce a temperature high enough to autoignite the gasoline. It would just boil away into vapor.

Characteristics like this can be a real "foolers" on the incident scene, particularly when dealing with flammable liquids where the boiling points and flash points are high. Responders and industry personnel sometimes become complacent when dealing with flammable liquids with high boiling and flash points. They think that because the boiling and flash points are high, the danger of fire is low. If, however, ignition sources are not controlled, what little vapor is present above the liquid can ignite if the temperature of the ignition source is above the ignition temperature of the liquid (Table 5.5). For example, a lighted cigarette, with no drafts, has a surface temperature of about 550°F (287°C), and Number 1 fuel oil has an ignition temperature of 444° F (228°C). A cigarette can be an ignition source for a combustible liquid because of the low ignition temperatures. A cigarette cannot be an ignition source for gasoline because the temperature of the lighted cigarette, without a draft, is below gasoline's ignition temperature of approximately 800°F (426°C). Table 5.6 provides the temperatures of common ignition sources.

Table 5.5 Ignition Temperatures of Common Combustible Materials

Wood	392°F
#1 Fuel oil	444°F
Paper	446°F
60 Octane gas	536°F
Acetylene gas	571°F
Wheat flour	748°F
Corn	752°F
Propane gas	871°F

Table 5.6 Temperatures of Common Ignition Sources

Lighted cigarette, no drafts	550°F
Lighted cigarette, with drafts	1350°F
Struck match	2000°F+
Electric arc	2000°F+

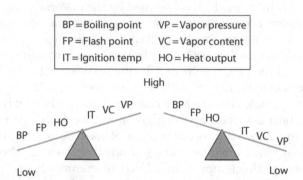

Figure 5.10 Physical relationships of parameters of combustion.

Referring back to the seesaw (Figure 5.10), there is an opposite relationship between boiling/flash point and ignition temperature. Flammable liquids that have low boiling points and flash points have high ignition temperatures. Liquids that have high boiling points and flash points have low ignition temperatures. High-boiling-point and flash-point flammable liquids generally are more difficult to ignite than low-boiling-point and flash-point flammable liquids. However, once they are ignited, they have a much higher heat output than the lower-boiling-point and flash-point flammable liquids. Because of this high heat output, high-boiling-point and high-flash-point flammable liquid fires are much more difficult to extinguish than low-boiling-point and flash-point flammable liquid fires. There is a parallel relationship between boiling point and heat output. Materials that have low boiling points have low heat outputs.

Flammable range

Flammable range is the last physical characteristic that must be met for combustion to occur. If a material has reached its flash point temperature and if and the ignition source is at the ignition temperature of the liquid, the vapor from the liquid must be within the flammable range of the material for combustion to finally occur. Flammable range is defined as the percent of vapor in air necessary for combustion to occur and is referred to as the explosive limit. It is expressed on a scale from 0% to 100% (Figure 5.11). There is an upper explosive limit (UEL) and a lower explosive limit (LEL) between the two. Additionally, a proper mixture of vapor (fuel) and air must occur in order to have combustion. Above the UEL, there is too much vapor and not enough air; in other words, the mixture is too rich to burn. Below the LEL, there is enough air, but too little vapor; therefore, the mixture is too lean to burn. Most hydrocarbon flammable liquids have explosive limits between 1% and 12%. They have a narrow flammable range; all of the conditions must be just right for combustion to occur. The

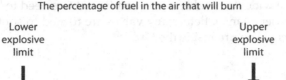

Figure 5.11 Flammable range (explosive limit).

Table 5.7 Flammable Ranges
of Flammable Liquid Families

Fuel family	1%–8%
Aromatic hydrocarbons	1%–7%
Ketones	2%–12%
Esters	1%–9%
Amines	2%–14%
Alcohols	1%–36%
Ethers	2%–48%
Aldehydes	3%–55%
Acetylene	2%–85%

liquid must be at its flash point, the air–vapor mixture must be within its flammable range, and the ignition source temperature must be above the ignition temperature of the liquid. Table 5.7 provides typical flammable ranges of flammable liquid families.

Some families of hydrocarbons and hydrocarbon derivatives have wide flammable ranges. Wide flammable-range materials are dangerous because they can burn inside a container since they burn rich. Alcohols, ethers, and aldehydes are families of flammable liquids that have wide flammable ranges and should be addressed with extreme caution (Figure 5.12).

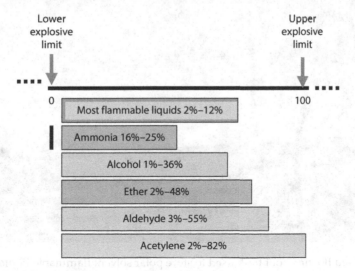

Figure 5.12 Flammable ranges of common flammable materials.

Acetylene also has a wide flammable range, and firefighters need to be careful when fighting fires involving acetylene tanks. Before any valves are turned to shut off the source of acetylene, make sure there is no fire inside the tank.

Vapor pressure

Boiling point is related to vapor pressure and vapor content, although the relationship is opposite in nature. Materials with low boiling points and flash points will have high vapor pressure and high vapor content. The vapor pressure of a liquid is defined in the *Condensed Chemical Dictionary* as "the characteristic at any given temperature of a vapor in equilibrium with its liquid or solid form. This pressure is often expressed in millimeters of mercury, mm Hg."

In simple terms, vapor pressure is the pressure exerted by the liquid against atmospheric pressure. When the pressure of the vapor is greater than atmospheric pressure, vapor will spread beyond an open container or an open spill. If liquid is in a container, vapor pressure is the pressure exerted by the liquid vapors on the sides of the container (Photo 5.2). For example, when a gas has been liquefied, the only thing keeping it a liquid is the pressure in the tank; the liquid is already above its boiling point. The pressure inside the container is the atmospheric pressure in that container. It can be much higher than outside atmospheric pressure. For example, if the pressure in the tank is 50 psi, the atmospheric pressure in that tank is 50 psi regardless of conditions outside the tank.

Photo 5.2 Closed floating-roof tank used to store polar solvent flammable liquids, such as alcohols and ketones.

Vapor content

Vapor content is the amount of vapor that is present in a spill or open container. The lower the boiling point and the flash point of a liquid, the more vapor there will be. The parallel relationship of boiling point and flash point is comparable to the opposite relationship of vapor pressure and vapor content shown in the diagram, using a seesaw to illustrate the up-and-down and opposing relationship. As shown in the previous illustration, when the boiling point and flash point are low, vapor content and vapor pressure are high. When boiling point and flash point are high, vapor pressure and vapor content are low. The lower the boiling point or flash point of a liquid, the higher the vapor content at a spill and the higher the vapor pressure inside a container. If a liquid is above its boiling point temperature, there is likely to be more vapor moving farther away from a spill. If the liquid is below its boiling point temperature, there will be some vapor above the surface of the liquid, but it will not travel far. If the flammable liquid in a container is below its boiling point, the vapor content and vapor pressure in the container will be low. If the flammable liquid in a container is above its boiling point, the vapor content and vapor pressure in the container will be high.

Vapor density

Vapor density is a physical characteristic that affects the travel of vapor; it is the weight of a vapor compared to the weight of air (see Figure 5.13). Vapor density is usually determined in the reference books by dividing the molecular weight of a compound by 29, which is the assumed molecular weight of air. Air is given a weight value of 1, which is used to compare the vapor density of a material. If the vapor of a material has a density greater than 1, it is considered heavier than air. Heavier-than-air vapor will lie low to the ground and collect in confined spaces and basements. This can cause problems because many ignition sources are in basements, such as hot-water heaters and furnace pilot lights. If the vapor density is less than 1, the vapor is considered to be lighter than air, so it will move up and travel farther from the spill.

Another term associated with vapor is volatility. It is the tendency of a solid or a liquid to pass into the vapor state easily. This usually occurs with liquids that have low boiling points. A volatile liquid or solid will produce significant amounts of vapor at normal temperatures, creating an additional flammability hazard. The vapor produced by a volatile liquid is affected by wind, vapor pressure, temperature, and surface area. Temperature

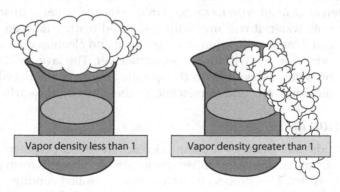

Figure 5.13 Vapor density.

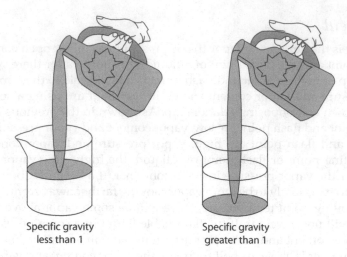

Specific gravity
less than 1

Specific gravity
greater than 1

Figure 5.14 Specific gravity.

always causes an increase in vapor pressure and vapor content in an incident. The more vapor pressure in a container, the greater the chance of container failure. The more vapor content, the farther the vapor may travel away from a spill.

Specific gravity

Specific gravity is to water what vapor density is to air. Specific gravity is the relationship of the weight of a liquid to water or another liquid (see Figure 5.14). Like air, water is given a weight value of 1. If a flammable liquid has a specific gravity greater than 1, it is heavier than water and will sink to the bottom in a water spill. If a flammable liquid has a specific gravity less than 1, it will float on top of the water. The specific gravity of a flammable liquid is important in a water spill because it will determine what tactics are necessary to contain the spill. Specific gravity is the theory behind the construction of overflow and underflow dams, which are used to stop the flow of hazardous materials in water spills. Overflow dams are constructed for liquids heavier than water. The liquid sinks to the bottom of the water, and the water flows over the dam, while the hazardous liquid is stopped by the dam. Underflow dams are constructed for liquids lighter than water (Photo 5.3). The liquid floats on the surface of the water and is stopped by the top of the dam, while the water continues to flow through a pipe at the bottom of the dam.

A term often associated with hazardous materials and water is miscibility. If a chemical is miscible with water, it will mix with water and neither float on top or sink to the bottom, which could make mitigation of the incident and cleanup difficult. If the chemical is not miscible with water, it will form a separate layer. The layer will form on top or on the bottom of the water, depending on the specific gravity of the liquid. Most flammable liquids are lighter than water and immiscible, so they float on the surface.

Polymerization and plastics

Flammable liquids may undergo a chemical reaction called polymerization, in which a large number of simple molecules, called monomers, combine to form long-chained molecule called a polymer. This process is used under controlled conditions to create plastics (Table 5.8). Alkene hydrocarbon compounds and hydrocarbon derivatives, such as aldehydes,

Photo 5.3 Underflow dams are used to stop the flow of flammable liquids that are lighter than water.

Table 5.8 Types of Plastics

Thermoplastics	Thermosets
ABS	Polyurethane
Acrylics	Amino resins
Nylons	Epoxy resins
Polycarbonate	Phenolic resins
Polyesters	Polyesters
Polyethylene	
Polypropylene	
Polystyrene	
Polyurethane	
Polyvinyl chloride	

alkyl halides, and esters, and the aromatic hydrocarbon styrene may undergo polymerization. There are other monomers that are flammable and can polymerize, but their primary hazard is poison. Monomers can be flammable liquids, flammable gases, and poisons.

When a monomer, such as styrene, is transported or stored, an inhibitor is included in solution to keep the styrene from polymerizing. An inhibitor, usually an organic compound, retards or stops an unwanted polymerization reaction. If an accident should occur, this inhibitor can become separated from the monomer, and a runaway polymerization may occur. Phenol—a deadly poison—is used as an inhibitor for vinyl chloride. Dibutylamine is used as an inhibitor for butadiene.

During a normal chemical reaction to create a particular polymer from a monomer, a catalyst is used to control the reaction. A catalyst is any substance that in a small amount noticeably affects the rate of a chemical reaction, without itself being consumed or undergoing a chemical change. For example, phosphoric acid is used as a catalyst in some polymerization reactions. Once an uncontrolled polymerization starts at an incident scene, it will not be stopped until it has completed its reaction, no matter what responders may try to

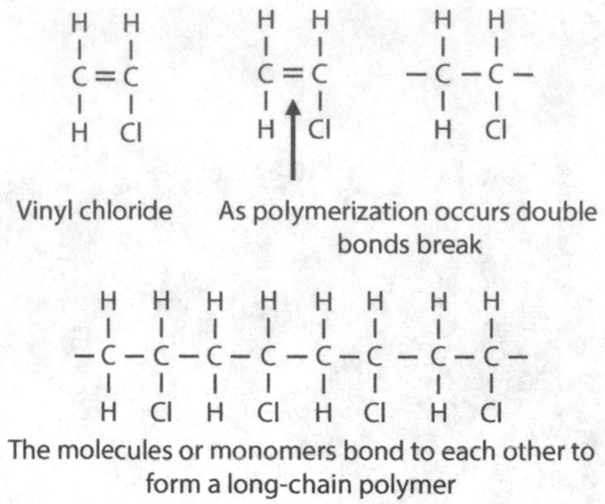

Figure 5.15 Polymers.

do. Bringing in a tanker full of inhibitor to apply to the reaction will do no good at this point even if it was available. If the polymerization occurs inside a tank, the tank may rupture violently. If a container of a monomer is exposed to fire, it is important to keep the container cool. Heat from an exposure fire may start the polymerization reaction. In Figure 5.15, the monomer vinyl chloride is shown along with the process of polymerization of the vinyl chloride molecules. This has been an abbreviated explanation of polymers and plastics, which are fairly complicated subjects. An entire book could be written on them. The most important thing for responders to understand about monomers and polymers is to be able to recognize them and the danger they present in the uncontrolled conditions of the incident scene.

Animal and vegetable oils

Some combustible liquids, such as animal and vegetable oils, have a hidden hazard: they may burn spontaneously when improperly handled. They have high boiling and flash points, narrow flammable ranges, low ignition temperatures, and are nonpolar. Examples of these liquids are linseed oil, cottonseed oil, corn oil, soybean oil, lard, and margarine. These unsaturated materials can be dangerous when rags containing residue are not properly disposed of or they come in contact with other combustible materials.

A double bond exists in the chemical makeup of animal and vegetable oils that reacts with oxygen in the air. This reaction causes the breakage of the double bond, which creates heat. If the heat is allowed to build up in a pile of rags, for example, spontaneous combustion will occur over a period of hours.

In Verdigris, Oklahoma, where the author was the assistant fire chief, a fire occurred in an aircraft hangar at a small airport. The owner's living quarters were on the second level of the hangar. Workers had been polishing wooden parts of an airplane with linseed oil in the afternoon. Rags used to apply linseed oil were placed in a plastic container inside a storage room in the hangar, just below the living quarters.

Around 2 a.m., the rags with the linseed oil spontaneously ignited and the fire traveled up the wall into the living quarters. Fortunately, the owner had smoke detectors; the family was awakened, and the fire department was called promptly. The fire was quickly extinguished with a minimum of damage. The V-pattern on the wall in the storage room led right back to the plastic container where the linseed oil-soaked rags had been placed. While investigating the fire cause, there was little doubt in the author's mind what had happened; the confinement of the pile allowed the heat to build up as the double bonds were broken in the linseed oil, which combined with oxygen in the air, and spontaneous combustion occurred.

A fire occurred on February 23, 1991, at the One Meridian Plaza Building in Philadelphia, resulting in the deaths of three firefighters (Photo 5.4). The fire was started by spontaneous combustion in linseed oil-soaked rags that were improperly disposed of after use. The fire occurred on the 22nd floor of the 30-story building. PCB contamination from the fire made the building uninhabitable, and it is currently being torn down.

While the author was gathering information for a Firehouse Magazine article on the Jacksonville, Florida, fire department, an alarm came in for an odor in a restaurant around noon (Photo 5.5). The author was invited by the battalion chief to ride along to the incident. Upon arrival, the hazmat team found a 5 gal pail of a corrosive liquid they thought was the problem. They removed it from the building and the odor inside continued. Further investigation led to a set of shelves where grill rags were placed that had just been laundered, dried in a commercial dryer. The odor was coming from those rags. As it turns out, each rag in the 1 ft tall stack had a burn pattern around the edge. Fire investigators first thought

Photo 5.4 A fire at the One Meridian Plaza building in Philadelphia that killed three firefighters, caused by linseed oil-soaked rags that were not properly stored and disposed of after use.

Photo 5.5 Upon arrival, the hazmat team found a 5 gal pail of a corrosive liquid they thought was the problem. They removed it from the building and the odor inside continued.

a liquid had been poured on them and ignited. The author was allowed to examine the rags and noticed the pattern was very symmetrical and not on the center of the rags as you would expect if a liquid had been poured on them. It was the opinion of the author that spontaneous combustion had occurred in the rags. They had animal vegetable oil on them before washing. The washing process did not remove all of the oil. When placed in the dryer, the heat from the dryer elevated the temperature of the oil and slow spontaneous combustion began to occur: another example of the dangers of animal vegetable oils.

Ordinary petroleum products, such as motor oil, grease, diesel fuel, and gasoline, to name a few, do not have a double bond in their chemical makeup. Therefore, those materials *do not undergo spontaneous combustion*! This fact may come as a surprise to some people because the author knows there have been numerous fires blamed on soiled rags with those products on them. The fact is that those types of flammable liquids do not spontaneously ignite and cannot start to burn without some other ignition source.

Fire-extinguishing agents

The theory behind fire extinguishing was first represented by the fire triangle and more recently by the fire tetrahedron. The triangle (Figure 5.16) represents the three components that were thought to be necessary for fire to occur: heat, fuel, and oxygen. If any of the components were removed, the fire would go out. The current theory (Figure 5.17) uses a four-sided geometric figure called a tetrahedron, representing the four components necessary for fire to occur: the original three, plus a chemical chain reaction. It is believed that fire is a chemical chain reaction.

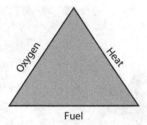

Figure 5.16 Fire triangle.

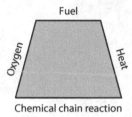

Figure 5.17 Fire tetrahedron.

Extinguishing agents, such as foam, act to eliminate one of the four components. Foam excludes oxygen from the fuel by blanketing the surface of the liquid and can also cool the material (removing heat). There are two general categories of firefighting foam: chemical and mechanical. Chemical foam was developed in the late 1800s. The foam bubble was produced by a chemical reaction of sodium bicarbonate powder, aluminum sulfate powder, and water. The reaction produces carbon dioxide, which is then encapsulated in the interior of a bubble. Chemical foams are expensive to produce and use and create a rigid foam blanket that does not reseal well when disrupted. Presently, chemical foams have been replaced by more effective and economical mechanical foams.

Four types of mechanical firefighting foams are currently available: protein, fluoroprotein, aqueous film forming (AFFF) and polar solvent (alcohol-type) foam. Different foams have different physical characteristics that affect their ability to form a foam blanket and extinguish a fire (Table 5.9). The type of foam selected will depend on the makeup of the fuel feeding the fire, the type of foam available, and the firefighting tactics chosen to extinguish the fire. Make sure that whatever extinguishing agent is used, it is the proper agent for the type of fire. Be sure there is enough agent on-scene to extinguish the fire before fire suppression efforts are started. Fires require a certain volume of water to effect extinguishment. Flammable liquid fires also require a certain volume of foam to extinguish the fire. If that amount of foam is not available, putting a lesser amount of foam on the fire is not going to extinguish it and is just a waste of time and foam.

Table 5.9 Comparison of Foam Types

	Protein	Fluoroprotein	AFFF
Drain time	Long	Shorter	Rapid
Viscosity	High	Low	Very low
Flow rate	Low	Moderate	Very fast
Polar compatibility	None	None	Varies by supplier

Photo 5.6 Portable fire extinguishers are the first line of defense for firefighting and can be effective if personnel are trained in their proper use.

In addition to foam, fires involving flammable liquids can be extinguished by using dry chemical, dry powder, halon, Purple K®, carbon dioxide, and water (Photo 5.6). Fire-extinguishing agents are rated according to the class of fire that they are effective in extinguishing (Table 5.10). Dry powder acts primarily to exclude atmospheric oxygen. Dry chemical, dry powder, halon, and Purple K® interrupt the chemical chain reaction. Dry-chemical fire extinguishers use sodium bicarbonate and monoammonium phosphate as agents. Purple K® extinguishers use potassium bicarbonate, which is where the purple color comes from. Dry-powder extinguishers use sodium chloride and graphite as agents. Water acts by cooling the fire or, in other words, removing the heat. Water may be ineffective against some flammable liquid fires. This precaution usually applies to materials with flash points below 100°F. When water is used, it should be applied in the form of a water

Table 5.10 Classes of Fire

Class A	Water
Ordinary combustibles	All purpose dry chemical
Class B	All purpose dry chemical
Flammable liquids	Foam
	Purple K™
	Halon
	Carbon dioxide
	Sodium bicarbonate
Class C	All purpose dry chemical
Electrical	Carbon dioxide
	Halon
Class D	Dry powder
Combustible metals	Graphite
	Sodium chloride

spray. Halon and carbon dioxide displace the oxygen needed for the fire to burn, so if there is not enough oxygen, the fire goes out. Because of their effect on the ozone layer above the earth, halon fire-extinguishing agents are being phased out, along with other gaseous halogenated chemicals.

Carbon tetrachloride was once used as a fire-extinguishing agent (see Chapter 8 alkyl halide compounds) (Photo 5.7). In fact, it was one of the first halons, halon 1040. It was discovered, however, that carbon tetrachloride has a hidden hazard. When carbon tetrachloride contacts fire or a hot surface, it gives off phosgene gas. As a result, carbon tetrachloride is no longer an approved extinguishing agent, but this does not mean that you will not see carbon tetrachloride extinguishers still in use. As recently as 1987, during routine fire inspections, the author found them in use in a rural school, an apartment building laundry room, and in the basement of a drugstore. They may also be found in antique stores and homes of collectors.

Soda-acid fire extinguishers are also obsolete and no longer approved in fire extinguishment (Photo 5.8). They are the familiar copper, brass, or stainless steel extinguishers that have become collectors' items. A screw-on cap on top exposes a glass bottle of sulfuric acid when removed. The rest of the tank is filled with soda water. When a soda-acid extinguisher is inverted, acid and soda water mix, expelling the mixture through a rubber hose with a nozzle on the end. After 10–15 years of service, these extinguishers became

Photo 5.7 Carbon tetrachloride fire extinguishers were one of the first halons. They are obsolete and dangerous to use as carbon tetrachloride produces phosgene gas when in contact with heat.

Photo 5.8 Soda-acid fire extinguishers are obsolete. They operated by mixing an acid with a base that created pressure expelling the agent from the container. Sometimes the pressure was too great for the container or the hose was blocked and the container exploded.

dangerous. Pressure that builds up to expel the acid–soda water mix is around 100 psi; however, if the hose or nozzle is plugged or the hose is kinked, the pressure can reach 300 psi. Many of these extinguishers failed when used, causing injury to operators. As recently as the late 1980s, this type of extinguisher was found by the author still in use in rural schools, old hotels, and main-street businesses. If these obsolete extinguishers are encountered, they should promptly be removed from service and properly disposed of.

Hydrocarbons

Hydrocarbon families of flammable materials were introduced in Chapters 2 and 4 in "Nonmetal compounds" and "Flammable gases" sections. Three hydrocarbon and one aromatic hydrocarbon families, which include flammable liquids, will be presented here (Photo 5.9). First is the alkane family, which has all single bonds and is considered saturated. Lighter alkanes (methane, ethane, propane, and butane) are all gases under normal conditions. All of these gases, however, may be encountered as liquefied gases (some cryogenic or very cold and some ambient temperature liquids under pressure). The alkane family is recognized by the "ane" suffix of the compound names and the single bonds between the carbons in the structures of the compounds. Starting with pentane, a five-carbon alkane, the remaining alkanes are flammable liquids. They are all naturally occurring

Photo 5.9 Lightning during a thunderstorm caused this fire in a 1.2-million gallon tank of unleaded gasoline at the Magellan petroleum storage facility in Kansas City, Kansas. (Courtesy of Chris Phillips.)

and are by products of crude oil and include pentane, hexane, heptane, octane, nonane, and decane. Other hydrocarbon compounds exist with more than 10 carbons, but the trivial naming system only works for the first 10. Other hydrocarbon compounds are named using the IUPAC naming system. IUPAC naming conventions are listed in the Appendix. As a group, pentane through decane have varying boiling points, flash points, narrow flammable ranges, high ignition temperatures (depending on the number of carbons in the compound), and are all nonpolar.

Isomers

Isomers occur more often in the hydrocarbon gas compounds and in hydrocarbon derivatives. It is possible that they can occur with hydrocarbon compounds with five or more carbons, but it is not common. Hydrocarbon structures may be altered so that their physical characteristics make them more economically valuable. One such alteration is called branching, or isomers. An *isomer* is a hydrocarbon that has the same molecular formula, that is, the same number of hydrogen atoms and carbons, but a different structural form.

The physical effect that branching has on a hydrocarbon is the lowering of the boiling point of that material. The straight-chained hydrocarbon is sometimes referred to as the "normal" configuration. When a material is listed in a reference book, you may see a small "n-" in front of the name or molecular formula; this indicates that it is the "normal," or straight-chained, form of the compound. For example, pentane can be normal,

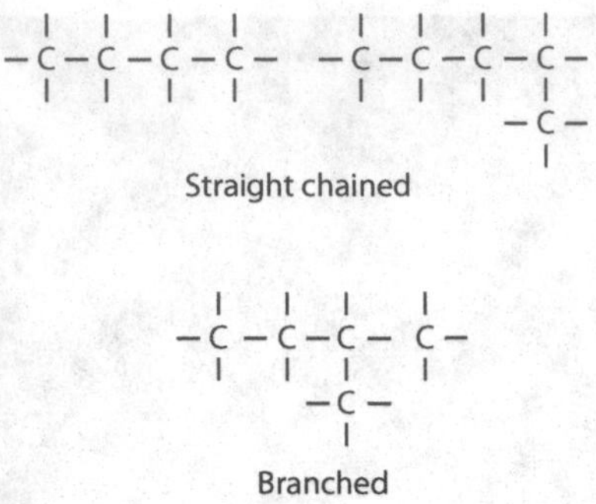

$$-C-C-C-C- \quad -C-C-C-C-$$
$$\qquad\qquad\qquad\qquad -C-$$

Straight chained

$$-C-C-C-\ C-$$
$$\qquad\quad -C-$$

Branched

Figure 5.18 Structures of branched compounds.

which would be written as normal pentane, or n-pentane. Pentane has three forms, the normal or straight-chained, the iso branch, and the neo branch. The iso-branched form of pentane would be written isopentane. The molecular formula would be C_5H_{12} or n-C_5H_{12} for the normal and i-C_5H_{12} for the branched compound (Figure 5.18). The neo branch of pentane would be written neopentane. The molecular formula would be the same neo-C_5H_{12} (Figure 5.19).

Branching has the effect of lowering the boiling point. Normal pentane has a boiling point of 97°F (36°C), iso pentane has a boiling point of 83°F (28°C), and neopentane has a boiling point of 50°F (10°C). Normal is not used all of the time with the normal form of the compound; it is found mostly in reference books and on laboratory containers. If no prefixes appear in the name, it is understood to be the normal form. "Iso" or the small "i" must be used to designate the branched form of the compound. The small "i" is used in front of the molecular formula and the prefix "iso" is used in front of the name. Without this designation, there is no way to determine the branching from the name or the molecular formula.

There are four forms or isomerization, the iso branch, the secondary branch, the tertiary branch, and the neo branch. Isomerization changes the chemical, physical, and at times the toxicological properties of the material. If the material is the isomer, or branched form, a small "i" will be placed in front of the molecular formula and the prefix "iso" in front of the name (Figure 5.20). The secondary branch will have a small "s" in front of the molecular formula and the word secondary in front of the name. Neo branches will have the "neo" in front of the molecular formula and the name. One way of determining whether the structure is branched or straight-chained is to try to draw a line through all of the carbons connected together in a chain without lifting the pencil or having to backtrack to reach another carbon. Branching can also occur in the alkene family as well. In order for branching to occur in hydrocarbons, there must be at least a four-carbon compound. See Figure 5.56.

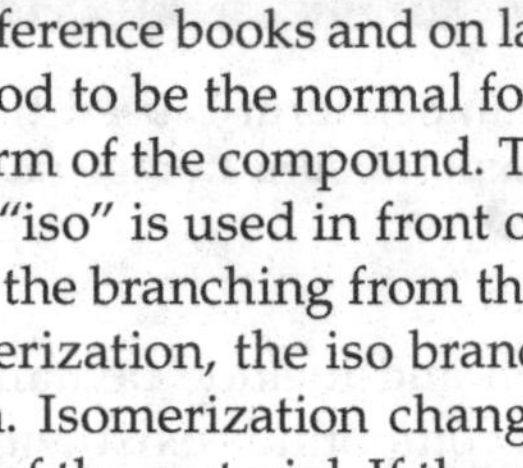

$$\begin{array}{c} H \\ | \\ H-C-H \\ \end{array}$$

$$\begin{array}{ccc} H & H & \\ | & | & \\ H-C-C-C-H & \\ | & | & \\ H & H & \end{array} \quad C_5H_{12}$$

$$\begin{array}{c} H-C-H \\ | \\ H \end{array}$$

Figure 5.19 Structure of neopentane.

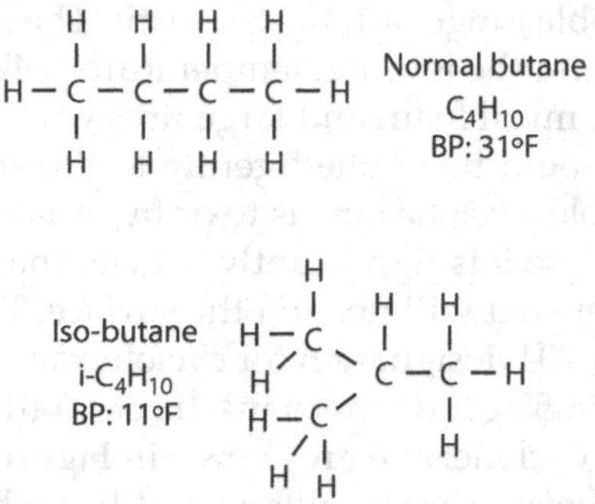

$$H-\overset{\overset{\displaystyle H}{|}}{C}-\overset{\overset{\displaystyle H}{|}}{C}-\overset{\overset{\displaystyle H}{|}}{C}-\overset{\overset{\displaystyle H}{|}}{C}-H$$

Normal butane
C_4H_{10}
BP: 31°F

Iso-butane
i-C_4H_{10}
BP: 11°F

Figure 5.20 Normal and isobutane.

Cyclic alkanes

Cyclic alkanes are circular in shape with all of the carbons connected together in a circle. Because of all the carbons being connected together, there are two less hydrogen atoms in a cyclic alkane of the same number of carbons as the straight-chained form. Cyclic forms exist for five, six, seven, and eight carbons. Formulas for cyclic compounds will have a small c in front of the formula to indicate the compound is a cyclic compound. The word cyclo will appear in front of the alkane name. Physical properties change when you change the shape of an alkane from the normal or straight-chained form. Some cyclic compounds have anesthetic properties and are flammable. See Figure 5.9 for the structure of cyclo compounds.

Pentane (IUPAC) is a five-carbon alkane and has a boiling point of 97°F (36°C), a flash point of −40°F (−40°C), a flammable range of 1.5%–7.8%, and an ignition temperature of 500°F (260°C). As the carbon content of a compound increases, so do the boiling point and flash point. Pentane has an iso branch and a neo branch and is found in a cyclic form as well. The structure and molecular formula for pentane are shown in Figure 5.21.

Hexane (IUPAC) is a six-carbon alkane and has a boiling point of 156°F (68°C), a flash point of −7°F (−21°C), a flammable range of 1.1%–7.5%, and an ignition temperature of 437°F (225°C). Hexane has an iso and neo branch and is found in a cyclic form as well. The structure and molecular formulas are shown in Figure 5.22.

Cyclohexane (IUPAC) is a cyclic alkane hydrocarbon with single bonds between the carbons. The liquid is colorless, nonpolar, and immiscible with water. Cyclohexane is highly

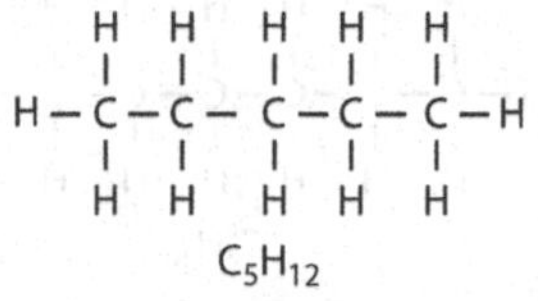

$$H-\overset{\overset{\displaystyle H}{|}}{C}-\overset{\overset{\displaystyle H}{|}}{C}-\overset{\overset{\displaystyle H}{|}}{C}-\overset{\overset{\displaystyle H}{|}}{C}-\overset{\overset{\displaystyle H}{|}}{C}-H$$
C_5H_{12}

Figure 5.21 Pentane.

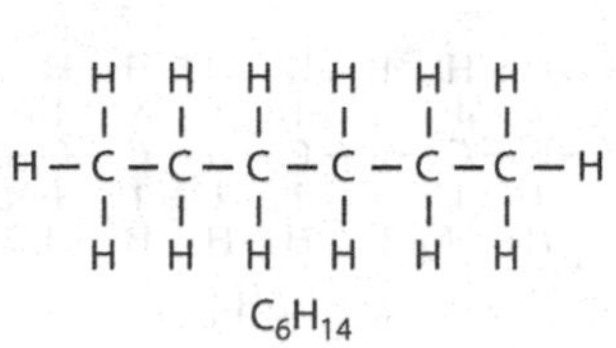

$$H-\overset{\overset{\displaystyle H}{|}}{C}-\overset{\overset{\displaystyle H}{|}}{C}-\overset{\overset{\displaystyle H}{|}}{C}-\overset{\overset{\displaystyle H}{|}}{C}-\overset{\overset{\displaystyle H}{|}}{C}-\overset{\overset{\displaystyle H}{|}}{C}-H$$
C_6H_{14}

Figure 5.22 Hexane.

flammable, with a flammable range of 1.3%–8% in air. The boiling point is 179°F (81°C), the flash point is −4°F (81°C), and the ignition temperature is 473°F (245°C). Small fires may be extinguished with dry-chemical foam and large fires with hydrocarbon-type foam. Water may be ineffective and should be applied gently to the surface of the liquid if used. In addition to being flammable, cyclohexane is toxic by inhalation, with a TLV of 300 ppm in air. The vapor density is 29, so it is significantly heavier than air. The specific gravity is 0.8, which is lighter than water so it will float on the surface. The four-digit UN identification number is 1145. The NFPA 704 designation for cyclohexane is health 1, flammability 3, and reactivity 0. It is shipped in 55 gal drums, tank trucks, railcars, and barges. The structure and molecular formula of cyclohexane are shown in Figure 5.23.

Heptane (IUPAC) is a seven-carbon alkane and has a boiling point of 208°F (98°C), a flash point of 25°F (−3°C), a flammable range of 1.05%–6.7%, and an ignition temperature of 433°F (222°C). The structure and molecular formula are shown in Figure 5.24. Heptane also has an iso branch.

Octane (IUPAC) is an eight-carbon alkane and has a boiling point of 259°F (125.6°C), a flash point of 56°F (13°C), a flammable range of 1.0%–6.5%, and an ignition temperature of 428°F (220°C). The structure and molecular formula are shown in Figure 5.25. Octane also has an iso branch.

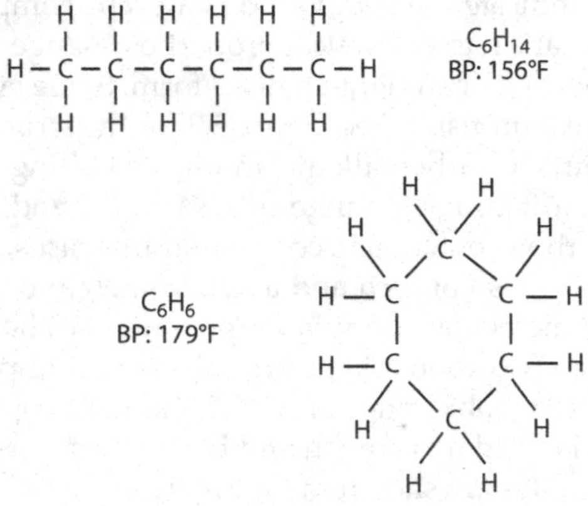

Figure 5.23 Hexane and cyclohexane.

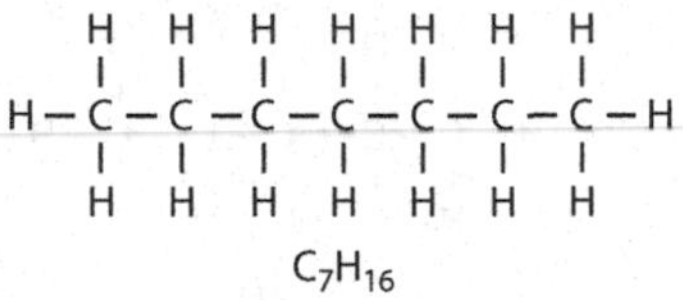

Figure 5.24 Heptane.

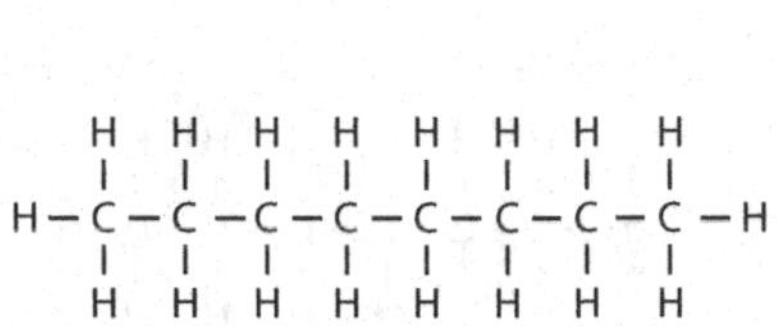

Figure 5.25 Octane.

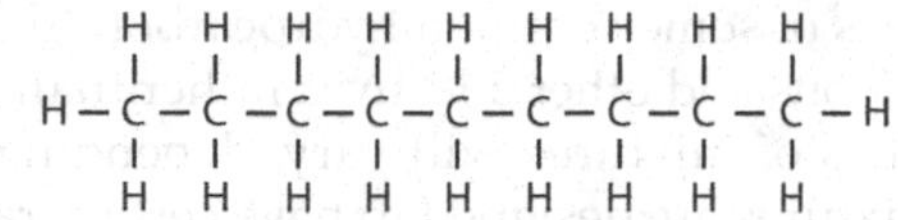

$$
\begin{array}{c}
\text{H} \quad \text{H} \quad \text{H} \quad \text{H} \quad \text{H} \quad \text{H} \quad \text{H} \quad \text{H} \quad \text{H} \\
| \quad | \quad | \quad | \quad | \quad | \quad | \quad | \quad | \\
\text{H}-\text{C}-\text{C}-\text{C}-\text{C}-\text{C}-\text{C}-\text{C}-\text{C}-\text{C}-\text{H} \\
| \quad | \quad | \quad | \quad | \quad | \quad | \quad | \quad | \\
\text{H} \quad \text{H} \quad \text{H} \quad \text{H} \quad \text{H} \quad \text{H} \quad \text{H} \quad \text{H} \quad \text{H}
\end{array}
$$

$$C_9H_{20}$$

Figure 5.26 Nonane.

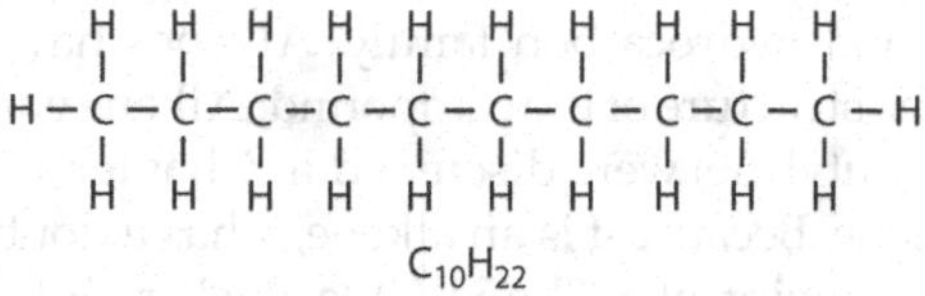

$$
\begin{array}{c}
\text{H} \quad \text{H} \quad \text{H} \quad \text{H} \quad \text{H} \quad \text{H} \quad \text{H} \quad \text{H} \quad \text{H} \quad \text{H} \\
| \quad | \quad | \quad | \quad | \quad | \quad | \quad | \quad | \quad | \\
\text{H}-\text{C}-\text{C}-\text{C}-\text{C}-\text{C}-\text{C}-\text{C}-\text{C}-\text{C}-\text{C}-\text{H} \\
| \quad | \quad | \quad | \quad | \quad | \quad | \quad | \quad | \quad | \\
\text{H} \quad \text{H} \quad \text{H} \quad \text{H} \quad \text{H} \quad \text{H} \quad \text{H} \quad \text{H} \quad \text{H} \quad \text{H}
\end{array}
$$

$$C_{10}H_{22}$$

Figure 5.27 Decane.

Nonane (IUPAC) is a nine-carbon alkane and has a boiling point of 315°F (150.7°C), a flash point of 86°F (30°C), a flammable range of 0.8%–2.9%, and an ignition temperature of 403°F (206°C). The structure and molecular formula are shown in Figure 5.26.

Decane (IUPAC) is a 10-carbon alkane and has a boiling point of 345°F (173d°C), a flash point of 115°F (46°C), a flammable range of 0.8%–5.5%, and an ignition temperature of 410°F (210°C). The structure and molecular formula are shown in Figure 5.27.

While it is certainly possible to encounter these compounds in transportation and fixed facilities as individual chemicals, they are more likely be found in mixtures with other compounds. Mixtures do not involve chemical reactions or bonding. They are pure chemical compounds that have been mixed together to form a solution without losing their individual chemical makeup. However, the overall physical characteristics of the mixture will change based on the physical characteristics of the compounds' makeup. Boiling and flash points of the mixture will fall somewhere between the boiling and flash points of the components. In simple terms, a pure compound is octane, pentane, or isooctane. It is possible to draw structures or write formulas for these compounds. When pure compounds are mixed together, you cannot draw a structure or write a formula for the mixture. These mixtures include gasoline, diesel fuel, jet fuel, fuel oil, petroleum, ether, and others. Gasoline has a boiling point between 100°F and 400°F, depending on the contents of the mixture; its flash point is −36°F to −45°F (−37°C to −42°C); the flammable range is 1.4%–7.6%; and the ignition temperature ranges from 536°F to 853°F (280°C–456°C). Diesel fuel has a flash point range of 100°F–130°F (37°C–64°C). Hazardous materials reference books do not list a boiling point or ignition temperature for diesel fuel. Table 5.11 shows

Table 5.11 Physical Characteristics of Common Fuels

	Gasoline	Kerosene	Fuel oil #4	Jet fuel
Boiling point	100–400°F	338–572°F		250°F
Flash point	−45°F	100–150°F	130°F	100°F
Ignition temp	536–853°F	444°F	505°F	435°F
Vapor density	3.0–4.0	4.5		1
Specific gravity	0.8	0.81	<1	0.8
LEL	1.4%	0.7%		0.6%
UEL	7.6%	5%		3.7%

the physical characteristics of some common hydrocarbon fuels. Remember that the fuels are mixtures of hydrocarbons and other additives rather than pure compounds.

Physical characteristics of mixtures will vary, depending on the components of the mixture. Many times, mixtures are designed to have certain ranges of flash points so they will perform a particular function. If a high-boiling-point and flash-point liquid is mixed with a low-boiling-point and flash-point liquid, the boiling and flash points of the mixture will be somewhere between the two liquids that were mixed together. Mixtures will be discussed further in "Hydrocarbon derivatives" section.

Alkenes are the second hydrocarbon family. Alkenes have one or more double bonds between the carbons in the structure of the compound. Alkene gases ethene (ethylene), propene (propylene), and butene (butylene) were discussed in Chapter 4. The first flammable liquid in the alkene family is pentene. Because it is an alkene, it has a double bond in the compound.

Pentene has a boiling point of 86°F (54°C), a flash point of 0°F (−17°C), a flammable range of 1.5%–8.7%, and an ignition temperature of 527°F (275°C). The structure and formula for pentene are shown in Figure 5.28.

Hexene has a boiling point of 146°F (63°C), a flash point of less than 20°F (−6°C), and an ignition temperature of 487°F (252°C). The structure and molecular formula for hexene are shown in Figure 5.29.

Just like the alkane family, as the carbon content increases within the alkene family, so do the boiling point and flash point temperatures. Some alkene-family flammable liquids have more than one double bond. *Pentadiene*, also known as 1,3-pentadiene and piperylene is a five-carbon compound with two double bonds. Pentadiene is a highly flammable liquid with an NFPA 704 designation for flammability of 4. It has a boiling point of −45°F (−42°C) and a flash point of 112°F (44°C). *Hexadiene* is a six-carbon liquid compound that is highly flammable with a boiling point of 147°F (63°C), a flash point of −6°F (−21°C), and a flammable range of 2%–6.1%. The structures and molecular formulas of pentadiene and hexadiene are shown in Figure 5.30.

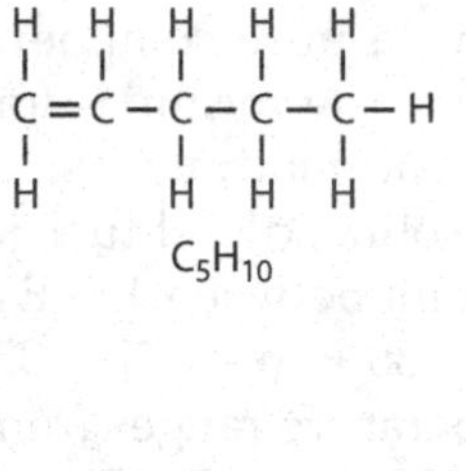

$$
\begin{array}{c}
\text{H \quad H \quad H \quad H \quad H} \\
| \quad | \quad | \quad | \quad | \\
\text{C} = \text{C} - \text{C} - \text{C} - \text{C} - \text{H} \\
| \quad \quad | \quad | \quad | \\
\text{H \quad \quad H \quad H \quad H} \\
C_5H_{10}
\end{array}
$$

Figure 5.28 Pentene.

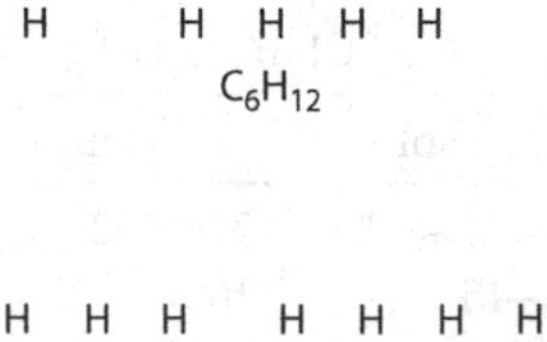

$$
\begin{array}{c}
\text{H \quad H \quad H \quad H \quad H \quad H} \\
| \quad | \quad | \quad | \quad | \quad | \\
\text{C} = \text{C} - \text{C} - \text{C} - \text{C} - \text{C} - \text{H} \\
| \quad \quad | \quad | \quad | \quad | \\
\text{H \quad \quad H \quad H \quad H \quad H} \\
C_6H_{12}
\end{array}
$$

Figure 5.29 Hexene.

$$
\begin{array}{cc}
\text{H \quad H \quad H \quad H \quad H} & \text{H \quad H \quad H \quad H \quad H \quad H} \\
| \quad | \quad | \quad | \quad | & | \quad | \quad | \quad | \quad | \quad | \\
\text{C} = \text{C} - \text{C} - \text{C} = \text{C} & \text{C} = \text{C} - \text{C} - \text{C} - \text{C} = \text{C} \\
| \quad | \quad \quad | & | \quad | \quad | \quad | \quad \quad | \\
\text{H \quad H \quad \quad H} & \text{H \quad H \quad H \quad \quad H} \\
C_5H_8 & C_6H_{10}
\end{array}
$$

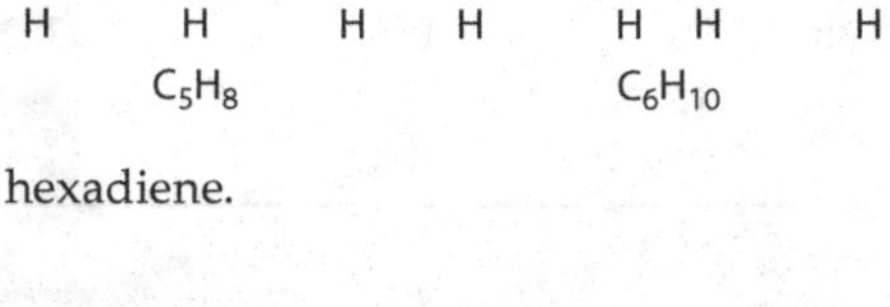

Figure 5.30 Pentadiene and hexadiene.

The third of the hydrocarbon families is the alkyne family. Since there are no commercially valuable alkyne flammable liquid compounds that are likely to be encountered by emergency responders, the alkyne family will not be covered here. Acetylene, the alkyne gas, and the characteristics of the alkyne family were covered in Chapter 4.

Finally, the fourth and final hydrocarbon family to be discussed is known as the aromatic hydrocarbons, sometimes referred to as the BTX fraction (benzene, toluene, xylene). One additional aromatic beyond the BTX fraction is called styrene and will be covered as the fourth aromatic compound. Aromatics as a group are toxic and flammable. They have moderate boiling and flash points, narrow flammable ranges, high ignition temperatures, and are nonpolar.

Benzene is a known carcinogen. Toluene and xylene have not yet been found to be carcinogenic, but certainly should be suspected since they come from benzene. Parent member of the family is benzene, which has a molecular formula of C_6H_6. There is a ratio that can be used to recognize the benzene ring in a formula: there has to be a minimum of six carbons to have a benzene ring and the number of carbon atoms to hydrogen atoms is almost in a 1:1 ratio. None of the other hydrocarbon families have that kind of carbon-to-hydrogen ratio. Benzene, toluene, and xylene have a ringed structure with six carbons, sometimes referred to as the benzene ring.

Aromatics were thought at one time to be unsaturated because the structure was thought to have double bonds (see the first-theory benzene structure in Figure 5.31). The structure appeared to have three double bonds to satisfy the octet rule of bonding. However, in reality, aromatics do not behave like unsaturated compounds. They burn with incomplete combustion. Aromatics are unreactive, so it is theorized that instead of three double bonds, they have a unique structure where the six extra electrons are in a state of resonance within the benzene ring. They are not attached to any one of the carbons, but rather go from one to another at a speed faster than the speed of light, much the same way a rotor works inside a distributor in an automobile. Bonding in the aromatic hydrocarbons is called a resonant bond. Resonant bonding is represented by a large circle located within the six-carbon benzene ring.

Benzene is the backbone of three other aromatic compounds that will be presented here. The first is toluene, with the molecular formula $C_6H_5CH_3$. Sometimes, all of the carbon and hydrogen in the compound are combined and the formula is shown as C_7H_8. Toluene is a benzene ring with one hydrogen atom removed to attach a methyl radical CH_3. Xylene is another aromatic with a molecular formula of $CH_3C_6H_4CH_3$. If all of the carbon and hydrogen were added together, the formula would be C_8H_{10}. The benzene ring is the

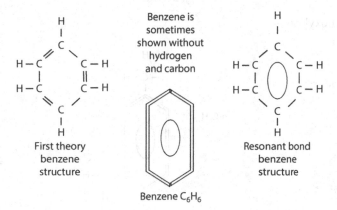

First theory benzene structure

Benzene is sometimes shown without hydrogen and carbon

Benzene C_6H_6

Resonant bond benzene structure

Figure 5.31 History of benzene structure.

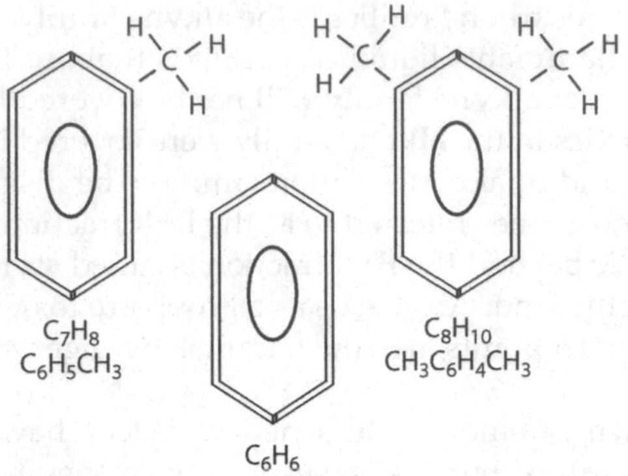

Figure 5.32 Toluene, benzene, and xylene.

backbone for xylene with two hydrogen atoms removed. Two methyl radicals replace two hydrogen atoms in the ring to make xylene. Structures of benzene, toluene, and xylene are shown in Figure 5.32.

The positioning of the methyl radicals on the xylene ring is of particular importance. There are names for the different positions around the benzene ring. If the methyl radicals are placed on the top and first side position, it is referred to as the "ortho" position; "ortho" refers to "straight ahead." The ortho structure and formula of ortho xylene (1,2 dimethyl benzene) are shown in Figure 5.33.

When methyl radicals are placed on the top and on the second side position, it is referred to as the "meta" position; "meta" translates to "beyond." The structure and formula for meta xylene (1,3 dimethyl benzene) are shown in Figure 5.34.

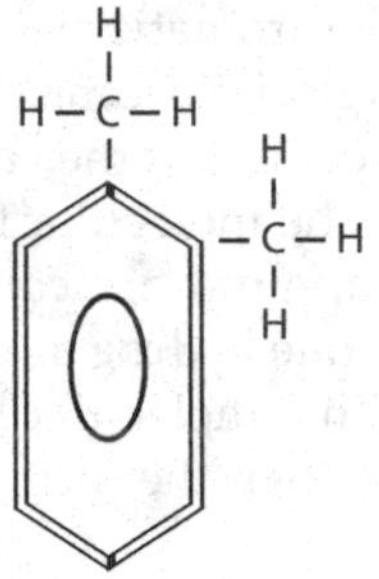

Figure 5.33 Orthoxylene.

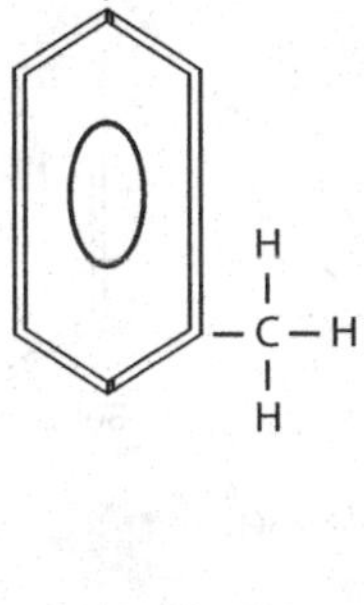

Figure 5.34 Metaxylene.

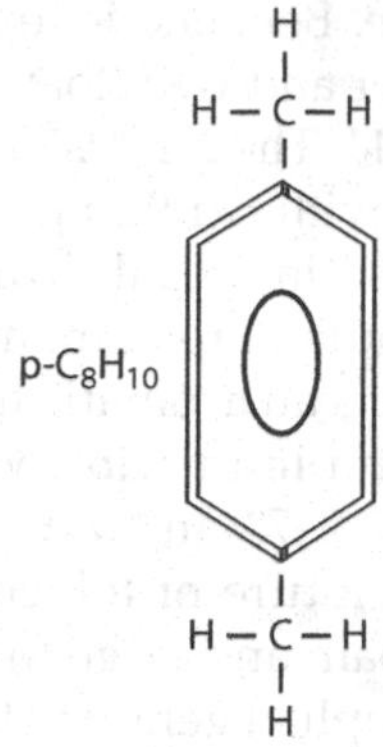

Figure 5.35 Paraxylene.

If the methyl radicals are on the top and bottom of the ring, it is referred to as the "para" position; "para" means "opposite." The structure and molecular formula for para-xylene (1,4-dimethyl benzene) are shown in Figure 5.35.

Effects that positioning has on a compound are changes in toxicity and some physical characteristics, such as melting point. Changes to boiling point, flash point, and ignition temperature are insignificant for these isomers of xylene. It is important when looking up these materials in reference sources that you make sure which xylene is involved in an incident. The "para," "meta," and "ortho" structural isomers of xylene are often used in the making of pesticides.

Finally, the last aromatic compound we will discuss in this book is styrene. It is, however, quite different from the other members of the aromatic family. Styrene is a monomer used in the manufacture of polystyrene. It has a vinyl radical attached to the benzene ring. The double bond in the vinyl radical is reactive. A reaction can occur with the oxygen in the air, with an oxidizer, or it can self-react in storage. The structure and molecular formula are shown in Figure 5.36.

Benzene (IUPAC) is the parent member of the aromatic hydrocarbon family. It is a colorless to light yellow liquid with a characteristic aromatic odor. Benzene is nonpolar and burns with incomplete combustion, producing a smoky fire. Flammable range is 1.5%–8% in air. Benzene is also toxic, with a TLV of 10 ppm, and is a known carcinogen. Concentrations of 8000 ppm for 30–60 min are fatal. Its boiling point is 176°F (80°C) and the flash point is 12°F (–11°C). The ignition temperature is 928°F (497°C). Its vapor density is 2.8, making it heavier than air. Small fires involving benzene should be fought with dry-chemical foams and large fires with hydrocarbon foams. Water may be ineffective and, if used, should be

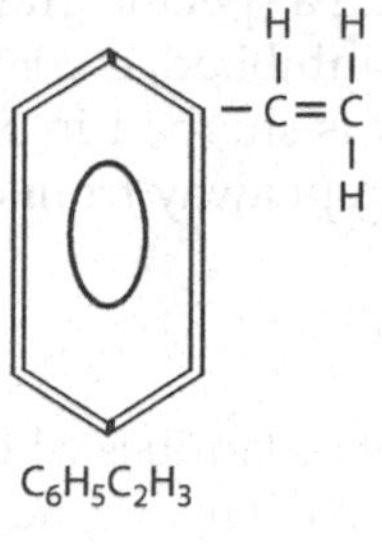

Figure 5.36 Styrene.

applied gently to the liquid surface. Benzene is immiscible with water and has a specific gravity of 0.9. It is lighter than water and will float on the surface. The four-digit UN identification number for benzene is 1114. The NFPA 704 classification is health 2, flammability 3, and reactivity 0. Benzene reacts with oxidizing agents and should be stored away from them in fixed facilities. It is shipped in 55 gal drums, highway tank trucks, railcars, and barges. Benzene is used in the manufacture of many other chemicals and as a solvent.

Toluene (methyl benzene IUPAC) is an aromatic hydrocarbon and a colorless liquid with an aromatic odor. It is nonpolar and immiscible with water. Toluene is a dangerous fire risk, with a flammable range of 1.27%–7% in air. Boiling point is 231°F (110°C), with a flash point of 40°F (4°C). Ignition temperature of toluene is 896°F (480°C). Small fires may be extinguished with dry-chemical foam and large fires with hydrocarbon-type foam. Water may be ineffective and should be applied gently to the surface of the liquid if used. In addition to flammability, toluene is toxic by ingestion, inhalation, and skin absorption, with a TLV of 100 ppm in air. Its vapor density is 3.1, which is heavier than air. The specific gravity is 0.9, so it is lighter than water and will float on the surface. The four-digit UN identification number is 1294. The NFPA 704 designation is health 2, flammability 3, and reactivity 0. It is shipped in 55 gal drums, tank trucks, railcars, and barges.

Xylene and *paraxylene*, sometimes referred to as dimethylbenzene and xylol, are aromatic hydrocarbons. *Condensed Chemical Dictionary* refers to xylene as "a commercial mixture of the three isomers: "ortho," "meta," and "para." Xylene is a clear liquid that is nonpolar and immiscible with water. It is a moderate fire risk, with a flammable range of 0.9%–7% in air. Boiling point is between 281°F and 292°F (138°C–144°C), depending on the mixture. Flash point ranges from 81°F to 90°F (27°C) and ignition temperature ranges from 867°F to 984°F (463°C–528°C). Small fires may be extinguished with dry-chemical foam and large fires with hydrocarbon-type foam. Water may be ineffective and should be applied gently to the surface of the liquid if used. In addition to flammability, xylene is toxic by ingestion and inhalation, with a TLV of 100 ppm in air. Vapor density is 3.7; therefore, it is heavier than air. Specific gravity is 0.9, so the xylenes will float on top of water. The four-digit UN identification number is 1307. The NFPA 704 designation for xylene is health 2, flammability 3, and reactivity 0. It is shipped in 55 gal drums, tank trucks, railcars, and barges.

Styrene (phenylethene IUPAC) is a colorless, oily liquid aromatic hydrocarbon with a characteristic odor. It is sometimes called vinyl benzene or phenyl ethylene. Styrene is a monomer and must be inhibited during transportation and storage to prevent polymerization. It is a moderate fire risk, with a flammable range of 1.1%–6.1%. Boiling point is 295°F (146°C), with a flash point of 88°F (31°C) and an ignition temperature of 914°F (490°C). Small fires may be extinguished with dry-chemical foam and large fires with hydrocarbon-type foam. Water may be ineffective and should be applied gently to the surface of the liquid if used. In addition to flammability, styrene is toxic by ingestion and inhalation, with a TLV of 50 ppm in air. The vapor density is 3.6, which is heavier than air. Styrene is nonpolar and it is immiscible with water, with a specific gravity of 0.9. The four-digit UN identification number for styrene monomer, inhibited, is 2055. The NFPA 704 designation is health 2, flammability 3, and reactivity 2. It is shipped in 55 gal drums, tank trucks, railcars, and barges. When stored, it should be kept away from oxidizers.

Hydrocarbon derivatives

There are seven hydrocarbon-derivative families whose primary hazard is flammability: alkyl halide, amine, ether, alcohol, ketone, aldehyde, and ester. Alkyl halides, amines, and ethers are nonpolar. Ethers, alcohols, and aldehydes are polar and have wide flammable ranges.

Some organic acids are flammable; inorganic acids do not burn. However, flammability is not the primary hazard of most organic acids. They will be discussed in detail in Chapter 10.

Some of the alkyl halides are also flammable, but their primary hazard is toxicity, and they will be presented in Chapter 8. Hydrocarbon derivatives are manmade materials. They are made from hydrocarbons by removing hydrogen and adding some other element. The primary elements used in making hydrocarbon derivatives, in addition to carbon and hydrogen, are oxygen, nitrogen, chlorine, fluorine, bromine, and iodine. Once a hydrogen atom is removed from a hydrocarbon, the hydrocarbon becomes a radical. The same prefixes are used for single-bonded hydrocarbon radicals as for the hydrocarbons; however, a "yl" is added to the prefix, indicating that it is a radical. For example, a one-carbon radical is called "methyl," two carbons "ethyl," three carbons "propyl," etc. Remember that hydrogen has been removed, so the radical is not a complete compound. It must be attached to a hydrocarbon-derivative functional group to be complete (Table 5.12). If a compound has more than one radical of the same type (except for ether), prefixes are used to indicate the number of radicals present. The prefixes are "di" for two, "tri" for three, and "tetra" for four. Radicals for single-bonded hydrocarbons are shown in Figure 5.37.

Table 5.12 Hydrocarbon-Derivative Families

Family	General formula	Hazard
Alkyl halide	R–X	Toxic/flammable
	X=F,Cl,Br,I	
Nitro	R–NO2	Explosive
Amine	R–NH2	Toxic/flammable
Nitrile	R–CN	Toxic/flammable
Isocyanate	R–NCO	Toxic/flammable
Sulfide	R–S–R	Toxic/flammable
Thiol	R–SH	Toxic/flammable
Ether	R–O–R	Flammable, WFR
Peroxide	R–O–O–R	Oxidizer, explosive
Alcohol	R–OH	Flammable/toxic, WFR
Ketone	R–CO–R	Flammable/toxic
Aldehyde	R–CHO	Flammable/toxic, WFR
Ester	R–COOH	Corrosive, flammable

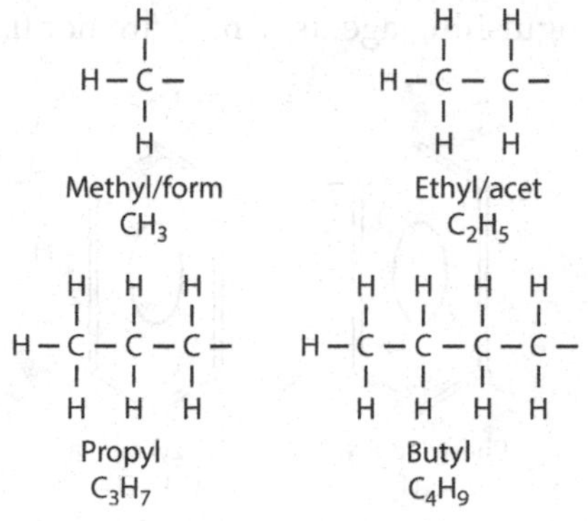

Figure 5.37 Alkane hydrocarbon radical structures.

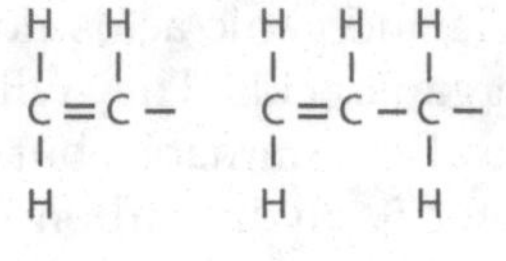

$$\begin{array}{ccc} \text{H} & \text{H} & \quad \text{H} \quad \text{H} \quad \text{H} \\ | & | & \quad | \quad | \quad | \\ \text{C} = \text{C} - & \quad \text{C} = \text{C} - \text{C} - \\ | & | & \quad | \quad | \\ \text{H} & & \quad \text{H} \quad \text{H} \end{array}$$

Figure 5.38 Vinyl and acryl radical structures.

Hydrocarbon compounds with double bonds can also be made into radicals. In order to have a double bond, there must first be at least two carbons; there are no double-bonded radicals with only one carbon. Only two double-bonded hydrocarbon radicals are important here. They are two-carbon and three-carbon radicals with double bonds between the carbons. Since the prefixes for two and three carbons have been used up in the single-bonded compounds, the names for these double-bonded radicals are different from the others: a two-carbon compound with a double bond is called vinyl, which is actually a radical of ethene or ethylene; the three-carbon compound with one double bond is called "acryl," which is a radical of propene or propylene. The structures for the vinyl and acryl radicals are shown in Figure 5.38.

Hydrogen can be removed from aromatic compounds to make radicals. If one carbon is removed from benzene, the radical is called "phenyl." If one carbon is removed from toluene, the radical is called "benzyl." The structures and molecular formulas for the phenyl and benzyl radicals are shown in Figure 5.39.

With certain hydrocarbon-derivative functional groups, alternate names for one- and two-carbon single-bonded radicals are used. This occurs when the radicals are used with the aldehydes, esters, and organic acids. A one-carbon radical for aldehydes, esters, and organic acids is called "form" and the two-carbon radical is called "acet." Additionally, when naming the radical for these compounds, the carbon in the functional group is counted as part of the total number of carbons when choosing the prefix. Figure 5.40 shows the structures, molecular formulas, and names for the one- and two-carbon compounds of aldehydes, esters, and organic acids. (Just a hint for future reference: when naming ester functional groups, nothing is named ester.) There are some alternate naming rules for esters based on which radical is attached to the carbon in the ester functional group.

Alkyl halide

First of the hydrocarbon-derivative flammable liquid families we will discuss is alkyl halide. As a family, alkyl halides vary widely in hazards. Some are flammable, some are toxic, and some are used as fire-extinguishing agents, which are not flammable or overly toxic. They

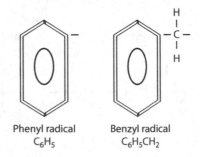

Phenyl radical
C_6H_5

Benzyl radical
$C_6H_5CH_2$

Figure 5.39 Benzene family radical structures.

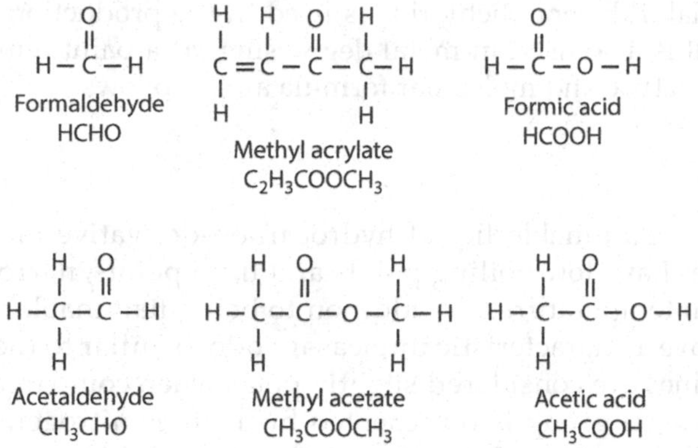

Figure 5.40 One- and two-carbon compounds of aldehydes, esters, and organic acids.

generally have moderate boiling and flash points and high ignition temperatures. Alkyl halides are nonpolar. The alkyl halide group is represented by a single hydrocarbon radical or hydrocarbon backbone, to which is attached one or more of the halogen family of elements in place of hydrogen. The general formula is a radical and one of the halogens: *R-X*. Halogens are represented in the general formula by X, because it can be any of the halogens. Generally, fluorine, chlorine, bromine, and iodine may be found in the alkyl halides. They may, by themselves, have more than one atom of an element or multiple atoms of different elements.

Ethylene dichloride (1,2 dichloroethane IUPAC) is an alkyl halide hydrocarbon derivative. Ethylene dichloride is a colorless oily liquid with a chloroform-like odor and a sweet taste. It is a dangerous fire risk, with a flammable range of 6%–16% in air. The boiling point is 183°F (83°C), flash point is 56°F (13°C), and the ignition temperature is 775°F (412°C) (Figure 5.41). Small fires involving ethylene dichloride should be fought with dry-chemical foam and large fires with hydrocarbon foam. Water may be ineffective and, if used, should be applied gently to the surface of the liquid. Water is generally ineffective against flammable liquid fires where the liquid has a flash point below 100°F (37°C). The farther below 100°F (37°C) the liquid's flash point is, the less effective water will be.

In addition to flammability, ethylene dichloride is toxic by ingestion, inhalation, and skin absorption; it is also a known carcinogen, with a TLV of 10 ppm in air. The vapor density is 3.4, which is heavier than air, so the vapors will stay close to the ground. The specific gravity is 1.3, which makes it heavier than water, so it will sink to the bottom. Alkyl halides are nonpolar, and ethylene dichloride is only slightly miscible with water. The four-digit UN identification number is 1184. The NFPA 704 designation for ethyl dichloride is health 2, flammability 3, and reactivity 0. Ethylene dichloride is shipped in metal cans, drums, tank trucks, railcars, and barges. It is usually packaged under nitrogen gas, which

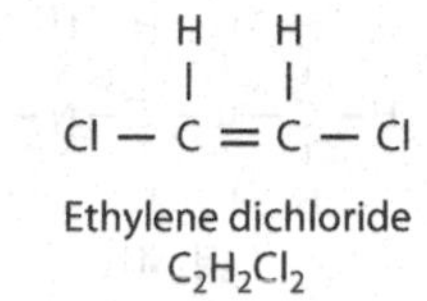

Figure 5.41 Ethylene dichloride.

is an inert material. Ethylene dichloride is used in the production of vinyl chloride and trichloroethane. It is also used in metal degreasing, as a paint remover, a solvent, and a fumigant. The structure and molecular formula are as follows.

Amines

Amine is the next flammable liquid hydrocarbon-derivative family we will discuss. Generally, amines have low boiling points and flash points, narrow flammable ranges, and high ignition temperatures. In addition to being flammable, amines are toxic and irritants. They have a characteristic unpleasant odor, similar to the odor of the bowel or rotten flesh. Amines are considered slightly polar when compared to nonpolar materials. Amine functional group is represented by a single nitrogen atom surrounded by two or fewer hydrogen atoms. Nitrogen requires three bonds to satisfy the octet rule of bonding. General formulas for amines are $R-NH_2$, R_2-NH, and R_3-N. Nitrogen identifies the amine group, not the numbers of hydrogen attached to the nitrogen. There may be one, two, or three radicals connected to the nitrogen. $R-NH_2$ indicates that one radical and two hydrogen atoms are attached to the nitrogen. R_2-NH indicates that two radicals and one hydrogen atom are attached to the nitrogen. R_3-N indicates that three radicals and no hydrogen atoms are attached to the nitrogen. To name amines, start with the smallest radical and proceed through, however, many more radicals there are, in order of size, and end with the word amine. Types of radicals attached to the nitrogen may be the same or different.

Propylamine (propan-1-amine, IUPAC) is a colorless liquid that is slightly soluble in water. Specific gravity is 0.7, which is lighter than water. Propylamine is flammable, with a flammable range of 2%–10% in air. Boiling point is 120°F (48°C) and flash point is −35°F (−37°C). Ignition temperature is 604°F (317°C). Vapor density is 2, which is heavier than air. In addition to being flammable, propylamine is corrosive and is a strong irritant to skin and tissue. The four-digit UN identification number is 1277. The NFPA 704 designation is health 3, flammability 3, and reactivity 0. It is shipped in glass bottles, cans, drums, and tank cars. The primary uses are as a chemical intermediate and as a lab reagent. The structure and molecular formula for propylamine is shown in Figure 5.42.

Butylamine (butan-1-amine, IUPAC) is a colorless, volatile liquid with an amine-like odor. It is miscible with water and has a specific gravity of 0.8, which is lighter than water. It is a dangerous fire risk, with a flammable range of 1.7%–9.8% in air. Flash point is 10°F (−12°C), with a boiling point of 172°F (77°C). Ignition temperature is 594°F (312°C). Vapor density is 2.5, which is heavier than air. Butylamine is also a skin irritant, with a TLV ceiling of 5 ppm in air. The four-digit UN identification number is 1125. The NFPA 704 designation is health 3, flammability 3, and reactivity 0. It is shipped in glass bottles, cans, drums, and tank cars. The primary uses are in the manufacture of pharmaceuticals, insecticides, dyes, and rubber chemicals. Structure and molecular formula for butylamine are shown in Figure 5.43.

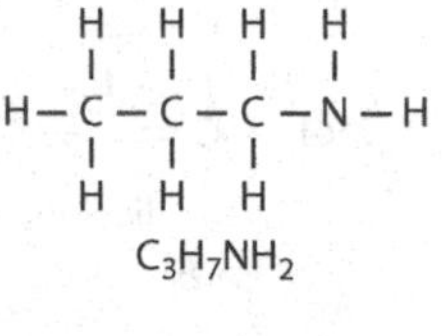

```
     H   H   H   H
     |   |   |   |
 H — C — C — C — N — H
     |   |   |
     H   H   H
         C₃H₇NH₂
```

Figure 5.42 Propylamine.

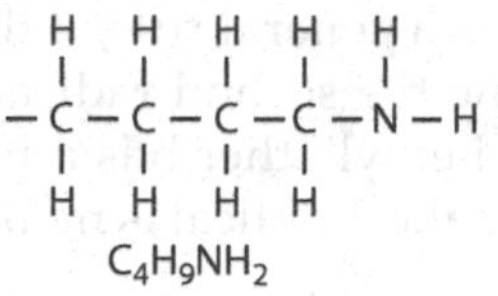

$$
\begin{array}{ccccc}
H & H & H & H & H \\
| & | & | & | & | \\
H-C & -C & -C & -C & -N-H \\
| & | & | & | \\
H & H & H & H \\
\end{array}
$$

$$C_4H_9NH_2$$

Figure 5.43 Butylamine.

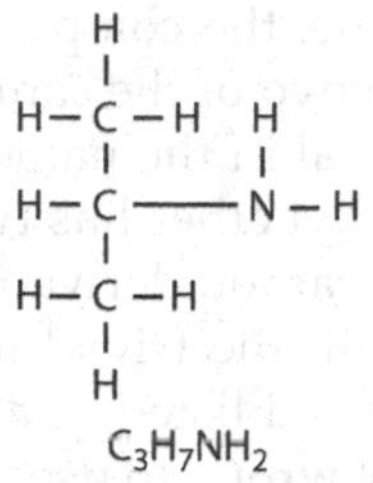

$$
\begin{array}{c}
H \\
| \\
H-C-H \quad H \\
| \quad\quad | \\
H-C \text{———} N-H \\
| \\
H-C-H \\
| \\
H \\
\end{array}
$$

$$C_3H_7NH_2$$

Figure 5.44 Isopropylamine.

Isopropylamine (propan-2-amine, IUPAC) is a colorless, volatile liquid. It is highly flammable, with a flammable range of 2%–10.4% in air. Boiling point is 93°F (33°C), flash point is −15°F (−26°C), and ignition temperature is 756°F (402°C).

It is miscible with water, with a specific gravity of 0.69, which is lighter than water. Vapor density is 2.04, which is heavier than air. In addition to flammability, isopropylamine is a strong irritant to tissue and has a TLV of 5 ppm in air. The four-digit UN identification number is 1221. The NFPA 704 designation for isopropylamine is health 3, flammability 4, and reactivity 0. Primary uses for isopropylamine are pharmaceuticals, dyes, insecticides, and as a dehairing agent. Structure and formula for isopropylamine are shown in Figure 5.44.

Ethers

Ether is the next flammable liquid hydrocarbon-derivative family. Primary hazard of ether is flammability. Ethers have low boiling and flash points, low ignition temperatures, and are nonpolar. In addition to being flammable, ethers are anesthetic; they have wide flammable ranges, from 2% to 48% in air; and can form explosive peroxides as the ether ages. Ethers are nonpolar compounds. When a container of ether is opened, oxygen from the air gets inside and bonds with the oxygen in the ether, forming an unstable peroxide. Heat, friction, or shock can cause the peroxide to explode. Oxygen has been known to permeate the soldered seam in a metal container even without the container being opened.

Ethers can become quite dangerous in storage. Most ether should not be stored longer than 6 months. If an aging container of ether is discovered, the nearest bomb squad should be called; the container should be treated as if it were a bomb. Ether is composed of a single oxygen atom with two hydrocarbon radicals, one on either side of the oxygen. It is expressed by the general formula *R-O-R*. Ether names do not always follow the trivial naming system. However, the formula will have a single oxygen atom, which indicates the ether family. Radicals on either side of the oxygen may be the same or they may be different. Ether compounds are named by identifying the two hydrocarbon radicals and ending with the word "ether." As a general rule, if the radicals are different, start with the smallest and name it, then name the second radical, and end with the word "ether." For example, the compound methyl ethyl ether has a methyl radical and an ethyl radical on either side of the oxygen. The smallest radical is methyl, the second radical is ethyl, and the name ends with "ether."

There may be occasions when a compound is looked up in a reference book, but it is not listed by the trivial naming system. When this happens, just look it up using the other radical name, such as ethyl methyl ether. When the radicals on each side of the oxygen in the ether compound are the same, the compound is named with just the one radical name. No prefix is used to indicate two of the same radical. Because ether must have two radicals, and there is only one radical in the name, it is understood that there are two of the same radical. For example, methyl ether has two methyl radicals, one on each side of the oxygen. Ether is the only hydrocarbon derivative where the prefix "di," indicating the number of radicals, is not used with the trivial naming system. However, even though uncommon, the "di" prefix is used and listed as a synonym for the common name of the ether compound. Therefore, it is not wrong to use "di" to identify the two radicals, it is just not common.

Ethyl methyl ether (*methoxyethane, IUPAC*) is a colorless liquid that is soluble in water. Specific gravity is 0.70, which is lighter than water. It is highly flammable, with a flammable range of 2%–10.1% in air. Boiling point is 51°F (10°C), flash point is −35°F (−37°C), and ignition temperature is 374°F (190°C). Vapor density is 2.07, which is heavier than air. In addition to flammability, ethyl methyl ether is an anesthetic and can form explosive peroxides as it ages. The four-digit UN identification number is 1039. The NFPA 704 designation is health 1, flammability 4, and reactivity 1. The primary use is in medicine as an anesthetic. The structure and molecular formula for ethyl methyl ether are shown in Figure 5.45.

Ethyl ether (diethyl ether) (*ethoxyethane, IUPAC*) is a colorless, volatile mobile liquid. It is slightly soluble in water, with a specific gravity of 0.7, which is lighter than water. It is a severe fire and explosion risk when exposed to heat or flame. The compound forms explosive peroxides from the oxygen in the air as it ages. Flammable range is wide, from 1.85% to 48% in air. Boiling point is 95°F (35°C), flash point is −49°F (−45°C), and ignition temperature is 356°F (180°C). Vapor density is 2.6, which is heavier than air. In addition to flammability, it is an anesthetic, which causes central nervous system depression by inhalation and skin absorption, with a TLV of 400 ppm in air. The four-digit UN identification number is 1155. The NFPA 704 designation is health 1, flammability 4, and reactivity 1. The primary uses are in the manufacture of smokeless powder, as an industrial solvent, in analytical chemistry, and as an anesthetic. Structure and molecular formula for ethyl ether are shown in Figure 5.46.

Isopropyl ether (diisopropyl ether) is a colorless, volatile liquid, which is slightly soluble in water. Specific gravity is 0.7, which is lighter than water. It is highly flammable, with a wide flammable range of 1.4%–21% in air. Boiling point is 156°F (68°C), flash point is −18°F

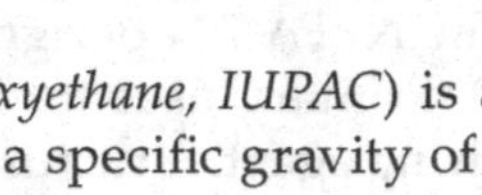

$$C_2H_5OCH_3$$

Figure 5.45 Ethyl methyl ether.

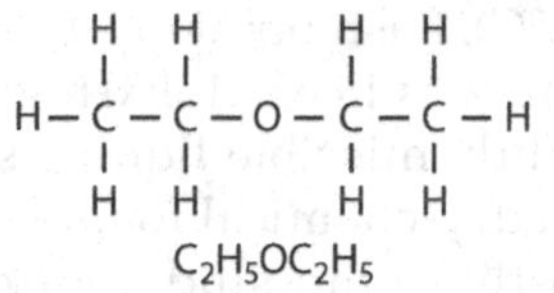

Figure 5.46 Ethyl ether.

(–27°C), and ignition temperature is 830°F (443°C). Vapor density is 3.5, which is heavier than air. In addition to flammability, isopropyl ether is toxic by inhalation and a strong irritant, with a TLV of 250 ppm in air. The four-digit UN identification number is 1159. The NFPA 704 designation is health 1, flammability 3, and reactivity 1. The primary uses are as a solvent and in rubber cements. Structure and molecular formula for isopropyl ether are shown in Figure 5.47.

Butyl ether (dibutyl ether) is a colorless, stable liquid, with a mild ether-like odor. It is immiscible with water, with a specific gravity of 0.8, which is lighter than water. Butyl ether is a moderate fire risk and will form explosive peroxides on aging. Flammable range is 1.5%–7.6% in air, with a boiling point of 286°F (141°C) and a flash point of 77°F (25°C). Ignition temperature is 382°F (194°C), and the vapor density is 4.5, which is heavier than air. In addition to flammability, butyl ether is toxic on prolonged inhalation. The four-digit UN identification number is 1149. The NFPA 704 designation is health 2, flammability 3, and reactivity 1. The primary use is as a solvent. The structure and molecular formula for butyl ether are shown in Figure 5.48.

Methyl tert-butyl ether (MTBE) is an ether hydrocarbon derivative and is highly flammable, with a wide flammable range. MTBE is not considered toxic; however, it is mildly irritating to the eyes and skin. If inhaled, it may cause suffocation. It is a gasoline additive, and there has been some controversy concerning potential health effects. MTBE is added to gasoline as an oxygenating compound that makes gasoline burn cleaner in the winter. The Centers for Disease Control conducted a study that was inconclusive: it did not vindicate the material; it just did not find enough evidence that the chemical is a health concern. MTBE is a colorless, nonpolar liquid with an anesthetic-like odor. It

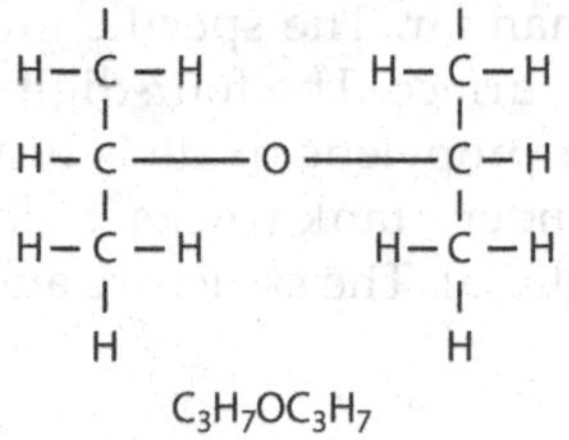

Figure 5.47 Isopropyl ether.

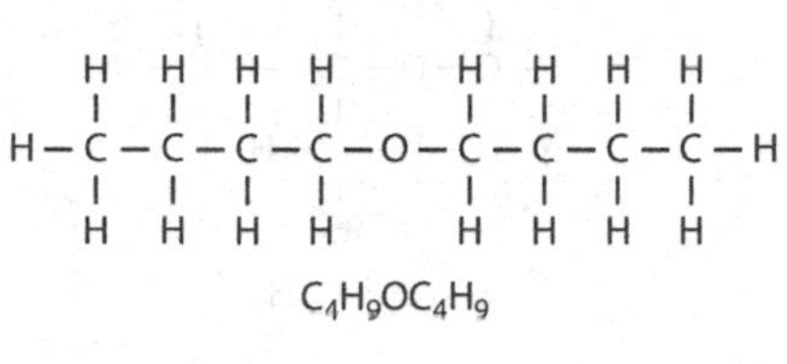

Figure 5.48 Butyl ether.

has a boiling point of 131°F (66°C), flash point of −14°F (−25°C), and ignition temperature of 707°F (375°C). Flammable range is 1.6–15.1. Extinguishing agents for ethers and other nonpolar, nonmiscible, or slightly miscible liquids should be selected carefully. Small fires can be extinguished with dry-chemical foam with some difficulty; remember that ether has an oxygen atom in the compound, so excluding atmospheric oxygen may not be effective. Water may also be ineffective. Alcohol-type foams may be effective against materials that are slightly miscible. The higher-molecular-weight liquids will attack the alcohol-type foam, and hydrocarbon foam will be needed. Vapor density is 3, which makes it heavier than air. Specific gravity is 0.74, which is less than the weight of water; therefore, it will float on top of water. MTBE has a four-digit UN identification number of 2398. It is used primarily as an octane booster for unleaded gasoline. There are no NFPA 704 data available for MTBE, and the reference information on the material is sketchy. The structure and molecular formula for MTBE are shown in Figure 5.49.

Ethylene oxide is an ether with the formula C_2H_4O. It is a cyclic ether. Ethylene oxide itself is a very hazardous substance; at room temperature, it is a flammable, carcinogenic, irritating, and anesthetic gas with a misleadingly pleasant aroma. As a poison gas that leaves no residue on items it contacts, pure ethylene oxide is an ether that is widely used in hospitals and the medical equipment industry to replace steam in the sterilization of heat-sensitive tools and equipment, such as disposable plastic syringes. Ethylene oxide is extremely flammable and explosive and is used as a main component of; therefore, it is commonly handled and shipped as a refrigerated liquid.

Propylene oxide (*epoxypropane, IUPAC*) is a cyclic ether hydrocarbon derivative, although it does not follow the trivial naming system for ethers. It has a cyclic structure between the oxygen atom and the two carbon atoms, and is a colorless liquid with an ether-like odor. Propylene oxide is nonpolar, partially soluble in water, and is highly flammable, with a wide flammable range of 2%–22% in air. Boiling point is 94°F (34°C), flash point is 35°F (1.66°C), and ignition temperature is 840°F (448°C). Small fires can be extinguished with dry-chemical foam with some difficulty. Remember that ether has an oxygen atom in the compound, so excluding atmospheric oxygen may not be effective. Water may also be ineffective. Alcohol-type foams may be effective against materials that are slightly miscible. The higher-molecular-weight liquids will attack the alcohol-type foam, and hydrocarbon foam will need to be used. In addition to flammability, propylene oxide is an irritant, with a TLV of 20 ppm in air, and it is also corrosive. The vapor density is 2, so it is heavier than air. The specific gravity is 0.83, which is lighter than water, and it will float on the surface. The four-digit UN identification number is 1280. The NFPA 704 designation for propylene oxide is health 3, flammability 4, and reactivity 2. It is shipped in steel cylinders, tank trucks, railcars, and barges under the cover of nitrogen, which is an inert material. The structure and molecular formula for propylene oxide are shown in Figure 5.50.

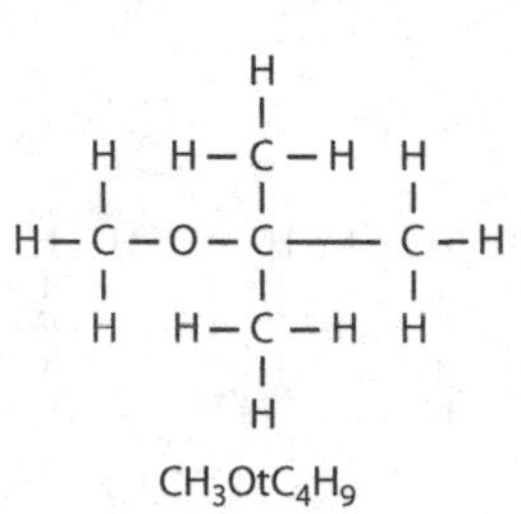

$$CH_3OtC_4H_9$$

Figure 5.49 MTBE.

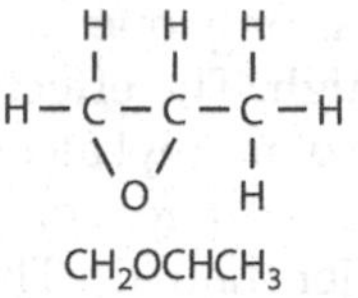

$$\begin{array}{c} \text{H} \quad \text{H} \quad \text{H} \\ | \quad | \quad | \\ \text{H}-\text{C}-\text{C}-\text{C}-\text{H} \\ \backslash \quad / \quad | \\ \text{O} \quad \text{H} \end{array}$$

CH₂OCHCH₃

Figure 5.50 Propylene oxide.

Alcohol

Alcohol is the next flammable liquid hydrocarbon-derivative family. In addition to being flammable, alcohols have wide flammable ranges from 1% to 36% in air and are toxic to some degree. They have high boiling points, moderate flash points, and high ignition temperatures. Small fires involving alcohols should be fought with dry-chemical fire extinguishers. Large fires should be fought with alcohol-type foam; water may be ineffective. Alcohols are miscible with water. Water, as it mixes with alcohol, will at some point raise the boiling and flash points of the alcohol until the mixture of the water and the alcohol are no longer flammable. The problem with this is that the container must be large enough to hold the mixture, or this method of extinguishment will not work. Therefore, care must be taken in choosing the method of extinguishing alcohol fires. Methyl alcohol, or methanol as it is sometimes called, is toxic by ingestion and can cause blindness or death, with a TLV of 200 ppm. Ethyl alcohol or ethanol, also referred to as grain alcohol, is consumed in alcoholic beverages. It is classified as a depressant drug; too much of it can produce toxic effects and can lead to liver damage.

The alcohol functional group is identified by the general formula *R-O-H*. There is one radical attached to the oxygen atom. Alcohols are polar liquids because they have hydrogen bonding. Alcohol is the second most polar functional group; the most polar are the organic acids. Because of polarity, alcohols are miscible with water and require the use of polar solvent or alcohol-type foam to extinguish fires. Because of polarity, alcohols as a family have high boiling and flash points.

An alcohol is named by identifying the radical attached to the oxygen. The radical is named first, and the compound ends in the word "alcohol." There are often various ways of naming the same chemical compounds. With alcohol, an "ol" ending may be added to the radical, indicating that it is an alcohol. For example, the radical "methyl" is attached to the oxygen in the alcohol functional group. The name of the compound is methyl alcohol or may be called methanol.

Methyl alcohol, also known as (*methanol, IUPAC*) wood alcohol, is an alcohol hydrocarbon derivative. It is a clear, colorless liquid that is highly polar and miscible with water. Alcohol has hydrogen bonding and is the second most polar material of the hydrocarbon derivatives after organic acids. It is a dangerous fire risk, with a wide flammable range from 6% to 36.5% in air. Fighting fires will require the use of alcohol-type foam. Boiling point is 147°F (63°C), and flash point is 52°F (11°C). By adding up the weights of the elements in methyl alcohol shown in Figure 5.51, you can determine that the molecular

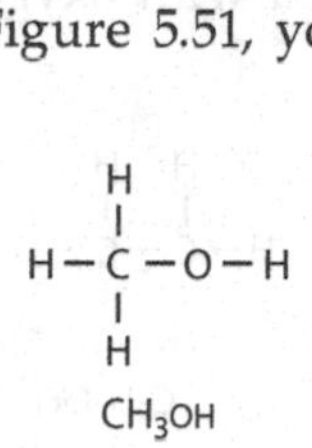

$$\begin{array}{c} \text{H} \\ | \\ \text{H}-\text{C}-\text{O}-\text{H} \\ | \\ \text{H} \end{array}$$

CH₃OH

Figure 5.51 Methyl alcohol.

weight of methyl alcohol is 32. It has only one carbon atom and one oxygen atom, yet the boiling point and flash point are high. The effects of polarity can readily be seen in this example. The ignition temperature of methyl alcohol is 867°F (463°C). In addition to being flammable, methyl alcohol is toxic by ingestion and has a TLV of 200 ppm in air. The vapor density is 1.1, which is heavier than air. The specific gravity is 0.8, which makes it lighter than water. Methyl alcohol is miscible with water, so it will mix rather than form defined layers. The four-digit UN identification number is 1230. The NFPA 704 designation is health 1, flammability 3, and reactivity 0. Methyl alcohol is shipped in glass bottles, 55 gal drums, tank trucks, and railcars. The structure and molecular formula for methyl alcohol are shown in Figure 5.51.

Ethyl alcohol (ethanol, IUPAC) is a colorless, clear, volatile liquid. It is polar and soluble in water, with a specific gravity of 0.8, which is lighter than water. It is highly flammable with a flammable range of 3.3%–19% in air. The boiling point is 173°F (78°C), and flash point is 55°F (12°C). Vapor density is 1.6, which is heavier than air. Ignition temperature is 685°F (362°C), and ethanol is classified as a depressant drug with a TLV of 1000 ppm in air. The four-digit UN identification number is 1170. The NFPA 704 designation is health 0, flammability 3, and reactivity 0. The primary uses are as a solvent and in beverages, antifreeze, gasohol, pharmaceuticals, and explosives. The structure and molecular formula for ethyl alcohol are shown in Figure 5.52.

Ethanol (ethyl alcohol) is the alcohol used to create beer, wine, and other alcoholic beverages. Ethanol is also produced as an alternative fuel to gasoline. Ethanol is a renewable energy source that can help reduce the U.S. dependence on foreign oil imports. In the United States, over 9 billion gallons of ethanol fuel are produced annually (Photo 5.10). Ethanol is typically produced from corn or sugarcane. The United States is the largest producer of ethanol in the world and primarily uses corn. Ethanol fuel does have a few disadvantages; it has a lower energy density than gasoline, so a tank of ethanol fuel will not go as far as a tank of gasoline, and ethanol fuel can be more difficult to start in very cold temperatures.

Over the past decade, ethanol has become an important source of fuel for automobiles in an effort to find a way to reduce our dependence on foreign oil. Ethanol distilling and blending plants have been popping up in many parts of the United States. Ethanol is primarily shipped by rail, and this has created additional problems for emergency responders because of the quantities being shipped. Hazards of ethanol are not as well understood by responders as gasoline and diesel fuel may be.

Gasoline and diesel fuel are primarily transported by pipeline and with highway transportation vehicles. There is some rail transportation of gasoline and diesel fuel, but it is not nearly the volume as pipelines and highway transportation. Ethanol on the other hand is largely transported by rail, supplemented by highway and minimal pipeline transportation. It is classified by the DOT as a Class 3 Flammable Liquid just like gasoline and diesel fuel, and it is marked in transportation with red placards and labels. Ethanol along with other alcohols is single bonded with oxygen bonded to a single hydrogen (–OH)

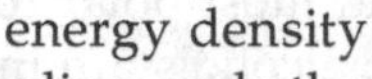

$$\text{H}-\underset{\underset{\text{H}}{|}}{\overset{\overset{\text{H}}{|}}{\text{C}}}-\underset{\underset{\text{H}}{|}}{\overset{\overset{\text{H}}{|}}{\text{C}}}-\text{O}-\text{H}$$

$$C_2H_5OH$$

Figure 5.52 Ethyl alcohol.

Photo 5.10 Ethanol plants are springing up all over the Midwest, where the majority of them utilize the renewable resource of corn to manufacturer fuel for automobiles.

added to the hydrocarbon compound. Ethanol is an alcohol with the –OH attached to a two-carbon backbone from the hydrocarbon ethane. It is a pure chemical where gasoline and diesel fuels are mixtures of chemicals. Mixtures that form gasoline and diesel fuel include members of the hydrocarbon family.

Alcohols including ethanol are members of a hydrocarbon-derivative subgroup known as polar solvents. Water, like alcohol, is also a polar compound. Water has a molecular weight of 18. Air has an average molecular weight of 29. With a molecular weight of 18, water should be a gas at normal temperatures and pressures. But as we know it is not, it is a liquid. It is polarity that allows water to exist as a liquid. Because water and alcohols are polar compounds, they are miscible. That is to say, they mix when placed together. (There are some experiments you can do at home that will prove this is true.) Because water and alcohol mix, it is one of the reasons that different types of fire-fighting foam are required for alcohol fires vs. gasoline or diesel fires. Ethanol is miscible with water, and if spilled in a water source, it would be difficult to clean up. Polarity is one of the major differences between ethanol and gasoline and diesel fuel. Flash point, boiling point, and ignition temperature of flammable liquids are affected by polarity. Compounds that are polar have a tendency to have higher boiling points and flash points than nonpolar compounds with similar molecular weights. Gasoline and diesel fuel have narrow flammable ranges, between 3 and 10. Ethanol, as with other alcohols, has a wider flammable range, from 3.3 to 19. That means that ethanol will burn within a greater percentage of mixtures of fuel and air than gasoline and diesel fuel. The wider the flammable range, the more likely it will be that the fuel may burn inside a container. Ethanol has flash point and ignition temperatures of 61.88°F (16.6°C) and 685°F (362°C) respectively. In terms of ignition temperature, gasoline and ethanol are closer than diesel fuel, which means the liquids must be heated to near the same temperature or the ignition source temperature would be close to ignite both liquids. Ethanol has a lower ignition temperature than most gasoline mixtures but is higher than diesel fuel. However, the flash point temperatures are much different between gasoline, diesel fuel, and ethanol. When exposed to water, gasoline and diesel fuel will float on top of the water. So the real major differences between gasoline, diesel fuel, and ethanol are polarity and miscibility. This results in different firefighting tactics in terms of foam use with ethanol.

Ethanol will mix with the water. For this reason, the foam used to fight fires involving gasoline, diesel fuel, and ethanol need to be different to be effective (Photo 5.11). If you use regular protein foam, AFFF, or fluoroprotein foam, it will not work on ethanol fires because the water from the foam will mix with the alcohol and the foam blanket will break down. When fighting ethanol fires, you will need to use alcohol-type or polar-solvent-type foams. Another possible tactic for extinguishing ethanol fires is using the issue of miscibility to your advantage. If there is enough room in a container to fill the container with water, at some point the ethanol will no longer be flammable and the fire will go out because of the percentage of water applied that mixes with the alcohol.

Ethanol burns with a bluish flame that may be difficult to view under certain light conditions. Ethanol fires also give off less carbon or black smoke than gasoline

Photo 5.11 Ethanol requires a special foam to extinguish fires. Alcohol-type or polar solvent foam is the only foam that will effectively extinguish a fire involving ethanol.

and diesel fuel fires. To sum up what we have discussed about the differences and similarities between gasoline and diesel fuel and ethanol, they are all DOT Class 3 Flammable Liquids. Gasoline and diesel fuel are nonpolar mixtures that are immiscible with water. Ethanol is a pure chemical that is polar and miscible with water. In transportation, you will likely encounter larger quantities in an accident because ethanol is largely shipped by rail. Once a fire has occurred in an accident at a plant or in storage or transportation, you need to use the proper type of foam in order to extinguish the fire.

During the manufacturing process and to a lesser degree in storage at the manufacturing facility, ethanol is pure 190 proof grain alcohol (Photo 5.12). As a pure alcohol, ethanol is placarded by the DOT as a flammable bulk liquid assigned the United Nations (UN) identification number of 1170. This material is also referred to as E100. When it is shipped from the manufacturing facility, ethanol is denatured with 2%–5% natural gasoline also known as E98 and E95 respectively. A blend of 95% ethanol and 5% gasoline has been assigned a DOT/North American (NA) identification number of 1987 for denatured alcohol or alcohol n.o.s. (not otherwise specified) and UN 1987. Mixtures of E95 through E99 are also assigned the 1987 UN identification number. Additionally, E95 may utilize the UN identification number 3475. Ethanol is ultimately blended with petroleum gasoline to form a motor fuel in various concentrations depending on whether it is used as an additive/oxygenator or blended motor fuel. Ethanol and gasoline mixtures are assigned

Photo 5.12 Ethanol manufactured and stored at plants is 190 proof grain alcohol. It is denatured with 2%–5% natural gasoline to prevent anyone from drinking it.

the identification number UN 3475 including E11 through E99. E1–E10 blends are assigned the UN number 1203, which is also used for gasoline. Pure ethanol (E100), E95, E90, E85, and gasoline are all assigned an NFPA 704 designation of flammability 3, health 1, and reactivity 0.

I have read where people have referred to explosions involving ethanol tank cars as BLEVEs. A BLEVE is a boiling liquid, expanding vapor explosion. Tanks that contain boiling liquids subject to a BLEVE are pressure tanks that contain materials like propane, butane, and liquefied petroleum gases (LPG). Ethanol tanks are liquid atmospheric pressure tanks. Under normal conditions, the tanks are at atmospheric pressure (Photo 5.13). The types of materials that are subject to a BLEVE are gases that have been liquefied in order to ship larger quantities. Liquefied gases remain a liquid in the tank, above their boiling point because of continued pressure in the tank. They continue to be liquids as long as the pressure tank car is intact. If a breach occurs to the container, all of the liquid in the tank car immediately turns back into gas at once. This occurs explosively, often rocketing parts of the disintegrating tank car over 1000 ft. Thus, the term BLEVE, boiling liquid in the tank, vapor expands as the tank is breached, occurring explosively. The main factor to bear in mind is that the power of a BLEVE lies in the release of *liquid* product that quickly (if not instantaneously) boils or evaporates into the vapor phase and combines with the surrounding atmosphere to form a flammable and explosive mixture. Tanks containing

Photo 5.13 Most petroleum products are shipped long distances by pipeline. There are no significant ethanol pipelines in place yet, so most ethanol is shipped by rail. Trains with large numbers of ethanol tank cars are not usual.

Photo 5.14 Because there are large numbers of ethanol tank cars on trains, and if an accident occurs, it seems like a bigger problem than it really is. Ethanol tank cars are liquid cars and do not BLEVE under fire conditions. (From *Rockford Register Star*, photographer Scott Morgan.)

ethanol and its mixtures with gasoline exposed to fire can rupture because of excess pressure built up by flame impingement, but they do not BLEVE.

Since 2000, there have been reports of at least 25 incidents involving ethanol and its blends at fixed facilities and in transportation. On June 17, 2009, 18 cars of a Canadian National Railway train containing denatured ethanol derailed and 14 of them caught fire in Cherry Valley, Illinois, near Rockford (Photo 5.14). Of the train with 114 cars, 74 contained denatured ethanol. A civilian setting in their vehicle at the railroad crossing was fatally burned by the fire that engulfed her vehicle. Six others were also injured and taken to area hospitals. Approximately 600 homes in the area of the derailment were evacuated. Firefighters from 26 local departments responded to the fire that was allowed to burn itself out over several days (Photo 5.15).

Propyl alcohol, 1-propanol (propan-1-ol, IUPAC), is a colorless liquid with an odor similar to ethanol. It is polar, soluble in water, and has a specific gravity of 0.8, which is lighter than water. It is a dangerous fire risk with a flammable range of 2%–13% in air. Vapor density is 2.1, which is heavier than air. Boiling point is 207°F (97°C), flash point is 74°F (23°C), and ignition temperature is 775°F (412°C). It is toxic by skin absorption with a TLV of 200 ppm in air. The four-digit UN identification number is 1274. The NFPA 704 designation is health 1, flammability 3, and reactivity 0. The primary uses are in brake fluid, as a solvent, and as an antiseptic. The structure and molecular formula for propyl alcohol are shown in Figure 5.53.

n-Butyl alcohol (butan-1-ol, IUPAC) is an alcohol hydrocarbon derivative and is a colorless liquid with a wine-like odor. Alcohols are highly polar and are miscible with water. Butyl alcohol is a moderate fire risk, with a flammable range of 1.4%–11.2% in air (Photo 5.16). Boiling point is 243°F (117°C), flash point is 98°F (36°C), and ignition temperature is 650°F (343°C). Fires should be fought with alcohol-type foams. In addition to being flammable, butyl alcohol is toxic when inhaled for long periods, is irritating to

Photo 5.15 Ethanol plant personnel are the best source of information during an emergency at a fixed facility.

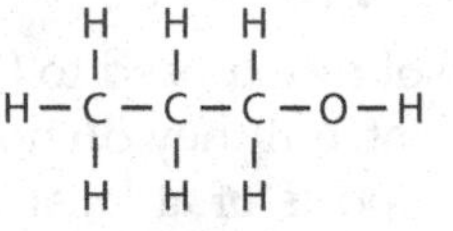

$$
\begin{array}{cccc}
& H & H & H \\
& | & | & | \\
H- & C- & C- & C-O-H \\
& | & | & | \\
& H & H & H \\
& & C_3H_7OH
\end{array}
$$

Figure 5.53 Propyl alcohol.

the eyes, and is absorbed through the skin. TLV is 50 ppm in air. Vapor density of butyl alcohol is 2.6, so it is heavier than air. Specific gravity is 0.8, which is lighter than water; however, it is miscible with water and will mix rather than form layers. The four-digit UN identification number is 1120. The NFPA 704 designation for butyl alcohol is health 1, flammability 3, and reactivity 0. It is shipped in glass bottles, pails, 55 gal drums, tank trucks, railcars, and barges. The structure and molecular formula for *n*-butyl alcohol are shown in Figure 5.54.

Isopropyl alcohol (*propan-2-ol, IUPAC*) is an alcohol hydrocarbon derivative. The liquid is colorless, with a pleasant odor, highly polar, and miscible with water. Firefighting will require polar solvent foam for extinguishment. Isopropyl alcohol is highly flammable, with a flammable range of 2%–12% in air. Boiling point is 181°F (82°C), flash point is 53°F (11°C), and ignition temperature is 750°F (398°C). In addition to flammability, isopropyl alcohol is toxic by ingestion and inhalation, with a TLV of 400 ppm in air. Vapor density is 2.1, which is heavier than air. Specific gravity is 0.8, which is lighter than water; however, it is miscible with water and will mix rather than form layers. The four-digit UN identification number is 1219. The NFPA 704 designation is health 1, flammability 3, and reactivity 0. It is shipped in glass bottles, pails, 55 gal drums, tank trucks, railcars, and barges. The structure and molecular formula for isopropyl alcohol are shown in Figure 5.55.

Photo 5.16 Cone-roof tank for flammable liquids.

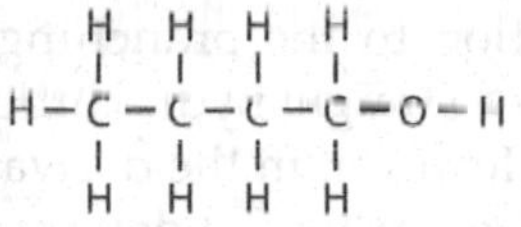

$$H-\overset{\overset{\displaystyle H}{|}}{\underset{\underset{\displaystyle H}{|}}{C}}-\overset{\overset{\displaystyle H}{|}}{\underset{\underset{\displaystyle H}{|}}{C}}-\overset{\overset{\displaystyle H}{|}}{\underset{\underset{\displaystyle H}{|}}{C}}-\overset{\overset{\displaystyle H}{|}}{\underset{\underset{\displaystyle H}{|}}{C}}=O-H$$

$$C_4H_9OH$$

Figure 5.54 Butyl alcohol.

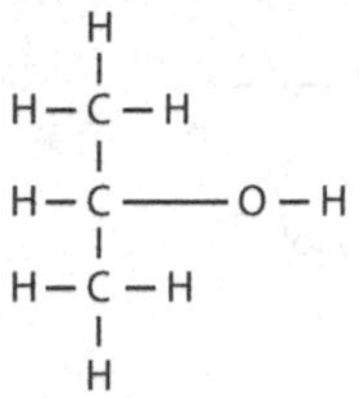

$$
\begin{array}{c}
H \\
| \\
H-C-H \\
| \\
H-C\!\!-\!\!-\!\!-\!\!O-H \\
| \\
H-C-H \\
| \\
H
\end{array}
$$

$$C_3H_7OH$$

Figure 5.55 Isopropyl alcohol.

Isomers

Isomers were introduced in Chapter 4. An isomer is a compound with the same formula as the "normal" compound, but a different structure. Isomers are sometimes referred to as branched compounds. Branching of a compound has the effect of lowering the boiling point. In "Hydrocarbon families" section of Chapter 4, only the "iso" branch was discussed. With the hydrocarbon-derivative functional groups, there will also be an "iso" branch and, in addition, there will be secondary and tertiary branches.

When dealing with the hydrocarbon compounds, the branch is a carbon atom attached to one of the center carbon atoms of the chain, but not on the end carbons as that would still be a straight chain. In the derivatives, the functional group is a part of that carbon chain and is considered when determining branching of the compound. The types of branches will be shown in this section because they occur commonly with the alcohol compounds. However, branching can occur in any of the hydrocarbon-derivative groups. Figure 5.56 shows examples of branches of a four-carbon alcohol. The branch is determined by the location of the functional group on the carbon chain. The first structure is the straight-chained compound butyl alcohol with a molecular formula of C_4H_9OH. Straight-chained compounds are sometimes referred to as the "normal" form. Normal butyl alcohol is represented by a small "*n*" in front of the molecular formula and the word normal in front of the name. The next structure is isobutyl alcohol. The "iso" branch is determined by locating the –O–H of the alcohol functional group and using it as an entry point into the structure. Then go to the first carbon that is attached to the –O–H. See how many carbon atoms are attached to the first carbon. In the case of the "iso" branch, only one carbon atom is attached. The third structure is secondary butyl alcohol. The –O–H is attached to a carbon that is attached to two other carbons. In the final structure, the compound is called tertiary butyl alcohol. The functional group is attached to a carbon atom, which is attached to three other carbon atoms. Notice that all of the compounds have the same molecular formula. To distinguish between them, it is necessary to include a small letter indicating which branch is in the structure in front of the molecular formula. A small "i" is used for "iso" branches, a small "s" for secondary, and a small "t" for tertiary branches. In front of the names, insert iso, sec or secondary, or tert or tertiary.

Propane is the one exception to the branching rules of hydrocarbon derivatives. Propane cannot be branched as a straight hydrocarbon because there are not enough carbon atoms to create a branch. However, in the derivatives, the functional group becomes a part of the carbon chain for the purpose of determining branching. It is possible to put

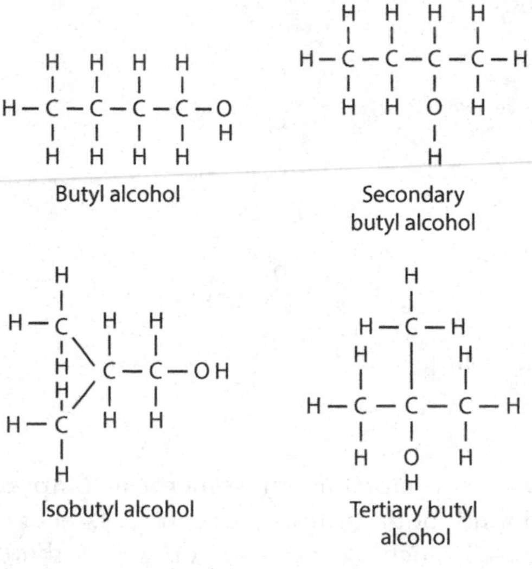

Figure 5.56 Four isomers of butyl alcohol.

a functional group on the center carbon atom of propane. The structure formed appears to be secondary, according to the examples shown. However, there is only one way to branch propane, so it is called the "iso" branch. That is something that has to be committed to memory as an exception to the branching rules. Another way to remember the exception is that secondary and tertiary are not used until there are four carbons in a compound.

Denatured alcohol is ethyl alcohol, or ethanol, to which another liquid has been added to make it unfit to use as a beverage. The primary reason for denaturing is for tax purposes. There are approximately 50 formulations of denatured alcohol. The hazards are the same as for ethanol. The primary uses for denatured alcohol are in the manufacture of acetaldehyde, solvents, antifreeze, brake fluid, and fuels. Ethanol manufactured as an additive to motor fuels is also denatured before it leaves the manufacturing facility to prevent consumption.

Ketone

Ketone is next in the hydrocarbon-derivative family. As a group, the ketones are flammable and narcotic. They have moderate boiling and flash points, narrow flammable ranges, and high ignition temperatures. Ketones are polar, and fires should be fought with alcohol- or polar-solvent-type foams because water may be ineffective. They are made up of a carbon atom double-bonded to an oxygen atom, with a radical on each side. The general formula is R-C-O-R. Ketone is the first of several compounds that are part of the carbonyl family. Carbonyl compounds have a carbon double-bonded to oxygen. Carbonyls are polar. The degree of polarity is less than that of the alcohols and organic acids. There are two radicals required in ketone compounds. The radicals may be the same, in which case the prefix "di" is used to indicate two, or the radicals may be different. When naming these compounds, the smallest is named first, then the second radical, ending in the word "ketone." Some ketones have trade names by which they are commonly known. There are no naming rules that can be used to determine trade names. There are some hints, such as with acetone, which is a three-carbon ketone, also called dimethyl ketone (Figure 5.58). The "one" ending indicates ketone, just as the "ol" ending indicates alcohol. The ending would be a tip-off that acetone may be a ketone. DMK is the trade name used for dimethyl ketone, which is also known as acetone. MEK is often used as a shortened name or trade name for methyl ethyl ketone (MEK). These are common ketones, and the more familiar you are with hazardous materials, the more familiar you will become with alternate names and trade names.

Acetone, also known as dimethyl ketone, is a ketone hydrocarbon derivative. It is a colorless, volatile liquid with a sweetish odor. Acetone is a carbonyl, is polar, and is miscible with water, which means it will require polar solvent foam for fire extinguishment. This compound is highly flammable, with a range of 2.6%–12.8% in air. Boiling point is 133°F (56°C), flash point is −4°F (−20°C), and ignition temperature is 869°F (465°C). Fighting fires will require the use of alcohol-type foam. In addition to flammability, acetone is a narcotic at high concentrations and is moderately toxic by ingestion and inhalation, with a TLV of 750 ppm in air. The vapor density is 2, which is heavier than air. The specific gravity is 0.8, so it is lighter than water, but is miscible and will mix with the water rather than form layers. The four-digit UN identification number is 1090. The NFPA 704 designation is health 1, flammability 3, and reactivity 0. It is shipped in pails, drums, tank trucks, railcars, and barges. Acetone is the solvent used to dissolve acetylene in cylinders to keep it stable. The structure and molecular formula are shown in Figure 5.57.

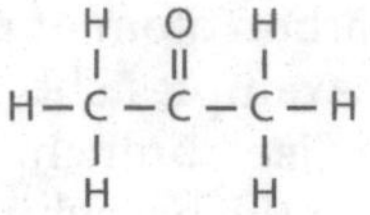

$$
\begin{array}{c}
\;\;\;\;\;\text{H} \;\; \text{O} \;\; \text{H} \\
\;\;\;\;\;| \;\;\; \| \;\;\; | \\
\text{H}-\text{C}-\text{C}-\text{C}-\text{H} \\
\;\;\;\;\;| \;\;\;\;\;\;\;\; | \\
\;\;\;\;\;\text{H} \;\;\;\;\; \text{H}
\end{array}
$$

DMK
acetone
CH_3COCH_3

Figure 5.57 Dimethyl ketone.

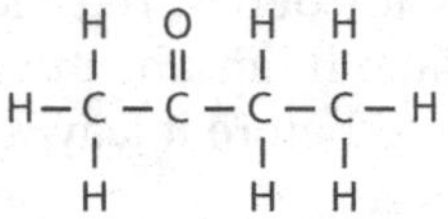

$$
\begin{array}{c}
\;\;\;\;\;\text{H} \;\; \text{O} \;\; \text{H} \;\; \text{H} \\
\;\;\;\;\;| \;\;\; \| \;\;\; | \;\;\; | \\
\text{H}-\text{C}-\text{C}-\text{C}-\text{C}-\text{H} \\
\;\;\;\;\;| \;\;\;\;\;\;\;\; | \;\;\; | \\
\;\;\;\;\;\text{H} \;\;\;\;\; \text{H} \;\; \text{H}
\end{array}
$$

MEK
$CH_3COC_2H_5$

Figure 5.58 Methyl ethyl ketone.

MEK (*butan-2-one, IUPAC*) is a colorless liquid with an acetone-like odor. MEK is polar, soluble in water, and has a specific gravity of 0.8, which is lighter than water. It is highly flammable, with a flammable range of 2%–10% in air, a flash point of 1°F (–17°C), and a boiling point of 176°F (80°C). Ignition temperature of MEK is 759°F (403°C), and the vapor density is 2.5, which is heavier than air. It is toxic by inhalation, with a TLV of 200 ppm in air. The four-digit UN identification number is 1193. The NFPA 704 designation for MEK is health 1, flammability 3, and reactivity 0. The primary uses of MEK are as a solvent and in the manufacture of smokeless powder, cleaning fluids, in printing, and acrylic coatings. The structure and molecular formula for MEK are shown in Figure 5.58.

Methyl vinyl ketone, MVK, or vinyl methyl ketone (*butenone, IUPAC*), is a colorless liquid that is soluble in water. It is polar, with a specific gravity of 0.8636, which is lighter than water. It is flammable, with a flammable range of 2.1%–15.6% in air, and the vapor density is 2.4, which is heavier than air. Boiling point is 177°F (80°C), flash point is 20°F (–6°C), and ignition temperature is 915°F (490°C). MVK is a skin and eye irritant. The four-digit UN identification number is 1251. The NFPA 704 designation is health 4, flammability 3, and reactivity 2. The primary uses are as a monomer for vinyl resins and as an intermediate in steroid and vitamin A synthesis. The structure and molecular formula for MVK are shown in Figure 5.59.

Methyl isobutyl ketone (*4-methylpentan-2-one, IUPAC*) is a colorless, stable liquid with a pleasant odor. It is slightly soluble in water, with a specific gravity of 0.8, which is lighter than water. The vapor density is 3.5, which is heavier than air. It is highly flammable, with a flammable range of 1.2%–8% in air. The boiling point is 244°F (117°C), flash point is 64°F (17°C), and ignition temperature is 840°F (448°C). It is toxic by inhalation, ingestion, and skin absorption, with a TLV of 50 ppm in air. The four-digit UN identification number is

$$
\begin{array}{c}
\;\;\;\;\;\text{H} \;\; \text{O} \;\; \text{H} \;\; \text{H} \\
\;\;\;\;\;| \;\;\; \| \;\;\; | \;\;\; | \\
\text{H}-\text{C}-\text{C}-\text{C}=\text{C} \\
\;\;\;\;\;| \;\;\;\;\;\;\;\;\;\;\;\;\;\; | \\
\;\;\;\;\;\text{H} \;\;\;\;\;\;\;\;\;\;\; \text{H}
\end{array}
$$

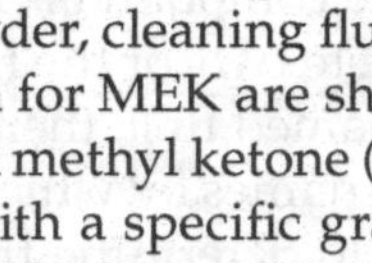

$CH_3COC_2H_3$

Figure 5.59 Methyl vinyl ketone.

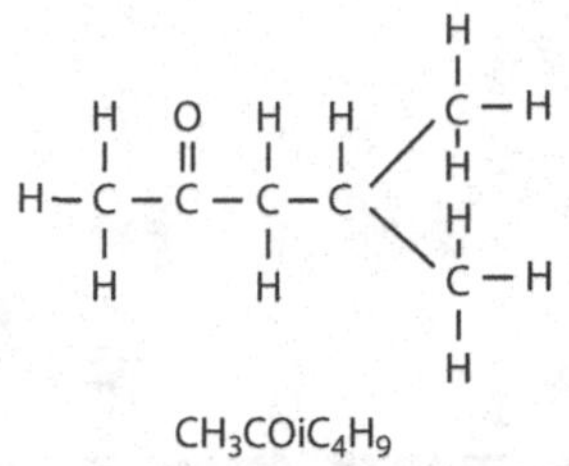

$CH_3COiC_4H_9$

Figure 5.60 Methyl isobutyl ketone.

1245. The NFPA 704 designation is health 2, flammability 3, and reactivity 1. The primary uses are as a solvent for paints, varnishes, and lacquers; in the extraction of uranium from fission products; and as a denaturant for alcohol. The structure and molecular formula for methyl isobutyl ketone are shown in Figure 5.60.

Aldehyde

Aldehydes have a wide flammable range from 3% to 55% in air; they are toxic and may polymerize. They have moderate boiling and flash points and high ignition temperatures. Fires involving aldehydes should be fought with polar-solvent-type foams because water may be ineffective. Aldehydes may also form explosive peroxides as they age, much the same way ethers do. Aldehydes are composed of a carbon atom double-bonded to an oxygen atom with a hydrogen atom on the other carbon connection. Aldehydes are carbonyls and, therefore, polar. The degree of polarity is much the same as ketone and ester, and much less than alcohol and organic acid. They are miscible with water and require the use of polar solvent foams to extinguish fires. Aldehydes have the general formula of *R-CHO*. There is one radical attached to the carbon atom of the aldehyde functional group. Aldehydes are one of the three derivatives in which the carbon atom in the functional group is counted when naming the compound. The alternate terms for one- and two-carbon radicals are also used with the aldehydes. A one-carbon aldehyde uses "form," and a two-carbon uses "acet." The aldehydes are named by identifying the radical, naming it, and ending with the word "aldehyde." Aldehydes may also be named in the same manner as the alternate names for alcohols; however, with the aldehydes, the ending is "al" instead of "ol." For example, a one-carbon aldehyde is called formaldehyde, with the alternate name of methanal. In the following examples, the structures, molecular formulas, and names are shown for one-, two-, and three-carbon aldehydes. The polarity of the carbonyls is somewhat less than that of the alcohols and the organic acids (Photo 5.17).

Acetaldehyde (ethanal, IUPAC) is a colorless liquid with a pungent, fruity odor. The odor is detectable at 0.07–0.21 ppm in air. It is highly flammable and a dangerous fire and explosion risk, with a wide flammable range of 4%–60% in air. Boiling point is 69°F (20°C), flash point is −36°F (−37°C), and ignition temperature is 374°F (190°C). Acetaldehyde is miscible with water, and the specific gravity is 0.78, which is lighter than water. The vapor density is 1.52, which is heavier than air. In addition to flammability, it is toxic (narcotic) and has a TLV of 100 ppm in air. Eye irritation occurs at 25–50 ppm in air. The four-digit UN identification number is 1089. The NFPA 704 designation is health 3, flammability 4, and reactivity 2. Its primary use is in the manufacture of other chemicals and artificial flavorings. The structure and molecular formula are shown in Figure 5.61.

Propionaldehyde, also known as *(propanal, IUPAC)* propyl aldehyde, is a water-white liquid with a suffocating odor that is water soluble. It is a dangerous fire and explosion risk, with a flammable range of 3%–16% in air. Boiling point is 120°F (48°C), flash point

Photo 5.17 Open floating-roof tank used for hydrocarbon fuels.

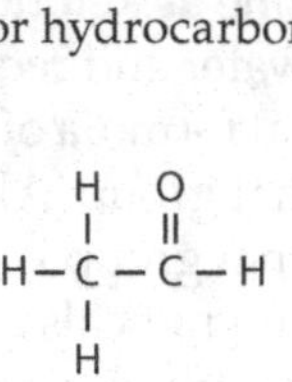

$$H-\overset{\displaystyle H}{\underset{\displaystyle H}{C}}-\overset{\displaystyle O}{C}-H$$

Ethanal
CH₃CHO

Figure 5.61 Acetaldehyde.

is 16°F (−8°C), and ignition temperature is 405°F (207°C). It is partially soluble in water and has a specific gravity of 0.81, which is lighter than water. Vapor density is 0.807, which is lighter than air. In addition to flammability, propionaldehyde is an irritant to the eyes, skin, and respiratory system. The four-digit UN identification number is 1275. The NFPA 704 designation is health 2, flammability 3, and reactivity 2. The primary uses are in the manufacture of other chemicals and plastics, as well as a preservative and disinfectant. The structure and molecular formula for propionaldehyde are shown in Figure 5.62.

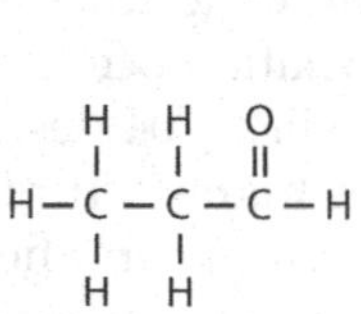

$$H-\overset{\displaystyle H}{\underset{\displaystyle H}{C}}-\overset{\displaystyle H}{\underset{\displaystyle H}{C}}-\overset{\displaystyle O}{C}-H$$

C₂H₅CHO

Figure 5.62 Propionaldehyde.

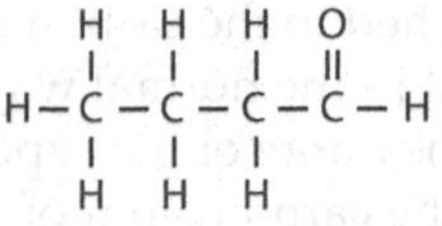

$$H-C-C-C-C-H$$

(structure with H H H O across top, H H H below)

$$C_3H_7CHO$$

Figure 5.63 Butyraldehyde.

Butyraldehyde (butanal, IUPAC) is a water-white liquid with a pungent aldehyde odor. Butyraldehyde is a dangerous fire risk, with a flammable range of 2.5%–12.5% in air. Boiling point is 168°F (75°C), flash point is 10°F (–12°C), and ignition temperature is 446°F (230°C). It is slightly soluble in water, with a specific gravity of 0.8, which is lighter than water. Vapor density is 0.804, which is lighter than air. In addition to flammability, butyraldehyde is corrosive and causes severe eye and skin burns. It may be harmful if inhaled. The four-digit UN identification number is 1129. The NFPA 704 designation is health 3, flammability 3, and reactivity 2. The primary uses of butyraldehyde are in plastics and rubber and as a solvent. The structure and molecular formula for butyraldehyde are shown in Figure 5.63.

Esters

The next flammable liquid hydrocarbon-derivative family is ester. In addition to being flammable, esters may polymerize. They have moderate boiling and flash points, narrow flammable ranges, and high ignition temperatures. Esters are made through a process referred to as esterification. An ester is formed when an alcohol is combined with an organic acid, with water as a by-product. This process is illustrated in Figure 5.64 by combining acrylic acid and methyl alcohol; the resulting ester compound is methyl acrylate.

Esters have a carbon atom double-bonded to one oxygen atom and a single bond with another oxygen atom. Esters are carbonyls and are polar. The degree of polarity is much less than organic acids and alcohols and is similar to ketones and aldehydes. Esters are miscible with water and require polar solvent foams when fighting fires. In this book, only three esters will be discussed because of their common commercial use. The general formula for ester is $R-C-O-O-R$ or $R-C-O_2-R$. There are two radicals in the ester compounds.

Because nothing is ever called ester, the word "ester" will not appear in the name of the compound. The radical that is attached to the carbon atom in the functional group determines which ester compound it will be. Esters are one of the functional groups in which all the carbon atoms are counted, including the carbon atom in the functional group, when naming the type of ester compound. In addition, esters use the alternate name for one- and

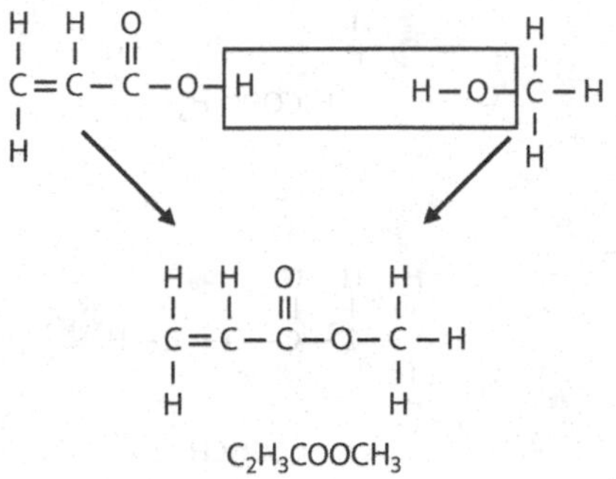

$$C_2H_3COOCH_3$$

Figure 5.64 Esterification process.

two-carbon radicals that are attached to the carbon atom in the functional group only. The radical on the other side is named in the normal way. The ester is named using the radical on the right first and ending in the name of the type of ester on the left.

Certain radicals attached to the carbon atom of the functional group will produce certain esters. For example, in a one-carbon ester, the carbon in the functional group is used as the one carbon. A single hydrogen atom is attached to the carbon atom in the functional group to complete the bonding requirements. A one-carbon ester uses the alternate name for one carbon, which is "form"; esters end in "ate," so the name of a one-carbon ester is formate.

Any radical can be attached to the oxygen atom in the functional group. If a methyl is added to the oxygen atom, the name of the ester compound is methyl formate. When a methyl radical is attached to the carbon atom in the functional group, it forms a two-carbon chain. The alternate name for two carbons, "acet," is used; the ending "ate" is used to indicate an ester, so a two-carbon ester is an acetate. The second radical added to the oxygen in the functional group determines what type of acetate the compound is. Theoretically, any radical can be used. If a vinyl radical were used, the compound would be vinyl acetate.

The last ester has a vinyl radical attached to the carbon in the functional group. This forms a three-carbon chain with one double bond in the chain. The name for a three-carbon radical with one double bond is "acryl." The ending for the ester is "ate"; so the ester is called acrylate. Any radical can be attached to the oxygen atom in the functional group. If a methyl is attached, the ester compound is called methyl acrylate. Figures 5.65 through 5.67 show the structures and molecular formulas for methyl formate, vinyl acetate, and methyl acrylate. There can be other radicals attached to the oxygen atom, which change the name of the compound.

Methyl formate is a colorless liquid with an agreeable odor. It is a dangerous fire and explosion risk, with a flammable range of 5%–23% in air. Boiling point is 89°F (31°C), flash point is –2°F (–18°C), and ignition temperature is 853°F (456°C). In addition to being flammable and a polymerization hazard, it is also an irritant, with a TLV of 100 ppm in air. It is water soluble and

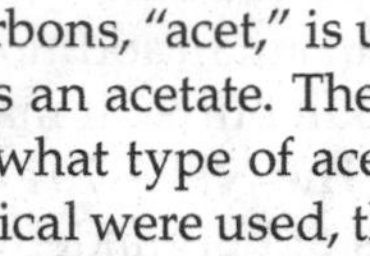

$HCOOCH_3$

Figure 5.65 Methyl formate.

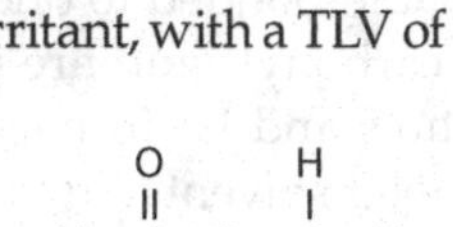

$CH_3COOC_2H_3$

Figure 5.66 Vinyl acetate.

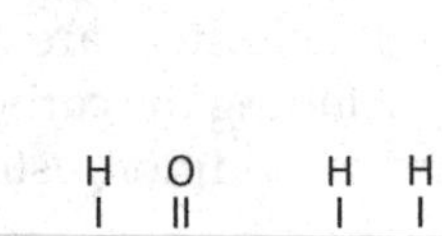

$C_2H_3COOCH_3$

Figure 5.67 Methyl acrylate.

has a specific gravity of 0.98, which is slightly lighter than water. Vapor density is 2.07, which is heavier than air. The four-digit UN identification number is 1243. The NFPA 704 designation is health 2, flammability 4, and reactivity 0. The primary uses of methyl formate are as a solvent, a fumigant, and a larvicide. The structure and molecular formula are shown in Figure 5.65.

Vinyl acetate (*ethenyl acetate*, IUPAC) is an ester hydrocarbon-derivative compound. It is a colorless liquid that has been stabilized with an inhibitor. Although it is a polar compound because of the carbonyl structure, it is only slightly miscible with water. Vinyl acetate is a highly flammable liquid, with a flammable range of 2.6%–13.4% in air, and it may polymerize without the inhibitor or when exposed to heat or an oxidizer during an accident. Boiling point is 161°F (71°C), flash point is 18°F (–7°C), and ignition temperature is 756°F (402°C). Fighting fires will require the use of alcohol-type foam. In addition to flammability, vinyl acetate is toxic by inhalation and ingestion, with a TLV of 10 ppm in air. Vapor density is 3, so it is heavier than air. The specific gravity is 0.9, which means it will float on water. The four-digit UN identification number is 1301. The NFPA 704 designation is health 2, flammability 3, and reactivity 2. It is shipped in 55 gal drums, tank trucks, railcars, and barges. It should be stored separately from oxidizing materials. The structure and molecular formula for vinyl acetate are shown in Figure 5.66.

Methyl acrylate (inhibited) (*methyl propenoate*, IUPAC) is a colorless volatile liquid. It is a dangerous fire and explosion risk, with a flash point of 2.8%–25% in air. Boiling point is 177°F (80°C), flash point is 27°F (–2.7°C), and ignition temperature is 875°F (357°C). It is immiscible with water and has a specific gravity of 0.96, which is lighter than water. Vapor density is 0.957, which is slightly lighter than air. In addition to flammability and polymerization hazards, methyl acrylate is toxic by inhalation, ingestion, and skin absorption. It is an irritant to skin and eyes, with a TLV of 10 ppm in air. The four-digit UN identification number is 1919. The NFPA 704 designation is health 3, flammability 3, and reactivity 2. The primary uses of methyl acrylate are in polymers, in vitamin B_1, and as a chemical intermediate. The structure and molecular formula are shown in Figure 5.67.

Sulfur is the next flammable hydrocarbon-derivative family that will be discussed in this chapter. Sulfur compounds have a general formula of *R-S-R*. This general formula applies to the thioethers or sulfides. *R-SH* is the general formula for the thiol/mercaptan compounds, which is the IUPAC suffix where the hydroxyl group has been replaced by sulfur. These compounds are also still called by the suffix mercaptan, which means mercury seizing, although this is an outdated naming convention. Thioethers look very much like ether and are named using the same rules as ether. You name the hydrocarbon radical(s) first and end in sulfide. If the prefix thio is used with two ethyl groups attached to the sulfur, it would be named diethyl thioether (IUPAC 1,1-thiobisethane). Thiol is used only for the mercaptans. For example, if two vinyl radicals were added one to each side of a sulfur, it would be called vinyl sulfide. Because the same naming rules are followed as for the ethers, you do not have to say divinyl even though there are two vinyls in the compound.

$$C_2H_3 + S + C_2H_3 = (C_2H_3)_2S, \text{ vinyl sulfide}$$

If a methyl radical is attached to a sulfur, it would be called methyl mercaptan. The hydrogen on the end of the sulfur is just a filler to complete the octet rule of bonding.

$$CH_3 + S + H = CH_3SH, \text{ methyl mercaptan}$$

All of the chemicals mentioned in this section have a primary hazard of flammability; however, because they are different in chemical makeup, some require different firefighting tactics. Note the differences in physical and chemical characteristics from compound to compound and family to family. The secondary or hidden hazards of flammable liquids vary widely from one chemical to another. Among the flammable liquids, many are also toxic, anesthetic, narcotic, and undergo polymerization. Many different hydrocarbon-derivative functional groups are represented in the flammable liquids. Earlier in this chapter, it was mentioned that almost 64% of all hazardous materials incidents involve flammable liquids. While many of those spills are hydrocarbon fuels, like gasoline and diesel fuel, they also involve industrial flammable liquids.

Organic acids

Organic acids are the final flammable hydrocarbon-derivative family to be discussed in this chapter. They are the only type of acid that can burn. Inorganic acids do not burn. Organic acids have a carbon atom double-bonded to one oxygen atom and a single bond with another oxygen atom that is attached to a single hydrogen. Organic acids are the most polar compound that is discussed in this book. Organic acids are miscible with water and require polar solvent foams when fighting fires. The general formula for organic acid is R-C-O-O-H. There is one radical in the organic acid compounds. They are named using the alternate naming convention for esters, aldehydes, and organic acids. Meth is form for one carbon, ethyl is acet for two carbons, and three carbons with a double bond are acryl. The prefix ends in "ic" and the compound ends in acid. When naming organic acids, the carbon in the functional group is also counted when choosing a name. For example, a one-carbon organic acid would utilize the carbon in the organic acid functional group, and an H would be placed where the radical would normally be placed.

$$H + COOH = HCOOH, \text{ formic acid}$$

If a one-carbon radical, methyl, was added to the organic acid functional group, it would create a two-carbon compound. Since you count both carbons for the name, you would use the alternate prefix for two carbons.

$$CH_3 + COOH = CH_3COOH, \text{ acetic acid}$$

As a family, organic acids are corrosive, toxic, and some may burn.

Formic acid (IUPAC), also methanoic acid, is a colorless fuming acid with a penetrating odor. It has a boiling point of 213°F (100.8°C), a flash point of 156°F (69°C), and an ignition temperature of 1114°F (600°C). It is a strong reducing agent. It is corrosive to skin and tissue with a TLV of 5 ppm in air (Figure 5.68). DOT classifies as corrosive liquid.

Acetic acid (IUPAC), also ethanoic acid, is a clear colorless liquid with a pungent odor like vinegar. Boiling point is 244°F (118°C), flash point is 110°F (43°C), and ignition temperature is 800°F (426°C). It is a moderate fire risk, moderately toxic by ingestion and inhalation. The TLV is 10 ppm in air (Figure 5.69) (Photo 5.18).

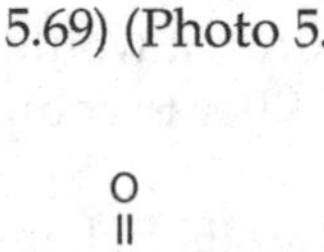

Figure 5.68 Formic acid.

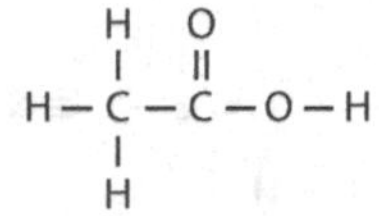

$$\begin{array}{ccc} & H & O \\ & | & || \\ H- & C-C-O-H \\ & | \\ & H \end{array}$$

CH$_3$COOH

Figure 5.69 Acetic acid.

Photo 5.18 Geodesic dome retrofitted as closed floating-roof tank.

Other flammable liquids

Acrylonitrile is sometimes referred to as vinyl cyanide. It is a colorless liquid with a mild odor, is nonpolar, and is partially miscible with water. Acrylonitrile is a dangerous fire risk, with a flammable range of 3%–17% in air. Small fires can be extinguished with dry-chemical foam with some difficulty. Water may also be ineffective. Alcohol-type foams may be effective. Boiling point is 171°F (77°C), flash point is 32°F (0°C), and ignition temperature is 898°F (481°C). In addition to flammability, acrylonitrile is toxic by inhalation and skin absorption. Acrylonitrile may polymerize because of the double bond in the vinyl and the triple bond between the carbon and nitrogen; it is always shipped with an inhibitor. It is a cyanide compound and highly toxic. The TLV is 2 ppm in air, and it is considered a human carcinogen. The vapor density is 1.8, which is heavier than air. Specific gravity is 0.8, which is lighter than water, so it will float on the surface. The four-digit UN identification number is 1093. The NFPA 704 designation is health 4, flammability 3, and reactivity 2. It is shipped in 55 gal drums, tank trucks, railcars, and barges. It should not be stored or shipped uninhibited. The structure and molecular formula are shown in Figure 5.70.

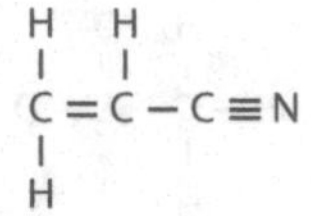

$$\begin{matrix} H & H \\ | & | \\ C = C - C \equiv N \\ | \\ H \end{matrix}$$

CH₂CHCN

Figure 5.70 Acrylonitrile.

Carbon disulfide is a nonmetal compound that is a clear, colorless, or faintly yellow liquid, and almost odorless. It is highly flammable, with a flammable range of 1.3%–50% in air, and can be ignited by friction. Boiling point is 115°F (46°C), flash point is –22°F (–5°C), and ignition temperature is 194°F (90°C). Contact with a steam pipe or a lightbulb could ignite carbon disulfide. It is slightly water soluble and has a specific gravity of 1.26, which is heavier than water. In addition to being highly flammable, carbon disulfide is also a poison and is toxic by skin absorption, with a TLV of 10 ppm in air. The four-digit identification number is 1131. The NFPA 704 designation is health 3, flammability 4, and reactivity 0. The primary uses of carbon disulfide are as a solvent and in the manufacture of rayon, cellophane, and carbon tetrachloride. The structure and molecular formula for carbon disulfide are shown in Figure 5.71.

Cumene, also known as isopropylbenzene, is a colorless liquid that is insoluble in water. The specific gravity is 0.9, which is lighter than water. It is a moderate fire risk, with a flammable range of 0.9%–6.5%. Boiling point is 306°F (152°C), flash point is 96°F (35°C), and ignition temperature is 795°F (423°C). Small fires may be extinguished with dry-chemical foam, and large fires with hydrocarbon-type foam. Water may be ineffective and should be applied gently to the surface of the liquid if used. Vapor density is 4.1, which is heavier than air. In addition to flammability, cumene is toxic by ingestion, inhalation, and skin absorption; it is also a narcotic. TLV is 50 ppm in air. The four-digit UN identification number is 1918. The NFPA 704 designation is health 2, flammability 3, and reactivity 0. The primary uses are in the production of phenol, acetone, and methylstyrene solvents. The structure and molecular formula for cumene are shown in Figure 5.72.

Ethyl benzene is a colorless aromatic hydrocarbon with a characteristic odor. It is a dangerous fire risk, with a flammable range of 0.8%–6.7% in air. Boiling point is 277°F (136°C),

$$S = C = S$$

CS₂

Figure 5.71 Carbon disulfide.

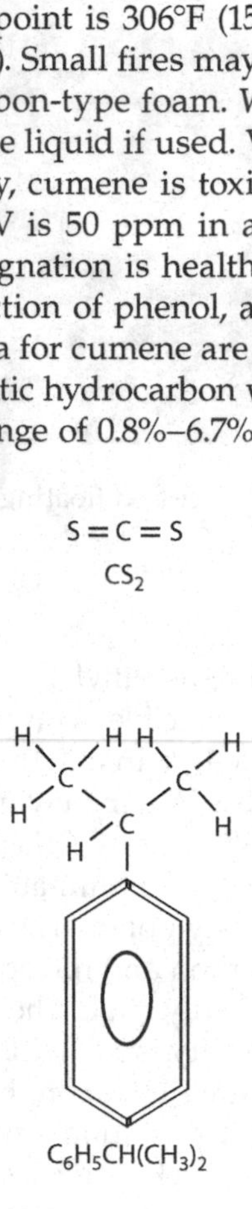

C₆H₅CH(CH₃)₂

Figure 5.72 Cumene.

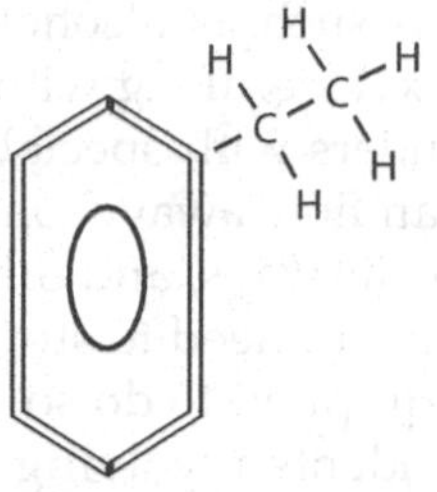

$C_6H_5CH_2CH_3$

Figure 5.73 Ethyl benzene.

flash point is 70°F (21°C), and ignition temperature is 810°F (432°C). Small fires may be extinguished with dry-chemical foam and large fires with hydrocarbon-type foam. Water may be ineffective and should be applied gently to the surface of the liquid if used. In addition to flammability, it is toxic by ingestion, inhalation, and skin absorption, with a TLV of 100 ppm in air. Vapor density is 3.7, which makes it heavier than air, and the vapors will tend to stay close to the ground. It is nonpolar, with a specific gravity of 0.9, which means it will float on water. Ethyl benzene is immiscible with water. The four-digit UN identification number is 1175. The NFPA 704 designation is health 2, flammability 3, and reactivity 0. It is shipped in cans, bottles, 55 gal drums, tank trucks, railcars, and barges. It should not be stored near oxidizing materials. The primary uses of ethylbenzene are as a solvent and as an intermediate in the production of styrene. The structure and molecular formula are shown in Figure 5.73.

Incidents

Most of the more commonly encountered flammable liquids are fuels: gasoline, diesel fuel, heating oil, and jet fuel. Motor vehicle accidents are the leading cause of flammable liquid spills (Photo 5.19).

Photo 5.19 Motor vehicle accidents are the leading cause of flammable liquid spills and fires.

Other spills include materials such as alcohols, ketones, aldehydes, paint thinners, pesticides, benzene, toluene, and xylene, along with other industrial solvents. These flammable liquids can present responders with special problems as well as hidden hazards. Because they are liquids, they can flow away from the scene, following the terrain into storm drains, sanitary sewers, waterways, and other low-lying areas. In addition to the flammability hazard, responders will need to stop the flow of the product in some situations, if properly trained and equipped to do so. Incidents range from leaks in vehicle fuel tanks, to transportation accidents involving tank trucks (Photo 5.20) resulting in leaks and fires, to large bulk-storage tank fires.

Kansas City, Kansas, Tuesday August 18, 1959, started out like any other summer day in the Kansas City metropolitan area, sunny with temperatures in the 1990s and a south wind of 13 mph. Before the day would end, five Kansas City, Missouri, firefighters and one civilian would die in an inferno of burning gasoline referred to by KMBC TV reporter Charles Gray as "when all hell broke loose" (Photo 5.21). Gray who has always been a strong supporter of the Kansas City Fire Department called it "one of the darkest days in modern history of Kansas City firefighting." It was the second largest loss of life in Kansas City Fire Department history.

At approximately 08:20 h, the Kansas City, Kansas, Fire Department received a report of a fire at the Continental Oil Company located at #2 Southwest Boulevard. The fire started on a loading rack at the combination bulk plant and service station located in Kansas City, Kansas, near the Kansas/Missouri State Line. The first alarm was dispatched at 08:33 and a second alarm was requested at 08:37 followed by a third at 08:45, fourth at 08:54, fifth at 08:59, and sixth at 10:00 (following the rupture of the tank). Despite the best efforts of firefighters,

Photo 5.20 MC/DOT 305/405 tanker used to transport flammable fuels, such as gasoline and diesel fuel.

Photo 5.21 A dozen streams are poured onto the raging oil and gasoline fire that engulfs the service station and two bulk oil plants at No. 2 Southwest Boulevard in Kansas City, Kansas. The scene is along Southwest Boulevard, which extends to the northeast diagonally toward the upper right of the picture. The intersecting street at the lower right is 31st Street. The two streets meet at approximately the Missouri–Kansas state line. (Courtesy of Kansas City Fire Department, Kansas City, MO; *The Kansas City Star.*)

the burning gasoline from the leaking fuel extended underneath four 11 ft-by-30 ft cylindrical horizontal storage tanks resting on concrete cradles, each with 21,000 gal of fuel capacity. Three contained gasoline and one kerosene. From left to right at the fire scene, tank 1 contained 6,628 gal of gasoline, tank 2 contained 15,857 gal of kerosene, tank 3 contained 3,000 gal of gasoline, and tank 4 contained 15,655 gal of premium gasoline. All of the tanks failed during the fire, but tanks 1–3 did not leave the concrete cradles they rested in. This lack of movement of the first three tanks may have given firefighters a false sense of security while fighting the fire involving tank 4. The tanks began to fail at approximately 10:00 or about 90 min after the fire started. Tank 4 was the last to fail, and when it did, it moved 94 ft from its cradles into Southwest Boulevard through a 13 in. brick wall, spreading burning gasoline and flying bricks in its path. Firefighter positions with 2½ in. hose lines were located just 74 ft from the tank when it ruptured, so their positions were overrun by the tank and burning gasoline that completely crossed Southwest Boulevard. Twenty-two firefighters were admitted to hospitals, five in critical condition along with civilian "firefighter" Rocky Tooms. All five critically injured firefighters and Rocky would die, the first at 2:45 p.m. on the day of the fire and the last on August 24. An additional 35 firefighters were treated at hospitals and released. Approximately 40 firefighters were given first aid at the scene. All suffered from burns caused by contact with the burning gasoline from tank 4. All five firefighters killed were from the Kansas City, Missouri, Fire Department and were from two companies, Pumper 19 and Pumper 25. Pumper 19 lost its entire crew and Pumper 25 lost two of its three crew members. Driver Earl Dancil scheduled to work that day but was off on sick leave. He had heard about the fire through radio and television and responded to the scene to relieve Driver Delbert Stone, who joined fellow crew members on the fire line behind Captain Sirna. Tony Valentini and civilian Francis "Rocky" Tooms were helping on the hose line behind Stone. Tooms, a civilian, had appeared on the scene and approached Captain Sirna and asked if he could help. The captain told him yes. According to KMBC Reporter Charles Gray, "Tooms had arrived on scene that day as a civilian but was a fire-fighter by the time he left." Because of his actions in assisting firefighters, he was honored by inclusion on the Southwest Boulevard Fire Memorial.

The irony for Pumper 25 was that they should not have even been at the fire that day. Pumper 25's crew had been dispatched earlier to a fire and saw the smoke from the Southwest Boulevard fire as they were finishing up. They went back to Station 25 to change out their wet and dirty hose with fresh dry hose, a common practice following fires. Radio calls from dispatch were piped through all stations and any dispatches were heard by all companies on the department. Pumper 2 had been dispatched to the fire scene to stage approximately two blocks from the fire scene in case they were needed. In route to the fire, they were involved in an accident at 11th and Broadway and taken out of service. Pumper 25 was sent to replace Pumper 2 and stage two blocks north of the fire scene. Pumper 11 was deployed on Southwest Boulevard in front of the fire sup-plying water to 2 ½ in. hose lines when its pump failed. Pumper 25 was sent to the scene from the staging area to replace Pumper 11. When they arrived, they placed a 2 ½ in. hose line in service from their pumper and began fighting the fire. This fire changed the way flammable liquids were stored at automotive service stations and how flammable liquid fires were fought involving horizontal storage tanks. National Fire Protection Association codes were changed to require flammable liquid storage tanks at automotive service sta-tions to be placed underground following the Southwest Boulevard fire. New procedures for fighting fires in horizontal flammable liquid storage tanks involved approaching the tanks from the sides and not the ends. Fire officers stressed the importance of wearing full turnout gear during all fires.

Firefighters killed at the "Southwest Boulevard Fire" (Photo 5.22)

Pumper 19
Captain George E. Bartels
FF Neal K Owen
Driver Virgil L. Sams

Pumper 25
Captain Peter T Sirna
Driver Delbert W. Stone

Civilian Firefighter
Rocky Tooms

Philadelphia, Pennsylvania, Sunday August 17, 1975, just before dawn. A seagoing tanker in the Schuylkill River at Girard Point was off-loading crude oil at the Gulf Oil Refinery in South Philadelphia (Photo 5.23). At the time of this fire, the refinery produced 180,000 bar-rels per day of refined petroleum products.

Suddenly and without warning, accumulated vapors from off-loading the tanker were ignited starting a fire that threatened 600 storage tanks at the refinery tank farm on shore,

Photo 5.22 Monument erected to the fallen firefighters and civilian at the Southwest Boulevard fire rededicated on August 18, 2009, 50 years to the day and hour from when the fire occurred.

many with a capacity of 80,000 gal of crude oil. Hydrocarbon vapors, emanating from tank 231, accumulated in the area of nearby Boiler House #4 and were ignited. A flame front followed the vapors back to tank 231 causing fire at the tank's vents and an explosion within the outer shell of the stack. These events began to unfold at 5:57 a.m. At approximately 6:02 a.m. in the wake of the first explosions and fire, the tanker terminated its pumping operations, left its Schuylkill River berth, and relocated to the Gulf piers at Hog Island.

Philadelphia's fire alarm office received the first report of the fire at approximately 6:04 a.m.: upon receiving the report, they transmitted the refinery's fire alarm box: Box 5988, Penrose and Lanier Avenues. Engine 60 was first due at the refinery, and as they left their station, firefighters could see fire and smoke conditions at a distance and, before arriving, requested a second alarm at 6:09 a.m. Before the fire was over, 500 firefighters would battle the blaze and 11 alarms would be transmitted, six Philadelphia firefighters would lose their lives, and an additional nine would be injured along with four Gulf firefighters. Two more Philadelphia firefighters would succumb to their injuries several days later. Not long after the original ignition of the fire, a second explosion occurred within tank 231. Burning petroleum spilled from the tank's vents into a diked area surrounding the tank. Within the diked area, a second tank (no. 114) just north of tank 231, containing no. 6 grade fuel oil, also ignited as pipelines within the diked area began to fail. The initial explosion damaged the pipe manifold outside of the dike wall and petroleum pouring out under pressure ignited. First arriving companies encountered large clouds of heavy black smoke billowing from tank 231, fire on top of tank 114, and fire showing from the 150 ft stack at Boiler House no. 4. The third and fourth alarms were ordered in quick succession by Battalion Chief 1, Arthur Foley, at 6:11 a.m. and 6:14 a.m. Acting Assistant Fire Chief Dalmon Edmunds ordered the fifth alarm at 6:34 a.m. Engine 33 and Foam 133 responded on the fifth alarm from their station on the North side of Philadelphia. Firefighter Hugh McIntyre of Engine 56 had been detailed to Engine 33 on the day of the fire. He was the oldest firefighter to die in the inferno. The sixth alarm was ordered

Photo 5.23 Gulf Oil Refinery fire in Philadelphia on August 17, 1975, was not the first at the facility, but was certainly the deadliest. Before the fire was over, 500 firefighters would battle the blaze, 11 alarms would be transmitted, 8 firefighters would lose their lives, and another 14 were burned or injured during rescue attempts. (Courtesy of Philadelphia Fire Department.)

by Fire Commissioner Joseph Rizzo at 6:52 a.m. Over the next several hours, firefighters utilized deluge guns and master streams to cool down surrounding exposures and applied foam directly to the burning tanks and piping in an effort to extinguish the fire. By 8:44 a.m., it appeared that the fire was well contained and the situation sufficiently stabilized to declare the fire under control. Throughout the day, Philadelphia's two foam pumpers, Foam Engines 160 and 133, along with the Gulf Refinery's foam pumper, continued to apply foam to the burning tank, piping, and manifolds. Additional foam to support the operation was acquired from the fire department's warehouse and the nearby Atlantic-Richfield refinery. It was also obtained from the National Foam Company in West Chester, PA.

Without warning at approximately 3:30 p.m. that afternoon, the accumulating liquid surrounding Engine 133 ignited, immediately trapping firefighters Campana, Fisher, and Andrews working at Engine 133. Instinctively and without hesitation, other nearby firefighters dove into the burning liquid to rescue their comrades, not aware of the danger to themselves. Five more firefighters would be consumed by the advancing fire. Firefighters McIntyre, Willey, and Parker died during the attempted rescue of their fallen comrades. Lieutenant Pouliot and Firefighter Brenek were gravely injured during the rescue efforts and would die several days later in the St. Agnes Hospital burn center in South Philadelphia.

At approximately 4:41 p.m., a fire storm was developing as the fire quickly spread eastward along Avenue "Y" toward 5th Street. Viewing the unfolding horror before him, Commissioner Rizzo ordered two more alarms, five additional rescue squads, and the recall of all companies that had previously been released from the fire ground throughout the day. On these orders, the fire alarm office transmitted the seventh and eighth alarms simultaneously. As the fire had been placed under control nearly 8 hours earlier, firefighters in stations across the city knew that the unthinkable had occurred as these additional alarms were struck. At 4:46 p.m., Commissioner Rizzo ordered the ninth alarm and notification of Philadelphia Managing Director Hillel Levinson as a major disaster was now unfolding at the Gulf Refinery. As the fire swept rapidly eastward along 5th Street, Philadelphia's foam pumpers 160 and 133, and the Gulf Refinery foam pumper, were rapidly destroyed in the fire's advance. At 5th Street, where Engines 16 and 40 had been assigned to improve drainage, their pieces were also destroyed in the fire's path, although their pump operators were able to escape. Upon reaching 5th Street, the fire traveled two city blocks north along 5th Street, now threatening four additional storage tanks and the 125 ft Penrose Avenue Bridge. At 5:37 p.m., Commissioner Rizzo ordered the 10th alarm as the fire was now traveling southward and engulfing the refinery's administration building, which was located on the south side of Avenue "Y" between 4th and 5th Streets. The 10th alarm companies were ordered to report to Gate 24 at Penrose and Lanier Avenues, to set up deluge guns and leave the area. As the situation continued to deteriorate at the Gulf Refinery, Commissioner Rizzo ordered all "D" platoon members from the day shift held over, and at 6:01 p.m., he ordered the 11th alarm. By seven o'clock, the involved tanks and pipelines were gushing flames, and nearby streets in the complex were burning streams of oil and other petroleum products. For a period of time, it was far from certain where the fire would be stopped. At 1:00 a.m., Commissioner Rizzo left the fire ground relinquishing command to Deputy Fire Commissioner Harry T. Kite, who placed the fire under control at 5:38 a.m. on Monday, August 18, 1975.

The original cause of the fire was the overfilling of tank 231. While no crude oil escaped from the tank as a result of being overfilled, large quantities of hydrocarbon vapors were trapped above the surface of the tank's crude oil. As the quantity of crude oil increased, these hydrocarbon vapors were forced out of the tank's vents and into the area of the no. 4 Boiler House, where the initial flash occurred. The overfilling of the tank, in turn, resulted from a failure of the tanker's personnel to properly monitor the quantity of crude oil being pumped to the tank. It is believed that the second fire and subsequent explosions were triggered when the hot muffler of Foam Engine 133 came in contact with the hydrocarbon vapors above the flammable liquids floating on the water under the truck.

Firefighters killed at the Gulf Oil Refinery fire (Photo 5.24)

Lt. James J Pouliot, 35
FF John Andrews, 49
FF Ralph J. Campana, 41
FF Robert J. Fisher, 43
FF Hugh McIntyre, 52
FF Roger Parker Jr., 28
FF Joseph R. Wiley, 33
FF Carroll K. Brenek, 33

Photo 5.24 Plaques honoring each of the firefighters who died at the Gulf Oil Refinery fire have been placed in the concrete in front of the Philadelphia Fire Museum.

Norfolk, Virginia, September 1984, explosion and fire occurred involving an MC 306 gasoline tanker truck that collided with a garbage truck. The garbage truck driver was charged with reckless driving. The Exon tanker was carrying 8500 gal of gasoline. When firefighters arrived, they found the tanker fully involved, with leaking, burning fuel flowing down the street into storm sewers leading to a retention pond. They had fires on multiple fronts, including the tanker, parked cars in an adjacent parking lot, on the street, in the sewer, and on the retention pond. Exposures included two apartment buildings, a 64-unit and a 168-unit senior-citizen apartment building next to the fire in the retention pond. Both apartment buildings were evacuated. The truck driver received first- and second-degree burns and was taken to a hospital in serious condition. Firefighters used foam to extinguish the fire, and it was brought under control within an hour of its beginning. During mop-up operations, two firefighters received minor injuries: one was injured when cut by a jagged edge on the burned-out shell of the tanker as he fell into a portion of the tank.

Review questions

5.1 The boiling point of a liquid is defined as the point at which one of the following occurs:

    A. Where the proper mixture of fuel and air occur

    B. Vapor pressure equals atmospheric pressure

    C. Vapor produced will be lighter than air

    D. Vapor produced will easily ignite

5.2 Flash point has to do with which of the following temperatures?

    A. Temperature of the air around the liquid

    B. Temperature of the flammable liquid

    C. Temperature required for ignition to occur

    D. None of the above

5.3 Liquids that have low boiling points will also have low
- A. Flash points
- B. Ignition temperatures
- C. Heat output
- D. Melting points

5.4 High flash point liquids have what type of vapor pressure?
- A. High
- B. Equal
- C. Low
- D. None

5.5 List the three factors that affect the boiling point of a liquid.

5.6 Match the following compounds with their degree of polarity.

Benzene   A. Polar
Propyl alcohol   B. Non polar
Ethyl ether   C. Super polar
Methyl ethyl ketone   D. Super-duper polar
Formic acid
Pentane

5.7 Polymers are long-chained molecules made up of individual building blocks called
- A. Links
- B. Monomers
- C. Elastomers
- D. Inhibitors

5.8 Identify the following hydrocarbon formulas as to whether they are alkanes, alkenes, alkynes, or aromatics.

$$C_6H_{14} \quad C_7H_8 \quad C_7H_{14} \quad C_6H_6 \quad C_8H_{16}$$

5.9 Provide structures, names, and hazards for the following hydrocarbon-derivative compounds.

$$HCHO \quad C_2H_3COOCH_3 \quad HCOOH \quad CH_3CHO \quad CH_3COOH$$

5.10 Provide the names, formulas, and hazards for the following structures:

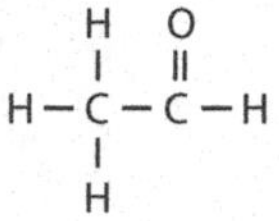

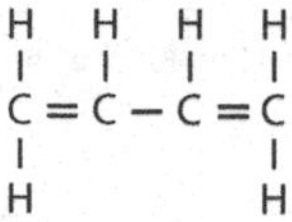

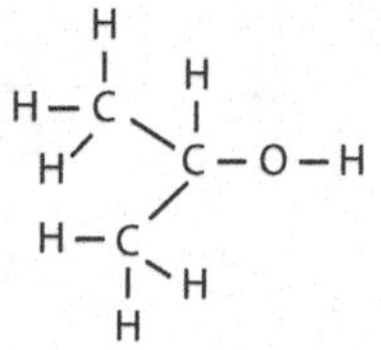

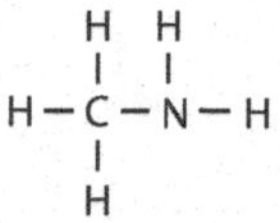

```
                          H
                          |     H
   H  O      H  H  H  H   H-C    |                H  H
   |  ||     |  |  |  |      \    \               |  |
 H-C-C-H   C=C-C=C        H´    C-O-H         H-C-N-H
   |        |     |             /                 |
   H        H     H       H-C                     H
                              |\
                              | `H
                              H
```

chapter six

Flammable solids

Hazard Class 4 is composed of flammable solids (Table 6.1). Additionally, some pyrophoric and water-reactive liquids are included under 4.2 Spontaneously Combustible and 4.3 Dangerous When Wet divisions because there are no such categories for just spontaneously combustible or dangerous when wet liquids. Flammable solids can be categorized into five groups according to the hazards of the materials: flammable metals, spontaneous combustibles, intensely burning or difficult-to-extinguish flash point solids, and water reactives. Solid materials may take on many different physical forms (not to be confused with physical states), including fine powder, filings, chips, and various-sized solid chunks (Photo 6.1). The smaller the solid particle, the more dangerous it becomes in terms of flammability and potential explosiveness. This is because the smaller the particle, the more surface area that is created to react.

Solid materials such as white phosphorus may be pyrophoric; that is, they spontaneously ignite upon exposure to air. Calcium carbide, which is a binary salt, must become wet before it creates a hazard. When in contact with water, calcium carbide releases the flammable gas acetylene. When wet, solids like the salt hydrides release flammable hydrogen gas. Certain solids, when they contact water, release poisonous gases. For example, binary phosphide salts release phosphine gas, binary nitrides release ammonia gas, and the oxysalt hypochlorites release chlorine. Peroxide salts when in contact with water release oxygen. Materials that release other gases when in contact with water can present an added danger to firefighters because the primary extinguishing agent they use is water. Water can cause the release of oxygen from water-reactive materials. If fire is present, the oxygen will accelerate the combustion process and make the fire more difficult to extinguish. Solid materials may spontaneously combust when exposed to water; the reaction is exothermic or heat-producing. If combustible materials are present, the heat of the reaction can cause combustion to occur. Some materials like phosphorus are shipped under water, whereas others such as picric acid are shipped with 10%–50% water in the container. Picric acid is classified as a "Wetted Explosive" when shipped in transportation. As long as the water is present, there is no danger of explosion. However, when picric acid dries out, particularly in storage, it is a high explosive similar to TNT. Wetted explosives present only a limited hazard as long as the water is present and, therefore, they are considered flammable solid materials.

The DOT divides the flammable solid hazard class into three divisions: 4.1 Flammable Solids, 4.2 Spontaneous Combustibles, and 4.3 Dangerous When Wet materials. Each division presents its own particular hazards when encountered in a hazardous material incident (Photo 6.2). It is important that emergency responders have a thorough understanding of each of the divisions, the types of materials that make them up, and the hazards posed in a release.

Table 6.1 DOT Hazard Class 4 Materials

Flammable solids	Spontaneously combustible	Dangerous when wet
Matches	Activated carbon	Alkaline earth metal alloys
Nitrocellulose membrane filters	Lithium alkyds	Aluminum powder
Pentaborane	Barium	Calcium hydride
Silicon powder	Phosphorus	Calcium
Wetted explosives	Potassium sulfide, anhydrous	Calcium carbide
Sulfur	Oily rags	Lithium
Zinc resinate	Sodium sulfide, anhydrous	Magnesium
Titanium powder, wetted	Seed cake	Sodium
Naphthalene	Butyl lythium	Sodium borohydride

Photo 6.1 Dry-bulk truck for transporting solid materials.

Class 4.1 flammable solids

Class 4.1 materials are flammable solids. The DOT defines 4.1 materials as "4.1. Wetted explosives that, when dry, are Class 1 Explosives. 4.2. Self-reactive materials that are liable to undergo a strongly exothermic decomposition at normal or elevated temperatures caused by excessively high transport temperatures or by contamination (Photo 6.3). 4.3. Readily combustible solids that may cause a fire through friction, such as matches, show a burning rate faster than 0.087 inches per second, or any metal powders that can be ignited and react over the whole length of a sample in 10 minutes or less."

Photo 6.2 Flares used by police officers are shipped as flammable solid materials.

Photo 6.3 Containers of spontaneously combustible materials in a warehouse.

Flammable solids may be elements, metals in various physical forms, or salts. Alkali metals located in column 1 on the periodic table are considered flammable solids dangerous when wet, particularly lithium, sodium, and potassium. They burn or in some cases explode depending on the size of the piece of metal when they come in contact with water. Even moisture in the air can cause the materials to react. Magnesium as a solid mass must already be burning before it reacts violently with water. Magnesium shavings, filings, and powder react explosively with water. With both the alkyl metal and alkaline earth metal families, the degree and intensity of water reaction combustion depends on the physical form of the solid material. Other metals, such as aluminum and titanium, can also be dangerously flammable and explosive as filings, shavings, or powders.

Flash-point solids/sublimation

A small group of flammable solid materials go through a process called sublimation at normal temperatures. Sublimation is a process by which solid materials go directly from a solid to a vapor without becoming a liquid. Even though these materials do not become liquids, they still have a flash point; hence, the term flash-point solid is given to the group. Flash points are generally above 100°F (37°C). Once ignition occurs, the solid material melts and flows like a flammable liquid. In addition to being flammable, flash-point solids may also be narcotic and toxic.

Two common flash-point solids are camphor and paradichlorobenzene (PDB), also known as mothballs or flakes. Mothballs are placed in areas where clothing is stored to prevent moths from doing damage to the clothing. The fact that the mothballs are flash-point solids allows them to pass from a solid to a vapor without becoming a liquid. Because of this feature, the vapor from the mothballs repels the moths without harming the clothing.

Camphor, also known as gum camphor and 2-camphanone, is a naturally occurring ketone that comes from the wood of the camphor tree. Camphor is composed of colorless or white crystals, granules, or easily broken masses. It has a penetrating aromatic odor that sublimes slowly at room temperature. Flash point is 150°F (65.5°C), and the autoignition temperature is 871°F (466°C). Flammable and explosive vapors are evolved when heated. While searching old newspapers for articles on fires, the author discovered an article from the late 1800s that told of a bottle of camphor placed on a wood stove. The bottle exploded from the increased vapor pressure from the heat and severely burned the resident. Camphor is slightly water-soluble, undergoes sublimation, and is a flash-point solid. The molecular formula and structure of camphor are shown in Figure 6.1.

Paradichlorobenzene, or *PDB*, also commonly known as mothballs, are white, volatile crystals with a penetrating odor. Boiling point is 345°F (173°C), flash point is 150°F (65.6°C), and the melting point is 127°F (52°C). PDB is insoluble in water and has a specific gravity of 1.5, which is heavier than water. Vapor density is 5.1, which is heavier than air. In addition to being flammable, it is also toxic by ingestion and an irritant to the eyes, with a TLV of 75 ppm in air. The four-digit UN identification number is 1592. The NFPA 704 designation for PDB is health 2, flammability 2, and reactivity 0. The primary uses are as a moth repellent, a general insecticide, a soil fumigant, and in dyes. The structure and molecular formula are shown in Figure 6.2. The following are some examples of 4.1 flammable solid materials, including wetted explosives and flammable metals.

Barium azide, $Ba(N_3)_2$, is a crystalline solid with not less than 50% water by mass that explodes when shocked or heated. Barium azide decomposes and gives off nitrogen at

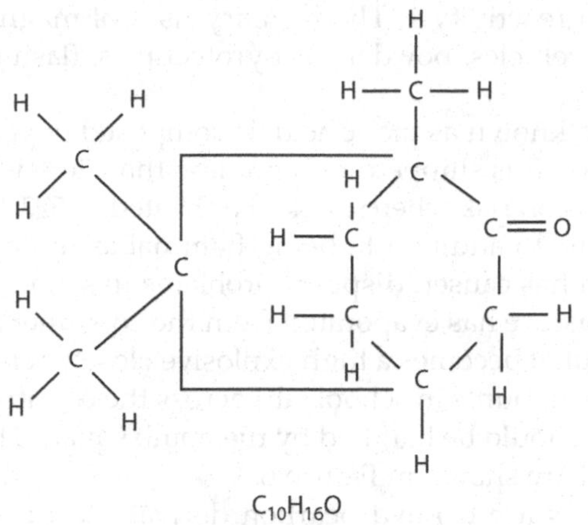

$C_{10}H_{16}O$

Figure 6.1 Camphor.

$C_6H_4Cl_2$

Figure 6.2 Paradichlorobenzene.

240°F (115.5°C) and is soluble in water. The four-digit UN identification number for barium azide is 1571. Its primary use is in high explosives.

Fusees are flares used on the highway or rail as warning devices and are considered flammable solids that are ignited by friction. They produce a temperature of 1200°F (648°C) and a 70-candle flame visible for over one-quarter of a mile. Fusees are primarily composed of strontium nitrate (72%), which produces the characteristic red color and provides oxygen; potassium perchlorate (8%), which is an oxidizing agent; and sulfur (10%), which is an oxidizer and combustion controller. Oil, wax, and sawdust act as binding agents that aid in the control of burning. Flares absorb water and should be stored in a dry place. They are said to have an indefinite shelf life.

Magnesium, Mg, is a metallic element of the alkaline-earth metal family. It is found as a silvery, soft metal, as a powder, or as pellets, turnings, or ribbons. Magnesium is flammable and a dangerous fire hazard, and has an ignition temperature of about 1200°F (648°C). It is insoluble in water; however, when burning, it reacts violently with water. In the form of a powder, a pellet, or as turnings, it can explode in contact with water. Dry sand or talc is used to extinguish small fires involving magnesium. The four-digit UN identification number for magnesium powder or magnesium alloys with more than 50% magnesium powder is 1418. The number for magnesium pellets, turnings, ribbons, or magnesium alloys with more than 50% magnesium powder is 1869. The NFPA 704 designation is health

0, flammability 1, and reactivity 1. The primary uses of magnesium are in die-cast auto parts, missiles, space vehicles, powder for pyrotechnics, flash photography, and dry and wet batteries.

Trinitrophenol, also known as picric acid, is composed of yellow crystals and is a nitro hydrocarbon derivative. It is shipped with not less than 10% water as a wetted explosive. There is a severe explosion risk when shocked or heated to 572°F (300°C), and it reacts with metals or metallic salts. In addition to being flammable and explosive, it is toxic by skin absorption. Picric acid has caused disposal problems in school and other chemistry laboratories where the moisture has evaporated from the container as the material ages. When the picric acid dries out, it becomes a high explosive closely related to TNT. Picric acid has been found in various amounts in school labs across the country. In a dry condition, picric acid is dangerous and should be handled by the bomb squad. The structure and molecular formula for picric acid are shown in Figure 6.3.

Ammonium picrate is a nitro hydrocarbon derivative. It is composed of yellow crystals with not less than 10% water by mass. Ammonium picrate is highly explosive when dry and a flammable solid when wet, and is slightly soluble in water. The four-digit UN identification number for ammonium picrate with not less than 10% water is 1310. The primary uses are in pyrotechnics and explosives. The structure and molecular formula are shown in Figure 6.4.

Matches are shipped as flammable solids and will ignite by friction. The "safety" match was invented in 1855. The primary composition is mostly potassium chlorate (an oxysalt), which is a strong oxidizer, antimony III sulfide (a binary salt), and glue on the match head. The striking surface is composed of a mixture of red phosphorus, antimony III sulfide, a little iron III oxide (a binary oxide), and powdered glass held in place by glue. The match functions when a strike produces heat that converts a tiny trace of red phosphorus to white phosphorus, which instantly ignites. The heat ignites the chemicals in the match head, and

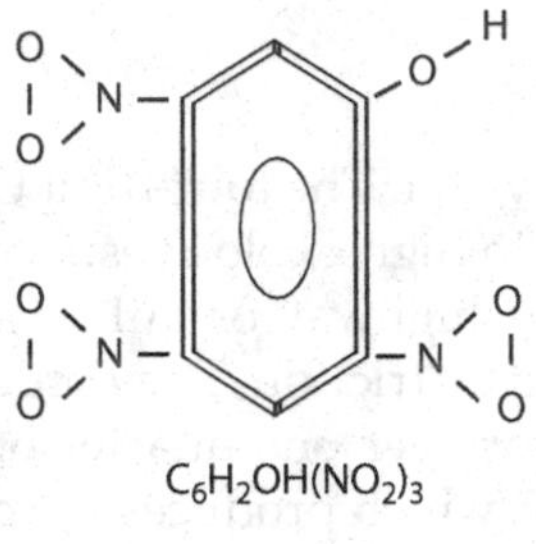

$C_6H_2OH(NO_2)_3$

Figure 6.3 Trinitrophenol (picric acid).

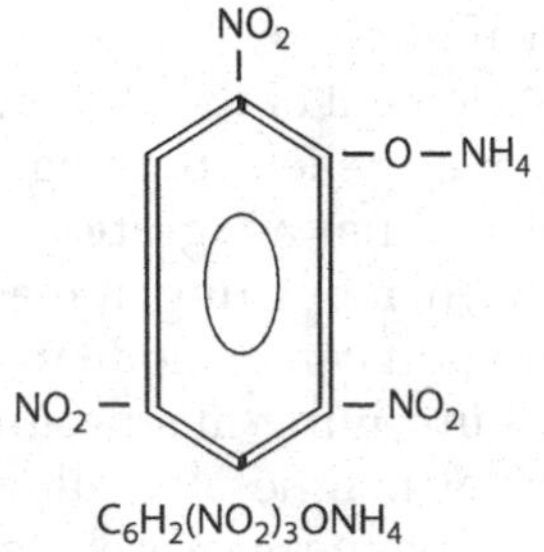

$C_6H_2(NO_2)_3ONH_4$

Figure 6.4 Ammonium picrate.

their short blaze ignites the wood or paper of the matchstick. The "strike-anywhere" match was invented in 1898. It is primarily composed of potassium chlorate, tetraphosphorus trisulfide, ground glass, and the oxides of zinc and iron. This match functions when the strike gives enough heat to initiate a violent reaction between the $KClO_3$ and the P_4S_3; the heat from the reaction ignites the matchstick. The four-digit UN identification numbers for matches are as follows: fusee matches 2254, safety matches 1944, strike-anywhere matches 1331, and wax (Vesta) matches 1945.

Sulfur, S, is a nonmetallic element composed of pale yellow crystals. It is a dangerous fire and explosion risk in a finely divided form. In the molten form, it has a flash point of 405°F (207°C), a boiling point of 832°F (444°C), and an ignition temperature of 450°F (232°C). Sulfur is insoluble in water, and its specific gravity is 1.8, which is heavier than water. The four-digit UN identification number as a dry solid is 1350. The NFPA 704 designation is health 2, flammability 1, and reactivity 0. The primary uses of sulfur are in the manufacture of sulfuric acid, carbon disulfide, and petroleum refining.

Titanium, Ti, is a metallic element, which is a silvery solid or dark-gray amorphous powder. Titanium is shipped and stored in a number of physical forms. These include powder, sheets, bars, tubes, wire, rods, sponges, and single crystals. Titanium is a dangerous fire and explosion risk, and will burn in a nitrogen atmosphere. It has an ignition temperature of 2192°F (1200°C) when suspended in air. Water and carbon dioxide are ineffective extinguishing agents for fires involving titanium. The four-digit UN identification number for dry titanium powder is 2546. The number for titanium powder with not less than 20% water is 1352. Titanium sponges, granules, or powder have a number of 2878. The primary uses of titanium are in the manufacture of structural material in aircraft, jet engines, missiles, marine equipment, surgical instruments, and orthopedic appliances.

Urea nitrate, $CO(NH_2)_2HNO_3$, is a colorless crystal shipped with not less than 20% water by mass. It is a dangerous fire and explosion risk, and is slightly soluble in water. It decomposes at 305.6°F (152°C). The four-digit UN identification number is 1357. It is used in the manufacture of explosives and urethane.

Zirconium, Zr, is a metallic element with a grayish, crystalline scale or gray amorphous powder form. It is flammable or explosive in the form of a powder or dust and as borings and shavings. The powder should be kept wet in storage. Zirconium is a suspected carcinogen, with a TLV of 5 mg/m^3 of air and it is insoluble in water. The four-digit UN identification numbers depend on the form and the amount of water present. The number for zirconium, dry, as wire, sheeting, or in the form of strips is 2009. The number for zirconium, dry, as a wire, sheeting, or as strips that are thinner than 254 μm, but not thinner than 18 μm, is 2858. Dry zirconium metal powder is 2008. Wet zirconium powder is 1358. Zirconium metal in a liquid suspension is 1308. The primary uses are as a coating on nuclear fuel rods, photo flashbulbs, pyrotechnics, explosive primers, and laboratory crucibles.

Flammable particles and dusts

Flammable solid materials that are not listed in a DOT hazard class exist, but may become flammable solids because of their physical state. These materials are flammable dusts, which are finely divided particles of some other ordinary Class A flammable material. The surface area of a solid material increases greatly when it becomes a dust (Figure 6.5). This increases the amount of material exposed to oxygen which makes it much more flammable. Included in this group are such ordinary materials as sawdust, grain dusts,

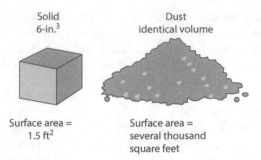

Figure 6.5 Surface area comparison.

flour, and coal dust. Although the DOT does not consider these materials a hazard class, they may, however, be shipped in dry-bulk transportation containers without placards. Flammable dusts become a problem when they are suspended in air in the presence of an ignition source. If this happens in a fixed facility, such as a grain elevator, flourmill, and woodworking operations in manufacturing plants, an explosion may occur. Flammable fine particulate materials may also be suspended in air in a transportation accident and, in the presence of an ignition source, may create an explosion. Even though flammable dusts are not considered in any of the DOT hazard classes, they present a significant fire and explosion hazard under certain conditions, and responders should be aware of this hazard.

For several years, Nebraska had the unwelcome distinction of having more deaths as a result of grain elevator explosions than any other state (Photo 6.4). The state fire marshal implemented an intensive grain elevator inspection program. The inspections focused

Photo 6.4 Open-and closed-leg grain elevators are a primary source of dust explosions.

on housekeeping, maintenance, and compliance with state and NFPA codes and regulations. The initial inspection of all grain elevators throughout the state took approximately 2 years to complete. Since that time, it is the author's understanding that there have not been any deaths from grain elevator explosions in the state. These types of explosions are also discussed in Chapter 3.

Class 4.2 spontaneous combustibles

Class 4.2 materials are spontaneously combustible. The DOT defines them as pyrophoric materials. Even though this hazard class is flammable solids, these materials may be found as solids or liquids. They can ignite without an external ignition source within 5 min after coming into contact with air. There are other 4.2 materials that may be self-heating; i.e., in contact with air and without an energy supply (ignition source), they are liable to self-heat, which can result in a fire involving the material or other combustible materials nearby.

Activated carbon is flammable, and the dust is toxic by inhalation. When some carbon-based materials, such as activated carbon or charcoal briquettes, are in contact with water, an oxidation reaction occurs between the carbon material, the water, and pockets of trapped air. The reaction is exothermic, which means heat is produced in the reaction and slowly builds up until ignition occurs spontaneously. Materials subject to spontaneous heating are listed in Table 6.2.

Carbon-based animal or *vegetable oils*, such as linseed oil, cooking oil, and cottonseed oil, can also undergo spontaneous combustion when in rags or other combustible materials (Figure 6.6). These materials have one or more reactive double bonds in these very large hydrocarbon compounds. This double bond reacts with the oxygen in the air and the breaking of the bonds is exothermic and if confined can build up heat until spontaneous ignition will occur over a period of hours (Photo 6.5). This type of spontaneous heating *cannot* occur in the case of petroleum oils and other petroleum-based products. Petroleum-based products do not have double bonds in the compounds. Oxidation reactions that occur with animal and vegetable oils are different from the reaction with the carbon-based materials. Oxygen from the air trapped in the mass reacts with double bonds present in the animal and vegetable oils. Breaking of the double bonds creates heat, which ignites the materials. Petroleum products do not contain these double bonds and, therefore, cannot undergo this type of spontaneous heating to cause a fire.

Fires started by this spontaneous heating process can be difficult to extinguish because they usually involve deep-seated fires. Materials involved in these fires are large hydrocarbon compounds with lots of heat output. In order for enough heat to be sustained to cause combustion, there must be insulation. This insulation can be the material itself or may be in the form of some other combustible material, such as rags.

Table 6.2 Materials Subject to Spontaneous Heating

Alfalfa meal	Hides	Peanut oil
Used burlap	Castor oil	Charcoal
Coal	Powdered eggs	Lanolin
Coconut oil	Lard oil	Linseed oil
Cottonseed oil	Manure	Soybean oil
Fertilizers	Metal powders	Fish meal
Fish oil	Olive oil	Whale oil

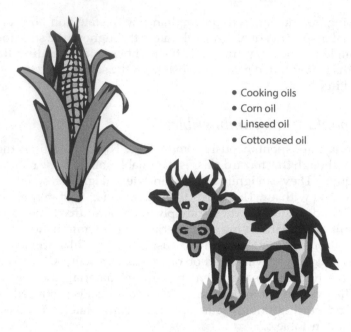

- Cooking oils
- Corn oil
- Linseed oil
- Cottonseed oil

Figure 6.6 Animal/vegetable oils.

Photo 6.5 Animal vegetable oils like linseed oil and turpentine are subject to spontaneous combustion when rags are not properly disposed of.

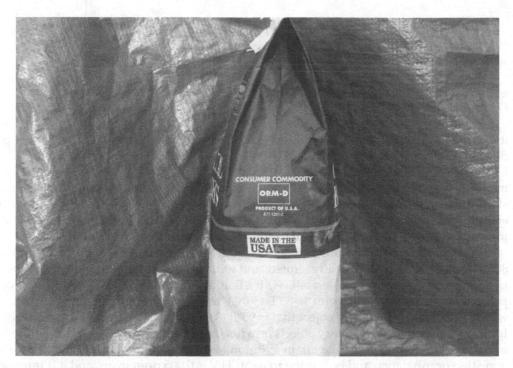

Photo 6.6 A bag of charcoal, an ORM-D material, that may undergo spontaneous heating upon contact with water. If the heat is confined, combustion may occur.

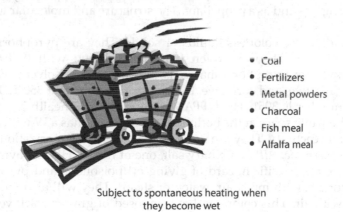

- Coal
- Fertilizers
- Metal powders
- Charcoal
- Fish meal
- Alfalfa meal

Subject to spontaneous heating when they become wet

Figure 6.7 Slow oxidation materials.

Charcoal briquettes are a dangerous fire risk (Photo 6.6). They may undergo spontaneous ignition when they become wet. However, this is a slow process, and the heat generated must be confined as it builds up and ignites (Figure 6.7).

Pyrophoric solids and liquids

Diethyl zinc is an organometal compound and is a dangerous fire hazard. It spontaneously ignites in air and reacts violently with water, releasing flammable vapors and heat. It is a colorless, pyrophoric liquid with a specific gravity of 1.2, which is heavier than water, so

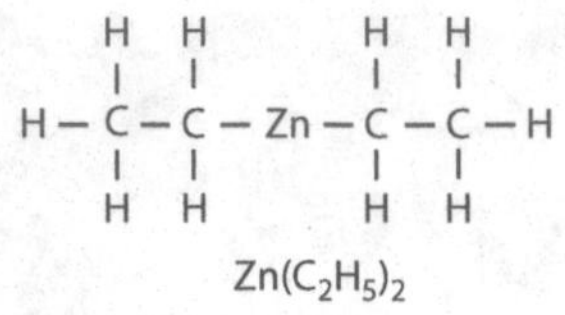

$$
\begin{array}{c}
\text{H H} \qquad\qquad \text{H H}\\
\text{| |} \qquad\qquad\ \ \text{| |}\\
\text{H}-\text{C}-\text{C}-\text{Zn}-\text{C}-\text{C}-\text{H}\\
\text{| |} \qquad\qquad\ \ \text{| |}\\
\text{H H} \qquad\qquad \text{H H}
\end{array}
$$

$Zn(C_2H_5)_2$

Figure 6.8 Diethylzinc.

it will sink to the bottom. It decomposes explosively at 248°F (120°C). It has a boiling point of 243°F (117°C), a flash point of −20°F (−28°C), and a melting point of −18°F (−27°C). The four-digit UN identification number is 1366. The NFPA 704 designation is health 3, flammability 4, and reactivity 3. The white space at the bottom of the diamond has a W with a slash through it to indicate water reactivity. Primary uses of diethyl zinc are in the polymerization of olefins, high-energy aircraft, and missile fuel and in the production of ethyl mercuric chloride. Molecular formula and structure are shown in Figure 6.8.

Pentaborane is a nonmetallic, colorless liquid with a pungent odor. It decomposes at 300°F (148°C), if it has not already ignited, and will ignite spontaneously in air if impure. It is a dangerous fire and explosion risk, with a flammable range of 0.46%–98% in air. Boiling point is 145°F (64°C), flash point is 86°F (30°C), and ignition temperature is 95°F (35°C), which is extremely low. Any object that is 95°F (35°C) or above can be an ignition source. Ignition sources can be ordinary objects on a hot day in the summer, such as the pavement, metal on vehicles, and even the air. In addition to extreme flammability, it is also toxic by ingestion or inhalation and is a strong irritant. TLV is 0.005 ppm in air, and it is immiscible in water. The four-digit UN identification number is 1380. The NFPA 704 designation for pentaborane is health 4, flammability 4, and reactivity 2. The primary uses are as fuel for air-breathing engines and as a propellant. The structure and molecular formula are shown in Figure 6.9.

Aluminum alkyls are colorless liquids or solids. They are pyrophoric and may ignite spontaneously in air. They are often found in solution with hydrocarbon solvents. Aluminum alkyls are pyrophoric materials in a flammable solvent. Vapors are heavier than air, water reactive, and corrosive. Decomposition begins at 350°F. The four-digit UN identification number is 3051. The NFPA 704 designation is health 3, flammability 4, and reactivity 3. The white space at the bottom of the diamond has a W with a slash through it, indicating water reactivity. They are used as catalysts in polymerization reactions.

Aluminum phosphide, AlP, is a binary salt, one of the NCHP acronyms (see Chapter 2). These salts have the specific hazard of giving off poisonous and pyrophoric phosphine gas when in contact with moist air, water, or steam. They will also ignite spontaneously upon contact with air. This compound is composed of gray or dark yellow crystals and is a dangerous fire risk. Aluminum phosphide decomposes upon contact with water and has a specific gravity of 2.85, which is heavier than water. The four-digit UN identification number is 1397. The NFPA 704 designation is health 4, flammability 4, and reactivity 2. The white section at the bottom of the diamond has a W with a slash through it,

$$
\begin{array}{c}
\text{H H H H H}\\
\text{| | | | |}\\
\text{H}-\text{B}=\text{B}=\text{B}=\text{B}=\text{B}-\text{H}\\
\text{|} \qquad\qquad\quad\ \text{|}\\
\text{H} \qquad\qquad\quad\ \text{H}
\end{array}
$$

B_5H_9

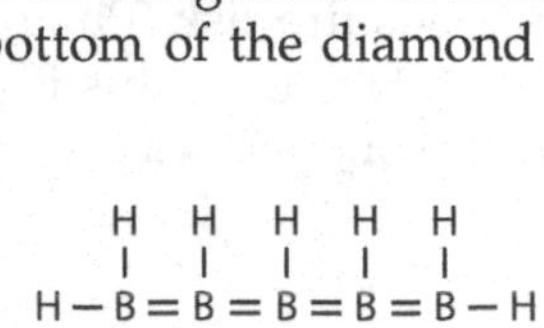

Figure 6.9 Pentaborane.

indicating water reactivity. Aluminum phosphide is used in insecticides, fumigants, and semiconductor technology.

Potassium sulfide, K_2S, is a binary salt. It is a red or yellow-red crystalline mass or fused solid. It is deliquescent in air, which means it absorbs water from the air, and it is also soluble in water. Potassium sulfide is a dangerous fire risk and may ignite spontaneously. It is explosive in the form of dust and powder. It decomposes at 1562°F (850°C) and melts at 1674°F (912°C). The specific gravity is 1.74, which is heavier than air. The four-digit UN identification number is 1382. The NFPA 704 designation is health 3, flammability 1, and reactivity 0. Potassium sulfide is used primarily in analytical chemistry and medicine.

Sodium hydride, *NaH*, is a binary salt that has a specific hazard of releasing hydrogen upon contact with water. It is an odorless powder that is violently water reactive. The four-digit UN identification number is 1427. The NFPA 704 designation is health 3, flammability 3, and reactivity 2. The white space at the bottom of the diamond has a W with a slash through it, indicating water reactivity.

White phosphorus, *P*, also known as yellow phosphorus, is a nonmetallic element that is found in the form of crystals or a wax-like transparent solid (Photo 6.7). It ignites spontaneously in air at 86°F, which is also its ignition temperature. White phosphorus should be stored under water and away from heat. It is a dangerous fire risk, with a boiling point of 536°F (280°C) and a melting point of 111°F (43°C). White phosphorus is toxic by inhalation and ingestion, and contact with skin produces burns. The TLV is 0.1 mg/m^3 of air, and it is insoluble in water, with a specific gravity of 1.82, which is heavier than air. White phosphorus is shipped and stored under water to keep it from contacting air. The four-digit UN identification number is 2447. The NFPA 704 designation is health 4, flammability 4, and reactivity 2. The primary uses are in rodenticides, smoke screens, and analytical chemistry.

Photo 6.7 Rail tank car of air-reactive phosphorous shipped under water.

Incidents

A series of fires have occurred in laundries around the country since 1989. One in six commercial, industrial, or institutional laundries reports a fire each year, which results in over 3000 fires. The primary cause is thought to be spontaneous combustion. Chemicals, including animal and vegetable oils, may be left behind in fabrics after laundering. The heat from drying may cause the initiation of the chemical reaction that causes spontaneous ignition. Spontaneous combustion, according to the *Handbook of Fire Prevention Engineering,* "is a runaway temperature rise in a body of combustible material, that results from heat being generated by some process taking place within the body." On June 16, 1992, a fire in a nursing home laundry in Litchfield, Illinois, caused $1.5 million in damage. The cause was determined to be the spontaneous ignition of residual chemicals in the laundered fabric reacting to heat from the dryer. In Findlay, Ohio, on July 2, 1994, a fire destroyed a commercial laundry, causing over $5 million in damage. Traces of linseed oil were found in a pile of clean, warm garments piled in a cart waiting to be folded.

Fires have also occurred involving residual animal or vegetable oils in cleaning rags in restaurants. The oils are never completely removed by laundering. When placed in the dryer, the rags are heated. When put away on the storage shelf, this heat can become trapped, along with the oil remaining on the rags when confined. The spontaneous combustion process begins slowly and the heat of the reaction increases.

In Jacksonville, Florida, hazmat teams were called to a popular local restaurant at noontime with the report of a chemical smell (Photo 6.8). Personnel first thought a 5-gal pail of corrosive liquid was responsible and removed it from the building. The odor still

Photo 6.8 Jacksonville firefighters responded with a full hazmat assignment to a restaurant at the height of the lunch hour for a reported chemical smell. It turns out to be a fire involving spontaneous ignition of animal vegetable grill rags that has been freshly laundered.

persisted. Odors were traced to the kitchen area and were coming from a shelf of grill rags that had just been laundered, dried, and placed on the shelf. A browning was found around the outer perimeter of each rag exactly in the same place. There were approximately 50 rags in the pile. Investigators first thought a flammable liquid was poured on the rags and ignited. However, the pattern was too symmetrical; if a liquid were poured, there would have been an irregular pattern and burning would not have occurred on each rag. It was finely determined animal/vegetable oil in the grill rags had spontaneously combusted causing the burn patterns and odor.

Another incident occurred in Gettysburg, Pennsylvania, on March 22, 1979, involving phosphorus shipped under water in 55-gal drums (phosphorus is air reactive). One of the drums developed a leak, and the water drained off. This allowed the phosphorus to be exposed to air, which caused it to spontaneously ignite. Fire spread to the other containers and eventually consumed the entire truck. The ensuing fire was fought with large volumes of water and in the final stages covered with wet sand. Cleanup created problems because as the phosphorus and sand mixture were shoveled into overpacked drums, the phosphorus was again exposed to air and reignited small fires. When phosphorus burns, it also gives off toxic vapors.

A train derailment in Brownson, Nebraska, resulted in a tank car of phosphorus overturning and the phosphorus igniting upon contact with air (Photo 6.9). Phosphorus is shipped under water, so there was water inside the tank car. CHEMTREC (Chemical Transportation Emergency Center) was called, and responders were told correctly that the phosphorous would not explode. However, the water inside the tank car turned to

Photo 6.9 Phosphorous fire in Brownson, Nebraska, during train derailment. Phosphorous is an air-reactive material and is shipped under water to keep the phosphorous from burning. During the derailment, the water drained off of the phosphorous and it ignited.

steam from the heat of the phosphorus fire. Pressure from the steam caused a boiler type of explosion that had nothing to do with the phosphorus. This is just another example of the hidden hazards that emergency responders must be aware of when dealing with hazardous materials. Not only do the hazardous materials have to be considered, but also the container and any "inert" materials that may be involved with the product.

Miamisburg, Ohio, experienced a train derailment involving a tank car of yellow phosphorous derailed along with 15 other cars. Phosphorus is air reactive and is shipped under water to keep it from igniting. During the derailment, the tank car overturned and the water leaked from the tank car exposing the phosphorus to the air, which ignited about 15 min later, creating thick clouds of phosphorous oxides, which are very toxic. Placards on the tank car were ripped off by the force of the derailment. In addition to the phosphorous, a tank car of molten sulfur also derailed and began to leak. Because of the potential toxic cloud that could be formed if the phosphorous and sulfur combined, and the cloud from the burning phosphorous, the fire chief ordered the evacuation of approximately 37,000 people or approximately one-third of the population of the area. Mutual aid was requested and received from 14 additional fire departments. County, regional, and state agencies also arrived and offered assistance. Fires were fought from a distance using large master streams. Approximately 3500 gal per minute were discharged on the burning phosphorous tank car. As specialists from the shipper, carrier, subcontractors, and state and federal agencies, several meetings were held to discuss tactics. Ideas put forth included direct hose stream attack, plugging the leak, water flooding of the interior of the tank car, foam application, burial, opening the manhole to allow air injection to accelerate the burn rate, and the use of explosive demolition. It was suggested that the Air Force from nearby Wight Patterson Air Force Base send a fighter to shoot a missile at the tank car. It was decided to proceed with the opening of the manhole to accelerate the burn rate. The incident went on for 5 days. No emergency response personnel were killed or seriously injured during the incident. There were no civilian deaths or serious injuries as well.

Class 4.3 dangerous when wet

Class 4.3 materials are dangerous when wet (Photo 6.10). The DOT definition is "a material that, by contact with water, is liable to become flammable or to give off flammable or toxic gas at a rate greater than 1 liter per kilogram of the material per hour." Examples of water-reactive materials include zinc powder; trichlorosilane; binary salt sodium phosphide; sodium aluminum hydride; metallic elements potassium, sodium, and lithium. Potassium, lithium, and sodium come from family 1 on the periodic table, known as the alkali metals (Photo 6.11). Located in the first column on the table and, as with other families on the chart, they have similar chemical characteristics. They are silvery soft metals that are reactive with air and violently reactive with water. Contact with water causes spattering, the release of free hydrogen gas, and the production of heat. The heat can be so great that it ignites the hydrogen gas (Table 6.3).

Metallic elements, such as *magnesium* and *calcium*, are from family 2 on the periodic table. These materials are known as the alkaline earth metals. Unlike the alkali metals, magnesium must be burning before it reacts with water or it must be in a finely divided form, such as filings and powder. Filings, flakes, dusts, and powders can ignite explosively upon contact with water to evolve flammable hydrogen gas and heat. Heat may be great enough to ignite the hydrogen gas. Magnesium is insoluble in water. The ignition temperature of magnesium is about 1200°F (648°C), as is the melting point. When magnesium ignites, the temperatures can reach 7200°F (3982°C). In contact with burning magnesium,

Photo 6.10 Sodium metal is a solid material that is very water reactive and can react with moisture in the air. It is shipped and stored under a petroleum-based product to prevent air contact.

water produces a violent explosion. Water in contact with magnesium fillings or powder can produce a spontaneous explosion. Talc, dry sand, Met-L-X, foundry flux, and G-1 powder should be used to extinguish small magnesium fires. Large fires should be fought with flooding volumes of water from unmanned monitors and aerial devices.

Calcium carbide, CaC_2, is a binary salt (Photo 6.12). It is a grayish-black hard solid that reacts with water to produce acetylene gas, a solid corrosive that is calcium hydroxide, and release heat. Acetylene gas is manufactured by reacting calcium carbide with water. Because acetylene is so unstable, it is not shipped in bulk quantities.

Calcium carbide is shipped to acetylene-generating plants where it is reacted with water in a controlled reaction. After the reaction process, the acetylene gas is placed into specially designed containers with a honeycombed mesh inside for shipment and use. It is dissolved in acetone for stability. Calcium carbide has a specific gravity of 2.22, which is heavier than water. The four-digit UN identification number for calcium carbide is 1402. The NFPA 704 designation is health 3, flammability 3, and reactivity 2. The white section at the bottom of the diamond contains a W with a slash through it, indicating water reactivity. It is shipped in metal cans, drums, and specially designed covered bins on railcars and trucks. When shipped and stored, it should be kept in a cool, dry place. Primary uses are in the generation of acetylene gas for welding, vinyl acetate monomer, and as a reducing agent.

Phosphorus pentasulfide, P_4S_{10}, is a nonmetallic inorganic compound (Photo 6.13). It is a yellow to greenish-yellow crystalline mass with an odor similar to hydrogen sulfide. It is a dangerous fire risk and ignites by friction or in contact with water. Boiling point is 995°F (535°C) and ignition temperature is 287°F (141°C). It decomposes upon contact with water

Photo 6.11 Alkali metals such as lithium, sodium, and potassium are water reactive and the reaction is exothermic. There can be enough heat produced to ignite flammable hydrogen gas released from the water during the reaction.

Table 6.3 Products of Water Reaction

Contact with water can cause any of the following reactions
Release heat
Release corrosive liquid
Release oxidizer
Release flammable gas
Release toxic gas
Explosion

or moist air, liberating toxic and flammable hydrogen-sulfide gas. Specific gravity is 2.09, so it is heavier than water. It is toxic by inhalation, with a TLV of 1 mg/m³ of air. The four-digit UN identification number is 1340. The NFPA 704 designation is health 2, flammability 1, and reactivity 2. Primary uses are in insecticides, safety matches, ignition compounds, and sulfonation.

Methyl dichlorosilane, CH_3SiHCl_2, is a colorless liquid with a sharp irritating odor. It is a dangerous fire risk, corrosive, and water reactive. The flammable range is wide, from 6% on the lower end to 55% on the upper end. Boiling point is 107°F (41°C), flash point is 15°F (–9°C), and ignition temperature is more than 600°F (315°C). Specific gravity is

Photo 6.12 Calcium carbide, a dry solid, is shipped to locations where it can be reacted with water, which produces acetylene gas.

Photo 6.13 Closed containers of water-reactive phosphorous pentasulfide, which produces toxic hydrogen-sulfide gas upon contact with water.

1.11, which is heavier than water. Vapors are heavier than air and will travel to ignition sources. It is immiscible in water and decomposes on contact to release hydrogen chloride gas. Methyl dichlorosilane is toxic by inhalation and skin absorption; it is irritating to the skin, eyes, and respiratory system. Contact with the material may cause burns to the eyes and skin. The four-digit UN identification number is 1242. The NFPA 704 designation is health 3, flammability 3, and reactivity 2. The white space at the bottom of the diamond has a W with a slash through it, indicating water reactivity. The primary use is in the manufacture of siloxanes, which are straight-chained compounds similar to paraffin hydrocarbons.

Potassium, K, is a metallic element, also known as kalium. It is an alkali metal that is soft, silvery, and rapidly oxidizes in moist air. Potassium is a combustible solid that may ignite spontaneously on contact with moist air. It is a dangerous fire risk and it reacts violently with water and moisture in the air to release hydrogen gas and form potassium hydroxide, which is a corrosive liquid. The boiling point is 1410°F (765°C), and the melting point is 146°F (63°C). The specific gravity is 0.86, which is lighter than water. The reaction with water is also exothermic, and the heat produced is enough to ignite the hydrogen gas that is released. Potassium metal is usually stored under kerosene to keep it from reaching the air. As it ages, it can form explosive peroxides, much like ethers do. When these peroxides are present, it may explode violently if handled or cut. Potassium that is coated with peroxides should be destroyed by burning. The four-digit UN identification number is 2257. The NFPA 704 designation is health 3, flammability 3, and reactivity 2. The white section at the bottom of the diamond has a W with a slash through it, indicating water reactivity.

Carbon black is a finely divided form of carbon. It may ignite explosively if suspended in air in the presence of an ignition source or slowly undergo spontaneous combustion upon contact with water. In addition, it is toxic by inhalation, with a TLV of 3.5 mg/m^3 in air. Primary uses are in the manufacture of tires, belt covers, plastics, carbon paper, colorant for printing inks, and as a solar-energy absorber.

Incidents

In Portland, Oregon, firefighters responded to a fire in an engraving company involving magnesium shavings (Photo 6.14). While in route, dispatch informed firefighters that flames were reported 4 ft high and several surrounding businesses were being evacuated. Hazmat was also dispatched to the scene. Firefighters used bags of sand to cover the burning magnesium and extinguish the fire. Magnesium when burning is violently water reactive and gives off toxic gases. Remaining magnesium was removed from the building and the cause of the fire was not determined. The incident was handled without injuries.

In New York City, three firefighters died and several others were injured in a fire involving sodium metal. Firefighters were extinguishing a fire in a 55-gal drum of molten sodium, when a small amount of water on a shovel came in contact with the sodium. This triggered a chemical reaction and explosion producing temperatures in excess of 2000°F, splattering the molten sodium on the firefighters. Sodium burned through their turnouts, station uniforms, and underwear. Contact with moisture on the skin caused burning of the tissue below. Water in contact with water-reactive and molten metals can produce violent reactions. Had the firefighters tried to extinguish the fire in the drum with a hose line, many more might have died.

Photo 6.14 Portland, Oregon, firefighters fight a fire involving magnesium shavings with sand. Magnesium when on fire is violently water reactive. (Courtesy of Portland Oregon Fire Department.)

Fire-extinguishing agents

Class 4.3 materials are water reactive. When large amounts of these materials are involved in fire, water is the only extinguishing agent available in quantities large enough to extinguish the fires; just understand that when water is used, there may be violent reactions and explosions. Preparations need to be made for the safety of personnel based upon the hazards of the materials. Small fires of water-reactive materials, especially metallic-based materials, can be extinguished with a dry-powder extinguishing agent. For other flammable solid materials, water is also the agent of choice in most cases. It is important, however, to make sure that there is a positive identification of the product, as with all hazardous materials. Once the product is identified, the proper extinguishing agent for any individual material can be identified through reference materials such as the Department of Transportation *Emergency Response Guidebook,* the CAMEO (Computer-Aided Management of Emergency Operations) computer database, CHEMTREC, or some other reference source.

Review questions

6.1   List the three subclasses of Class 4 hazardous materials.
6.2   Sublimation is the process in which a solid goes directly to which physical state?
  A. Liquid
  B. Slurry
  C. Particles
  D. Gas

6.3   Phosphorus is a Class 4.1 Flammable Solid and will do what when removed from its container?
      A. Dry out
      B. Spontaneously combust
      C. Oxidize
      D. Nothing

6.4   Which of the following may provide combustible dusts?
      A. Flour
      B. Coal
      C. Grain
      D. All of the above

6.5   Which of the following compounds or mixtures may undergo spontaneous heating?
      A. Fish oils
      B. Motor oils
      C. Linseed oil
      D. Cottonseed oil
      E. Gasoline
      F. Methyl ethyl ketone
      G. Isopropyl ether

6.6   Which of the following extinguishing agents would be appropriate for large fires involving water-reactive materials?
      A. Carbon dioxide
      B. Dry-chemical foam
      C. Dry powder
      D. Water

6.7   Which families from the periodic table are water-reactive materials in their elemental state?
      A. 1 and 7
      B. 1 and 2
      C. 2 and 8
      D. Only 1

6.8   When carbide salts contact water, what is produced?
      A. Phosgene
      B. Chlorine
      C. Carbon
      D. Acetylene

6.9   Class 4.2 Spontaneously Combustible contains which types of materials?
      A. Solids
      B. Liquids
      C. Gases
      D. Both A and B

6.10 Class 4.1 Flammable Solids contains which of the following materials?
      A. Wetted explosives
      B. Road flares
      C. Phosphorus
      D. All of the above

chapter seven

Oxidizers

Hazard Class 5 Oxidizers are separated into two divisions: 5.1 and 5.2. Class 5.1 materials are solids and liquids that, according to the DOT, "by yielding oxygen, can cause or enhance the combustion of other materials." Although that is not a technical definition from a chemistry standpoint, it gets right to the point for emergency response purposes. While oxidizers themselves do not burn, if present in a fire situation, they will make the fire burn faster and become more difficult to extinguish.

The NFPA classifies oxidizers into four groups. These groups are identified in Table 7.1. Common groups of oxidizers include oxysalts, inorganic peroxides (salt peroxides), certain acids, halogens Family 7 on the periodic table of elements, and organic peroxides. Examples of oxidizers from each of the NFPA classes are illustrated in Table 7.2. Oxygen is probably the most recognized oxidizer. Even though oxygen is essential for life to exist, it can be a dangerous hazardous material. Oxygen is found in transportation and storage as a compressed gas and as a cryogenic liquid with a temperature of $-183°F$ ($83°C$). If cryogenic oxygen leaks from a container and comes in contact with asphalt surfaces, it forms a contact explosive. Driving on it or even walking or dropping a tool can cause an explosion to occur. Oxygen does not burn, but in contact with organic materials, it can become explosive. Oxygen-enriched atmospheres can be deadly to emergency responders if there is a fire or heat source nearby. Many times, the enriched atmosphere is not visible or detectable to responders without the use of monitoring instruments. While the oxidizers in this chapter are solids and liquids, through physical and chemical reactions, they can release oxygen gas, which can cause some of the same problems as compressed oxygen.

Class 5.2 materials are organic peroxides. Unlike peroxide salts, which contain metals, these are organic compounds that contain carbon in their formula. These materials contain oxygen in the bivalent $-O-O-$ structure and may be considered a derivative of hydrogen peroxide. The structure and molecular formula for hydrogen peroxide are shown in Figure 7.1.

In organic peroxide compounds, both of the hydrogen atoms in hydrogen peroxide have been replaced by organic radicals. One of the major hazards of 5.2 organic peroxides is the instability of the compounds. The oxygen-to-oxygen single bond is an unstable bond. It is this same bond that is responsible for the explosiveness of the nitro compounds and peroxides formed in ether as it ages as discussed in Chapters 3 and 5. Oxidizers, especially the organic peroxides, should be treated with a great deal of respect. They can be just as dangerous and explosive as Class 1 compounds.

Class 5.1 oxidizers

Oxidizers may be elements, acids, or salts classified into families, with specific hazards associated with each family. There are elements found on the periodic table that are oxidizers in their elemental state. These include oxygen, chlorine, fluorine, bromine,

Table 7.1 NFPA Classes of Oxidizers

Class 1	Solid or liquid that readily yields oxygen or oxidizing gas or that readily reacts to oxidizer combustible materials.
Class 2	Oxidizing material that can cause spontaneous ignition when in contact with combustible materials.
Class 3	Oxidizing material that can undergo vigorous self-sustained decomposition when catalyzed or exposed to heat.
Class 4	Oxidizing material that can undergo an explosive reaction when catalyzed or exposed to heat, shock, or friction.

Table 7.2 Examples of Oxidizers in Four NFPA Classes

Class 1	Class 2	Class 3	Class 4
Aluminum nitrate	Calcium hypochlorite	Ammonium dichromate	Perchloric acid (60%–72%)
Calcium peroxide	Nitric acid (above 70%)	Calcium hypochlorite (over 50%)	Hydrogen peroxide (>90%)
Potassium persulfate	Sodium peroxide	Hydrogen peroxide (52%–91%)	Potassium superoxide
Sodium nitrite	Potassium permanganate		

$$H-O-O-H$$
$$H_2O_2$$

Figure 7.1 Hydrogen peroxide.

- Oxygen
- Halogens
- Oxysalts
- Peroxide salts
- Inorganic acids

Figure 7.2 Types of oxidizers.

and iodine. Peroxide salts and oxysalt families are oxidizers. Acids such as perchloric above 70%, which is an oxy-acid, and nitric above 40%, which is an inorganic acid, are oxidizers (Figure 7.2).

Oxygen, O_2, can be encountered as a gas, cryogenic liquid, and liquid or solid in compound with other materials. Although nontoxic, it is reactive with hydrocarbon-based materials. Oxygen is a strong oxidizer. Oxygen is a nonmetallic gaseous element.

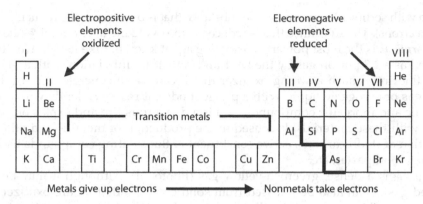

Figure 7.3 Periodic table of electropositive and electronegative elements.

Oxygen makes up approximately 21% of the atmosphere. The boiling point of oxygen is −297°F. It is nonflammable, but supports combustion. Liquid oxygen can explode when exposed to heat or organic materials.

Chlorine, fluorine, bromine, and iodine in their elemental forms are all strong oxidizers even though they are placarded and labeled as poisons (Figure 7.3). DOT placards the most severe hazard of materials with multiple hazards. Poison is considered more dangerous than the oxidizer hazard. Two terms commonly associated with oxidizers are oxidation (or oxidation reaction) and reduction. *Oxidation* is the loss of electrons by one reactant, and *reduction* is the gaining of electrons by another. Metals usually lose electrons, and nonmetals usually gain electrons. Elements in the upper-right corner of the periodic table are electronegative, or electron-drawing. Fluorine is the most electron-drawing element known. Chlorine and oxygen are also electron-drawing (Table 7.3). Oxidation and reduction always occur together. No substance is ever oxidized unless something else is reduced.

For example, when sodium and chlorine combine with an ionic bond, the electron of sodium is given to chlorine; sodium has been oxidized. Chlorine receives the electron of sodium; chlorine has been reduced. The substance that accepts the electrons is known as the oxidizing agent. Therefore, in the reaction between sodium and chlorine, chlorine is the oxidizing agent and sodium is the reducing agent. Chlorine is reduced to the chloride ion in the

Table 7.3 Strength of Oxidizing Agents

1. Fluorine
2. Ozone
3. Hydrogen peroxide
4. Metallic chlorates
5. Nitric acid
6. Chlorine
7. Sulfuric acid
8. Oxygen
9. Bromine
10. Iron III
11. Iodine
12. Sulfur

reaction with sodium. In summary, the substance that is oxidized is the reducing agent (gives up its electrons). The substance that is reduced is the oxidizing agent (receives electrons).

Fluorine, F, is the most powerful oxidizing agent known. Like its relative, chlorine, it is classified as a 2.3 poison gas by the DOT and is toxic by inhalation, with a TLV of 1 ppm. Liquid fluorine is such a strong oxidizer that it can cause concrete to burn. It is a pale-yellow gas or cryogenic liquid with a pungent odor. It reacts violently with a wide range of organic and inorganic compounds, and is a dangerous fire and explosion risk when in contact with these materials. It is used in the production of metallic and other fluorides, production of fluorocarbons, active constituent of fluoridating compounds used in drinking water, toothpastes, etc.

Chlorine is a dense, greenish-yellow gas (Photo 7.1). Although it may be a gas or a liquefied gas, it can also be released from solid compounds that are oxidizers. Chlorine is not combustible; however, it will support combustion just like oxygen. It is dangerous in contact with turpentine, ether, ammonia, hydrocarbons, hydrogen, powdered metals, and other reducing materials. Chlorine does not occur freely in nature. It is found in compounds within the minerals halite (rock salt), sylvite, and carnallite, and as the chloride ion in seawater. Chlorine is used in the manufacture of carbon tetrachloride, trichloroethylene, chlorinated hydrocarbons, polychloroprene (neoprene), polyvinyl chloride, hydrogen chloride, ethylene dichloride, hypochlorous acid, metallic chlorides, chloroacetic acid, chlorobenzene, chlorinated lime; water purification; shrink-proofing wool; in flame-retardant compounds; in special batteries with (lithium or zinc); and in processing of meat, fish, vegetables, and fruit.

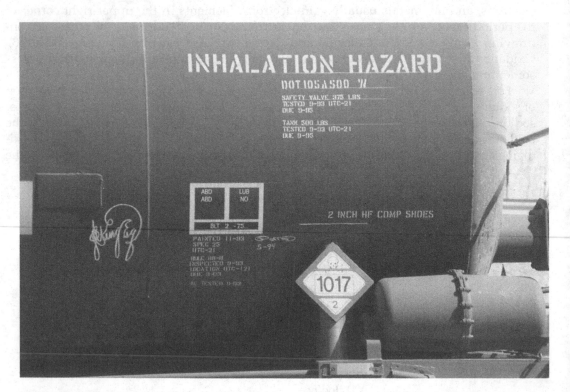

Photo 7.1 Chlorine, although placarded as a poison, is also a strong oxidizer and will support combustion just like oxygen.

Bromine is a dark, reddish-brown liquid with irritating fumes. It attacks most metals and reacts vigorously with aluminum and explosively with potassium. It is a strong oxidizing agent and may ignite combustibles on contact. It is used in the manufacture of ethylene dibromide (antiknock gasoline), organic synthesis, bleaching, water purification, solvent, intermediate for fumigants (methyl bromide), analytical reagent, fire-retardant for plastics, dyes, pharmaceuticals, photography, shrink-proofing wool.

Iodine is the least reactive of the halogens. It consists of heavy, grayish-black granules or solid plates, readily sublimed to a violet vapor. Iodine is insoluble in water and soluble in alcohol, carbon disulfide, chloroform, ether, carbon tetrachloride, and glycerol. It is toxic by ingestion and inhalation, strong irritant to eyes and skin. TLV: ceiling 0.1 ppm in air. Iodine is used as an antiseptic, germicide, x-ray contrast, water treatment, medicinal soaps, and food and feed additive. Fluorine, chlorine, bromine, and iodine will be discussed further in Chapter 8.

Oxysalts

Oxysalts are combinations of metals ionically bonded to a complex covalently bonded, nonmetal oxyradicals. They have a general formula (GF) of M + oxyradical or M + radical, ending in "ate" or "ite," and may have the prefixes "per" or "hypo." They are named by naming the metal first followed by the oxyradical and with a prefix indicating the oxygen content. Oxysalts have a hazard of oxidizers as a family. Eleven oxysalt radicals will be presented with this group (Table 7.4). As shown, the following oxysalts are in their base state. The first six all have -1 charges: FO_3 (fluorate), ClO_3 (chlorate), BrO_3 (bromate), IO_3 (iodate), NO_3 (nitrate), and MnO_3 (manganate). The next three have -2 charges: CO_3 (carbonate), SO_4 (sulfate), and CrO_4 (Chromate). The last two have -3 charges: PO_4 (phosphate) and AsO_4 (arsenate).

All of these radicals are considered to be in their base state, i.e., containing the "normal" number of oxygen atoms present in that oxyradical. When a metal is added to any oxyradical in the base state, the compound ends in "ate," such as sodium phosphate. Oxyradicals may be found with varying numbers of oxygen atoms. There may be more or less oxygen atoms in a compound than the base state. Regardless of the number of oxygen atoms on the oxyradical, the charge of the radical *does not* change. When naming compounds with one additional oxygen atom, the prefix "per" is used; to indicate excess oxygen over the base state, the ending is still "ate." When the number of oxygen atoms is one less than the base state of an oxyradical, the ending of the oxyradical name is "ite." An example is magnesium sulf*ite* ($MgSO_3$). Furthermore, an oxyradical can have two less oxygen atoms than

Table 7.4 Oxysalts

Any metal + Oxyradical		
-1	-2	-3
FO_3	CO_3	PO_4
ClO_3	CrO_4	BO_3
BrO_3	SO_4	AsO_4
IO_3		
NO_3		
MnO_3		

Table 7.5 Naming Oxysalts

+1 Oxygen prefix	Per-	____ate
Base state ending		____ate
−1 Oxygen ending		____ite
−2 Oxygen prefix	Hypo-	____ite

the base state; the oxyradical name has a prefix "hypo" and ends in "ite." An example is aluminum *hypo*phosph*ite* ($AlPO_2$) (Table 7.5).

All oxysalt compounds are salts and have the hazard of being oxidizers. They contain, among other things, fluorine, chlorine, and oxygen, which are strong oxidizers. Most oxysalts do not react with water, but are soluble in water. In the process of mixing with water, they may liberate oxygen, fluorine, or chlorine. Oxysalts have varying numbers of oxygens in their compounds. The common oxysalt compound has three or four oxygen atoms, which is known as the base state. Some oxysalts are loaded with oxygen, such as the "per-ate" compounds, which will have four or five oxygen atoms. The compounds that have two oxygen atoms will end in "ite." The "hypo-ite" compounds will have one oxygen atom. The fact that the "hypo-ites" have only one oxygen atom does not mean that they are not dangerous oxidizers. Some of the "hypo-ite" compounds, in addition to oxygen, also have chlorine atoms, which are oxidizers. Perchlorates and other oxyradicals in the "per-state" contain one more oxygen atom than the base-state chlorates. They are loaded with oxygen and want to give it up readily. Perchlorates can form explosive mixtures with organic, combustible, or oxidizable materials. Contact with acids, such as sulfuric acid, can form explosive mixtures.

Lithium perchlorate, $LiClO_4$, is an oxysalt that is a colorless, deliquescent crystal. Oxysalt "per-ate" compounds are loaded with excess oxygen and will readily give it up in a reaction. Lithium perchlorate is a powerful oxidizing agent. It has more available oxygen than does liquid oxygen on a volume basis. Lithium perchlorate has a specific gravity of 2.429, which is heavier than water, and is water soluble. It is a dangerous fire and explosion risk in contact with organic materials and is an irritant to skin and mucous membranes. The primary use of lithium perchlorate is as a solid rocket propellant. Chlorates are strong oxidizing agents. When heated, they give up oxygen readily. Contact with organic or other combustible materials may cause spontaneous combustion or explosion. They are incompatible with ammonium salts, acids, metal powders, sulfur, and finely divided organic or combustible substances.

Potassium chlorate, $KClO_3$, is a transparent, colorless crystal or white powder. It is soluble in boiling water and decomposes at approximately 750°F (398°C), giving off oxygen gas. Potassium chlorate is a strong oxidizer and forms explosive mixtures with combustible materials, such as sugar, sulfur, and others. Potassium chlorate is incompatible with sulfuric acid, other acids, and organic material. The four-digit UN identification number is 1485. Its primary uses are as an oxidizing agent in the manufacture of explosives and matches; in pyrotechnics; and as a source of oxygen. Sodium and potassium chlorates have similar properties. Chlorites are powerful oxidizing agents. They have one less oxygen than the base-state oxysalts. They form explosive mixtures with combustible materials, and in contact with strong acids, they can release explosive chlorine dioxide gas.

Calcium chlorite is an oxysalt with a molecular formula of $Ca(ClO_2)_2$. It is a white, crystalline material that is soluble in water. It is a strong oxidizer and a fire risk in contact with organic materials. The four-digit UN identification number is 1453. Hypochlorites have two less oxygen atoms than the base-state compounds. They can cause combustion in high concentrations when in contact with organic materials. When heated or in contact with water, they can give off oxygen gas. At ordinary temperatures, they can give off chlorine and oxygen when in contact with moisture and acids. They are commonly used as bleaches and swimming pool disinfectants.

Calcium hypochlorite, Ca(ClO)₂, is an oxysalt; it is a crystalline solid and an oxidizer that decomposes at 212°F (100°C). Calcium hypochlorite is a dangerous fire risk in contact with organic materials. It is also a common swimming pool chlorinator and decomposes in contact with water, releasing chlorine into the water. If a container of calcium hypochlorite becomes wet in storage, the result can be an exothermic reaction. If combustible materials are present, a fire may occur. The chlorine in the compound will be released by contact with the water and will then accelerate the combustion process. The four-digit UN identification number for dry mixtures with not less than 39% available chlorine (8.8% oxygen) is 1748; hydrated with not less than 5.5% and not more than 10% water, the number is 2880; mixtures that are dry, with not less than 10% but not more than 39% available chlorine, are numbered as 2208. The NFPA 704 designation for calcium hypochlorite is health 3, flammability 0, and reactivity 1. The white section at the bottom of the diamond has the prefix "oxy," indicating an oxidizer. The primary uses are as a bleaching agent, a swimming pool disinfectant, a fungicide, in potable-water purification, and as a deodorant.

Metal nitrates are oxysalts and, as a group, have a wide range of hazards. Common to many of them, however, is the fact they are oxidizers and are heat and shock sensitive. When heated, they will melt, releasing oxygen, which will increase the combustion process. Molten nitrates react violently with organic materials. When solid streams of water are used for fire suppression, steam explosions may occur upon contact with the molten materials. Nitrates can be dangerous oxidizers and will explode if contaminated, heated, or shocked. Most nitrates have similar properties.

Aluminum nitrate, Al(NO₃)₃, is a white, crystalline material that is soluble in cold water. It is a powerful oxidizing agent that decomposes at approximately 300°F. Aluminum nitrate should not be stored near combustible materials. The four-digit UN identification number is 1438. The primary uses are in textiles, leather tanning, as an anticorrosion agent, and as an antiperspirant.

Sodium nitrate, also known as Chile saltpeter and soda niter, has a molecular formula of *NaNO₃*. Sodium nitrate is a colorless, odorless, transparent crystal. It oxidizes when exposed to air and is soluble in water. This material explodes at 1000°F (537°C), much lower than temperatures encountered in many fires. Sodium nitrate is toxic by ingestion and has caused cancer in test animals. When used in the curing of fish and meat products, it is restricted to 100 ppm. Sodium nitrate is incompatible with ammonium nitrate and other ammonium salts. The four-digit UN identification number is 1498. Sodium nitrate is used as an antidote for cyanide poisoning and in the curing of fish and meat.

Potassium nitrate (saltpeter) has a molecular formula of *KNO₃*. It is found as a transparent to white crystalline powder and as crystals. Potassium nitrate is water soluble and is a dangerous fire and explosion risk when heated or shocked or in contact with organic materials. It is a strong oxidizing agent, with a four-digit UN identification number of 1486. Potassium nitrate is used in the manufacture of pyrotechnics, explosives, and matches. It is often used in the illegal manufacture of homemade pyrotechnics and explosives.

Persulfates are strong oxidizers and may cause explosions during fires. Oxygen may be released by the heat of the fire and cause explosive rupture of the containers. Explosions may also occur when persulfates are in contact with organic materials.

Potassium persulfate, K₂S₂O₈, is composed of white crystals that are soluble in water, and it decomposes below 212°F (100°C). Potassium persulfate is a dangerous fire risk in contact with organic materials. It is a strong oxidizing agent and an irritant, with a four-digit UN identification number of 1492. The primary uses are in bleaching, as an oxidizing agent, as an antiseptic, as a polymerization promoter, and in the manufacture of pharmaceuticals.

Permanganates mixed with combustible materials may ignite from friction or spontaneously in the presence of inorganic acids. Explosions may occur with either solutions or dry mixtures of permanganates.

Potassium permanganate, $KMnO_4$, is composed of dark purple, odorless crystals with a blue metallic sheen. It is soluble in water, decomposes at 465°F (240°C), and is a powerful oxidizing material. Potassium permanganate is a dangerous fire and explosion risk in contact with organic materials. Potassium permanganate is incompatible with sulfuric acid, glycerin, and ethylene glycol. The four-digit UN identification number is 1490. The primary uses of potassium permanganate are as an oxidizer, bleach, or dye; during radioactive decontamination of the skin; and in the manufacture of organic chemicals.

Some ammonium compounds are oxysalts. Although ammonia is not a metal, in the case of the ammonium ion, it acts like a metal when attached to the oxyradicals. When ammonia gas is added to water, it readily dissolves and remains as NH_3. One of the hydrogen atoms leaves water, but leaves its electrons behind. The protons of the hydrogen then attach to the unbonded electrons on nitrogen to complete its duet. This hydrogen is loosely held to the nitrogen and comes off easily. The hydrogen ions in the water are attracted to the negative side of the ammonia molecule where that unbonded pair of electrons is located (you can think of the ammonia as a slightly polar molecule). The ammonium ion is positive because the hydrogen ion contributes no electrons. It is not important to understand why this happens, but rather that the hazards of the compounds will be similar to the rest of the oxysalts, i.e., they are oxidizers. This is a complex covalent-sharing arrangement and is one of those chemistry concepts that should be accepted rather than explained for the purposes of emergency response. The ammonium ion is shown in (Figure 7.4).

Ammonium iodate, NH_4IO_3, is an oxidizing agent and a dangerous fire and explosion risk in contact with organic materials. It is a white, odorless powder.

Ammonium chlorate is an ammonium compound with a molecular formula of NH_4ClO_3. It is a colorless or white crystal that is soluble in water. Ammonium chlorate is a strong oxidizer and, when contaminated with combustible materials, can spontaneously ignite. It is shock sensitive and can detonate when exposed to heat or vibration. One of its primary uses is in the manufacture of explosives.

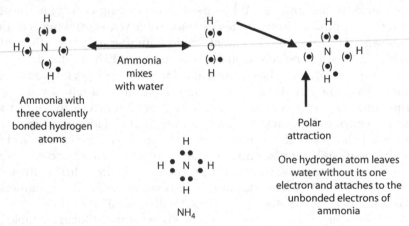

Figure 7.4 Ammonium ion.

Peroxide salts

Peroxide salts should not be confused with organic peroxides. Peroxide salts are inorganic salts containing a metal and a peroxide radical (O_2). Organic peroxides are organic compounds made up of nonmetal hydrocarbon radicals and the peroxide functional group –O–O–. Peroxide salts are composed of an M + ionically bonded to the covalently bonded complex radical O_2. They are named by naming the metal and ending with peroxide. Hazards are water reactive and give off oxygen, evolve heat, and produce a corrosive liquid in contact with water. They are oxidizers. When a corrosive liquid is created by the water reaction, the hydroxide of the metal in the compound is created as a by-product. Heat produced may be sufficient to ignite nearby combustible materials. Metal peroxides may also decompose when exposed to heat, with results similar to their reaction with water. Hazards and physical and chemical characteristics of metal peroxides are similar. Sodium peroxide and barium peroxide are common metal peroxides and are detailed in the following paragraphs. Keep in mind that they are not the only metal peroxides.

Sodium peroxide has a molecular formula of Na_2O_2 and is an inorganic peroxide salt. It is a yellowish-white powder that turns yellow when heated. Sodium peroxide absorbs water and carbon dioxide from the air and is soluble in cold water. It is a strong oxidizing agent, is corrosive and can cause burns to the eyes and skin, and is also toxic by ingestion and inhalation. Sodium peroxide is water reactive and a dangerous fire and explosion risk in contact with water, alcohol, or acids. Sodium peroxide forms self-igniting mixtures with powdered metals and organic materials. It is incompatible with ethyl or methyl alcohol, glacial acetic acid, carbon disulfide, glycerin, ethylene glycol, and ethyl acetate. The four-digit UN identification number is 1504. The NFPA 704 designation is health 3, flammability 0, and reactivity 1. The 704 diamond has the prefix "oxy" in the white space at the bottom. It is used as bleach and as an oxygen-generating material for diving bells and submarines.

Barium peroxide, BaO_2, is a grayish-white powder that is slightly soluble in water. Barium peroxide is a dangerous fire and explosion risk in contact with organic materials and decomposes around 1450°F (787°C). It is also toxic by ingestion, is a skin irritant, and should be kept cool and dry in storage. The four-digit UN identification number is 1449. The primary uses of barium peroxide are in bleaching, in thermal welding of aluminum, as an oxidizing agent, and in the dyeing of textiles.

Inorganic acid oxidizers

At higher concentrations, some acids can be strong oxidizers and cause combustion in contact with organic materials. Nitric acid above 40% is a dangerous oxidizing acid; it is also corrosive, and its vapors are toxic.

Nitric acid, HNO_3, is considered an oxidizer above 40% concentration. Nitric acid is a dangerous fire risk in contact with organic materials. It is shipped in bulk quantities in MC/DOT 312 tanker trucks. The MC 312 tanker is a small-diameter tank with reinforcing rings around the outside. The corrosive materials carried in these tanks are heavy. By recognizing the shape of MC 312, you will realize that there is a corrosive material in the tank even though it may be placarded with the oxidizer placard. It is used in the manufacturer of ammonium nitrate for fertilizer and explosives. Nitric acid will be discussed further in Chapter 10.

Chromic acid, CrO_3, is composed of dark, purplish-red, odorless crystals that are soluble in water. The specific gravity is 2.7, which is heavier than water. It is a powerful oxidizing agent and may explode on contact with organic materials. Chromic acid is a poison, corrosive to the skin, and has a TLV of 0.05 mg/m³ of air. Chromic acid is a known human carcinogen. The four-digit UN identification number is 1463. The NFPA 704 designation is health 3, flammability 0, and reactivity 1. The white section at the bottom of the 704 diamond has an "oxy" prefix, indicating that it is an oxidizer.

Perchloric acid has a molecular formula of $HClO_4$. At concentrations of more than 50% but less than 72%, by volume, it is placarded as an oxidizer. Concentrations above 72% are forbidden in transportation. It is a colorless, fuming liquid, and is unstable in its concentrated form. It is a strong oxidizing agent, is corrosive, and is placarded as a Class 8 corrosive in concentrations less than 50%. Perchloric acid will ignite vigorously upon contact with organic materials or detonate by shock or heat. It is toxic by inhalation and ingestion, and is a strong irritant. Perchloric acid is water soluble and has a specific gravity of 1.77, which means it is heavier than water. However, it is water soluble and will mix rather than form layers. Upon contact with water, heat is produced. Boiling point is 66°F and vapor density is 3.46, which is heavier than air. Perchloric acid is incompatible with acetic anhydride, bismuth and its alloys, alcohols, paper, wood, and other organic materials. The four-digit UN identification number for concentrations greater than 50% is 1873. Concentrations less than 50% have a four-digit identification number of 1802. The NFPA 704 designation is health 3, flammability 0, and reactivity 3. There is an "oxy" prefix in the white section of the 704 diamond, indicating an oxidizer.

Oxidizers, like many of the other hazard classes, can have more than one hazard. They can be corrosive, poisonous, and explosive under certain conditions. Many are water reactive, and some may react violently with other chemicals, particularly organic materials. Oxygen is one of the components of the original fire triangle and the more recent fire tetrahedron. The combustion process involves heat, oxygen, fuel, and a chemical chain reaction. Combustion is simply a rapid oxidation reaction that is accompanied by the emission of energy in the form of heat and light. When we think about combustion, we usually think about atmospheric oxygen allowing combustion to occur. In the combustion process, the fuel source is heated; the molecules start to vibrate rapidly, and if the vibrations are strong enough, the molecules break into small fragments with incomplete bonds.

These "free radical" fragments are unstable and cannot remain as fragments, so they want to bond with some other element to complete the octet rule of bonding and become stable. As the heat increases, the radical fragments rise and encounter oxygen from the air. The oxygen is electronegative (electron-drawing) and quickly attracts the electrons from the radical fragments. Bonding occurs between the radical fragments and the oxygen, forming a chemical bond. This bonding process is exothermic, and the heat energy is fed back into the combustion process, allowing the combustion to continue. Without oxygen or chemical oxidizers, most combustion cannot occur.

Materials that contain oxygen in their compounds can support combustion even in the absence of atmospheric oxygen. When extra oxygen is added, over and above atmospheric oxygen, the process of combustion is accelerated. The more oxygen that is present, the more bonds that can take place between the radical fragments and the oxygen. The higher the oxygen concentration, the more accelerated the combustion process. When oxysalts are dissolved in water from hose streams and the liquid comes in contact with turnouts, the turnouts become wet. As they dry out, the oxysalt impregnates the turnout fibers. Now the turnouts have an oxidizer in the fabric. When a firefighter is exposed to the heat of a fire, the oxidizer in the turnouts can cause the turnouts to ignite and burn vigorously. The turnouts must be decontaminated prior to being used again, or the firefighters will be in unnecessary danger when exposed to heat or fire.

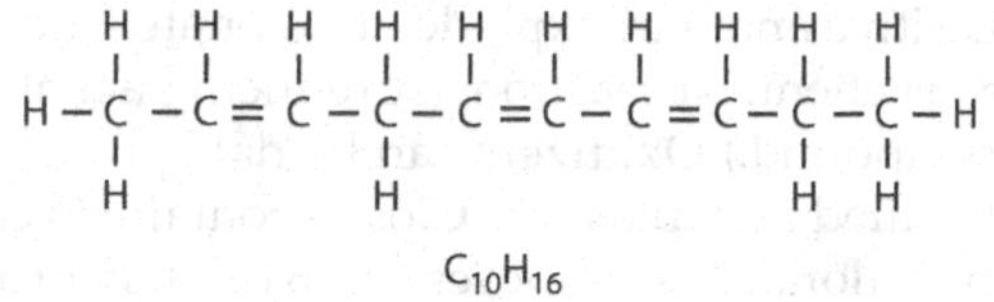

$$H - \overset{\displaystyle H}{\underset{\displaystyle H}{C}} - \overset{\displaystyle H}{C} = \overset{\displaystyle H}{C} - \overset{\displaystyle H}{\underset{\displaystyle H}{C}} - \overset{\displaystyle H}{C} = \overset{\displaystyle H}{C} - \overset{\displaystyle H}{C} = \overset{\displaystyle H}{C} - \overset{\displaystyle H}{\underset{\displaystyle H}{C}} - \overset{\displaystyle H}{\underset{\displaystyle H}{C}} - H$$

$$C_{10}H_{16}$$

Figure 7.5 Turpentine.

Oxidizers can undergo spontaneous combustion by three different means. First, there is slow oxidization. This occurs when oxygen comes in contact with a material that has double bonds (also known as Pi bonds), such as animal or vegetable oils and alkenes. Animal and vegetable oils are actually large esters, with the exception of turpentine, which is a pure hydrocarbon compound. The problem is the same, however; the double bonds in turpentine can be attacked by oxygen from the air and can undergo spontaneous ignition. The structure and molecular formula for turpentine are shown in Figure 7.5; notice the double bonds.

As oxygen breaks the double bonds, heat is produced. First, if there is enough insulating material present, the heat can build up to the point of spontaneous ignition. Second, the reaction of an inorganic oxidizer with water can be exothermic, i.e., it gives off heat. Oxidizers also generate oxygen gas in contact with water. The heat and oxygen represent two sides of the fire triangle or tetrahedron. All that is needed is fuel. If combustible materials are present, combustion can occur spontaneously. When that occurs, there is also excess oxygen present that accelerates the rate of combustion. Third, the corrosive action of a strong acid, such as nitric acid, can generate heat. The heat can be sufficient to ignite combustible materials to which the corrosive is exposed.

Some oxidizers, particularly salts, mix with water rather than react. The liquid mixture can impregnate combustible and noncombustible materials. When the water evaporates from the material, the oxidizer is left behind. The materials will now burn with great intensity if exposed to heat or fire because of the oxidizers present. Some compounds can undergo spontaneous combustion when they contact water. This is a slow process, as in the case of charcoal. Wetted charcoal is oxidized by oxygen in the air. If the heat produced by the water reaction is accumulated in the material, combustion may occur. This air oxidation can also occur with animal and vegetable oils when they are present in combustible materials, such as rags. The double bond, also referred to as a Pi bond, in the animal and vegetable oils is oxidized by the oxygen in the air. This action on the double bond creates heat. If the heat is confined, spontaneous combustion may occur. The corrosive action of strong acids can generate heat. The heat may be enough to ignite combustible materials. When the corrosive material is also an oxidizer, the oxidizer will contribute to the acceleration of the combustion process. An explosion is really nothing more than a rapid combustion. A chemical explosion, as discussed in Chapter 3, requires a chemical oxidizer to be present. Without the chemical oxidizer, the combustion does not accelerate fast enough to allow an explosion to occur. Oxidizers that allow for explosions can themselves explode when heated or shocked.

In Henderson, Nevada, a fire occurred in the Pepcon chemical plant. The fire heated ammonium perchlorate to the point that a detonation occurred. Ammonium perchlorate is an oxidizer used in the manufacture of solid rocket fuel, but it is not an explosive. However, this oxidizer detonated, creating a shock wave that blew the windshields out of responding fire apparatus and injuring several firefighters.

Ammonium perchlorate, NH_4ClO_4, is a white, crystalline material that is soluble in water. It is a strong oxidizing agent and a skin, eye, and respiratory irritant. Ammonium perchlorate is shock sensitive and may explode or detonate when exposed to heat-reducing agents or by spontaneous chemical reaction. (A reducing agent is a material that removes the oxygen from the compound.) Oxidizers can be dangerously explosive materials even though they are not classified as explosives. Closed containers can rupture violently when heated. Ammonium perchlorate decomposes at 464°F and produces oxides of nitrogen, hydrogen chloride, and ammonia. It is incompatible with acids, alkalis, powdered metals, and organic materials. It has a four-digit UN identification number of 1442. The NFPA 704 classification is health 1, flammability 0, and reactivity 4. The prefix "oxy" is placed in the white section at the bottom of the 704 diamond. Ammonium perchlorate is usually shipped in fiber drums, bags, steel drums, and tote bins. It is primarily used in the manufacture of explosives.

Other oxidizer compounds

Sodium carbonate, Na_2CO_3, also known as soda ash, is an oxysalt. It can be found naturally or can be synthetic. It is a grayish-white powder or lumps containing up to 99% sodium carbonate. Sodium carbonate is soluble in water. It is not a particularly hazardous material and is not regulated in transportation by the DOT. The primary uses are in the manufacture of other chemicals and products, including glass, paper, soaps, cleaning compounds, petroleum refining, and as a catalyst in coal liquefaction.

Ammonium nitrate, NH_4NO_3, is a colorless crystal ammonium salt that is soluble in water (Photo 7.2). It is a strong oxidizer. It should not be confused with the explosive ammonium nitrate that is mixed with a hydrocarbon fuel, commercially called ANFO.

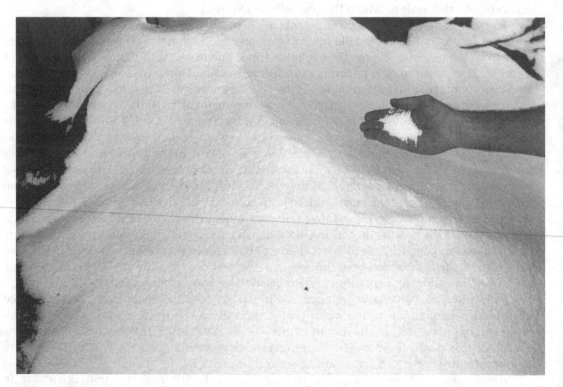

Photo 7.2 Commercial-grade ammonium nitrate, an oxidizer used to make blasting agents.

Table 7.6 Ammonium Nitrate Four-Digit Identification Numbers

More than 0.2% combustible material—0222
Not more than 0.2% combustible material—1942
With organic coating—1942
Ammonium nitrate fertilizer—2067
Fertilizer that is liable to explode—0223
Fertilizer with ammonium sulfate—2069
Fertilizer with calcium carbonate—2068
Fertilizer with not more than 0.4% combustible material—2071
Fertilizer with phosphate or potash—2070
Fertilizers—2071, 2072
Fertilizers N.O.S.—2072
Ammonium nitrate fuel oil mixture—0331
Mixed fertilizers—2069

Specific gravity of ammonium nitrate is 1.72, which is heavier than water. Ammonium nitrate is soluble in water and decomposes at 410°F (210°C), evolving nitrous oxide gas. It may explode under confinement and at high temperatures. Large amounts of water should be applied using unmanned appliances to fight fires, with all personnel evacuated to a safe distance. Ammonium nitrate is incompatible with acids, flammable liquids, metal powders, sulfur, chlorates, and any finely divided organic or combustible substance. The four-digit UN identification numbers are listed in Table 7.6. The NFPA 704 designation for ammonium nitrate is health 0, flammability 0, and reactivity 3. In the white space at the bottom of the diamond, the prefix "oxy" indicates it is an oxidizer. The primary uses of ammonium nitrate are in fertilizers, explosives, pyrotechnics, herbicides, and insecticides. It is also used as an oxidizer in solid rocket fuel (Photo 7.3).

Ammonium sulfate, $(NH_4)_2SO_4$, an ammonium compound, is a brownish-gray to white crystal. Ammonium sulfate is soluble in water and is nonflammable. This compound is an oxidizer with a specific gravity of 1.77, which is heavier than water. The primary uses are in fertilizers, water treatment, fermentation, fireproofing compositions, and as a food additive.

Incidents

Ammonium nitrate fertilizer, mixed with a hydrocarbon fuel, was used in the explosion that rocked the Federal Building in Oklahoma City in April 1995, killing 166 people, including 19 children, and injuring 450 others. The damage to the building was so extensive that it had to be demolished. In addition, several other buildings in the downtown area were damaged by the explosion. Ammonium nitrate was also used in the bombing of the World Trade Center in New York City in 1994 (Photo 7.4).

Commercial-grade ammonium nitrate was involved in the explosion in Kansas City, Missouri, in 1988 that killed six firefighters (Photo 7.5). The firefighters were responding to a construction site where explosives were in a box trailer used for storage. The storage trailer was on fire; the firefighters may have been unaware that the explosives were stored there and fought the fire. The resulting explosion totally destroyed one fire engine and damaged another beyond repair and, in addition, killed six firefighters. As a result of that

Photo 7.3 Hopper truck hauling ammonium nitrate.

tragic explosion in Kansas City, OSHA issued a new regulation involving the use of DOT placards and labels in fixed storage. All hazardous materials that require DOT placarding and labeling in transportation must continue to be placarded and labeled in fixed storage. The placards and labels must remain on the containers until the materials are used up and the containers have been purged or properly discarded (Photo 7.6).

Ammonium nitrate fertilizer can be made resistant to flame and detonation by an exclusive process involving the addition of 5%–10% ammonium phosphate.

Class 5.2 organic peroxides

The DOT assigns Class 5.2 organic peroxides to seven generic types (Table 7.7), and they are classified by the extent to which they will detonate or deflagrate. Organic peroxides are a hydrocarbon derivative. Organic peroxides are explosive. They are highly dangerous materials used as initiators and catalysts for polymerization reactions. Organic peroxides are highly reactive because of the presence of the oxidizer and the fuel within the formula. They can start their own decomposition process when contaminated, heated, or shocked. Organic peroxides are nonpolar and immiscible in water. Some examples of organic peroxide compounds and compounds subject to peroxide formation are listed in Table 7.8.

The organic peroxide functional group is composed of two oxygens single-bonded to each other. There is a hydrocarbon radical on each side of the single-bonded oxygens. Their GF is $R–O–O–R$. Peroxides are named much the same way as ethers and ketones. There must be two radicals, which are named smallest to largest with the word "peroxide" at the end of the name. If the radicals are the same, the prefix "di" is used to indicate two of the same radicals. For example, if the radicals "methyl" and "ethyl" are attached to the peroxide, the

Photo 7.4 New York City police and firefighters inspect the bomb crater inside the World Trade Center after the explosion that killed six and injured more than a thousand. (From The New York City Fire Department, Firefighter John Strandberg, FDNY Photo Unit. Used with permission.)

compound is named methyl ethyl peroxide ($CH_3O_2C_2H_5$). Shown in Figures 7.6 and 7.7 are the names, molecular formulas, and structures of two common organic peroxides: methyl ethyl ketone (MEK) peroxide and ditert butyl peroxide. Note that benzoyl peroxide does not follow the trivial naming system for peroxides; however, "peroxide" in the name provides the family information that will help in determining the hazard, which is explosive.

Organic peroxides are widely used in the plastics industry as polymerization reaction initiators. All organic peroxides are combustible. Many can be decomposed by heat, shock, or friction; some, such as MEK peroxide, can detonate. Organic peroxides can be liquids or solids, and are usually dissolved in a flammable or combustible solvent. Organic peroxides can be dangerously explosive materials. Organic peroxides have a self-accelerating decomposition temperature (SADT), and they are shipped and stored under refrigeration to keep them cool (Photo 7.7). The SADTs range from 0°F to 50°F and higher. SADT is the temperature at which the compounds will start to decompose. This decomposition reaction may result in a violent detonation that cannot be stopped by anything responders might try to do. The best bet is to make sure that the materials *do not* reach their SADT. Some organic peroxides are so unstable that they are forbidden in transportation. Organic peroxides exhibit the following hazards to emergency responders: they are unstable, flammable, and highly reactive; may explode in a fire; are

Photo 7.5 Pumper 41 was damaged beyond recognition as a piece of fire apparatus by the explosion. (From Kansas City, Missouri, Fire Department. Used with permission.)

Photo 7.6 Memorial located near the site of the Kansas City, Missouri, ammonium nitrate explosion that killed six Kansas City firefighters.

Table 7.7 Organic Peroxides and Peroxidizable Compounds

Methyl ethyl ketone peroxide	Dioxane
Benzoyl peroxide	Furan
Ether peroxides	Butadiene
Peracetic acid	Vinyl chloride
Potassium metal	Styrene
Vinylidene	Cyclohexane
Methyl acetylene	Tetrahydrofuran
Cyclopentane	Cumene

Table 7.8 Generic Types of Organic Peroxides

Type A	Organic peroxide that can detonate or deflagrate rapidly as packaged for transport. Transportation of type A organic peroxides is forbidden.
Type B	Organic peroxide as packaged for transport, neither detonates nor deflagrates rapidly, but can undergo a thermal explosion.
Type C	Organic peroxide as packaged for transport, neither detonates nor deflagrates rapidly and cannot undergo a thermal explosion.
Type D	Organic peroxide which (I) Detonates only partially, but does not deflagrate rapidly and is not affected by heat when confined; (II) Does not detonate, deflagrates slowly, and shows no violent effect if heated when confined; or (III) Does not detonate or deflagrate, and shows a medium effect when heated under confinement.
Type E	Organic peroxide which neither detonates nor deflagrates and shows low, or no, effect when heated under confinement.
Type F	Organic peroxide which will not detonate in a cavitated state, does not deflagrate, shows only a low, or no, effect if heated when confined, and has low, or no, explosive power.
Type G	Organic peroxide that will not detonate in a cavitated state, will not deflagrate, shows no effect when heated under confinement, has no explosive power, and is thermally stable.

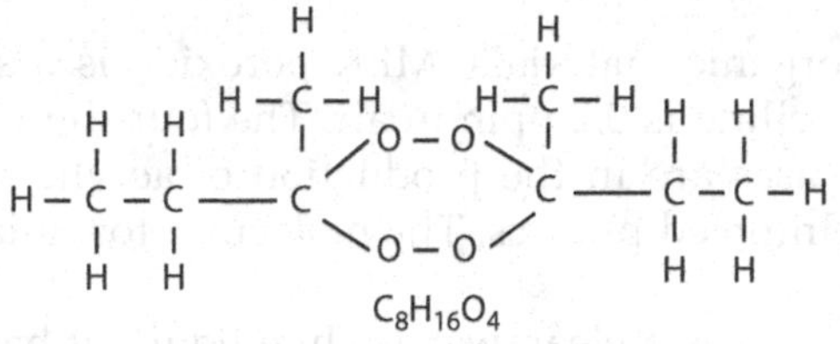

$C_8H_{16}O_4$

Figure 7.6 Methyl ethyl ketone peroxide.

corrosive; may be toxic; and are oxidizers. They all have SADTs, and once this reaction starts, there is little responders can do to stop it. Responders should withdraw and treat the material as Class 1 explosives.

MEK peroxide is an organic peroxide, even though there is also a ketone functional group in the compound. It is a colorless liquid and a strong oxidizing agent. It is a fire

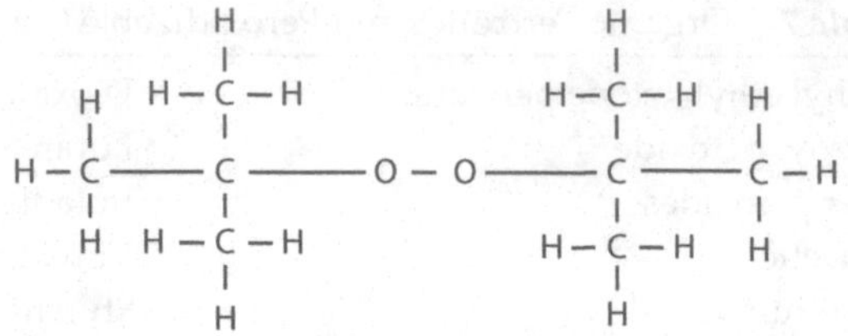

$CH_3OtC_4H_9$

Figure 7.7 Ditert butyl peroxide.

Photo 7.7 Organic peroxides have SADT and may decompose explosively upon exposure to heat.

risk in contact with organic materials. MEK peroxide is a strong irritant to the skin and tissues. The TLV ceiling is 0.2 ppm in air. The four-digit UN identification number is 2550. The primary uses are in the production of acrylic resins and as a hardening agent for fiberglass-reinforced plastics. The molecular formula and structure are shown in Figure 7.6.

Ditertiary butyl peroxide is a clear, water-white liquid. It has a specific gravity of 0.79, which is lighter than water, and it will float on the surface. It is nonpolar and insoluble in water. Ditertiary butyl peroxide is a strong oxidizer and may ignite organic materials or explode if shocked or in contact with reducing agents. In addition to being an oxidizer, ditertiary butyl peroxide is highly flammable. It has a boiling point of 231°F (110°C) and a flash point of 65°F (18°C). The NFPA 704 designation is health 3, flammability 2, and reactivity 4. The prefix "oxy" for oxidizer is placed in the white section at the bottom of the 704 diamond. The molecular formula and structure are shown in Figure 7.7.

Photo 7.8 MC 312 highway tanker hauling hydrogen peroxide, a strong oxidizer.

Hydrogen peroxide is a colorless liquid organic peroxide that is a powerful oxidizer. It is also a dangerous fire and explosion risk and is toxic in high concentrations, with a TLV of 1 ppm in air (Photo 7.8). It is soluble in water, with a boiling point of 25°F (25°C). Hydrogen peroxide is also corrosive and a strong irritant. It is shipped and stored in 40%–60% and in greater than 60% concentrations. Common commercial strengths are 27.5%, 35%, 50%, and 70%. Hydrogen peroxide is incompatible with most metals and their salts, alcohols, organic substances, and any flammable substances. The four-digit UN identification number for the 40%–60% concentrations is 2014. The NFPA 704 designation is health 2, flammability 0, and reactivity 1. The white section at the bottom of the 704 diamond has an "oxy" for oxidizer. The four-digit UN identification number for concentrations greater than 60% is 2015. The NFPA 704 designation is health 2, flammability 0, and reactivity 3. The white section at the bottom of the 704 diamond has an "oxy" for oxidizer.

In its pure form, *benzoyl peroxide* is a white, granular, crystalline solid that ignites easily, burns with great intensity, and may explode. It has a faint odor of benzaldehyde and is tasteless. Benzoyl peroxide is slightly water soluble. It may explode spontaneously when dry with less than 1% water. It decomposes explosively above 221°F (105°C). Its autoignition temperature is 176°F (80°C). It should not be mixed unless at least 33% water is present. Benzoyl peroxide is highly toxic by inhalation, with a TLV of 5 mg/m³ of air. The burning characteristics are similar to black powder. Benzoyl peroxide decomposes rapidly when heated, and if the material is confined, detonation will occur. The molecular formula and structure for benzoyl peroxide are shown in Figure 7.8.

Most ethers, when stored for more than 6 months, will form explosive *ether peroxides* in their containers (Photo 7.9). Peroxide formation can also occur in potassium metal and

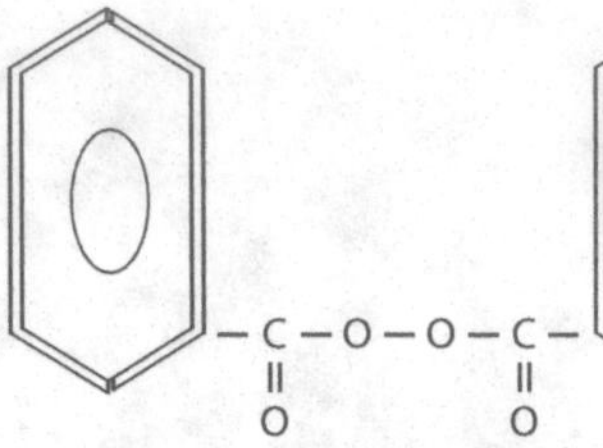

$$(C_6H_5CO)_2O_2$$

Figure 7.8 Benzoyl peroxide.

Photo 7.9 Ethers have a hidden hazard. They can form explosive peroxides as they age. Moving or shaking the container can cause an explosion.

aldehydes. These peroxides are organic peroxides. The primary ethers to be concerned about are ethyl ether, ethyl tertiary butyl ether, ethyl tertiary amyl ether, and isopropyl ether. Isopropyl ether is considered the worst hazard in storage.

Ether peroxides that form inside containers are organic peroxides and are sensitive to shock and heat. When these peroxides are concentrated or heated, they may detonate. Peroxide formation can be detected as early as 1 month in storage. Ethers are usually stored in amber glass bottles. Light and heat are important contributors to peroxide formation, although light seems to have more effect than heat. There is no effective means of inhibiting peroxide formation in ether containers. If aging containers of ethers are encountered, they should be treated like bombs. The bomb squad should be called for handling and disposal. There have been cases where employees and response personnel

Table 7.9 Hazards of Oxidizers

Intensification of combustion
Spontaneous combustion
Water reactive mechanism
Explosion
Toxic smoke and fumes

have been severely injured when an aging can of ether has exploded in their hands. Ethers may be found in high schools and college laboratories and throughout industry. Ether is also used in the processing of illegal drugs and may be encountered in clandestine drug lab operations.

The dangers presented by all oxidizers are similar. They are dangerous in contact with organic materials. They accelerate combustion. They can be a serious fire and explosion hazard. Even though some are water reactive, water is the extinguishing agent of choice for fires involving oxidizers. Extinguishing agents that work by excluding atmospheric oxygen will not always work with oxidizers. Oxidizers have their own oxygen supply within the compound and do not need atmospheric oxygen to support combustion. Emergency responders should treat oxidizers with the same respect as incidents involving explosives, because they may be just as dangerous (Table 7.9).

Incident

Two workers were killed and 13 people injured, including three firefighters, in an explosion and fire at a chemical plant. The facility manufactured organic peroxides, including MEK peroxide, which was involved in the fire. Also involved in the fire were bunkers for storing the MEK peroxide and benzoyl peroxide. Arriving companies were ordered to stage until the explosions subsided and an aggressive attack could be mounted safely.

Review questions

7.1 Name four elements from the periodic table that are oxidizers.
7.2 Which families of compounds are considered oxidizers?
7.3 Oxidizers may be shipped under which of the following placards?
   A. Dangerous
   B. Oxidizer
   C. Poison or poison gas
   D. All of the above
7.4 Name the following oxysalt compounds and balance the formulas if needed. (Any compounds with transition metals are balanced.)

$AlSO_5$ $LiClO_2$ $NaFO_3$ $MgPO_2$ $CuClO_3$

7.5 Provide formulas and structures for the following organic peroxide hydrocarbon-derivative compounds.

Methyl ethyl peroxide   Di-vinyl peroxide   Isopropyl butyl peroxide

7.6 When in contact with water, some peroxide salts may produce which of the following hazards?
   A. Release heat
   B. Release oxygen
   C. Impregnate other materials
   D. All of the above

7.7 Some organic peroxides may undergo polymerization if inhibitors are released during an accident. Which of the following best describes polymerization?
   A. Violent reaction
   B. Releases heat
   C. Self-reaction
   D. All of the above

7.8 Oxidizers, when present during a fire, will do which of the following?
   A. Help extinguish the fire
   B. Accelerate combustion
   C. Purify runoff water
   D. None of the above

chapter eight

Poisons

According to Paracelsus (1494–1541), "Too much of anything is toxic. All substances are poisons; there is none which is not a poison. The right dose differentiates a poison from a remedy." Why are some chemicals more harmful than others? Structure is the most important factor in toxicity. All matter is made up of protons, electrons, and neutrons. It is how they are arranged that determines if they are food, fuel, or a deadly poison. Hazard Class 6 materials are poisons that are solids and liquids. Some liquids are volatile and produce vapors, which are an inhalation hazard. Volatile poisons that are an inhalation hazard require the transport vehicle to be placarded regardless of the quantity.

Class 6 is divided into two subclasses: 6.1 and 6.2. The DOT defines a Class 6.1 poison as "a material, other than a gas, known to be so toxic to humans as to afford a health hazard during transportation or which, in the absence of adequate data, is presumed to be toxic to humans because it falls within any one of the categories shown in Figure 8.1, when tested on laboratory animals". DOT definitions of toxicity include the following:

Oral toxicity is a liquid with a lethal dose (LD) of 50 or not more than 500 mg/kg, or a solid with an LD_{50} of not more than 200 mg/kg of the body weight of the animal. (LD_{50} is the single dose that will cause the death of 50% of a group of test animals exposed to it by any route other than inhalation.)

Dermal toxicity is a material with an LD_{50} for acute dermal toxicity of not more than 1000 mg/kg of the body weight of the research animal.

Inhalation toxicity is a dust or mist with a lethal concentration (LC) of 50 for acute toxicity upon inhalation of not more than 10 mg/kg of body weight of the laboratory animal. (LC_{50} is the concentration of a material in air that, on the basis of laboratory tests through inhalation, is expected to kill 50% of a group of test animals when administered in a specific period.)

Simply stated, the DOT definition really says that a small amount of a poison is dangerous to life and that the material should be considered dangerous. A medical definition for a poison is "the ability of a small amount of a material to produce injury by a chemical action." A chemical definition is "the ability of a chemical to produce injury when it comes in contact with a susceptible tissue." Perhaps a better definition of a poison would be that "a poison is a chemical that, in relatively small amounts, has the ability to produce injury by chemical action when it comes in contact with a susceptible tissue." Corrosives are not usually thought of as poisonous materials. However, if the combined chemical definition is applied to corrosives, then, in fact, they are poisonous to the tissues they contact. Allyl alcohol (DOT/UN identification number 1098) is toxic by absorption, inhalation, and ingestion. However, like many other hazardous materials, it also has multiple hazards. Allyl alcohol is placarded as a 6.1 poison primary hazard, but it is also a flammable liquid.

It is placarded 6.1 Poison because in the DOT's hazard hierarchy, poison is more dangerous than flammability. Materials in subclass 6.2 are infectious substances, meaning "they are viable microorganisms or their toxins, which may cause diseases in humans or animals" (Photo 8.1). This section also includes regulated medical waste.

- Chlorine
- Phosgene
- Arsenic
- Cyanide
- Pesticides

Figure 8.1 Examples of poisons.

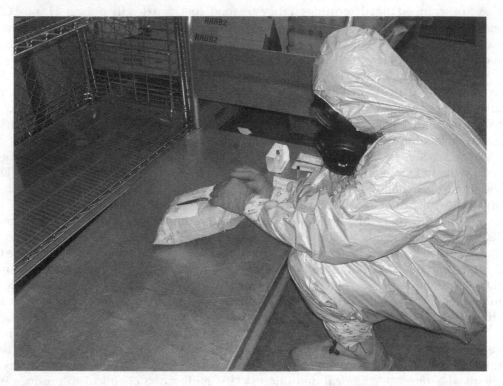

Photo 8.1 Emergency personnel have responded to thousands of "white powder" incidents involving potential biological terrorist agents. (Courtesy Durham NC Police Department, Durham, NC.)

Poisons are among the most dangerous materials for emergency responders (Figure 8.2). In particular, those that are inhalation hazards present the most serious danger to responders. Effects of poisons may not present themselves right away; in fact, the toxic effect may not appear for days, months, or years. Because the effects may not present themselves right away, responders may be led to believe that there is no danger. One of the main reasons decontamination is done for hazardous materials incidents is to prevent the spread of toxic materials (poisons) away from the "hot zone."

Toxicology is the science of the study of poisons and their effects on the human body. It is also the study of detection in the body systems and of antidotes to counteract poisonous effects. The science of toxicology is relatively new. It evolved out of the concern for worker health and safety that became a concern in the early twentieth century.

- Asphyxiants
- Corrosives
- Sensitizers
- Carcinogens
- Mutagens
- Teratogens
- Irritants

Figure 8.2 Types of poisons.

Trade unions raised some of the first concerns for worker health and safety in the workplace. Eventually, the Occupational Health and Safety Act was passed by Congress in the 1970s, creating the Occupational Safety and Health Administration (OSHA). These events led to the eventual formation of the science of industrial hygiene, which involved the protection of workers in the workplace. Industrial hygiene is the application of industrial hygiene concepts to the work environment. Many of the toxicological measuring terms found in reference sources come from the recommendations of safe workplace exposures (Table 8.1). When responding to poison incidents, it is important to remember that, first of all, the emergency scene is the workplace for emergency responders. Second, the concentrations that emergency responders will encounter at a spill will often be much higher than the "normal" or acceptable workplace measurements. Some of the toxic fire gases that will be encountered at fire scenes are shown in Table 8.2.

Table 8.1 OSHA PELs

Chemical	8 h PEL	Odor threshold
Ozone	0.1 ppm	0.1 ppm
Chlorine	0.5 ppm	0.3 ppm
Benzene	1 ppm	5–12 ppm
Hydrogen sulfide	10 ppm	Fatigues nose
Carbon monoxide	35 ppm	Odorless
Trichloroethylene	50 ppm	20 ppm
Toluene	100 ppm	2.0 ppm
Gasoline (TLV)	300 ppm	10 ppm
Freon 113	1000 ppm	350 ppm
Methane	Simple asphyxiant	Odorless

Table 8.2 Most Common Fire Gases

Carbon dioxide
Carbon monoxide
Sulfur dioxide
Nitric oxide
Nitrogen dioxide
Ammonia
Hydrogen sulfide
Hydrogen cyanide

Types of exposure

Types of exposure include the following:

Acute: A one-time, short-duration exposure. Depending on the concentration and duration, there may or may not be toxic effects. A one-time exposure can cause illness or death; however, it cannot be cumulative. In order for cumulative effects to occur, there must be multiple exposures. If multiple exposures occur, it is considered chronic rather than acute.

Subacute: Involves multiple exposures with a period of time between exposures. The effect is actually less than an acute exposure. The theory is that as long as there are periods of time between exposures, there will be no ill effects. There are no cumulative effects of subacute exposure because of the time between exposures. This concept is similar to the time factor when dealing with radioactive materials. Personnel can be exposed to certain levels of radioactivity for short periods of time without any ill effects.

Chronic: As with subacute, chronic exposures are multiple exposures; however, with chronic exposure, there can be cumulative effects. Cumulative effects are simply a buildup of poison in the body. After the first exposure, some or all of the toxic material stays in the body. The first exposure may not cause any illness or damage. As additional exposures occur and the poison builds up in the body, it can reach toxic levels where illness, damage, or death can occur.

All of these exposures are usually considered workplace events. The emergency responders' workplace is the incident scene. An emergency responder can have an acute exposure to a toxic material at the scene of a hazmat incident or other type of emergency response. A single exposure may not produce symptoms or illness, but multiple, or chronic, exposures may cause damage. It is important to monitor exposures of personnel to determine whether they experience any illnesses following a response to an incident. Sometimes, it takes several days for a toxic material to reach the susceptible target organ. The tendency is to assume that when a person is exposed to a poison at an incident scene, the ill effects, if there are going to be any, will occur right away. This is not always the case with many poisons. When an illness occurs following a hazardous materials incident, the personnel should be checked out just in case.

Once a poison has entered the body, it can behave in a number of ways. First, the effect on the body may be localized, i.e., it affects only the tissue that it has directly contacted. Second, the effect may be systemic or a whole-body effect. In this instance, the effect on the contact tissue is little, if any. The poison enters the bloodstream and travels throughout the body until a target organ is reached, or the material is secreted through the body's waste removal system. Last, the effects may be a combination of localized and systemic.

Routes of exposure

In order for a person to be affected by a poison, the poison must directly contact the body or enter into the body. There are four routes by which toxic materials can enter the body and cause damage (Figure 8.3): inhalation, absorption, ingestion, and injection.

Inhalation requires the poison gas or vapor to enter the body through the respiratory system, where most of the damage usually occurs (Photo 8.2). Once in the lungs, the poison can injure the respiratory tissues, enter the bloodstream, or both. Examples of toxicants that produce disease of the respiratory tract are shown in Table 8.3.

Absorption (skin and eye contact) occurs when solid, liquid, or gaseous poisons enter the body through the skin, eyes, or other tissues (Table 8.4). Damage may occur at the point

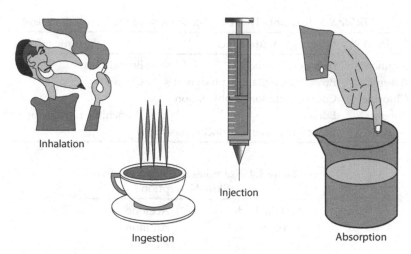

Figure 8.3 Routes of exposure.

Photo 8.2 Respiratory protection is the most important protection measure when dealing with military agents and other poisons. This can involve Positive Pressure Air Purifying Respirators (PAPR), cartridge respirators, and Self-Contained Breathing Apparatus (SCBA).

of contact, or the material may travel to the susceptible target organ and cause harm there. Effects may be local, such as irritation and death of body tissues through direct contact. For example, when naphthalene contacts the eyes, it can cause cataracts and retina damage. Phenothiazine (insecticide) damages the retina. Thallium causes cataracts and optic nerve damage. Methanol causes optic nerve damage and blindness.

Table 8.3 Toxicants That Produce Disease of the Respiratory Tract

Toxicant	Acute effect	Chronic effect
Ammonia	Irritation, edema	Bronchitis
Arsenic	Bronchitis, irritation, pharyngitis	Cancer
Chlorine	Cough, irritation, asphyxiation	
Phosgene	Edema	Bronchitis, fibrosis, pneumonia
Toluene	Bronchitis, edema, bronchospasm	

Table 8.4 Examples of Chemicals Toxic by Skin Absorption

Acetaldehyde	Acetone
Acrolein	Ammonia
Aniline	Arsenic
Benzene	Barium
Camphor	Carbon disulfide
Carbon tetrachloride	Chlordane
Butyric acid	Chlorine
Cumene	Bromine

Ingestion occurs when a solid, liquid, or gaseous poison enters the body through the mouth and is swallowed. Damage may occur to the tissues contacted, or the poison may enter the bloodstream. Absorption can also occur after ingestion. A poisonous material may be absorbed through the tissues in the mouth, stomach, intestines, or other tissues it contacts after ingestion occurs.

Injection involves a jagged or sharp object that has been contaminated with a toxic material creating or entering an open wound in the skin. The poison enters the bloodstream once injected into the skin. The exposure method may greatly impact the severity of the damage produced. A chemical that is extremely poisonous by one route of exposure may have little, if any, effect by other routes. For example, carbon monoxide is toxic by inhalation. Just a 1% concentration in air, if inhaled, is fatal in 1 min. However, you could stay in a 100% concentration indefinitely provided there was an outside air supply, such as self-contained breathing apparatus (SCBA). Carbon monoxide is not absorbed through the skin. Rattlesnake venom is poisonous if it gets in the bloodstream; it damages the cells. It must be injected to cause damage. If ingested, it may cause nausea, but it will not enter the bloodstream.

Effects of exposure

Toxicology relates to the physiological effect, source, symptoms, and corrective measures for toxic materials. Poisons can be divided into several general categories: asphyxiants, corrosives, sensitizers, carcinogens, mutagens, teratogens, and irritants.

Short-term effects

When a poison contacts a tissue and a chemical action occurs, the poison will produce an injury to that tissue. The effects of an exposure to a poison, however, will differ based on the poison involved, the type of exposure, and the method of exposure. There are three types of effects that can occur as a result of an exposure to any given poison: immediate, long-term, and etiologic. Immediate effects depend on the dosage received by the person exposed to the poison. Dosages can be large or small, short-term or long-term, and can be acute, subacute, or chronic. Effects from the type of exposure can vary from none, to slight discomfort, to illness, and even death. Effects may also depend on the person exposed. Not all people are affected in the same way by the same poison, the same dosage, and the same exposure. Differences are based upon individual body chemistry, age, health, sex, and size. Immediate effects can include asphyxiation, corrosive damage, sensitizing, and irritation.

Asphyxiants act upon the body by displacing oxygen. Asphyxiants can be simple or chemical. Gases, such as nitrogen, hydrogen, helium, methane, and others, are known as simple asphyxiants, because they dilute the oxygen in the air below the level required for life to exist. Victims die because there is not enough oxygen in the air they breathe. In the case of Class 6 poisons, however, asphyxiants act by interfering with the blood's ability to convert or carry oxygen in the bloodstream. These are known as chemical asphyxiants. Death may also result from a lack of oxygen, but it is not a result of a lack of oxygen in the air. Examples of chemical asphyxiants include hydrogen cyanide, benzene, toluene, aniline, and others.

Corrosives are a skin contact hazard. They are acids or bases that, in small amounts, can cause damage to tissue. Tissue is damaged in much the same way as a thermal burn; however, the burn is much more damaging. The type of damage is the same whether exposed to acids or bases. Examples of corrosive materials include nitric acid, sulfuric acid, phosphoric acid, sodium hydroxide, and potassium hydroxide. They are covered in detail in Chapter 10.

Sensitizers, on first exposure, cause little or no harm in humans or test animals, but on repeated exposure, they may cause a marked response not necessarily limited to the contact site. This response is similar to the process that occurs in allergies that humans develop. It is a physiological reaction to a sensitizing material. For example, a person who moves into an area that has high pollen counts and other airborne allergens may not experience any effects at first, but the longer the exposure occurs, the more symptoms that develop. Examples of sensitizers are isocyanates and epoxy resins.

Irritants are materials that cause irritation to the respiratory system, body organs, or the surface of the skin. The irritation may be a corrosive action, localized irritation, inflammation, pulmonary edema, or a combination. Effects may include minor discoloration of the skin, rashes, or tissue damage. Examples of irritants are tear gas and some disinfectants.

Several factors influence toxic effects, one of which is the concentration of a material (Table 8.5). Concentrations may be expressed in terms of percentage or parts per million (ppm) or billion per milligram or kilogram; the higher the concentration, the more serious the effect. (Figure 8.4) illustrates some parts-per-million approximations. There are also materials that can have an anesthetic effect. This may range from loss of feeling and sensation to unconsciousness. Anesthetics include nitrous oxide, ethers, and hydrocarbons.

Table 8.5 Expressions of Concentration

Percentages
Parts per million (ppm) or billion (ppb)
Milligrams per cubic meter, foot, kilogram, and cubic liter

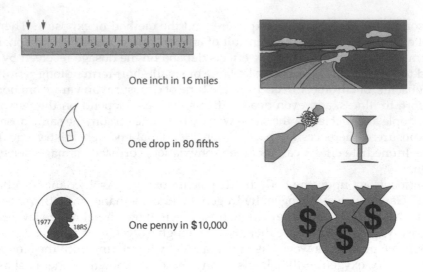

One inch in 16 miles

One drop in 80 fifths

One penny in $10,000

Figure 8.4 Equivalents of parts per million.

Long-term effects

Long-term effects can be divided into three types: carcinogenic, teratogenic, and mutagenic. All three types of effects take, from the time of exposure, up to 15 years or longer to develop symptoms. Effects can be aggravated by exposures to other materials or unrelated health problems.

Carcinogens are materials that cause cancer. This is a chronic toxic effect. Information about cancer-causing agents has been obtained by studying populations exposed to chemicals for long periods, usually in the workplace. Data have also been obtained through tests using laboratory animals. This does not mean that a material will actually cause cancer in humans, but it is a good indication. Examples of known carcinogens include benzene, asbestos, arsenic, arsenic compounds, vinyl chloride, and mustard gas.

Mutagens cause mutations or alterations in genetic material, thereby altering human DNA. Changes may be chromosomal breaks, rearrangement of chromosome pieces, gain or loss of entire chromosomes, or damage within a gene. Effects may involve the current generation that was exposed or future generations. Examples of mutagens include arsenic, chromium, dioxin, mercury, ionizing radiation from x-rays or other radioactive material, caffeine, LSD, marijuana, and nitrous oxide. Additionally, ethylene oxide, ethyleneimine, hydrogen peroxide, benzene, and hydrazine are also mutagens.

Teratogens cause one-time birth defects in offspring resulting from maternal or paternal exposure to toxic materials. The word "teratology" is derived from the Latin meaning "the study of monsters." It is actually the study of congenital malformations, which started with the study of the correlation of German measles to birth defects. Later, there was an industrial link to teratogens discovered involving the chemical methyl mercury. Results of exposure to teratogenic chemicals on living organisms include the alteration of developing cells, leading to improper functioning of these cells. This may result in the death of the embryo or fetus. Specific types of birth defects can be caused by specific types of chemicals. These chemicals do not permanently damage the reproductive system, and normal children can be produced as long as repeated exposure does not occur.

Examples of teratogens include thalidomide, ethyl alcohol, and *o*-benzoic sulfimide (the artificial sweetener saccharin). Thalidomide is a drug that was used in the 1960s as a treatment for morning sickness in pregnant mothers. However, the toxic effects were discovered too late for 10,000 babies who were born with various malformations. In Japan, where fish is a staple food, mothers ate fish contaminated with mercury compounds, causing children to be born with cerebral palsy.

Ethyl alcohol is found in alcoholic beverages and is a known teratogenic material, which is the reason doctors tell pregnant women not to drink alcohol. When a mother drinks, the unborn child drinks as well. Ethyl alcohol causes growth failure and impaired brain development. Unborn children exposed to alcohol may suffer the effects of fetal alcohol syndrome when they are born. Symptoms of fetal alcohol syndrome include sleep disturbance, jitteriness, a higher incidence of impaired vision and hearing, lack of motor coordination, balance problems, abnormal thyroid function, and a decrease in immune system effectiveness. Additional teratogens include heavy metals, methyl mercury, mercury salts, lead, thallium, selenium, penicillin, tetracyclines, excess vitamin A, and carbon dioxide.

Etiologic effects

Class 6.2 infectious substances are the last type of toxic effect that will be presented here. Etiological toxins are among the most poisonous materials known. For example, the bacterium *Clostridium botulinum*, the cause of botulism, is a single cell that can release a toxin so potent that four-hundred thousandths of an ounce is enough to kill 1,000,000 laboratory guinea pigs.

Variables of toxic effects

Not all people are affected by toxic materials in the same way. Variables include age, genetic makeup, sex (in terms of the genetic difference between males and females, most of which are related to reproduction and nurturing functions), weight, general health, body chemistry (the chemicals produced in each individual are unique, not only by form, but by quantity as well), and physical condition. These variables can all affect the way individuals respond to any toxic material. Infants and children are often more sensitive to some toxic materials than younger adults. Elderly persons have diminished physiological capabilities to deal with toxic materials. This age group may be more susceptible to toxic effects at relatively lower doses. Chemicals may be more toxic to one sex than the other. Males may be affected by chemicals that do not affect females; some chemicals can affect both. Chemicals may affect the reproductive system of either the male or female.

Females who are pregnant may be affected by a toxic material that will cause damage to the fetus, whereas a male may not be affected at all. Damage to the male reproductive system may include sterility, infertility, abnormal sperm, low sperm count, or reduced hormonal activity. Chemicals that cause these types of effects include lead; mercury; PCBs; 2,4,D; paraquat; ethanol; vinyl chloride; and DDT. Females are affected by DDT, parathion, PCBs, cadmium, methyl mercury, and anesthetic gases.

Dose/response

Dose can be expressed in three ways: the amount of the substance actually in the body, the amount of material entering the body, and the concentration in the environment. Response is the health damage resulting from a specified dose. All chemicals, no matter what their makeup, are toxic if taken in a large enough dose. Even water can be toxic if too much is ingested at one time. Paracelsus (1493–1541) observed that "all substances are poisons; there is none which is not a poison. The right dose differentiates a poison and a remedy." There is a no-adverse-effects level for almost all materials. Dose is more than just the amount of a toxic material that has caused the exposure. More correctly, dose is related to the weight of the individual exposed. This is the basis for many of the toxicological terms that have been developed based upon tests on laboratory animals. For example, a 1 lb rat given 1 oz of a toxic material should have the same response as a 2000 lb elephant given 2000 oz of the same toxic material. In both cases, the amount per weight is the same: 1 oz for each pound of animal weight. This amount per weight is known as the dose. Response of an individual plant or animal species to a chemical is based on concentration; length, type, and route of exposure; and susceptible target organ. Additionally, other health-related variables include age, sex, physical condition, and size (mass). This relationship is referred to as dose/response or, in other words, the amount of the exposure and the resulting biological effect. Each plant or animal species has its own individual response to a given chemical. A substance administered at a dose large enough to be lethal to rabbits may have a lesser effect on rats or dogs.

Susceptible target organs

When a poison enters the body through one of the four routes of exposure, it will cause damage to some bodily function. This is referred to as the susceptible target organ (Figure 8.5). Depending on the dose, it may take a toxic material several hours or several

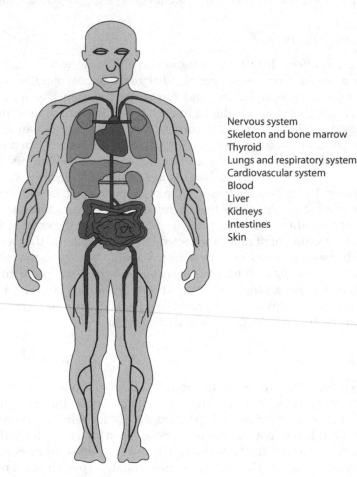

Nervous system
Skeleton and bone marrow
Thyroid
Lungs and respiratory system
Cardiovascular system
Blood
Liver
Kidneys
Intestines
Skin

Figure 8.5 Susceptible target organs.

Table 8.6 Target Organs and Chemicals That Affect Them

Lungs	Halogens, hydrogen sulfide
Liver	Vinyl chloride, aromatics, chlorinated HC
Kidneys	Mercury, calcium, carbon tetrachloride
Blood	Carbon monoxide, chlorinated HC
Neurologic	Organophosphates, carbon monoxide
Skeletal	Fluorides, selenium
Skin	Arsenic, chromium, beryllium

days to reach a susceptible target organ and cause damage. The target organ should not be confused with routes of exposure, which are the manner in which the poison enters the body. The susceptible target organ is the organ or system to which the poison does its damage once it enters the body. For example, a pesticide may enter the body through inhalation, but may have an effect on the central nervous system. In this instance, the route of exposure is inhalation through the lungs; the target organ is the central nervous system. Target organs in the body include the respiratory system, liver, kidneys, central nervous system, blood, bone marrow, skin, cardiovascular system, and other body tissues. The skin is the largest single organ of the body. When unbroken, it provides a barrier between the environment and other organs, except for the lungs and eyes, and is a defense against many chemicals. Studies show that 97% of chemicals to which the body is exposed are deposited on the skin. Chemicals that affect specific target organs are shown in Table 8.6.

Exposure rate

Exposure rate is the measurement of workplace exposure of hazardous materials based upon tests conducted on laboratory animals. Values are translated to humans based on the weight ratio between an animal and a human. These are only estimations and should be used with caution. Toxicology information is expressed in ppm or billion (ppb) and milligrams per cubic meter (mg/m^3), which are terms indicating the concentration of the toxic material (Table 8.7).

These terms are not related, but they do have one thing in common: the smaller the numerical value, the more toxic the material measured. The emergency responder's workplace is the incident scene, and the concentrations of toxic materials here are usually much higher than the values indicated through the following terms:

Table 8.7 Toxicity Comparisons of Some Hazardous Materials

Common Name	TLV ppm	LC_{50}
Hydrogen cyanide	10	300 ppm
Hydrogen sulfide	10	600 ppm
Sulfur dioxide	5	1,000 ppm
Chlorine	1	1,000 ppm
Carbon monoxide	50	1,000 ppm
Ammonia	50	10,000 ppm
Carbon dioxide	5,000	10%
Methane	90,000	Simple asphyxiant

TLV-TWA: Threshold limit value, time-weighted average is the average concentration for a normal 8 h workday and a 40 h workweek, to which nearly all workers can be exposed repeatedly, day after day, without adverse effect and without protective equipment.

TLV-STEL: Threshold limit value, short-term exposure limit is the maximum concentration averaged over a 15 min period to which healthy adults can be exposed safely. Exposures should not occur more than four times a day, and there should be at least 60 min between exposures.

TLV-C or TLV-ceiling: The concentration that should not be exceeded during any part of the workday. *This is the only reliable measurement that should be used by emergency responders on incident scenes.*

PEL: The permissible exposure limit is the maximum concentration averaged over 8 h, to which 95% of healthy adults can be repeatedly exposed for 8 h/day, 40 h/week.

IDLH: Immediately dangerous to life and health, the maximum amount of a toxic material that a healthy adult can be exposed to for up to 15 min (National Institute for Occupational Safety and Health [NIOSH]) and escape without irreversible health effects. There are also IDLH values established for oxygen-deficient atmospheres and explosive or near-explosive atmospheres (above, at, or near the LELs). At the scene of a hazardous materials incident, it should be assumed that the concentrations present are above the TLV-ceiling, and only responders wearing SCBA and proper chemical protective clothing should be allowed near the incident scene.

LD_{50} is the lethal dose by ingestion or absorption for 50% of the laboratory animals exposed. (There is a wide variation among species. The LD_{50} for one type of animal could be thousands of times less than for another type.) Examples of LD_{50} values for common chemicals are shown in Table 8.8.

LC_{50} is the lethal concentration by inhalation for 50% of the laboratory animals exposed.

NOAEL is no-observable-adverse-effect level.

GRAS is generally recognized as safe.

Table 8.8 LD_{50} Values for Some Common Chemicals

Sucrose (table sugar)	29,000 mg/kg
Ethyl alcohol	14,000
Sodium chloride	3,000
Vitamin A	2,000
Vanillin	1,580
Aspirin	1,000
Chloroform	800
Copper sulfate	300
Caffeine	192
Phenobarbital, sodium salt	162
DDT	113
Sodium nitrate	85
Nicotine	53
Sodium cyanide	6.4
Strychnine	2.5

Toxicity measures are based on dose and size or weight of the test animal. This is then projected to humans, based on the weight ratio of the dose, the weight of the animal, and the weight and health of the human. All of the values should be considered nothing more than an educated guess. Many factors can influence the way any given individual will react if exposed to a toxic material. Some factors may be allergies and previous illnesses or operations. Individuals that have had a splenectomy are much more susceptible to poisons than those who have not. Individuals with such a history should be aware that they are at additional risk whenever they are exposed to toxic chemicals. Consider all toxicity data as an estimate; do not stake your life on it.

Defense mechanisms for toxic materials

There are three types of defenses against toxic materials: internal, antidotal, and external. Three things can happen once a chemical is taken into the body: metabolism, storage, and excretion. Internal defenses are the ability of the body to get rid of a toxic material, sometimes referred to as metabolism. The body normally excretes waste materials through the feces or urine. Additionally, women can also excrete through the ova and breast milk. In these instances, the excretions from the mother represent exposure to the offspring.

Two main organs in the body filter materials and remove them through excretion: the liver and the kidneys. All of the blood in the body passes through these two organs and is filtered. Filtered material is passed from the liver to the intestine through the gallbladder or to the kidney through the blood, after some degree of chemical breakdown (metabolizing). If the liver is unable to break down a poison, it may store the toxic material within its own tissue. Toxic materials, such as lead, can be stored in the bone tissue. Some may be bound by blood proteins and stored in the blood. It is this storage of the toxic material that causes most of the long-term damage. Kidneys also filter poisons from the blood and may incur damage in the process. Materials filtered are then passed on to the urine, which is held in the bladder. This can lead to bladder cancer through chronic exposure to the toxic materials.

Toxic materials that enter the body cannot always be excreted and may be stored in the fatty tissues. Examples include the organochlorines, DDT, PCBs, and chlordane. When a person loses weight, for whatever reason, toxic materials that have been stored in the fatty tissues are released back into the body as the levels of fat decrease. When this occurs, the body is reexposed to the chemicals and illness can occur. Chemicals that cannot be excreted from the body may also be bound by blood proteins and stored in the blood. Lead is an example of a material that cannot be excreted from the body and is stored in the bones.

Polarity has an effect on the body's ability to excrete toxic materials. If a poison is polar, it is usually soluble in water, because water is also polar and is more easily removed from the body. The body also has a system of converting nonpolar compounds so that they can also be removed; however, the body cannot convert all nonpolar compounds.

Therefore, some poisons are difficult to convert and may stay in the body for long periods of time. For example, table salt (sodium chloride) is polar and is easily excreted. DDT, a now-banned pesticide, is nonpolar and is not easily excreted, so it stays in the body for long periods. This is one of the main reasons DDT is no longer allowed to be used on food crops. Characteristics of polarity or nonpolarity can have a crucial impact on the effect a toxic material has on the body.

Two other mechanisms by which the body defends against toxic materials is through breathing and sweating. The lungs are also able to remove materials from the blood.

Table 8.9 Antidotes for Poisons

Cyanide	Amyl nitrate, sodium nitrate, sodium thiosulfate
Organophosphate pesticides	Atropine, pralidoximine
Methanol or ethylene glycol	IV ethanol, hemodialysis
Nitrites	Oxygen, methylene blue
Hydrocarbons	Oxygen

This can be noted by the odor of alcohol on the breath of a person who has been drinking. Odor detected from a person who is sweating is toxic material being removed from the body.

While all three methods contribute to the removal of toxic materials from the body, the last two are of only minor significance. Internal defenses of the body against toxic materials do a good job against many types of chemicals. However, not all toxic materials can be removed by the body's systems. It is best not to rely on the body to remove toxic materials, but rather to take precautions to ensure that the toxic materials do not enter the body in the first place.

Antidotes are administered to counteract the effects of some toxic materials (Table 8.9). The definition of an antidote is "any substance that nulls the effects of a poison on the spot and prevents its absorption or blocks its destructive action once absorbed." The problem with antidotes is that there are only a few in existence and they do not work on all types of poisons (see Table 8.9). Another problem with antidotes is availability—an antidote must be given immediately after exposure to a poison or the victim may die anyway. Most EMS units do not carry antidotes for toxic exposures; by the time a person is taken to a medical center where the antidote may be available, it may be too late.

External defenses against toxic materials are by far more effective than the internal defenses of the body or antidotes. What it amounts to in simple terms is do not let toxic materials into the body to begin with; the idea is to place a barrier between responders and the poison. These barriers include chemical protective clothing and SCBA. Chemical protective clothing provides protection against absorption and contact tissue damage. SCBA prevents inhalation and ingestion of toxic materials. Firefighter turnouts do not provide any type of chemical protection from toxic materials. Turnouts may prevent some types of injection; however, the toxic material will contaminate the turnouts. There are also some preventive measures that can be taken to prevent ingestion of toxic materials. Decontamination is also an external defense. The primary reason for conducting decontamination is the removal of toxic materials. Do not eat, drink, smoke, or place anything in your mouth until decontamination has been completed.

Toxic elements

A number of elements are naturally toxic. These include arsenic; mercury; heavy metals, such as lead and cadmium; and the halogens: fluorine, chlorine, bromine, and iodine. Fluorine and chlorine are covered in "Poison gas" section in Chapter 4. It is important to note the uses of elements, because they give an indication where these materials may be found in storage and manufacturing.

Arsenic, As, is a nonmetallic element that is a silver-gray, brittle, crystalline solid that darkens in moist air. Arsenic is acutely toxic, depending on the dose, and is a carcinogen and mutagen. It is insoluble in water and reacts with nitric acid. Primary routes of exposure are through inhalation absorption, skin or eye contact, and ingestion. The exposure limit is 0.002 mg/m³. The IDLH is 100 mg/m³ in air. The target organs are the liver, kidneys, skin, lungs, and lymphatic system. Arsenical compounds are incompatible with any reducing agents. The four-digit UN identification number is 1558. The primary uses are as alloying additives for metals, especially lead and copper, and in boiler tubes, high-purity semiconductors, special solders, and medicines.

Mercury, Hg, is a liquid metallic element that is silvery in color and heavy. It is insoluble in water and has a specific gravity of 13.59, which is heavier than water. Mercury is highly toxic by skin absorption and inhalation of fumes or vapor. The TLV is 0.05 mg/m³ of air, and the IDLH is 10 mg/m³. All inorganic compounds of mercury are toxic by ingestion, inhalation, and absorption. Most organic compounds of mercury are also highly toxic. The target organs affected are the central nervous system, kidneys, skin, and eyes. Mercury is incompatible with acetylene and ammonia. The four-digit UN identification number is 2809. The uses of mercury are in electrical appliances, instruments, mercury vapor lamps, mirror coating, and as a neutron absorber in nuclear power plants.

Bromine, Br_2, is a nonmetallic, fuming liquid element of the halogen family on the periodic table. It is dark reddish-brown in color with irritating fumes. Bromine is slightly soluble in water and attacks most metals. The boiling point is 138°F (58°C), and the specific gravity is 3.12, which is heavier than water. Bromine is toxic by ingestion and inhalation, and is a severe skin irritant. The TLV is 0.1 ppm in air, and the IDLH is 10 ppm. The target organs are the respiratory system, the eyes, and the central nervous system. It is also a strong oxidizing agent and may ignite combustible materials on contact. The DOT lists it as a Class 8 corrosive; however, it carries the corrosive and poison label. The four-digit UN identification number for bromine is 1744. The NFPA 704 designation for bromine and bromine solutions is health 3, flammability 0, and reactivity 0. The white section at the bottom of the 704 diamond has the prefix "oxy," indicating an oxidizer. The primary uses are in antiknock compounds for gasoline, bleaching, water purification, as a solvent, and in pharmaceuticals.

Iodine, (I), is a nonmetallic element of family seven, the halogens. It is heavy, grayish-black in color, has a characteristic odor, and is readily sublimed to a violet vapor. It has a vapor density of 4.98, which is heavier than air. It melts at 236°F (113.5°C), has a boiling point of 363°F (184°C), and is insoluble in water. Iodine is toxic by ingestion and inhalation, and is a strong irritant to eyes and skin. The TLV ceiling is 0.1 ppm in air. Iodine is used for antiseptics, germicides, x-ray contrast material, food and feed additives, water treatment, and medicinal soaps. The four-digit UN identification number for iodine is only for the compounds iodine monochloride and iodine pentafluoride, and they are 1792 and 2495, respectively. The DOT lists iodine monochloride as a Class 8 corrosive, and iodine pentafluoride carries an oxidizer and poison label. Iodine does not have an NFPA 704 designation.

Toxic salts

Binary salts are composed of a metal and a nonmetal except oxygen. M + NM except oxygen = binary salt. They have varying hazards, one of which is toxicity. Some of the binary salts are highly toxic, such as sodium fluoride, calcium phosphide, and mercuric chloride. Cyanide salts are composed of a metal and the cyanide radical (CN), M + CN = cyanide salt. They are also highly toxic, such as sodium cyanide and potassium cyanide. The remaining salts, binary oxides, peroxide, hydroxides, and oxysalts are generally not considered toxic.

Calcium phosphide, Ca_3P_2, is a binary salt made up of reddish-brown crystals or gray, granular masses. It is water-reactive and evolves phosphine gas in contact with water, which is highly toxic and flammable. The four-digit UN identification number is 1360. The primary uses are as signal fires, torpedoes, pyrotechnics, and rodenticides.

Sodium fluoride, NaF, is a binary salt that is a clear, lustrous crystal or white powder. The insecticide grade is frequently dyed blue. It is soluble in water and has a specific gravity of 2.558, which is heavier than water. Sodium fluoride is highly toxic by ingestion and inhalation, and is also strongly irritating to tissue. The TLV is 2.5 mg/m^3 of air. The four-digit UN identification number is 1690. The primary uses are fluoridation of municipal water at 1 ppm, as an insecticide, rodenticide, and fungicide, and in toothpastes and disinfectants.

Mercuric chloride (mercury II chloride), HgCl$_2$, is a binary salt composed of white crystals or powder. It is odorless and soluble in water. It is highly toxic by ingestion, inhalation, and skin absorption. The TLV is 0.05 mg/m^3 of air. The four-digit UN identification number is 1624. The primary uses of mercuric chloride are in embalming fluids, insecticides, fungicides, wood preservatives, photography, textile printing, and dry batteries.

Sodium cyanide, NaCN, is a cyanide salt that is a white, deliquescent, crystalline powder and is soluble in water. The specific gravity is 1.6, which is heavier than water. Sodium cyanide is toxic by inhalation and ingestion, with a TLV of 4.7 ppm and 5 mg/m^3 of air. The target organs are the cardiovascular system, central nervous system, kidneys, liver, and skin. Reactions with acids can release flammable and toxic hydrogen cyanide gas. Cyanides are incompatible with all acids. The four-digit UN identification number is 1689. The NFPA 704 designation is health 3, flammability 0, and reactivity 0. The primary uses are in gold and silver extraction from ores, electroplating, fumigation, and insecticides.

Potassium cyanide, KCN, is a cyanide salt that is found as a white, amorphous, deliquescent lump or crystalline mass with a faint odor of bitter almonds. It is soluble in water and has a specific gravity of 1.52. It is a poison that is absorbed through the skin. Target organs are the same as for sodium cyanide. Reaction with acids releases flammable and toxic hydrogen cyanide gas. The four-digit UN identification number is 1680. The NFPA 704 designation is health 3, flammability 0, and reactivity 0. The primary uses are in gold and silver ore extraction, insecticides, fumigants, and electroplating.

Toxic hydrocarbons

Most of the alkane, alkene, and alkyne hydrocarbon compounds are considered to be flammable as their major hazard and the toxicity is considered as moderate to low. The vapors are more likely to be asphyxiant than toxic. TLVs range from 50 ppm for hexane to 300 ppm for octane. Decane is listed as having a narcotic effect. Many of these hydrocarbons are found in mixtures, and it will be necessary to look at the Material Safety Data Sheets (MSDSs) to obtain toxicity information on particular mixtures. Benzene, toluene, and xylene are aromatic hydrocarbons. They are considered highly toxic and human carcinogens. Benzene has a TLV of 0.1 ppm in air, according to the *NIOSH Guide 1997 Addition*, and an STEL of 1 ppm. The OSHA STEL is 5 ppm and a PEL of 1 ppm. Toluene is toxic by ingestion, inhalation, and skin absorption. The TLV for toluene is 100 ppm in air. Xylenes are toxic by inhalation and ingestion, with a TLV of 100 ppm. The target organs are the blood, skin, bone marrow, eyes, central nervous system, and respiratory system.

Toxic hydrocarbon derivatives

Several hydrocarbon derivatives are toxic as a primary hazard. While some compounds in each of the groups are toxic, not all compounds are toxic. It is important, however, to consider them all toxic within a group until the specific material can be researched and the exact hazards verified. Families with toxicity as a primary hazard are the alkyl halides, cyanides, isocyanates, and amines. Other groups that are toxic in certain concentrations and doses, although they may not be the primary hazard, include the alcohols, sulfides, aldehydes, and organic acids. Ethers are considered anesthetic; however, there are some ethers that the DOT lists as Class 6.1 poisons. They are compounds that have chlorine added to the ether. The toxicity comes from the chlorine. For example, 2,2 dichlorodiethyl ether and dichlorodimethyl ether have NFPA 704 health designations of 3 and 4, respectively. Epichlorohydrin, also known as chloropropylene oxide, is a Class 6.1 poison and is an ether with chlorine added to the compound. Methyl chloromethyl ether is another compound with chlorine added. The NFPA 704 designation for health is 3.

Ketones as a group are considered narcotic. The primary hazard of the esters is polymerization. Many of them are flammable liquids and polymers. The DOT does list some ester compounds as Class 6.1 poisons and, again, these compounds have chlorine added, which accounts for their toxicity. For example, ethyl chloroformate is an ester that has an NFPA health designation of 4. Remember that too much of any chemical can be toxic, so too much of an anesthetic or a narcotic can be toxic.

Alkyl halides

The alkyl halide functional group is composed of a hydrocarbon radical and some combination of halogens from family seven on the periodic table. The halogens are all toxic and, therefore, it is not difficult to see that the alkyl halides are also going to be toxic. The general formula for alkyl halide is R–X. The "R" represents one or more hydrocarbon radicals, and the "X" represents one or more halogens. The "X" can be replaced by fluorine (F), chlorine (Cl), bromine (Br), iodine (I), or combinations of two or more. It is important to remember that hydrocarbon derivatives started out as hydrocarbons before hydrogen was removed and other elements were added. Many of the hydrocarbon names are still used in the naming process with the alkyl halides.

There are three ways alkyl halides can be named. They are all correct naming conventions, and the compounds may be listed under any one of the possibilities. When researching the compounds in reference books, you may have to look under the alternate names to find information on the compound. The first naming convention is the one in which the radical is named first, the "ine" is dropped from the halogen, and an "ide" ending is added. For example, if the compound has one carbon, the radical for one carbon is methyl. If there is chlorine attached to the methyl radical, the alkyl halide compound is named methyl chloride. The structure, molecular formula, and names for it are shown in Figure 8.6.

If fluorine is attached to a one-carbon radical, the name is methyl fluoride, and so on. The second convention is to name the halogen first and then the hydrocarbon radical. In this case, the "ine" ending is dropped from the halogen and an "o" is added to the abbreviated name for the halogen. In the case of chlorine, it is "chloro." The radical is on the end

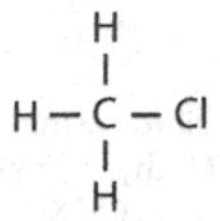

$$
\begin{array}{c}
H \\
| \\
H - C - Cl \\
| \\
H
\end{array}
$$

CH_3Cl

Figure 8.6 Chloromethane (methyl chloride).

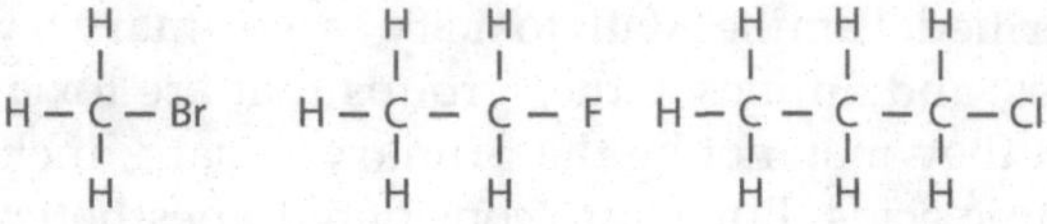

$$H-C-Br \qquad H-C-C-F \qquad H-C-C-C-Cl$$

Methyl bromide   Ethyl fluoride   Propyl chloride
Bromomethane   Fluoroethane   Chloropropane
   CH_3Br C_2H_5F C_3H_7Cl

Figure 8.7 One-, two-, and three-carbon alkyl halides.

of the name. When the radical is on the end, the name reverts back to the hydrocarbon that was used to form the radical. For example, methyl is a radical of the one-carbon alkane, methane. So if the halogen chlorine is added to methane, and the halogen is named first, the name is chloromethane. If bromine is the halogen, the name is bromomethane. If the radical is a two-carbon radical and the halogen is fluorine, the name of the compound is fluoroethane, and so on. Illustrated in Figure 8.7 are the names and structural formulas for some one, two-, and three-carbon alkyl halides.

It is possible to use more than one halogen to form alkyl halide compounds. If the multiple halogens are the same type, the prefix "di" is used for two, "tri" for three, and "tetra" for four. Some chemicals, as has been previously mentioned, have trade names. There are also trade names for some of the alkyl halides. For example, a one-carbon radical with three chlorine atoms attached is called trichloro methane, or methyl trichloride; however, the trade name for the compound is chloroform. A methyl radical with four chlorines attached is named tetrachloro methane, or methyl tetrachloride. The trade name for the compound is carbon tetrachloride, a material that was used as a fire-extinguishing agent. It is no longer approved as an extinguishing agent because when it contacts a hot surface, it decomposes to phosgene gas. Shown in Figure 8.8 are the names, molecular formulas, and structures for some alkyl halides with multiple numbers and combinations of halogens in the compounds.

Some alkyl halide compounds also have double bonds: dichloroethylene, dichloropropene, dichlorobutene, and trichloroethylene.

Trichloroethylene (IUPAC), $CHClCCl_2$, is a stable, low-boiling, colorless liquid with a chloroform-like odor. It is not corrosive to the common metals even in the presence of moisture. It is slightly soluble in water and is nonflammable. It is toxic by inhalation, with a TLV of 50 ppm and an IDLH of 1000 ppm in air. The FDA has prohibited its use in foods, drugs, and cosmetics. The four-digit UN identification number is 1710. The NFPA 704 designation is health 2, flammability 1, and reactivity 0. Its primary uses are in metal degreasing, dry cleaning, as a refrigerant and fumigant, and for drying electronic parts.

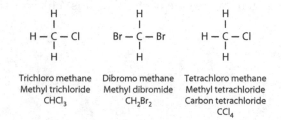

Trichloro methane   Dibromo methane   Tetrachloro methane
Methyl trichloride   Methyl dibromide   Methyl tetrachloride
   $CHCl_3$ CH_2Br_2 Carbon tetrachloride
                                               CCl_4

Figure 8.8 Alkyl halides with multiple numbers of halogens in the compounds.

Amines

The next toxic hydrocarbon-derivative family is the amines. Amines are toxic irritants, in addition to being flammable. They are considered slightly polar when compared to nonpolar materials. The amine functional group is represented by a single nitrogen surrounded by two or fewer hydrogen atoms. The general formulas for the amines are $R-NH_2$, R_2NH, and R_3N. It is the nitrogen that identifies the amine group, not the number of hydrogen atoms attached to the nitrogen. The amines are covered in detail in "Hydrocarbon derivatives" section in Chapter 5. The degree of toxicity of amines varies from compound to compound. Many of them are strong irritants. TLV values range in the low double digits from 5 to 10 ppm. Diethylamine is toxic by ingestion and is a strong irritant. It has a TLV of 10 ppm in air. Butyl amine is a skin irritant with a TLV of 5 ppm in air. It is important to obtain further information when dealing with amines. Look the materials up in reference books and MSDSs to determine the toxic characteristics of a given amine compound.

Aniline, $C_6H_5NH_2$, also known as phenylamine, is a colorless, oily liquid with a characteristic amine odor and taste. It rapidly turns brown when exposed to air. It is soluble in water, with a specific gravity of 1.02, which is slightly heavier than water. Aniline is an allergen and is toxic if absorbed through the skin. The TLV is 2 ppm in air, and the IDLH is 100 ppm. The target organs are the blood, cardiovascular system, liver, and kidneys. Aniline is incompatible with nitric acid and hydrogen peroxide. The four-digit UN identification number is 1547. The NFPA 704 designation is health 3, flammability 0, and reactivity 1. The primary uses are in dyes, photographic chemicals, isocyanates for urethane foams, explosives, herbicides, and petroleum refining. The structure and molecular formula for aniline are shown in Figure 8.9.

Cyanides and isocyanates

Cyanides and isocyanates are derivative families that are extremely toxic. Industry does not usually call any of its compounds cyanide as that would cause some concern if this material was shipped through or stored in a community. So they use the alternate naming system that uses the ending nitrile. Few people would have any idea what nitrile was. The cyanide general formula is $R-CN$ and the isocyanate general formula is $R-NCO$. Examples of cyanides vinyl cyanide (acrylonitrile) and methyl cyanide (acetonitrile). When naming cyanides, you name the radical hooked to the functional group and then the word cyanide. When using the nitrile ending for cyanide, you also name the radical first followed by the nitrile ending. Just as with the aldehydes, esters, and organic acids, you count the carbons, including the one in the cyanide functional group, when naming the compounds.

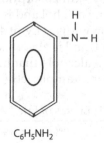

$C_6H_5NH_2$

Figure 8.9 Phenylamine (aniline).

Vinyl is added to the cyanide radical and you get the compound vinyl cyanide or the prefix for a three-carbon double bond, which is acryl, and the nitrile name for cyanide and you get acrylonitrile, which is toxic by inhalation and skin absorption. It is also a carcinogen, flammable and dangerous fire risk. Explosive limits are wider than most hydrocarbon compounds at 3%–17% in air and the TLV is 2 ppm in air. NFPA 704 lists the hazards as flammability 4, health 3, and reactivity 2.

$$CH_2 + CH + CN = CH_2CHCN, \text{ acrylonitrile}$$

When a methyl radical is added to the cyanide functional group, you call it methyl cyanide, or when using the nitrile ending, you have acetonitrile.

$$CH_3 + CN = CH_3CN, \text{ acetonitrile}$$

Just as with the aldehydes, esters, and organic acids, you count the carbons, including the one in the cyanide functional group when naming the compounds. Organic cyanides are not as toxic as the ionic bonded cyanide salts; however, they should be treated with respect in terms of toxicity. Acetonitrile is flammable and a dangerous fire risk and toxic by inhalation and skin absorption. The TLV is 40 ppm in air. It has an NFPA 704 listing of flammability 3, health 2, and reactivity 0. Isocyanates have a general formula of *R–NCO*. Isocyanates are named with the radical first and ending in the word isocyanate. An example would be to add the methyl radical to the isocyanate functional group, which would yield methyl isocyanate (MIC).

$$CH_3 + NCO = CH_3NCO, \text{ methyl isocyanate}$$

MIC is a flammable, dangerous fire risk, toxic by skin absorption, and a strong irritant. It is the chemical that was released from the chemical plant in India, which killed to date over 13,000 people. Many more will die of the chronic effects of the compound.

Alcohols

All alcohols are toxic to some degree. Ethyl alcohol is the drinking alcohol and, when consumed in moderation, has limited toxic effects, as shown in Table 8.10. However, ethyl alcohol, if taken in excess, can have toxic effects. In the short term, the effects of ethyl alcohol are sedative and depressant; in the long run, cancer or liver damage may occur. Methyl alcohol is toxic by ingestion and can cause blindness. It has a TLV of 200 ppm. Propyl alcohol is toxic by skin absorption. It has a TLV of 200 ppm. However, isopropyl alcohol is used as rubbing alcohol and is applied to the skin. Isopropyl alcohol is toxic by ingestion and inhalation with a TLV of 400 ppm. Butyl alcohol is toxic by prolonged inhalation, is an eye irritant, and is absorbed through the skin. The TLV ceiling is 50 ppm in air. There are also some alcohols that have chlorine added to the compound, which increases their toxicity. For example, ethylene chlorohydrin is an alcohol that

Table 8.10 Toxic Effects of Ethyl Alcohol in the Blood

0.08%	2.0%	3.0%	4.0%	5.0%
Happy	Very happy	Drunk	Falling down drunk	Death

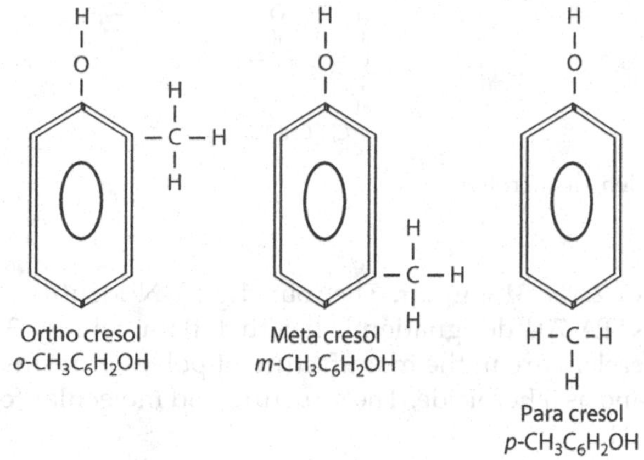

Ortho cresol
$o\text{-}CH_3C_6H_2OH$

Meta cresol
$m\text{-}CH_3C_6H_2OH$

Para cresol
$p\text{-}CH_3C_6H_2OH$

Figure 8.10 Cresol.

has an NFPA 704 designation for health of 4. As you can see, the hazards and routes of entry vary widely among the alcohols. It should be assumed that the alcohol in a spill is the worst-case scenario until the exact hazard of the compound can be researched. The alcohols are covered in detail in Chapter 4 under Hydrocarbon Derivatives.

Cresols (o-,m-,p-,), $CH_3C_6H_4OH$, are alcohol hydrocarbon derivatives. They are colorless to yellowish or pinkish liquids.

They are found in the "ortho," "meta," and "para" isomers, like the xylenes in Chapter 5. Cresol has a characteristic phenolic odor and is soluble in water, with a specific gravity of 1.05, which is slightly heavier than water. It is an irritant, corrosive to the skin and mucous membranes, and is absorbed into the skin. The TLV is 5 ppm in air, and the IDLH is 250 ppm. The target organs are the central nervous system, respiratory system, liver, kidneys, skin, and eyes. The four-digit UN identification number is 2076. The NFPA 704 designation is health 3, flammability 2, and reactivity 0. The primary uses are as a textile scouring agent, herbicide, phenolic resins, and ore flotation; the para isomer is used in synthetic food flavors. The structures and molecular formulas for the isomers of cresol are shown in Figure 8.10.

Aldehydes

Aldehydes are highly toxic compounds that are known to be human carcinogens. Formaldehyde is toxic through inhalation and is a strong irritant. The TLV is 1 ppm in air. Acetaldehyde is toxic, with narcotic effects. The TLV is 100 ppm in air, and the IDLH is 10,000 ppm. Propionaldehyde is an irritant. The toxicity hazards of the aldehydes also vary from one compound to another. Care should be taken to determine the exact hazards of any given compound.

Acrolein, CH_2CHCHO, also known as acrylaldehyde, is an aldehyde hydrocarbon derivative and it is a colorless or yellowish liquid with a disagreeable, suffocating odor. Acrolein is soluble in water, with a specific gravity of 0.84, which is lighter than water. It polymerizes readily, is reactive, and is not shipped or stored without an inhibitor. Acrolein is toxic by inhalation and ingestion, and is a strong irritant to the skin and eyes. The TLV is 0.1 ppm in air, and the IDLH is 5 ppm. The target organs are the heart, eyes, skin, and respiratory system. Acrolein is a dangerous fire and explosion risk, with a wide

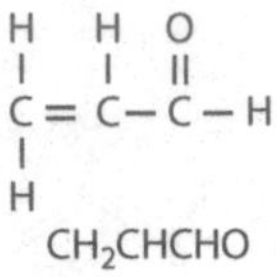

$$\begin{array}{ccccc} H & & H & & O \\ | & & | & & \| \\ C & = & C & - & C & - & H \\ | & & & & \\ H & & & & \end{array}$$

CH₂CHCHO

Figure 8.11 Acrylaldehyde (acrolein).

flammable range of 2.8%–31% in air. The four-digit UN identification number for acrolein is 1092. The NFPA 704 designation is health 4, flammability 3, and reactivity 3. The primary uses of acrolein are in the manufacture of polyester resins, polyurethane resins, pharmaceuticals, and as a herbicide. The structure and molecular formula for acrolein are shown in Figure 8.11.

Organic acids

Organic acids are toxic and corrosive. Corrosivity is a form of toxicity to the tissues that the acid contacts. However, the organic acids have other toxic effects. Formic acid is corrosive to skin and tissue. It has a TLV of 5 ppm in air and an IDLH of 30 ppm. Pure acetic acid is toxic by ingestion and inhalation. It is a strong irritant to skin and tissues. The TLV is 10 ppm in air, and the IDLH is 1000 ppm. Propionic acid is a strong irritant, with a TLV of 10 ppm in air. Butyric acid is a strong irritant to skin and tissues. The degree of toxicity varies with the different organic acid compounds. Review reference sources and MSDSs to determine the exact hazards of specific acids.

Phosphoric esters

Phosphoric esters are primarily military nerve agents and organophosphate pesticides and examples are presented in "Pesticides" and "Military and terrorist chemical agents" sections of this chapter. Organophosphate insecticides contain phosphorus. They are phosphorus-based esters with pentavalent phosphorus. Phosphorus forms five bonds in the insecticides and military nerve agents. They are derived from phosphoric acid. Carbamate insecticides are derived from carbamic acid. Primary hazard is toxicity.

Miscellaneous toxic materials

A wide variety of toxic liquids and solids do not fit neatly into any of the families discussed so far. Some of the most dangerous and more common ones will be listed here, but this will, by no means, be a comprehensive listing. The intent is to foster familiarity with the many different types of toxic chemicals that may be encountered in the real world, both in transportation and in fixed facilities. The chemicals presented will be drawn from the DOT Hazardous Materials Tables and the NFPA Hazardous Chemicals Data listing.

  Chloropicrin, CCl_3NO_2, is a slightly oily, colorless, refractive liquid. It is relatively stable and slightly water soluble. The specific gravity is 1.65, which is heavier than air. It is toxic by ingestion and inhalation, and is a strong irritant. The TLV is 0.1 ppm in air. The four-digit UN identification number is 1580. The NFPA 704 designation is health 4, flammability 0, and reactivity 3. The primary uses of chloropicrin include dyestuffs, fumigants, fungicides, insecticides, rat exterminator, and tear gas. The structure and molecular formula for chloropicrin are shown in Figure 8.12.

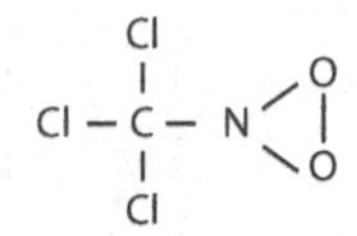

$$Cl - \underset{\underset{Cl}{|}}{\overset{\overset{Cl}{|}}{C}} - N \overset{O}{\underset{O}{<}}$$

CCl₃NO₂

Figure 8.12 Chloropicrin.

Epichlorohydrin, $ClCH_2CHOCH_2$, an epoxide, is a highly volatile, unstable liquid, with a chloroform-like odor. It is slightly water soluble and has a specific gravity of 1.18, which is heavier than water. Epichlorohydrin is toxic by ingestion, inhalation, and skin absorption. It is a strong irritant and a known carcinogen. The TLV is 2 ppm in air, and the IDLH is 250 ppm. The target organs affected are the respiratory system, skin, and kidneys. The four-digit UN identification number is 2023. The NFPA 704 designation is health 3, flammability 3, and reactivity 2. The primary uses are as a raw material for epoxy and phenoxy resins, in the manufacture of glycerol, and as a solvent. The structure and molecular formula for epichlorohydrin are shown in Figure 8.13.

Hydrogen cyanide, HCN, also known as hydrocyanic acid, is a water-white liquid with a faint odor of bitter almonds. The odor threshold is 0.2–5.1 ppm in air; however, if the odor of hydrogen cyanide is detected, it has already exceeded the allowable amount. The immediately fatal concentration is usually 250–300 ppm in air. Hydrogen cyanide blocks the uptake of oxygen by the cells. It is soluble in water, with a specific gravity of 0.69, which is lighter than water. Hydrogen cyanide is highly toxic by inhalation, ingestion, and skin absorption. The TLV is 4.7 ppm in air, and the IDLH is 50 ppm. The target organs are the central nervous system, cardiovascular system, kidneys, and liver. The four-digit UN identification number is 1051 for anhydrous (without water) and 1614 when it is absorbed in a porous material. The NFPA 704 designation is health 4, flammability 4, and reactivity 2. The primary uses are in the manufacture of acrylonitrile, acrylates, cyanide salts, rodenticides, and pesticides. The structure and molecular formula for hydrogen cyanide are shown in Figure 8.14.

Methyl hydrazine, CH_3NHNH_2, is a colorless, hygroscopic liquid with an ammonia-like odor. It is soluble in water, with a specific gravity of 0.87, which is lighter than water. Methyl hydrazine is toxic by inhalation and ingestion, and is a suspected human carcinogen. The TLV ceiling is 0.2 ppm in air, and the IDLH is 50 ppm. The target organs are the central nervous system, respiratory system, liver, blood, eyes, and cardiovascular system. The four-digit UN identification number is 1244. The NFPA 704 designation is health 4, flammability 3, and reactivity 2. The primary uses are as a missile propellant and a solvent. The structure and molecular formula for methyl hydrazine are shown in Figure 8.15.

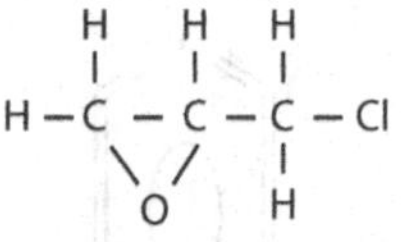

$$H - \underset{\underset{O}{\diagdown}}{\overset{\overset{H}{|}}{C}} - \underset{\underset{/}{}}{\overset{\overset{H}{|}}{C}} - \underset{\underset{H}{|}}{\overset{\overset{H}{|}}{C}} - Cl$$

CH₂CHOCH₂Cl

Figure 8.13 Epichlorohydrin.

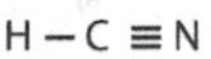

$$H - C \equiv N$$

HCN

Figure 8.14 Hydrogen cyanide.

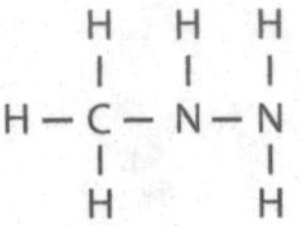

$$
\begin{array}{ccc}
H & H & H \\
| & | & | \\
H-C-N-N & & \\
| & & | \\
H & & H \\
\end{array}
$$

$$CH_3NHNH_2$$

Figure 8.15 Methyl hydrazine.

Methyl isocyanate (MIC), CH_3NCO, is a colorless liquid. It is water reactive, with a specific gravity of 0.96, which is lighter than water. This is the chemical that was released in Bhopal, India, that killed over 3000 people in 1984. Methyl isocyanate is toxic by skin absorption and a strong irritant. The TLV is 0.02 ppm in air, and the IDLH is 20 ppm. The target organs are the respiratory system, eyes, and skin. The four-digit UN identification number is 2480. The NFPA 704 designation is health 4, flammability 3, and reactivity 2. The white section at the bottom of the diamond has a W with a slash through it, indicating water reactivity. The primary use of methyl isocyanate is as a chemical intermediate. The structure and molecular formula are shown in Figure 8.16.

Toluene diisocyanate (TDI), $(OCN)_2C_6H_3CH_3$, is a water-white to pale-yellow liquid with a sharp, pungent odor. It reacts with water to release carbon dioxide. The specific gravity is 1.22, which is heavier than water. TDI is toxic by inhalation and ingestion, and is a strong irritant to skin and other tissue, particularly the eyes. The TLV is 0.005 ppm in air, and the IDLH is 10 ppm. The target organs are the respiratory system and the skin. The four-digit UN identification number is 2078. The NFPA 704 designation is health 3, flammability 1, and reactivity 3. The white section at the bottom of the diamond has a W with a slash through it, indicating water reactivity. The primary uses of TDI are in the manufacture of polyurethane foams, elastomers, and coatings. The structure and molecular formula for TDI are shown in Figure 8.17.

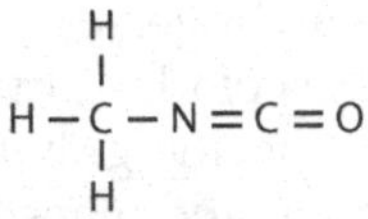

$$
\begin{array}{c}
H \\
| \\
H-C-N=C=O \\
| \\
H \\
\end{array}
$$

$$CH_3CNO$$

Figure 8.16 Methyl isocyanate.

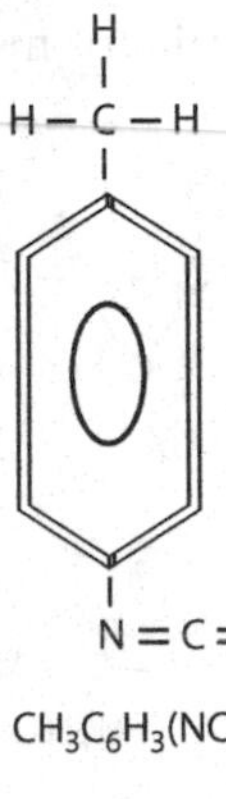

$$CH_3C_6H_3(NCO)_2$$

Figure 8.17 Toluene diisocyanate (TDI).

Pesticides

Pesticides can be found in manufacturing facilities, commercial warehouses, agricultural chemical warehouses, farm supply stores, nurseries, farms, supermarkets, discount stores, hardware stores, and other retail outlets.

A pesticide is a chemical or mixture of chemicals used to destroy, prevent, or control any living thing considered to be a pest, including insects (insecticides), fungi (fungicides), rodents (rodenticides), or plants (herbicides). The definition of a pesticide from the Federal Insecticide, Fungicide, and Rodenticide Act (FIFRA) is "a chemical or mixture of chemicals or substances used to repel or combat an animal or plant pest. This includes insects and other invertebrate organisms; all vertebrate pests, e.g., rodents, fish, pest birds, snakes, gophers; all plant pests growing where not wanted, e.g., weeds; and all microorganisms which may or may not produce disease in humans. Household germicides, plant-growth regulators, and plant-root destroyers are also included" (Photo 8.3).

Restricted-use pesticides are those regulated by EPA and only allowed for commercial or agricultural use by trained and licensed applicators. Restricted-use pesticides are usually more toxic, may be more environmentally sensitive, found in greater quantities in storage, and packaged in larger containers. When a pesticide is classified as restricted, the label will state "Restricted Use Pesticide" in a box at the top of the front panel. A statement may also be included describing the reason for the restricted-use classification. Usually, another statement will describe the category of certified applicator that can purchase and use the product. Restricted-use pesticides may only be used by applicators certified by the state or by the EPA. Home yard work often involves the use of general pesticides to

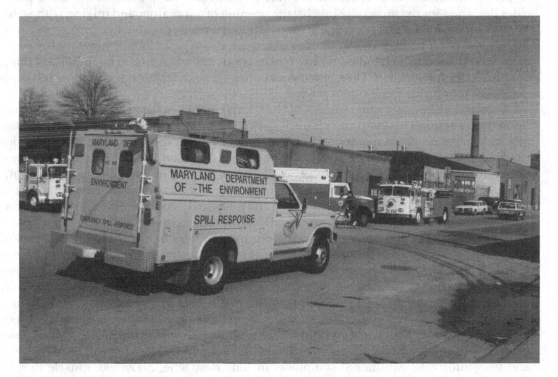

Photo 8.3 When there is a threat of air, water, or ground contamination from a poison incident, the local or state department of environment should be notified.

control weeds, insects, and fungus in lawns, and ornamental trees and bushes. Chemicals approved by the EPA for homeowner use do not require any special training or licensing by the homeowner to be used. Container size is usually small, and they can be safely used by following label directions.

The EPA estimates that there are 45,000 accidental pesticide poisonings in the United States each year, where more than 1 billion pounds are manufactured annually. Pesticides can be found in manufacturing facilities, commercial warehouses, agricultural chemical warehouses, farm supply stores, nurseries, farms, supermarkets, discount stores, hardware stores, and other retail outlets. They may also be found in many homes, garages, and storage sheds across the country.

More than 1,000 basic chemicals, mixed with other materials, produce about 35,000 pesticide products. However, when an emergency occurs involving different groups of pesticides, chemicals may become mixed that are not normally found together. This mixing of chemicals may provide toxicology and cleanup problems for emergency responders. Great care should be taken during firefighting operations including overhaul when pesticides and other chemicals are involved. If the fire is allowed to burn off, care should be taken to avoid smoke or fumes that evolve. Runoff can become contaminated with toxic materials and damage firefighting protective clothing, equipment, apparatus, and the environment. If a decision is made to fight a pesticide fire, runoff should be kept to a minimum. If possible, route runoff water to a holding area. Hazardous materials teams are called upon to respond to numerous incidents each year involving pesticides. Care should be taken so as not to overreact to a pesticide spill. A pint bottle of a pesticide broken on the display floor of a lawn and garden center does not necessarily require a full-blown hazmat response and a multi-hour operation to effectively mitigate. Keep in mind that those pesticides designed for consumer use involve the opening of a container and mixing with water for application by the end user, many times without a need for extensive protective clothing. On the other hand, larger quantities of restricted-use pesticides or unrestricted pesticides would need to be handled as any other serious chemical spill. The important thing is to evaluate the incident. Do a risk–benefit analysis to determine the level of response necessary to mitigate the incident safely.

Pesticides like many other groups of chemicals can be segregated into families based upon their chemical makeup and characteristics, including toxicity (Photo 8.4). Common pesticide families are organophosphates, carbamates, chlorophenols, and organochlorines. *Organophosphates* are derivatives of phosphoric acid and are acutely toxic, but are not enduring (Photo 8.5). They are generally known as all insecticides, which contain phosphorus. The chemical formulas of the organophosphates contain carbon, hydrogen, phosphorus, and at least one sulfur atom, and some may contain at least one nitrogen atom. Organophosphates break down rapidly in the environment and do not accumulate in the tissues. They are generally much more toxic to vertebrates than other classes of insecticides and are therefore associated with more human poisonings than any other pesticide (Table 8.11). They are also closely related to some of the most potent nerve agents including sarin and VX. Organophosphates function by overstimulating and then inhibiting neural transmission, primarily in the nervous, respiratory, and circulatory systems. Signs and symptoms of exposure include pinpoint pupils, blurred vision, tearing, salivation, and sweating. Pulse rate will decrease, and breathing will become labored. Intestines and bladder may evacuate their contents. Muscles will become weak and uncomfortable. Additional symptoms include headache, dizziness, muscle twitching, tremor, or nausea. Examples of organophosphate pesticides include malathion, methyl parathion, thimete, counter, lorisban, and dursban (Figure 8.18). The chemical

Photo 8.4 Large and multiple containers of pesticides should be expected to be found at the home base of agricultural spraying aircraft.

Photo 8.5 A tank car of organophosphate pesticide, the same type of material used as a military nerve agent.

Table 8.11 Toxicity of Organophosphorous Pesticides

Name	LD_{50} mg/kg	TLV mg/kg
Diazonon	80	0.1
Malathion	1000	10.0
Parathion	8	0.1
Tetraethylpyrophosphate (TEPP)	0.5	0.05
Dimethyldichlorovinyl phosphate (DDVP)	56	1.0

- Malathion
- Methyl parathion
- Thimete
- Counter
- Lorisban
- Dursban

Figure 8.18 Examples of organophosphates.

formulas of the organophosphates contain carbon, hydrogen, phosphorus, and at least one sulfur atom, and some may contain at least one nitrogen atom. Antidotes are available for organophosphate pesticide poisonings. Many hospitals in rural areas where organophosphates are used by farmers and others will have extra stocks of atropine, which is used to counter the effects of organophosphate pesticides (Photo 8.6).

Photo 8.6 Agricultural spray plane used to apply pesticides to food crops.

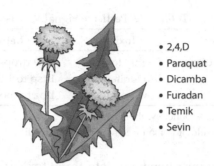

- 2,4,D
- Paraquat
- Dicamba
- Furadan
- Temik
- Sevin

Figure 8.19 Examples of carbamates.

Carbamate pesticides are derivatives of carbamic acid and are among the most widely used pesticides in the world. Most are herbicides and fungicides, such as 2,4,D, paraquat, and dicamba, and function by inhibiting nerve impulses. The formula will contain carbon, hydrogen, nitrogen, and sulfur. Other examples include furadan, temik, and sevin (Figure 8.19).

Organochlorines are chlorinated hydrocarbons. The formula contains carbon, hydrogen, and chlorine. They are neurotoxins, which function by overstimulating the central nervous system, particularly the brain. Examples are aldrin, endrin, hesadrin, thiodane, and chlordane (Figure 8.20). The best-known organochlorine is DDT, which has been banned for use in the United States because of its tissue accumulation and environmental persistence.

Chlorophenol pesticides contain carbon, hydrogen, oxygen, and chlorine. They affect the central nervous system, kidneys, and liver.

Pesticide labels contain valuable information for the emergency responder and medical personnel treating a patient exposed to pesticides. This information includes product name, "signal word," a statement of practical treatment, EPA registration number, a note to physician, and a statement of chemical hazards. Other information includes active and inert ingredients. "Inert" does not necessarily mean that the ingredients do not pose a danger; it means only that the inert ingredients do not have any action on the pest for which the pesticide was designed. Many times, the inert ingredient is a flammable or combustible liquid. The label also contains information about treatment for exposure. This information should be taken to the hospital when someone has been contaminated with a pesticide. Do not, however, take the pesticide container to the hospital. Take the label, or write the information down, take a polaroid picture of the label, or use a pesticide label book. (Label books are available from agricultural supply dealers.)

- DDT
- Aldrin
- Endrin
- Hesadrin
- Thiodane
- Chlordane

Figure 8.20 Examples of organochlorines.

Table 8.12 Pesticide Signal Words

Signal word	Toxicity	Lethal dose
Danger/poison[a]	Highly toxic	Few drops to 1[b] tsp
Warning	Moderately toxic	1 tsp to 1 tbsp
Caution	Low toxicity	1 oz to more than a pint

[a] Skull and crossbones symbol included.
[b] Less for a child or person weighing less than 160 lb.

Pesticides can be grouped generally into three toxicity categories: high, moderate, and low. You can tell the degree of toxicity by the signal word on the label. Three signal words indicating the level of toxicity of a pesticide are "Danger," "Caution," and "Warning" (Table 8.12). Highly toxic materials bear the word "Danger," with a skull-and-crossbones symbol and the word "Poison" printed on the label. The lethal dose may be a few drops to 1 tsp. Moderately toxic pesticides have the word "Warning," and the lethal dose is 1 tsp–1 tbsp. Low-toxicity pesticides carry the word "Caution," and the lethal dose is 1 oz–1 pt.

Pesticides may poison or cause harm to humans by entering the body in one or more of these four ways: through the eyes, through the skin, by inhalation, and by swallowing. As with most any chemical, exposure to the eyes is the fastest way to become poisoned. Whenever anyone is exposed to a pesticide, it is important to recognize the signs and symptoms of poisoning so that prompt medical help can be provided. Any unusual appearance or feeling of discomfort or illness can be a sign or symptom of pesticide poisoning. These signs and symptoms may be delayed up to 12 h. When they occur and pesticide contact is suspected, get medical attention immediately. (The National Pesticide Network, located in Texas, provides emergency information through a toll-free telephone number: 1-800-858-7378, from 8 a.m. to 6 p.m. Central Standard Time.) Information can also be obtained from the National Poison Control Center by calling 1-800-222-1222.

While emergency responders should never risk life to protect the environment, they certainly should not do anything to damage the environment further than if they had not responded at all (Photo 8.7). Some pesticide fires may need to be left to burn out on their own to avoid contaminated runoff, which could damage the environment. Many times fire can break down pesticides into less harmful chemicals when they burn. Pesticide labels contain information about what type of environmental damage can occur from the uncontrolled release of the pesticide. Care should be taken to protect the environment in consultation with state and local environmental officials.

Other toxic materials

Ethylene glycol, CH_2OHCH_2OH, an alcohol hydrocarbon derivative, is a clear, colorless, syrupy liquid with a sweet taste. Ethylene glycol is soluble in water and has a specific gravity of 1.1, which is slightly heavier than water. It is toxic by inhalation and ingestion. The lethal dose is reported to be 100 cm^3, and the TLV is 50 ppm. Ethylene glycol has not been assigned a four-digit UN identification number. The NFPA 704 designation is health 1, flammability 1, and reactivity 0. The primary uses are as coolants and antifreeze, brake fluids, low-freezing dynamite, a solvent, and a deicing fluid for airport runways. The structure and molecular formula for ethylene glycol are shown in Figure 8.21.

Photo 8.7 MC 312 tanker of pesticide designed to kill weeds or trees.

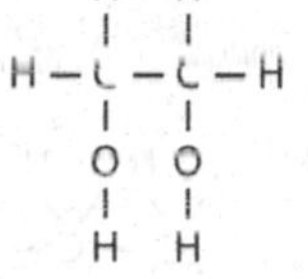

$$H - \overset{\overset{\displaystyle \|}{\underset{\displaystyle |}{}}}{C} - \overset{\overset{\displaystyle H}{\underset{\displaystyle |}{}}}{C} - H$$

CH₂OHCH₂OH

Figure 8.21 Ethylene glycol.

Phenol, C_6H_5OH, also known as carbolic acid, is a white, crystalline mass that turns pink or red if not perfectly pure or if exposed to light. Phenol absorbs water from the air and liquefies; it may also be found in transport as a molten material. It has a distinctive odor and a sharp, burning taste, but in a weak solution, it has a slightly sweet taste. Phenol is soluble in water, with a specific gravity of 1.07, which is heavier than water. It is toxic by ingestion, inhalation, and skin absorption, and is a strong irritant to tissues. The TLV is 5 ppm in air, and the IDLH is 250 ppm. The target organs are the liver, kidneys, and skin. The four-digit UN identification number is 1671 for solids, 2312 for molten materials, and 2821 for solutions. The NFPA 704 designation is health 4, flammability 2, and reactivity 0. The primary uses are phenolic resins, epoxy resins, 2-4-D herbicides, solvents, pharmaceuticals, and as a general disinfectant. The structure and molecular formula for phenol are shown in Figure 8.22.

Caprolactam, $CH_2CH_2CH_2CH_2CH_2NHCO$, is a solid material composed of white flakes. Caprolactam is soluble in water and has a specific gravity (in a 70% solution)

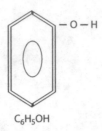

C₆H₅OH

Figure 8.22 Phenol.

of 1.05, which is heavier than water. It may also be encountered as a molten material. Caprolactam is toxic by inhalation, with a TLV of (vapor) 5 ppm in air and (dust) 1 mg/m³ of air. The primary uses are in the manufacture of synthetic fibers, plastics, film, coatings, and polyurethanes. The structure and molecular formula for caprolactam are shown in Figure 8.23.

Nitrobenzene is a greenish-yellow crystal or yellow oily liquid, and is slightly soluble in water. The primary hazard of nitrobenzene is toxicity; however, it is also combustible. The boiling point is about 410°F, the flash point is 190°F, and the ignition temperature is 900°F. The specific gravity is 1.2, which is heavier than water, and the material will sink to the bottom. The vapor density is 4.3, which is heavier than air. Nitrobenzene is toxic by ingestion, inhalation, and skin absorption, with a TLV of 1 ppm in air. The four-digit UN identification number is 1652. The NFPA 704 designation is health 3, flammability 2, and reactivity 1. Nitrobenzene is a nitro hydrocarbon derivative, but it is not very explosive. The primary uses are as a solvent, an ingredient of metal polishes and shoe polishes, and in the manufacture of aniline. The structure and molecular formula are shown in Figure 8.24.

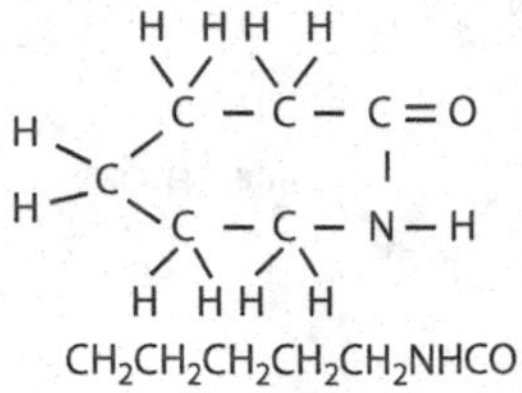

CH₂CH₂CH₂CH₂CH₂NHCO

Figure 8.23 Caprolactam.

C₆H₅NO₂

Figure 8.24 Nitrobenzene.

Military and terrorist chemical agents

Chemical agents

Chemical agents can be divided into groups, which include nerve agents, vesicants (blister agents), blood agents, choking agents, and riot-control agents (Figure 8.25). *Nerve agents* function by inhibiting the enzyme acetylcholinesterase, resulting in an excess of acetylcholine in the body. This excess results in the characteristic uncontrolled muscle movements associated with exposures to nerve agents. Nerve agents presented in this section will include GA (tabun), GB (sarin), GD (soman), GF, and VX. *Vesicants,* or blister agents, include mustard agents; sulfur mustard (H); (HD)(Agent "T"); nitrogen mustards (HN), (HN-2), and (HN-3); lewisite (L); and phosgene oxime (CX). Vesicants produce vesicles (blisters) when in contact with the human body, hence, their name blister agents. Mustard agents also cause damage to the eyes and respiratory system by direct contact and inhalation. *Blood agents* (cyanide compounds) at high concentrations kill quickly. Common forms of cyanide include hydrocyanic acid (AC) and cyanogen chloride (CK). *Choking agents* (lung-damaging agents) include phosgene (CG), diphosgene (DP), chlorine, and chloropicrin (PS). HC smoke (a smoke that contains zinc) and oxides of nitrogen (from burning munitions) also produce lung hazards when exposed. *Riot-control agents* are incapacitating compounds. CS, which is used by law enforcement agents and the military, and CN (Mace 7), which is sold in devices for self-protection, are the primary types of riot-control agents.

Like all other hazardous materials, chemical agents can exist as solids, liquids, or gases, depending on the existing temperatures and pressures. The only exception is riot-control agents, which exist as aerosolized solids at normal temperatures and pressures. Tear gas really is not a gas; it is actually a solid that is aerosolized. Mustard and nerve agents likewise are not gases unless boiled above 212°F, which is the boiling point of water at sea level. In fact, sulfur mustard has a boiling point of 422°F, nitrogen mustard 495°F, and lewisite 375°F. These agents vaporize much like boiling water. Chlorine, hydrogen cyanide, and phosgene are common industrial chemicals. Chlorine and phosgene are gases and hydrogen cyanide is a liquid at normal temperatures and pressures. Nerve and mustard agents and hydrogen cyanide are liquids under the same conditions in which chlorine and phosgene are gases. Some mustard agents are frozen solid at 57°F.

- Nerve agents
- Blister agents
- Blood agents
- Choking agents
- Irritants

Figure 8.25 Types of chemical warfare agents.

When in the liquid state, they evaporate at a rate similar to that of water. Chemical agent evaporation occurs not only because of its chemical makeup, but also because of the temperature, air pressure, wind velocity, and nature of the surface the agent comes in contact with. Water, for example, evaporates at a slower rate than gasoline, but at a faster rate than motor oil at a given temperature and pressure. Mustard is less volatile than the nerve agent sarin, but more volatile than the nerve agent VX. Volatility is the liquid's ability to produce a vapor at normal temperatures and pressures. Thus, a liquid that is said to be volatile is producing a lot of vapor; one that is not volatile will not be producing much vapor.

Evaporation rates of all chemicals mentioned are accelerated by increases in temperature and wind speed or when they are resting on a smooth surface rather than a porous one. Volatility has an inverse relationship to persistence. The more volatile a substance is, the more quickly it evaporates, and the less it tends to stay or persist as a liquid. Because of the relative low volatility of persistent-liquid chemical agents, the liquid hazard is generally more significant than the danger from the small amounts of vapor that may be generated. The reverse is true of the nonpersistent agents. They evaporate quickly enough so as not to present a liquid hazard for an extended period of time. Generally, the division between persistence and nonpersistence is related to the amount of the material left after 24 h. The nonpersistent agents are usually gone after 24 h.

The military has developed specialized toxicology terms for exposures to chemical agents in addition to the LD_{50} and LC_{50} already discussed. They are the ED_{50} and the ID_{50}. ED_{50} is the dose (D) of a liquid agent that will predictably cause effects (E) to anyone exposed. ID_{50} is the dose (D) that will cause the person to become incapacitated (I). When applying the LD_{50} to military agents, the lower the LD_{50} value, the less the amount of agent that is required to cause harm and the more potent is the agent. There is a difference in absorption rates for chemical agents and, therefore, the ED_{50} and LD_{50} for a particular agent are specific to the site of entry into the body. For example, the LD_{50} for mustard absorbed through dry intact skin is much higher than agent absorbed through the eyes.

The military uses a term called the concentration–time product, or Ct, which is the concentration of the agent present (usually expressed in terms of milligrams per cubic meter in air (mg/m^3)) multiplied by the time (usually expressed in minutes) of exposure to the agent. For example, exposure to a concentration of soman (GD) vapor for 10 min results in a Ct of 40 mg-min/m³. Exposures of 8 mg/m³ for 5 min result in the same Ct (40 mg-min/m³). This result is true for most of the agents except for cyanide. The Ct associated with a biological effect remains relatively constant even though the concentration and length of time of exposure may vary within certain limits. For example, a 10 min exposure to 4 mg/m³ of soman causes the same effects as a 5 min exposure to 8 mg/m³ of the agent or to a 1 min exposure to 40 mg/m³. When the exposure threat is in the form of a vapor or gas, the "E" for effect is attached to the "Ct," which is the agent concentration and time of exposure. The result is the effect that will occur from the given time/concentration (ECt_{50}) exposure by inhalation of the vapor or gas to 50% of those exposed. The same holds true for the "I" indicating incapacitation. Fifty percent of those exposed to a certain concentration/time by inhalation will be incapacitated by the exposure, which is expressed by the value ICt_{50}. Lethal concentrations over a given time to 50% of those exposed by inhalation are represented by LCt_{50}. When the exposure is to a liquid agent, the terms used to identify the exposure are ED_{50} for the effects-dose by ingestion or skin absorption, ID_{50} for the incapacitation effect by ingestion or skin absorption, and LD_{50} for the lethal dose by ingestion or skin absorption.

Nerve agents

The most common nerve agents are tabun (GA), sarin (GB), soman (GD), GF, and VX. "G" agents are volatile and penetrate the skin well, in seconds or minutes. "V" agents are less volatile and penetrate the skin well in minutes and hours. All nerve agents are related to organophosphate pesticides and in pure form are colorless. However, nerve agents may also be light brown in color when contaminated and vary in their degree of volatility. Some of the agents have the volatility of motor oil, while others have volatility similar to water. Compared to other liquids that are considered to be volatile, such as gasoline, none of the nerve agents are significantly volatile. When comparing the nerve agents as a group, sarin would be considered the most volatile. VX, on the other hand, is the least volatile of the nerve agents, but is the most toxic. G-agents, such as sarin, are considered to be nonpersistent, and V-agents, like VX, are considered persistent agents. Thickened nonpersistent agents may present a hazard for a longer period of time.

Most nerve agents are odorless; however, some may have a faint fruity odor when contaminated. Nerve agents are highly toxic and quick-acting. They enter the body through inhalation or skin absorption. Poisoning can occur also with ingestion of the agents placed in food or drink. Nerve agents work the quickest if inhaled. Inhalation of high concentrations can cause death within minutes. The LC_{50} for sarin is 100 mg/min/m^3, while for VX is 50 mg/min/m^3. Primary target organs for nerve agents are the respiratory and central nervous systems. Entry into the body through skin absorption requires a longer period of time for symptoms to develop. First symptoms from skin contact may not appear for 20–30 min after exposure. However, if the dose of the nerve agent is high, the poisoning process may be rapid. Eye exposure is extremely dangerous, and eyes should be flushed immediately with copious amounts of water if exposed. Liquid nerve agent splashed into the eyes is absorbed faster into the body than through skin contact.

Acetylcholine is an important neurotransmitter, which is essential to complete the transmission of neural impulses from one neuron (fibers that convey impulses to the nerve cell) to another. Without acetylcholine, the body cannot function normally. When a message is sent from the brain for a muscle to move or some other bodily function to activate, acetylcholine is released. It then binds to the postsynaptic membrane, which starts and continues the movement or action. When it is time for the movement to stop, acetylcholinesterase is released to remove the acetylcholine from the synapse, so it can be used again.

Nerve agents are acetylcholinesterase enzyme inhibitors and can affect the entire body. Acetylcholinesterase enzymes normally act upon acetylcholine when it is released. Nerve agents inhibit this action, and an accumulation of acetylcholine occurs. Initially, it results in an overstimulation of the nervous system. The system then becomes fatigued and paralysis results. Paralysis of the diaphragm muscle is the primary cause of death from nerve agent poisoning. This results in the cession of breathing. Accumulation of acetylcholine causes increased nerve and muscle activity, overfunctioning of the salivary glands, secretory glands, and sweat glands. Muscular twitching, fatigue, mild weakness, cramps, and flaccid paralysis, accompanied by dyspnea and cyanosis, result from the excess acetylcholine. Accumulation of excess acetylcholine in the brain and spinal cord results in central nervous system symptoms. Unless the accumulation of acetylcholine is reversed by the use of antidotes, the effects become irreversible and death will occur.

Symptoms of nerve agent poisoning resulting from a low dose include increased saliva production, runny nose, and a feeling of pressure on the chest. Pupils of the eyes exhibit pinpoint constriction (miosis), short-range vision is impaired, and the victim feels

Table 8.13 General Symptoms of Chemical
Agent Poisoning

Salivation
Lacrimation
Urination
Defecation
Gastric distress
Emesis
Miosis

pain when trying to focus on a nearby object. Headache may follow with tiredness, slurred speech, and hallucinations. Exposures to higher doses present more dramatic symptoms. Bronchoconstriction and secretion of mucus in the respiratory system lead to difficulty in breathing and coughing. Discomfort in the gastrointestinal (GI) tract may develop into cramps and vomiting. Involuntary discharge of urine and defecation may also occur. Saliva discharge is powerful and may be accompanied by running eyes and sweating. Muscular weakness, local tremors, or convulsions may follow in cases of moderate poisoning. Exposure to a high dose of nerve agent may lead to more pronounced muscular symptoms. Convulsions and loss of consciousness may occur (Table 8.13). Those individuals most sensitive to nerve agents will experience a lethal dose at about 70 mg/min/m^3. More resistant persons require about 140 mg/min/m^3. The amount of nerve agent for a dermal exposure that would result in a lethal effect is small. In fact, if you were to look at the back of a Lincoln penny and find the Lincoln Memorial, the amount of liquid that it would take to cover one of the columns on the memorial would be a toxic dermal exposure!

Sarin (GB) is a fluorinated organophosphorous compound with the chemical name of phosphonofluoridic acid, methyl-, isopropyl ester. The chemical formula is $CH_3PO(F)OCH(CH_3)_2$. The structure and molecular formula of the compound are shown in Figure 8.26.

Sarin has a vapor density that is about five times heavier than air and about the same specific gravity as water. It is a colorless and odorless liquid in the pure form. To date, OSHA has not identified a permissible exposure concentration for sarin. Sarin is listed by the American Conference of Governmental Industrial Hygienists (ACGIH) and OSHA as a carcinogen. Sarin is stable when in the pure state. While sarin will burn, it has a high flash point and would be difficult to ignite under normal circumstances. Sarin reacts with steam or water to produce toxic and corrosive vapors.

Tabun (GA) is an organophosphorus compound with a chemical name of ethyl N, N-dimethylphosphoramidocyanidate. The chemical formula is $C_2H_5OPO(CN)N(CH_3)_2$. The structure and molecular formula of the compound are shown in Figure 8.27.

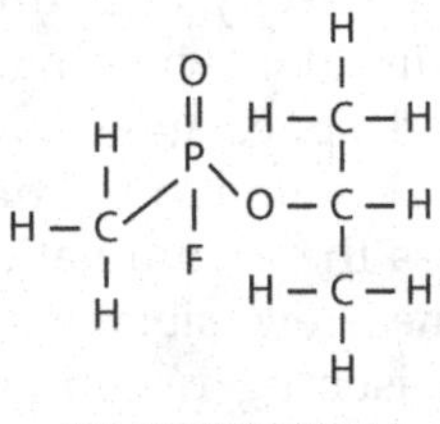

$$CH_3PO(F)OCH(CH_3)_2$$

Figure 8.26 Sarin (GB).

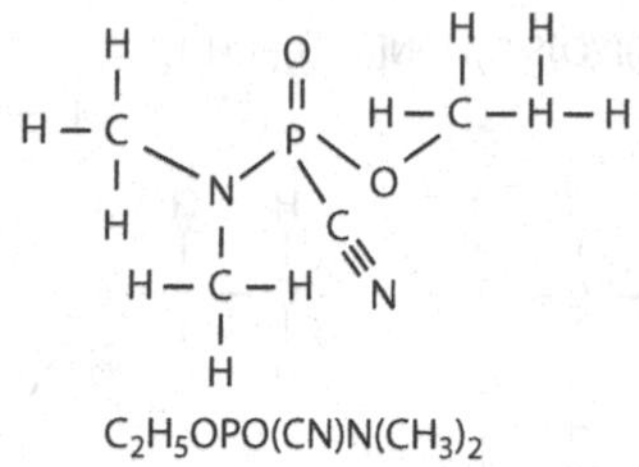

C₂H₅OPO(CN)N(CH₃)₂

$C_2H_5OPO(CN)N(CH_3)_2$

Figure 8.27 Tabun (GA).

Tabun is a colorless to brown liquid with a faint fruity odor. In its pure form, it does not have any odor. The boiling point of tabun is 475°F, which is approximately 159° higher than sarin. Tabun has a vapor density higher than sarin and about 5.5 times heavier than air. Its specific gravity is slightly heavier than water. Tabun has a flash point lower than sarin. There are no explosive limits available. Contact with the agent liquid or vapor can be fatal. Tabun is a lethal cholinesterase inhibitor similar to sarin in the way it affects the human body. It is only about half as toxic by inhalation as sarin, but in low concentrations, it is more irritating to the eyes. Symptoms presented by GA depend on the concentration and rate of entry into the body. Small dermal exposures may cause local sweating and tremors, with few other effects. Symptoms for larger doses are much the same as for sarin, regardless of the route of exposure to the body. They include, in order of appearance, runny nose; tightness of the chest; dimness of vision and pinpoint pupils (miosis); difficulty breathing; drooling and excessive sweating; nausea; vomiting; cramps and involuntary defecation and urination; twitching, jerking, or staggering; headache; confusion; drowsiness; coma; and convulsion. These symptoms are followed by cessation of breathing and death. First symptoms appear more slowly from skin contact than through inhalation. Skin absorption of a dose great enough to cause death can occur in 1–2 min; however, death may be delayed for 1–2 h. The inhalation lethal dose can kill in 1–10 min, and liquid splash in the eyes is almost as fast.

Soman (GD) is a fluorinated organophosphorus compound with a chemical name of pinacolyl methylphosphonofluoridate, and it is a lethal nerve agent. The chemical formula is $CH_3PO(F)OCH(CH_3)C(CH_3)_3$. The structure and molecular formula of the compound are shown in Figure 8.28. Soman has a vapor density more than six times heavier than air. It is a colorless liquid with a fruity odor, when pure, and an amber or dark brown color with an oil of camphor odor, when impure. Doses of soman that can cause death may be only slightly larger than those that produce symptoms. The median

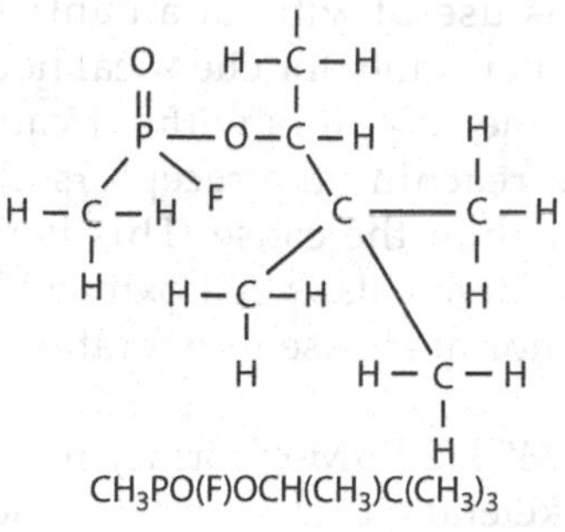

CH₃PO(F)OCH(CH₃)C(CH₃)₃

$CH_3PO(F)OCH(CH_3)C(CH_3)_3$

Figure 8.28 Soman (GD).

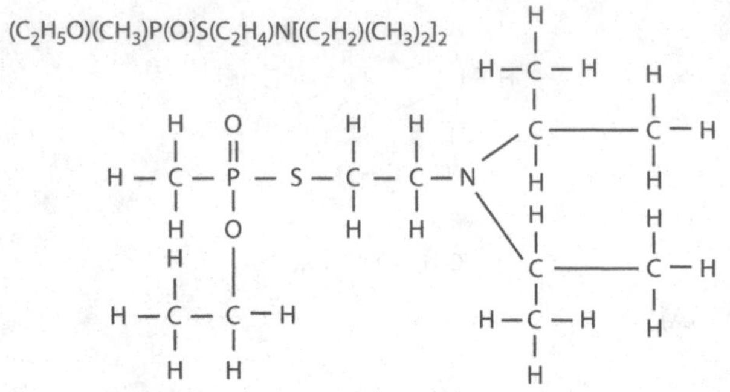

$(C_2H_5O)(CH_3)P(O)S(C_2H_4)N[(C_2H_2)(CH_3)_2]_2$

Figure 8.29 Lethal agent (VX).

incapacitation dose for soman is unknown. In contact with water, it will hydrolyze to form hydrogen fluoride (HF).

Lethal nerve agent (VX) is a sulfonated organophosphorous compound with a chemical name of O-ethyl-S-(2-iisopropylaminoethyl) methyl phosphonothiolate. The chemical formula is $(C_2H_5O)(CH_3O)P(O)S(C_2H_4)N[(C_2H_2)(CH_3)_2]_2$. The structure and molecular formula of the compound are shown in Figure 8.29. VX has a vapor density more than nine times heavier than air. VX is a colorless to straw-colored liquid with little, if any, odor. It has the consistency and appearance of motor oil. VX is a lethal cholinesterase inhibitor, much more toxic than sarin or any of the other nerve agents. Life-threatening doses may be only slightly larger than those producing symptoms. A drop of VX the size of a pinhead on the skin will kill in 5–15 min.

Lethal nerve agent (GF) is a fluorinated organophosphate compound. Limited information is available on this nerve agent. The chemical formula is $CH_3PO(F)OC_6H_{11}$. GF has a sweet, musty odor of peaches or shellac. It has a boiling point of 239°C (463°F); a vapor pressure of 0.044 mmHg at 20°C; a vapor density of 6.2, which is over six times heavier than air; a volatility of 438 mg/m^3 at 20°C; a specific gravity of 1.1327 at 20°C; and a freezing/melting point of −30°C. The flash point of GF is approximately 94°C (200°F), and the flammable limits are not available. GF has an MCt$_{50}$ of less than 1 mg-min/m^3 and an LD$_{50}$ on the skin of 30 mg. All nerve agents are 6.1 poisons, with a UN identification number of 2810. Their NFPA 704 designation would likely be health 4, flammability 1, reactivity 1, and no special information.

Antidotes

No chemical warfare agent is useful without an antidote to protect your own personnel. Atropine is the universal antidote for chemical nerve agent exposures. It functions by binding to the acetylcholine receptors without causing excitement. This keeps the excess of acetylcholine from reaching the receptors. Atropine treats the symptoms of nerve agent exposure rather than the cause. This is why atropine is administered in conjunction with a group of chemicals called oximes. They treat the cause of the problem by restoring the acetylcholinesterase to operation, by breaking the enzyme–nerve agent bond.

Protopam chloride, 2-PAMCl (2-PAM-chloride), may also be used in conjunction with atropine. Its effect is on the skeletal muscles, and it causes an increase in muscle strength. Protopam chloride should be administered in all cases of nerve agent poisoning unless the agent is GD, which is highly unlikely. Officials at the Philadelphia Fire Department have decided to place autoinjectors containing atropine and 2-PAMCl on every department vehicle for the crews in case of chemical nerve agent exposure. Material presented on the administration of atropine and other antidotes for nerve agents is provided only for informational purposes. Administration of atropine or any other medication should be undertaken only upon the advice and guidance of a physician and should be based upon local medical protocols. The material provided here is meant for informational purposes only.

Mustard agents (vesicants)

Blister agents, or vesicants as they are also known, include four basic types: sulfur mustard, nitrogen mustard, lewisite, and phosgene oxime. Like the nerve agents, mustard agents are sometimes incorrectly referred to as gases, when, in fact, they are not. All of the blister agents have the same basic characteristics and will be referred to as mustard, blister agents, or vesicants, interchangeably (Photo 8.8). Mustard agents are persistent by nature. They are made in varying formulations, including distilled, nitrogen, sulfur, and thickened mustards. Military designations for mustard agents are as follows: mustard (H),

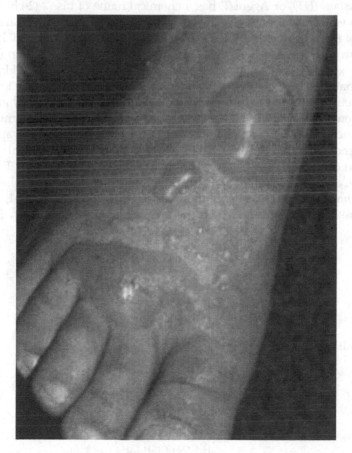

Photo 8.8 Blister agents or vesicants are named because they on contact form blisters on the skin. You cannot be exposed to blister agents by contacting the liquid in the blisters.

distilled mustard, (HD) or (HS), and thickened mustard (HT). Nitrogen mustards have the military designation of (HN$_1$), (HN$_2$), and (HN$_3$). Unlike the nerve agents, mustard agent exposure is rarely fatal. Symptoms for mustard agents do not appear for several hours after exposure. Mustard is designed to incapacitate, not to kill. It can, however, be fatal if a large enough dose is administered or a person has other health problems that will be compounded by mustard exposure.

Sulfur mustard (H), Distilled mustard (HD), is a liquid that has a consistency of motor oil and appears in colors from light yellow to brown and black, based upon the number and type of impurities present. Pure mustard is clear like water. The odor of mustard ranges from onions or garlic to horseradish or mustard, which is where it gets its name. It is a chlorinated sulfur compound. H or HD mustard has a chemical name of bis-(2-chloroethyl) sulfide. Mustard is only slightly soluble in water and is miscible with organophosphate nerve agents. It has a freezing point of approximately 57°F and, if stored outside, would remain frozen for a large part of the year. Mustard has a specific gravity that is heavier than water. The vapor density of mustard is about 5.5 times heavier than air. While mustard is flammable, it has a fairly high flash point. Explosive charges may be used to ignite mustard agents. Flammable limits or explosive ranges for mustards have not been identified. The chemical formula for sulfur/distilled mustard is $(ClH_2CH_2)_2S$, and the structure and molecular formula are shown in Figure 8.30.

Sulfur mustard (HT), or Agent T, has a chemical name of bis-(2-(2-chloroethylthio)ethyl) ether. It has a yellow color with a garlic-like odor, similar to other mustard agents. The specific gravity of HT is heavier than water and similar to distilled mustard. It has a vapor density more than nine times heavier than air and four times heavier than distilled mustard. No airborne exposure limits have been identified for sulfur mustard. Flash point or flammable limits have not been established, and it has a boiling point slightly lower than distilled mustard. Specific health-hazard data, such as airborne exposure limits, have not been established for Agent T; however, it has a toxicity similar to that of distilled mustard, and under no circumstances should anyone be allowed to be exposed to direct vapor or skin or eye contact. The median lethal dose (LCt$_{50}$) of Agent T in laboratory animals is 1650–2250 mg-min/m^3 based on 10 min of exposure. The chemical formula for Agent T sulfur mustard is $C_2H_4S_2C_2H_4OC_2H_4C_2H_4Cl_2$, and the structure and molecular formula are shown in Figure 8.31.

Nitrogen mustard (HN) was developed in three formulations: HN-1, HN-2, and HN-3. HN-1 was the first to be produced in the late 1920s and early 1930s. Originally, it was

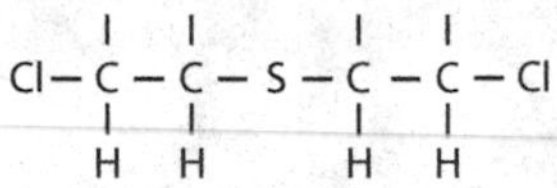

$(ClH_2CH_2)_2S$

Figure 8.30 Sulfur mustard.

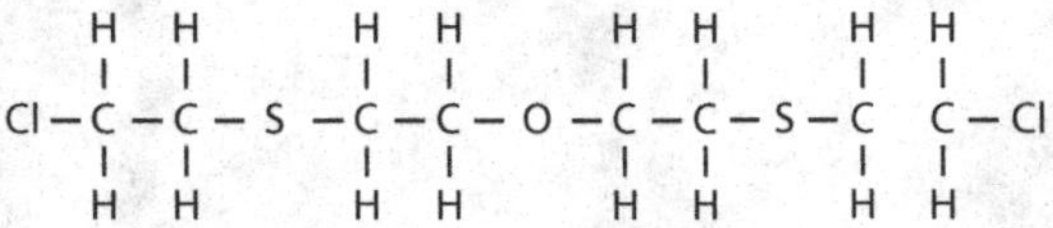

$C_2H_4S_2C_2H_4OC_2H_4C_2H_4Cl_2$

Figure 8.31 Agent T (sulfur mustard).

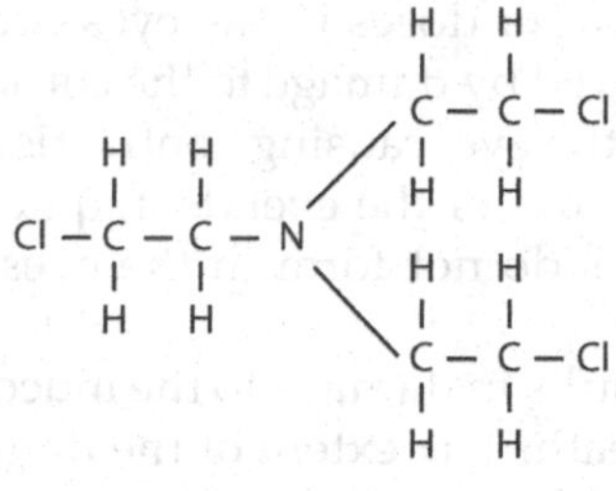

$$
\begin{array}{c}
& & & & \overset{\displaystyle H}{\underset{\displaystyle |}{}} \ \ \overset{\displaystyle H}{\underset{\displaystyle |}{}} \\
& & & & C - C - Cl \\
& \overset{\displaystyle H}{\underset{\displaystyle |}{}} \ \ \overset{\displaystyle H}{\underset{\displaystyle |}{}} & \nearrow & \overset{\displaystyle |}{H} \ \ \overset{\displaystyle |}{H} \\
Cl - C - C - N & & \\
& \overset{\displaystyle |}{H} \ \ \overset{\displaystyle |}{H} & \searrow & \overset{\displaystyle H}{\underset{\displaystyle |}{}} \ \ \overset{\displaystyle H}{\underset{\displaystyle |}{}} \\
& & & C - C - Cl \\
& & & \overset{\displaystyle |}{H} \ \ \overset{\displaystyle |}{H}
\end{array}
$$

$$N(CH_2CH_2Cl)_3$$

Figure 8.32 Nitrogen mustard.

developed as a pharmaceutical and used to remove warts before it became a military agent (Photo 8.8). Agent H-2 was developed as a military agent and became a pharmaceutical. HN-3 was designed as a military mustard agent and is the only one that remains in military use. Therefore, this section will only cover the characteristics of HN-3 mustard agent. HN-3 is colorless to pale yellow with a butter-almond odor. The chemical formula for nitrogen mustard agent HN-3 is $N(CH_2CH_2Cl)_3$, and the structure and molecular formula are shown in Figure 8.32. Mustard is effective in small doses and affects the lungs, skin, and eyes. Symptoms are likely to appear first on delicate tissues, such as the soft membranes surrounding the eyes. This will be followed by tissues of the throat, lungs, nose, and mouth. Mustard agents have the greatest effects on warm, moist areas of the body, such as eyes, respiratory tract, armpits, groin, buttocks, and other skin folds. Because of the delayed onset of symptoms, those exposed often do not know they have contacted mustard. Mustard agents can be an inhalation or skin absorption hazard. The symptoms of the agents do not appear for 2–24 h after the exposure, and the usual onset in most people is between 4 and 8 h. Liquid exposures to the skin and eyes produce an earlier onset of symptoms than do vapor exposures. When mustard agent contacts the skin, it does not produce pain or sensation. Those exposed would not likely be aware of the contact. Once mustard contacts the skin, the damage has already occurred. If the exposure is known at that point, decontamination will, at best, only reduce the amount of material still on the skin. It will otherwise be ineffective against stopping the damage to the body.

Because mustard is a persistent agent, it will remain on the clothing of victims and anything else that it touches for a long period of time. Decontamination, while it will not provide much help to the victim, is important to prevent secondary contamination of responders and others. Hypochlorite solution or flooding amounts of water are necessary to remove mustard agents. Unlike nerve agents, there is no known antidote for mustard exposure. It enters the body through the skin or mucous membrane and triggers biochemical damage within seconds to minutes, and no known medical procedures can stop or minimize the damage.

First symptoms from mustard exposure involve erythema, which is a redness of the skin similar to sunburn, followed by itching, burning, or stinging pain. The time between exposure and onset of symptoms may be affected by the dose of mustard, the air temperature, the amount of moisture on the skin, and the area of exposure on the skin. Within the vicinity of the reddened skin, vesicles form and eventually become blisters filled with a yellowish, translucent fluid. The eyes are sensitive to mustard agent, and exposure will produce the most rapid onset of symptoms. Irritation, accompanied by a "gritty" sensation of something in the eye, is followed in some cases by inflammation of the eyes (conjunctivitis). Larger doses in the eyes produce edema of the eyelids (fluid buildup), which may be followed by damage to the cornea. The eyelids may become swollen, accompanied by pain in the eye, causing contraction of the muscles surrounding the eye and, finally, complete closure of the eyelids. Liquid mustard causes the most severe eye damage; however, blisters do not form in the eyes. Almost all of the eye exposure symptoms are self-reversing.

Respiratory exposure results in damage to the mucosa, or cell lining, followed by cellular damage and cellular death. The extent of the damage is dependent on the amount of the dose during exposure. Mustard enters the respiratory system through inhalation, and the greater the amount inhaled, the more severe the symptoms and damage. Small amounts of mustard result in nasal irritation and possibly bleeding. Sinus passages may be irritated along with the throat, which will produce a scratchy sore throat type of pain. Laryngitis may result from larger exposures and can include complete loss of voice for a period of time. Shortness of breath and productive coughs may be present in the respiratory system resulting from large doses. When respiratory symptoms begin within 4 h of the exposure, it is usually an indication of a large dose through inhalation.

When a large dose of mustard agent has entered the body, it may be carried to the bone marrow. Damage occurs here to the precursor cells, which is followed by a decrease in the white blood cell count. Over a period of days, red blood cells and platelets are also diminished. The GI tract is also a target organ for mustard agent. Nausea and vomiting are not unusual within 12–24 h after exposure. Last, the central nervous system may be affected by exposure to mustard agents. Generally, the symptoms are not well recognized and may involve sluggish, apathetic behaviors. Symptoms have been recorded for as much as a year after exposure. While death from mustard exposure is rare, lethal doses will produce convulsions and severe pulmonary damage that is usually associated with infection. In large dose exposures, death can occur within 24–36 h.

Lewisite (L) is a vesicant from the arsenical (vesicant) chemical family. Lewisite causes many of the same types of damage to skin, eyes, and the respiratory system as do the mustard agents. The chemical name is dichloro-(2-chlorovinyl) arsine. It has not knowingly been used on the battlefield, and human exposure data are limited. The major difference between lewisite and mustard is that lewisite causes pain immediately upon exposure, whereas the mustards have a delayed onset. Visible tissue damage will occur quickly in the form of a grayish-appearing area of dead skin. Blisters also develop more quickly than with mustard, but may take up to 12 h for full blistering effects to develop.

Lewisite is also a systemic poison, which can result in pulmonary edema, diarrhea, restlessness, weakness, subnormal temperature, and low blood pressure. Severity of symptoms, in order of appearance, are blister agent, toxic lung irritant, tissue absorption, and systemic poison. If inhaled in high enough concentrations, lewisite can cause death in as little as 10 min. Common routes of exposure into the body are through the eyes, skin, and inhalation. Lewisite is an oily colorless liquid when pure. "War gas" is amber to dark brown in color with a geranium-like odor; pure lewisite has little, if any, odor. Lewisite is much more volatile than mustard agents. Often, lewisite is mixed with mustard to lower the boiling point of the mixture. The military designation for the mustard/lewisite mixture is (*HL*). Lewisite is insoluble in water and has a specific gravity, which makes it heavier than water. It has a vapor density more than seven times heavier than air. Lewisite has a low level of flammability, with no flash point or flammable range identified. The molecular formula for lewisite is $C_2H_2AsCl_3$, and the structure and molecular formula are shown in Figure 8.33.

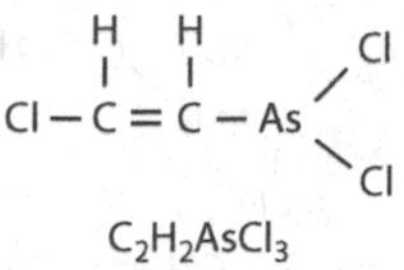

$$H \quad H$$
$$| \quad |$$
$$Cl - C = C - As \diagup^{Cl}_{\diagdown Cl}$$

$$C_2H_2AsCl_3$$

Figure 8.33 Lewisite.

Lewisite is a vesicant and toxic lung irritant that is absorbed into tissues. If inhaled in high concentrations, it can be fatal in as little as 10 min; the body is unable to detoxify itself from lewisite exposure. Routes of entry into the body include the eyes, skin absorption, and inhalation. Eye contact results in pain, inflammation, and blepharospasm (spasms of the muscles of the eyelid), which leads to closure of the eyelids, corneal scarring, and iritis (inflammation of the iris). If decontamination of the eyes occurs quickly after exposure, damage may be reversible; however, permanent injury or blindness can occur within 1 min of exposure.

Lewisite is irritating to the respiratory tract, producing burning, profuse nasal discharge, and violent sneezing. Prolonged exposure results in coughing and large amounts of froth mucus. Vapor exposure to the respiratory tract produces much the same symptoms as mustard. The main difference is that edema of the lungs is more exceptional and may be accompanied by pleural fluid. Lewisite does not affect the bone marrow, but does cause an increase in capillary permeability, with ensuing plasma leakage into tissues. This results in sufficient fluid loss to cause hemoconcentration, shock, and death. Exposures that do not result in death can cause chronic conditions, such as sensitization and lung impairment. Lewisite is a suspected carcinogen.

An antidote is available for lewisite exposure. BAL (British-Anti-Lewisite; dimercaprol) was developed by the British during World War II. The antidote is produced in oil diluent for intramuscular administration to counter the systemic effects of lewisite. There is no effect, however, on the skin lesions (eyes, skin, and respiratory system) from the antidote. Mustard agents (II), (HD), (HS), and (HT), like nerve agents, would be classified as Class 6.1 poisons by the DOT and would have NFPA 704 designations of health 4, flammability 1, reactivity 1, and special 0. The UN four-digit identification number for mustard agents would be 2810, as it also is for all of the mustard and nerve agents.

Phosgene oxime (CX) is also a vesicant, or blister agent, but it should not be confused with phosgene, which is a lung-damaging agent. Phosgene oxime causes urticaria (eruptions on the skin), rather than the fluid-filled blisters that occur with other vesicants. CX is a solid (powder) at temperatures below 95°F, but can be considered a flash-point solid, because of the high vapor pressure of 11.2 at 25°C. The boiling point is 53°C–54°C, and it is not considered flammable. Not much is known about the mechanical action or biological activity of phosgene oxime. It is absorbed through the skin, and contact with the material causes extreme pain immediately, much like lewisite, and tissue effects occur quickly. Irritation begins to occur in about 12 s at a dose of 0.2 mg-min/m³. When the agent has been in contact with the skin for 1 min or more, the irritation becomes unbearable at a dose of 3 mg-min/m³. It is also irritating to the respiratory system and causes the same types of eye damage as lewisite. Phosgene oxime remains persistent in the soil for about 2 h and is considered nonpersistent on other surfaces. The agent is known to be corrosive to most metals. There is no known antidote for phosgene oxime, and exposures should be treated symptomatically after decontamination, which should occur immediately. The structure and molecular formula are shown in Figure 8.34.

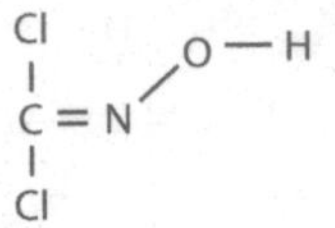

$$\begin{array}{c} Cl \\ | \\ C = N \\ | \\ Cl \end{array} \raise1ex\hbox{$\diagup$} O - H$$

CCl$_2$NOH

Figure 8.34 Phosgene oxime.

Blood agents (cyanogens)

Blood agents are common industrial chemicals that have been used on the battlefield to produce casualties. Two primary types of blood agents used by the military are hydrogen cyanide (AC) and cyanogen chloride (CK). Poisonous effects of cyanide have been well known since ancient times. Cyanide was the first blood agent used as a chemical warfare agent. Cyanides are salts with the metals potassium, sodium, and calcium most commonly used to form the compounds. The term "blood agent" was used because at the time cyanide was introduced as a warfare agent, it was thought to be the only agent that was transported via blood to the target organ. It is now known that other agents, such as nerve and vesicants, are also carried by the blood. Therefore, realistically, the term has become obsolete.

Cyanide is found naturally in many foods, such as lima beans, cherries, apple seeds, and the pits of peaches and similar fruits. It is reported to have an odor of burnt almonds or peach pits. However, 40% of the population cannot detect the odor of cyanide. Blood agents function by interfering with the body's ability to use oxygen at the cellular level. They are considered chemical asphyxiants, because death occurs from a lack of oxygen in the body. Inhalation is the primary route of exposure for blood agents. Unlike nerve agents and vesicants, blood agents are gases or volatile liquids. They produce vapors easily, which dissipate quickly in air and are, therefore, considered nonpersistent agents. Antidotes are available for cyanide poisoning, but must be administered quickly after exposure to be effective.

Hydrogen cyanide (HCN), (AC), also known as hydrocyanic acid or prussic acid, is a colorless to water-white or pale-blue liquid at temperatures below 80°F. It has an odor of peach kernels or bitter almonds at 1–5 ppm. HCN may not have a detectable odor in lethal concentrations. The acute toxicity of hydrogen cyanide is high. There is rarely a reported chronic exposure to HCN because you either get better quickly from an exposure or you die. It is toxic by inhalation, skin absorption, and ingestion, with a TLV of 10 ppm in air. The LD$_{50}$ through ingestion is 10 mg/kg of body weight. Skin absorption LD$_{50}$ is estimated to be 1500 mg/kg of body weight. Inhalation of HCN has an LC$_{50}$ of 63 ppm for 40 min. OHSA has established a PEL of 10 ppm (11 mg/m^3) for skin contact. TLV-TWA values are listed by ACGIH at 10 ppm (11 mg/m^3) for skin absorption. Inhalation of 18–36 ppm over a period of several hours can produce weakness, headache, confusion, nausea, and vomiting. Inhalation of 270 ppm can cause immediate death, and 100–200 ppm over a period of 30–60 min can also be fatal. Absorption of 50 mg through the skin can be fatal. Ingestion of 50–100 mg of HCN can also be fatal.

HCN is highly soluble and stable in water. It is extremely flammable, with a flash point of 0°F and an explosive range of 6%–41% in air. Hydrogen cyanide has a boiling point of 25.7°C (78°F). Its vapor density is 0.990, which is slightly lighter than air, and it has a liquid density of 0.687, which is lighter than water. The autoignition temperature for HCN is 1000°F. Liquid HCN contains a stabilizer (usually phosphoric acid) and, as it ages, may explode if the acid stabilizer is not maintained in the solution at a sufficient concentration. HCN can polymerize explosively if heated above 120°F or if contaminated with any

$$H - C \equiv N$$
HCN

Figure 8.35 Hydrogen cyanide.

alkali materials. The structure and molecular formula of hydrogen cyanide are shown in Figure 8.35.

Hydrogen cyanide is volatile, and the DOT lists it as a 6.1 poisonous liquid with an inhalation hazard from the vapor. The NFPA 704 designation for AC is health 4, flammability 4, reactivity 2, and special 0. Vapors from hydrogen cyanide are highly toxic. The four-digit UN identification number is 1613 for less than 20% hydrogen cyanide and 1051 for greater than 20% hydrogen cyanide. The Chemical Abstract Service (CAS) number is 74-90-8. Hydrogen cyanide is thought to act on the body by blending with cytochrome oxidase (an enzyme essential for oxidative processes of the tissues) and blocking the electron carrier system. This results in loss of cellular oxygen use. The central nervous system, and particularly the respiratory system, are notably sensitive to this effect, and respiratory failure is the usual cause of death.

Cyanogen chloride (CK) is a colorless compressed gas or liquefied gas with a pungent odor and is slightly soluble in water. It is not flammable; however, containers exposed to radiant heat or fire may explode and give off toxic or irritating fumes. CK is highly toxic by ingestion or inhalation, and is an eye and skin irritant. The TLV is 0.3 ppm, and the ceiling is 0.75 mg/m^3 in air. Because it is a gas, it is considered a nonpersistent agent. The vapor density is 2.16, which is heavier than air, and the specific gravity is 1.186, which is heavier than water. It has a boiling point of approximately 58°F, and rapid evaporation can cause frostbite. Cyanogen chloride has a freezing/melting point of 44°F (−6.9°C).

CK may polymerize violently if contaminated with hydrogen chloride or ammonium chloride. Upon heating, it decomposes, producing toxic and corrosive fumes of hydrogen cyanide, hydrochloric acid, and nitrogen oxides. Hazardous polymerization can occur. Cyanogen chloride will react slowly with water to form hydrogen chloride gas. It acts on the body in two ways. Systemic effects of CK are much like those of hydrogen cyanide. Additionally, it causes irritation of the eyes, upper respiratory tract, and lungs. Eye irritation results in tearing. CK, like AC, stimulates the respiratory system and rapidly paralyzes it. Exposure is followed by immediate, intense irritation of the nose, throat, and eyes, with coughing, tightness in the chest, and lacrimation. This is followed by dizziness and increasing difficulty breathing. Unconsciousness comes next, with failing respiration and death within a few minutes. Convulsions, retching, and involuntary urination and defecation may occur. If these effects are not fatal, the signs and symptoms of pulmonary edema may develop. There may be repeated coughing, with profuse foamy sputum, rales in the chest, severe dyspnea, and distinct cyanosis. Recovery from the systemic effects is usually as prompt as in AC poisoning. However, a higher incidence of residual damage to the central nervous system should be expected. Based upon the concentration of the cyanogen chloride to which the victim has been exposed, the pulmonary effects may evolve instantly or may be delayed until the systemic effects have subsided. Consequently, early prognosis must be cautious. The structure and the molecular formula for cyanogen chloride are shown in Figure 8.36.

Cyanogen chloride becomes volatile as temperatures increase, and the DOT lists it as a 2.3 poison gas. The NFPA 704 designation for CK is estimated to be health 4, flammability 0,

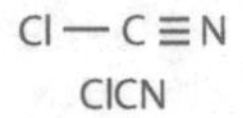

$$Cl — C \equiv N$$
ClCN

Figure 8.36 Cyanogen chloride.

reactivity 2, and special −0. Cyanogen chloride vapors are highly toxic. It has a four-digit UN identification number of 1589 (inhibited). Treatment for either AC or CK poisoning is to follow the treatment protocols for airway, breathing, and circulation (ABCs) and administer oxygen to assist breathing. Instructions for administration and dosage should be based on local protocols and with the advice of a physician. Sodium nitrate is administered to produce methemoglobin, thus seizing the cyanide on the methemoglobin. The sodium thiosulfate combines with the confiscated cyanide to form thiocyanate, which is then excreted from the body.

Choking agents (lung-damaging agents)

Choking agents are known to cause pulmonary edema, which is the accumulation of fluid in the lungs. Examples of lung-damaging agents include phosgene (CG), diphosgene (DP), chlorine, and chloropicrin (PS). Phosgene is the best known of the choking agents. It is produced by heating carbon tetrachloride, which was once used as a fire-extinguishing agent. It was discontinued as a fire-extinguishing agent because when it came in contact with hot surfaces, it released phosgene gas. Choking agents irritate the bronchi, trachea, larynx, pharynx, and nose, which may result in pulmonary edema and contribute to a choking sensation.

Phosgene (CG) (carbonyl chloride) is produced as a liquid or liquefied gas, which is colorless to light yellow. At ordinary temperatures and pressures, CG is a colorless gas. Odors range from strong and stifling, when concentrated, to the smell of freshly mowed hay in lower concentrations. Phosgene has a boiling point of 8.2°C (45.6°F) and is noncombustible. The vapor density is 3.4, which is heavier than air, and the specific gravity is 1.37 at 68°F (20°C), which is heavier than water. The primary route of exposure is through inhalation, by which it is highly toxic. It is also a strong eye irritant and has a TLV of 0.1 ppm or 0.40 mg/m³ in air. General population limits are 0.0025 mg/m³. In addition to mild conjunctival irritation, direct effects of exposure to phosgene result in damage to the lungs. The primary effect of exposure is pulmonary edema (fluid in the lungs). Death can occur within several hours after an exposure to a high concentration. Most fatalities, however, reach a maximum effect from the pulmonary edema in about 12 h, and death occurs within 24–48 h after exposure. Symptoms include coughing, choking (thus the term choking agent), tightness in the chest, nausea, and possibly vomiting, headache, and lacrimation. There is no real relationship between the symptoms and the prognosis. The structure and molecular formula for phosgene are shown in Figure 8.37. The DOT lists phosgene as a 2.3 poison gas. The NFPA 704 designation for CG is health 4, flammability 0, reactivity 1, and special 0. Vapors of phosgene are highly toxic. It has a four-digit UN identification number of 1076.

Diphosgene (DP), trichloromethyl chloroformate, is a clear, colorless liquid with an odor similar to phosgene. It is noncombustible, a strong irritant to the eyes and tissues, and is toxic by inhalation and ingestion. DP has a boiling point of 127°C–128°C (263°F) and a vapor pressure of 4.2 at 68°F (20°C). The liquid density is 1.65, which is heavier than water, and a melting/freezing point of 314°F (157°C). Inhalation LC$_{50}$ is 3600 mg/m³ for 10 min. Effects of exposure are quite similar to phosgene gas. Its molecular formula is

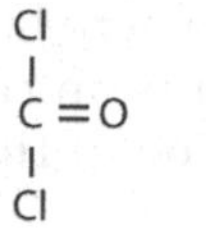

$$Cl$$
$$|$$
$$C = O$$
$$|$$
$$Cl$$

COCl$_2$

Figure 8.37 Phosgene.

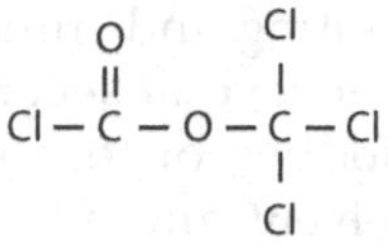

$$O \quad\quad Cl$$
$$\|\quad\quad |$$
$$Cl - C - O - C - Cl$$
$$|$$
$$Cl$$

ClCOOCCl

Figure 8.38 Diphosgene.

ClCOOCCl$_3$, and the structure and molecular formula are shown in Figure 8.38. The DOT lists diphosgene as a 6.1 poison liquid. The NFPA 704 designation for CG is estimated to be health 4, flammability 0, reactivity 1, and special 0. It has a four-digit UN identification number of 2972.

Riot-control agents (irritant agents and vomiting agents)

Riot-control agents are also referred to as tear gas, irritants, and lacrimators. They are local irritants, which in low concentrations act essentially on the eyes, resulting in intense pain and profuse tearing. Higher concentrations irritate the upper respiratory tract, skin, and sometimes cause nausea and vomiting. Exposure to these materials is rarely serious and may not require medical attention. Tear gases are not really gases at all. They are fine particulate smoke or aerosolized solid materials that can cause contamination of those exposed. Riot-control agents are commonly used by law enforcement officers. (CS) and (CN) are the primary agents used by law enforcement agencies today. However, it may be possible to encounter an older riot-control agent, (DM), which is a "vomiting gas." CN, chloroacetophenone, is also known as Mace. It has been replaced today by CS, O-chlorobenzylidene malononitrile, which was developed in the 1950s and has become the agent of choice of law enforcement. Other irritant agents include (CNC), chloroacetophenone in chloroform; (CA), bromobenzyl cyanide; and (CR), dibenz-(b,f)-1,4-oxazepine. OC is the designation for pepper spray. CK and PS are also considered to be lacrimators.

The biological mechanism by which riot-control agents work has not been well studied or documented compared to the other types of chemical agents. Generally, riot-control agents cause pain, tearing, and conjunctivitis in the eyes, which can be accompanied by spasms of the muscles around the eyes. A burning sensation occurs in the nose and respiratory tract. This is followed by sneezing and a large volume of nasal discharge. The chest may feel tight, and there may be a shortness of breath accompanied by coughing and secretions from the bronchial tubes when inhaled. Contact with the skin produces tingling, a burning feeling, and redness similar to sunburn. If conditions of high temperature and humidity are present along with a high concentration of agent, blisters may occur within 8–12 h, which are similar to those produced by vesicants. Blisters have been reported by firefighters resulting from exposure when entering buildings where law enforcement personnel have discharged CS. Death, although rare, may occur from riot-control agents when exposure occurs in confined areas for an extended period of time. However, there are no reported deaths from open-air use.

Agent CS, O-chlorobenzylidene malononitrile, is a white crystalline solid with a boiling point of 590°F–599°F (310°C–315°C), a flash point of 386°F (197°C), and a melting point of 194°F (87°C). It has a vapor pressure of 3.4×10^{-5} at 20°C and a vapor density several times heavier than air. CS is immiscible in water. Volatility is 0.71 mg/m^3 at 77°F (25°C). It has an odor that is biting and similar to pepper spray. Clouds produced by CS are white near the source for several seconds following the release. Usually, CS is disseminated by burning, exploding, or forming an aerosol. It may also be used in liquid form if dissolved into a solvent and aerosolized. CS acts rapidly compared to CN, and it is 10 times more potent, while it is also much less toxic. The LD$_{50}$ for CS is approximately 200 mg/kg of body weight, or 14 g for a 70 kg person. CS has a TLV-TWA of 0.4 mg/m^3, an ICt$_{50}$ of 10–20 mg-min/m^3, and an LCt$_{50}$ of 61,000 mg-min/m^3. The chemical formula for CS is $C_{10}H_5ClN_2$, and the structure and molecular formula are shown in Figure 8.39.

Agent CR, dibenz-(b,f)-1,4-oxazepine, is a pale-yellow, crystalline solid, which has a melting point of 163°F (72°C). It has a pepper-like odor and is only used in solution for dissemination in liquid dispensers. Solutions consist of 0.1% of CR in 80 parts of propylene glycol and 20 parts water. CR is an eye irritant in organic solutions at concentrations of 0.0025% or lower. Agent CR is less toxic when inhaled, but has more profound skin effects, which are longer lasting. It is a persistent agent when released in the environment and deposited on clothing.

Agents CN, chloroacetophenone, and *CA*, bromobenzylcyanide, are both white, crystalline solids with boiling points of 478°F (247°C) and 468°F (242°C), respectively, and freezing points of 129°F (53°C) and 77°F (25°C), respectively. Vapor density for CN is 5.3, and for CA it is 4.0, both of which are heavier than air by four and five times, respectively. CN has an odor similar to apple blossoms, and CA smells like sour fruit. Agent CA is normally found as a liquid, and CN can also be used as a liquid in suitable solvents. When released, these agents produce a bluish-white cloud at the time of release.

Riot-control agents CA and CN are dispersed as minute particulate smoke and as a vapor resulting from burning munitions, such as lacrimator candles and grenades. Liquid agents may be dispersed from aircraft spray or exploding munitions. The OSHA PEL for agent CN is 15 mg/m^3, and the TLV is 10 mg/m^3. ICt$_{50}$ for CN is 30 mg-min/m^3, and the LCt$_{50}$ is estimated to be 8,000–11,000 mg-min/m^3. For CA, the ICt$_{50}$ is 80 mg/m^3, and the LCt$_{50}$ is 7,000 mg-min/m^3 from a solvent and 14,000 mg-min/m^3 from a grenade. CN has a chemical formula of $C_6H_5COCH_2Cl$, and CA has a chemical formula of C_8H_6BrN. The structures and molecular formulas for CA and CN are shown in Figure 8.40.

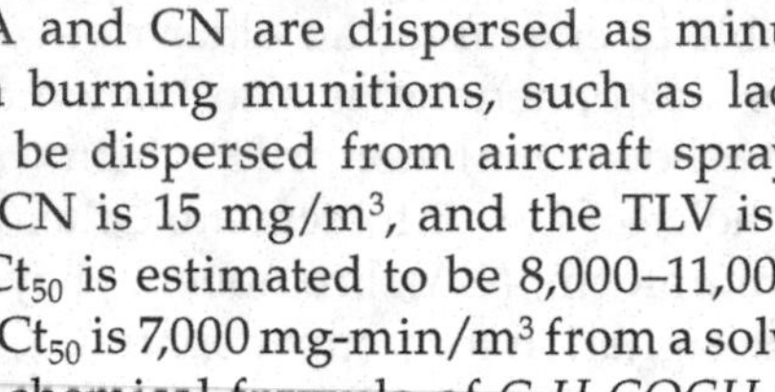

$C_{10}H_5ClN_2$

Figure 8.39 Riot control agent (CS).

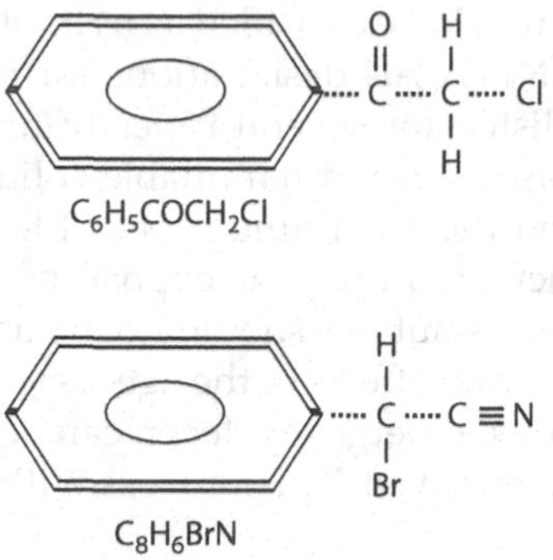

C6H5COCH2Cl

C8H6BrN

Figure 8.40 Riot control agents (CA) and (CN).

Chloropicrin (PS), nitrotrichloromethane, trichloronitromethane, nitrochloroform, is a slightly oily, colorless, pale to transparent liquid that is nearly stable. It is nonflammable, with a boiling point of approximately 235°F (112°C) and slight water solubility. The vapor density is 5.7, which is heavier than air. PS is a strong irritant, highly toxic by inhalation and ingestion, with a TLV of 0.1 ppm (0.7 mg/m^3) in air. When heated, it decomposes to form hydrogen chloride, phosgene, carbon monoxide, and oxides of nitrogen. Decomposition can be explosive, and it can become shock sensitive. The LC$_{50}$ for inhalation is 66 mg/m^3 over 4 h, and the oral dose, LD$_{50}$, is 250 mg/kg of body weight. Chloropicrin is incompatible with strong oxidizers, alcohols, sodium hydroxide, and aniline. The DOT lists chloropicrin as a 6.1 poison liquid, and it has a UN four-digit identification number of 1580. The NFPA 704 designation for chloropicrin is flammability 0, health 4, reactivity 3, and special 0. The CAS number is 76-06-2.

Vomiting agents

Vomiting agents create a strong pepper-like irritation of the upper respiratory tract along with irritation of the eyes and profuse tearing. This may be accompanied by intense, uncontrollable sneezing, coughing, nausea, vomiting, and an overall feeling of malaise (nonspecific feeling of illness). Agents primarily associated with this group include (DA) diphenylchloroarsine, (DM) diphenylaminochloroarsine (Adamsite), and (DC) diphenylcyanoarsine. Inhalation of the fine aerosol mist is the principal route of entry into the body, along with irritation and symptoms resulting from direct eye contact. Unlike other riot-control agents, vomiting agents generally do not produce any discomfort when in contact with skin.

These agents are all crystalline solids, which are released as smoke when heated. Smoke produced by DM is a light or canary yellow, while DA and DC produce a white smoke. These agents are highly effective in low concentrations, which may not be detectable during exposure. Symptoms produced include a feeling of pain and sense of fullness in the nose and sinuses, accompanied by a severe headache, intense burning in the throat, and tightness and pain in the chest. Coughing is uncontrollable, and sneezing is violent and persistent. Nasal secretion is greatly increased, and quantities of ropy saliva flow from the mouth. Nausea and vomiting are prominent. Mental depression may occur during the progression of symptoms. There are usually no long-term effects from vomiting agent exposures, and symptoms generally disappear within 20 min to 2 h. Like other riot-control agents, deaths have occurred from exposure to high concentrations in enclosed areas.

Most riot-control agents would be classified as irritants under DOT hazard class 6.1. No information is available on NFPA 704 designations for riot-control agents. UN four-digit identification numbers are listed for several types of tear gas and containers. Grenades and tear gas candles are poison 6.1 and flammable solids. Tear gas devices are assigned the number 1700; devices and the remaining types of tear gas are listed as 1693. Because riot-control agents are particulate in nature, exposures will require decontamination of victims and responders. Bleach solutions should not be used for decontamination, because they will make the symptoms and effects of the agents worse. Plain soap and water are the best decontamination solutions. Emergency decon can be performed using the water from hose lines. Clothing must be removed. This process will remove the majority of the agent particles off of the person.

Miscellaneous chemical agents

Psychedelic agent 3 (BZ), 3-quinuclidinyl benzilate, also known as "agent buzz," is a potent psychoactive chemical that affects the central nervous system as well as the circulatory, digestion, salivation, sweating, and vision systems. Those exposed experience hallucinations, and it acts as a sedative. The experience is much the same as that encountered by the narcotics amphetamines and cocaine. Three to four days after exposure, full recovery is expected from BZ intoxication. The TLV-TWA for BZ is 0.004 mg/m^3, and the general population limits are 0.0001 mg/m^3. BZ has an ICt$_{50}$ value of 101 mg-min/m^3 (15 L/min) and an estimated LCt$_{50}$ of 200,000 mg-min/m^3.

Infectious substances

Infectious substances are microorganisms or toxins derived from living organisms that produce death or disease in humans, animals, or plants. There are both good and bad bacteria present in the environment and living organisms. The bad bacteria are referred to as pathogens, because they can cause death to a living organism. Disease-causing microorganisms (pathogens) are classified as 6.2 Infectious Substances under the DOT hazard class system. This section will also cover potential terrorist biological agents, including bacteria, viruses, and toxins (Figure 8.41).

Toxins, which are poisons produced by microorganisms or plants, are classified as 6.1 Poisons. DOT defines an infectious substance or etiologic agent as "a viable (live) microorganism, or its toxin, that is capable of producing disease in humans." Included in the DOT regulation are agents listed in 42 CFR 72.3 of the regulations of the DSHS or

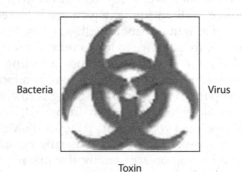

Figure 8.41 Groups of infectious substances/biological agents.

Table 8.14 Characteristics of Biological Agents

Some similarities to chemical agents
Onset of symptoms is the big difference
Route of entry
Sensitive to environmental conditions
Toxicity comparisons
Invisible to our senses, difficult to detect

any other agent that causes or may cause severe, disabling, or fatal disease. Many materials shipped as infectious substances are diagnostic specimens. A diagnostic substance is any human or animal material, including blood, and its components, tissue, and tissue fluids being shipped for purposes of diagnosis. Medical wastes are also included in this hazard subclass.

The main reason this group is included with poisons is that the living organisms produce toxins (poisons). These toxins are biological wastes produced by the microorganisms. In small doses, the body's defense systems can handle the toxins. However, as the volume of toxin increases, it overwhelms the body's ability to defend against it. Toxins travel through the body until they reach a susceptible tissue and cause damage. Examples of typical infectious substances include *Bacillus anthracis* (anthrax), *Botulinum toxin* (botulism), *Francisella tularensis* (tularemia/rabbit fever/deerfly fever), *Coxiella burnetii* (rickettsia/Q fever), *Venezuelan equine encephalitis* (VEE), *Brucella suis* (brucellosis/undulant fever/Bang's disease), smallpox, ricin, mycotoxins, and *Staphylococcal enterotoxin B* (SEB). Infectious substances also include bloodborne pathogens HIV and HBV, as well as any other substance that meets the DOT definition (Table 8.14).

Biological agents can be subdivided into several related groups. These include *bacteria* and *rickettsia*, *viruses*, and *toxins*. Bacteria and rickettsia are single-celled, microscopic organisms that can cause disease in plants, animals, and humans (Figure 8.42). Some of the diseases caused by bacteria include anthrax, botulism, plague, cholera, diphtheria, tuberculosis, typhoid fever, typhus, Legionnaire's disease, Lyme disease, and strep infections. Bacterial organisms have a nucleus, intracellular nonmembrane-bound organelles (a specialized cellular part that resembles an organ), and a cell wall. Rickettsia are pleomorphic (come in varying sizes), parasitic microorganisms that live in the cells of the intestines of arthropods (invertebrate animals), such as insects, spiders, and crabs, which have segmented bodies and jointed limbs. Some are pathogenic to mammals and man, where they are known to cause the typhus group of fevers. *Rickettsia* are smaller than bacteria, but

- Anthrax
- Plague
- Tularemia
- Q Fever

Anthrax bacillus

Figure 8.42 Examples of bacteria and rickettsia.

Photo 8.9 Toxic materials are the primary reason decontamination is conducted at hazmat incidents.

larger than viruses. Like the viruses, rickettsia are obligate (they cannot exist on their own or in any other form); they are considered intracellular parasites. *Viruses* are submicroscopic organisms, smaller than bacteria, and are unable to live on their own. They must invade a host cell and make use of its reproductive mechanism to multiply. *Toxins* are poisons produced by living organisms, including plants, bacteria, and animals. They would be classified as 6.1 Poisons rather than 6.2 Infectious Substances because they are toxic materials rather than disease-causing agents (Photo 8.9).

Bacterial agents

Bacteria are single-celled organisms that range in shape and size from cocci (spherical cells), with a diameter of 0.5–1.0 m (μm), to long rod-shaped organisms known as bacilli, which may be from 1 to 5 m in size. Chains of bacilli have been known to exceed 60 m in size. Some bacteria have the ability to turn into spores. In this form, the bacteria are more resistant to cold, heat, drying, chemicals, and radiation than the bacterial form would be. When in the spore form, bacteria are inactive, or dormant, much like the seeds of a plant. When conditions are favorable, the spores will germinate just like seeds.

Bacteria have two methods by which they can cause disease in humans and animals. The first is by attacking the tissues of the living host. Second, all living organisms produce waste. Bacteria may produce a toxic or poisonous waste material that causes disease in the host. Some bacteria attack using both methods.

Anthrax spores are highly persistent and resist adverse environmental elements and remain viable for hundreds of years in soil and in dried or processed hides. Spores are

- Primarily an inhalation hazard
- Subcutaneous through broken skin
- Ingestion
- Some are bloodborne pathogens

Inhalation   Broken skin   Blood

Figure 8.43 Routes of entry.

resistant to drying, heat, and sunlight. They have been known to survive in milk for up to 10 years, on dried filter paper for 41 years, on dried silk threads for up to 71 years, and in pond water for 2 years. Anthrax is a disease that is present in the soil and occurs naturally in livestock, including pigs, horses, goats, cattle, and sheep. Farmers who work with these animals have been known to contract anthrax. Human-to-human contact has not been documented and is thought to be unlikely. Infection occurs from *skin contact* with infected animal tissue and possibly from biting fleas feeding on the animals. The bacterium enters a cut or abrasion on the skin while handling contaminated products from infected animals. Most commonly, the disease appears on the hands and forearms of people working with infected animals (Figure 8.43).

Symptoms include the formation of carbuncles (inflammation of hair follicles and surrounding subcutaneous tissue) and swelling at the location of the infection (Photo 8.10). Scabs form over the lesion and turn a coal-black. Anthrax is the Greek term for coal, and so the name is derived from the coal-black scabs on the lesions. This localized infection can also become systemic and transformed to the inhalation form of the disease.

Inhalation exposure occurs from spores in contaminated soil areas or from fried or processed skins and hides of infected animals that become airborne. Symptoms from inhalation exposure, depending on the concentration and length of time, will present in two distinct phases. First, spores are carried to the lungs, specifically the avolai within the lungs. This is followed by a pus-producing infection with edema (fluid buildup) and

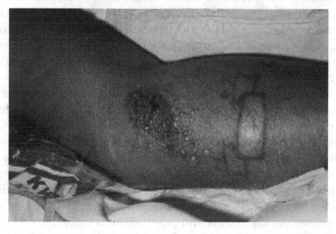

Photo 8.10 Anthrax skin exposure and ingestion cases have been successfully treated when antibiotic therapy begins quickly.

hemorrhage in the lungs. Symptoms are flu-like, including mild fever, malaise, fatigue, myalgia (muscle pain), a dry cough, and a feeling of pressure in the chest. It is believed that the anthrax exposure can be successfully treated at this stage if diagnosed in time. Depending on the dose, the first phase can last for several days or as little as 24 h. The first phase can be followed by a period when the victim is symptom-free.

Onset of the second phase can be sudden, with the evolution of shortness of breath and cyanosis. *Ingestion* of contaminated meat may also cause infection, although this is a rare occurrence. Symptoms from ingestion exposure to anthrax include acute inflammation of the intestinal tract, nausea, loss of appetite, vomiting, and fever, which is followed by abdominal pain, vomiting of blood, and severe diarrhea. Approximately 25%–60% of those victims who ingest anthrax will die.

The primary cause of human outbreaks is from skin contact with contaminated hides, leather, and animals. Natural infection from inhalation or ingestion is rare. The Canadian Center for Disease Control reports that the infectious dose in humans is 1300 organisms through inhalation. This equates to one-billionth of a gram (about the size of a speck of dust), which would be lethal to a single person. The U.S. military reports that the lethal dose is 8,000–10,000 spores. Once infected, the incubation period for anthrax is 7 days and, on average, symptoms begin within 2–5 days. When inhaled at the proper dose, the disease runs a swift sequence of events, and death can occur within 24–48 h. Spore size is also an important factor in determining their effectiveness. Particles from 2 to 5 μm (one-millionth of a meter, 25,400th of 1 in.) in size are considered to be the most efficient in causing infection through inhalation. Particles larger than 5 μm would tend to be filtered out by the upper airway. It is expected that heavy smokers might be more susceptible to large particles.

Skin exposure and ingestion cases have been successfully treated when antibiotic therapy begins quickly. Penicillin, tetracycline, ciprofloxacin, doxycycline, or other broad-spectrum antibiotics have proven effective, although inhalation exposure is almost always fatal, even with treatment. Penicillin has been the antibiotic of choice in the past, and dosage is usually 2 million units given intravenously every 2 h. The FDA has approved Cipro as the antibiotic of choice, but the drug has significant side effects. Naturally occurring strains of anthrax may be resistant to penicillin. Victims who do recover from the cutaneous form of the disease may develop immunity. Injections of anthrax vaccine occur in six steps: it is administered at 0, 2, and 4 weeks, followed by 6, 12, and 18 months, followed by an annual booster. It is believed that after the first three doses, protection against cutaneous anthrax is achieved. Little data is available on the length of protection against inhalation exposure, although tests in primates suggest good protection can be attained after 2 doses, with protection up to 2 years. It is possible that the vaccine could be overwhelmed by an extremely high dose of spores.

Plague (*Yersinia pestis*) is a zoonotic bacterium that is normally spread among rodents by infected fleas. Zoonotic bacteria are capable of being transmitted from lower animals to man under natural conditions. Three forms of the disease can affect humans: bubonic, pneumonic, and primary septicemic. Another type of plague, called pharyngeal, resembles acute tonsillitis. Periodic outbreaks of the plague occur naturally in rodent populations, which may result in a high death rate. Fleas that have lost their usual hosts pursue alternative sources of blood. When this happens, the risk to humans and other animals is increased. Epidemics of plague in humans commonly involves house rats and their fleas. The disease is passed on to humans and other animals when they are bitten by a flea that has bitten an infected rat or other living thing. Animals prone to be carriers in the United States include the rock squirrels, prairie dogs, and other burrowing rodents. During 1924 and 1925, the last epidemic in the United States occurred. Since that time, there have only been isolated cases reported, usually in rural areas from wild rodents.

Plague cases in the United States during the 1980s averaged around 18 per year, mostly in the Southwestern states of New Mexico, Arizona, Colorado, and California. Highest rates of infection occur among Native Americans, particularly the Navajos. Others at risk include hunters, veterinarians, pet owners, campers, and hikers. Of those, most cases involved persons that were under the age of 20, with a fatality rate of one in seven. Death rates from bubonic plague can reach as high as 50%–60% if not treated. When treated, the death rate is reduced to about 15%. If treatment is not begun within 24 h after symptoms develop, pneumonic plague has a near 100% death rate. Plague can also be transmitted when the organism enters the body through a break in the skin. This type of exposure occurs from direct contact with tissue or body fluids of a plague-infected animal, such as skinning a rabbit or other animal. This is, however, a rare occurrence. Plague can also be transmitted through inhalation by contacting infected droplets from a person or domestic animal coughing. Plague that develops from this type of exposure is called pneumonic. This infection involves the lungs as a result of inhalation of organisms, which results in primary pneumonic plague. Secondary pneumonic plague results from septicemia (blood infection) when the organisms spread to the lungs.

Symptoms from plague exposure usually develop within 2–6 days following the exposure. Pneumonic plague occurs a little faster, from 1 to 3 days, which is also dependent on the amount of organisms inhaled. Symptoms of bubonic plague include enlarged lymph nodes, fever, chills, and prostration. Pneumonic plague symptoms are similar to bubonic and include high fever and chills, which are accompanied by cough and difficulty breathing, production of a bloody sputum, and toxemia. Symptoms may be followed by rapid shock and death if treatment is not begun early. Death results from respiratory failure, circulatory collapse, and a predisposition toward bleeding. Bubonic plague can progress spontaneously to septicemic plague accompanied by fever, chills, prostration, abdominal pain, shock, and bleeding into the skin and other organs. It can also affect the central nervous system, lungs, and other parts of the body. As many as 80% of bubonic plague victims have positive blood cultures for septicemia. Approximately 25% of the patients will also have various types of skin lesions. Pustules, vesicles, eschars (dead tissue separating from living tissue), or papules (small elevation of the skin) containing leukocytes and bacteria may also be present near the site of the fleabite.

Antibiotic treatment should begin as soon as possible. Streptomycin is the drug of choice for treatment, but others, such as tetracyclines, chloramphenicol, gentamicin, or one of the sulfonamides, may also be effective. A vaccine for the plague is available; however, it is only effective as a preventative measure; once someone is exposed, the vaccine will not help. The initial dose of vaccine is followed by a second one 1–3 months later and a third one 3–6 months later. Booster shots are administered at 6, 12, and 18 months and then every 1–2 years. As with *all vaccines*, the level of protection is related to the size of the dose; vaccination defense could be overwhelmed by extremely high doses of the bacteria.

Tularemia (Francisella tularensis), also known as rabbit fever, deerfly fever, and Ohara's disease, like the plague, is a bacterial infection that can occur naturally from the bite of insects, usually ticks and deerflies. The disease can also be acquired from contact with infected rabbits, muskrats, and squirrels, ingestion of contaminated food, or inhalation of contaminated dust. Once contracted, it is not directly spread from human to human. Tularemia remains infectious in the blood for about 2 weeks and in lesions for a month. It remains ineffective in deerflies for 14 days and ticks throughout their lifetime (about

2 years). The disease can occur at anytime of the year, but is most common in the early winter during rabbit hunting season and in the summer when tick and deerfly activity is at its peak. Tularemia contracted naturally has a death rate of approximately 5%.

Tularemia can appear in several different forms in humans, depending on the route of exposure. The usual presentation is ulceroglandular, typhoidal, or septicemic. In humans, as few as 10–50 organisms can cause disease if inhaled or injected, but over 108 would be required for the disease if ingested. Ulceroglandular tularemia is acquired naturally from dermal or mucous membrane exposures of blood or tissue fluids of infected animals. The typhoidal form makes up 5%–15% of naturally occurring cases, which result from inhalation of infectious aerosols. Pneumonia can result from any of the forms of tularemia, but is most prominent in typhoidal. Incubation periods range from 2 to 10 days, depending on the dose and route of exposure. The average incubation period occurs within 3 days.

Symptoms of ulceroglandular disease include lymphadenopathy, fever, chills, headache, and malaise. About 90%–95% of patients may present cutaneous ulcers. When ulcers are absent, it is referred to as glandular tularemia. When symptoms are confined to the throat, it is called primary ulceroglandular disease. Oculoglandular tularemia results from contact to the eyes from an infected fluid or blood. Typhoidal or septicemic tularemia produces fever, prostration, and weight loss, without adenopathy (any disease of the gland, especially a lymphatic gland). After exposure, the usual treatment is 2 weeks of tetracycline. Streptomycin is given for more severe exposures. Aminoglycosides, genatamycin, kanamycin, and chloramphenicol are also effective antibiotics. Once a person recovers, he or she has permanent immunity from the disease. A vaccine is under development and has been successful during tests on more than 5000 persons, without significant adverse reactions.

Cholera (*Vibrio cholerae*) is a bacterial disease that is contracted by ingestion of contaminated water or food. However, it does not spread easily from person to person. Cholera occurs naturally in many underdeveloped countries and has caused widespread outbreaks in South America, with over 250,000 cases reported just in Peru. It can be spread through ingestion of food or water contaminated with feces or vomitus of patients, by dirty water, hands contaminated with feces, or flies. Cholera is an acute infectious disease, represented by a sudden onset of symptoms. Victims may experience nausea, vomiting, profuse watery diarrhea with "rice water" appearance, and the rapid loss of body fluids, toxemia, and frequent collapse. Not everyone exposed will show symptoms. In some cases, there may be as many as 400 people without symptoms for every patient showing symptoms. Where cases go untreated, the death rate can be as high as 50%. With treatment, the death rate drops to below 1%.

Cholera itself is not lethal, but the breakdown of medical treatment systems in large outbreaks can result in many deaths from dehydration, hypovolemia (loss of body fluid), and shock. Fluid loss can be as much as 5–10 L/day, and IV fluids used to replenish fluids can be in short supply. The incubation period varies from 12 to 72 h, depending on the dose of ingested organisms. An infectious dose is greater than 108 organisms to a healthy individual through ingestion. Treatment involves antibiotic and IV fluid therapy using tetracycline (500 mg every 6 h for 3 days). Doxycycline (300 mg once, or 100 mg every 12 h for 3 days), IV solutions of 3.5 g NaCl, 2.5 g $NaHCO_3$, 1.5 g KCl, and 20 g glucose/L are also appropriate treatments. Cholera shows a significant resistance to tetracycline and polymyxin antibiotics. Ciprofloxacin (500 mg every 12 h for 3 days) or erythromycin (500 mg every 6 h for 3 days) can be used as substitute. A vaccine is available for prevention; however, it has not proven very effective. It offers only 50% protection for a period of up to 6 months.

Brucella (brucellosis), also known as undulant fever or Bang's disease, is a bacterial disease caused by any one of four species of coccobacilli. They are naturally occurring diseases in cattle, goats, pigs, swine, sheep, reindeer, caribou, coyotes, and dogs. The organisms can be contracted by humans through ingestion of unpasteurized milk and cheese, or from inhalation of aerosols generated on farms and in slaughterhouses. Skin lesions on persons who have come in contact with infected animals can also spread infection. If these bacteria were used by terrorists, it would have to be aerosolized for inhalation by victims or be used to contaminate food supplies. Brucella is present worldwide among the animal populations, but is especially prevalent in the Mediterranean countries of Europe and in Africa, India, Mexico, and South America. The disease is also common in populations who eat raw caribou. When the animal population has a high rate of infection, the rate of disease occurrence is also higher in humans. It is unknown what the infectious dose of brucella is. Symptoms are nonspecific and insidious upon onset (the disease is well established when the symptoms appear). Symptoms include intermittent fever, headache, weakness, profuse sweating, chills, and arthralgia, and localized suppurative (pus-forming) infections are frequent. Incubation periods vary from 5 to 30 days, and in some cases, many months. Evidence of human-to-human transmission of the disease has not been documented. Side effects include depression and mental status changes. Osteoarthritic complications involving the axial skeleton are also common. Death is uncommon even in the absence of treatment. Treatment of brucellosis involves the administration of the antibiotics tetracycline and streptomycin or TMP-SMX. The disease is resistant to penicillin and cephalosporin. There is no vaccine available for use in humans.

Q fever (*Coxiella burnetii*), also known as Query fever and rickettsia, is a bacterial disease that occurs naturally in sheep, cattle, and goats. It is present in high concentrations in the placental tissues of these animals. Incidence of the disease is worldwide, and it is likely that more cases occur than those reported. Many epidemics occur in stockyards, meat-packing plants, and medical labs using sheep for research. Transmission occurs from airborne dissemination of rickettsiae in dust from contaminated premises. Organisms can be carried in the air over half a mile downwind. Infections are also contracted from contact with infected animals, their birth products (especially sheep), wool from sheep, straw, fertilizer, and laundry of exposed persons. The disease has also been traced to unpasteurized milk from cows. Transmission from human to human is rare. Several varieties of ticks may also carry the disease and transmit it from animal to animal. Mortality rates from this disease are low, from 1% to 3%. It would, however, be an effective incapacitating agent because it is highly infectious when delivered through inhalation; as little as one organism can cause clinical symptoms.

The usual infectious dose is considered to be 10 organisms through inhalation. Symptoms are not specific to the disease, and it may be mistaken for a viral illness or atypical pneumonia. The incubation period is from 10 to 20 days. Patients may experience fever, cough, and chest pain as soon as 10 days after exposure. Although somewhat rare, other symptoms that may appear include chills, headache, weakness, malaise, severe sweats, hepatitis, endocarditis, pericarditis, pneumonitis, and generalized infections. Patients are not critically ill and, in most cases, the illness lasts from 2 days to 2 weeks.

Q fever is generally a self-limiting illness and will clear up without treatment. Antibiotics given during the illness can shorten the period of incapacitation. Tetracycline is the antibiotic of choice and when given during the incubation period may delay the onset of symptoms. Usual dosage is 500 mg every 6 h or doxycycline, 100 mg every 12 h. Antibiotic treatment should be continued for 5–7 days. The disease is remarkably resistant to heat and drying, and is stable under diverse environmental conditions. Vaccines for humans are still in the development stages, although tests have shown promise.

Salmonella spp. (Salmonellosis) is a naturally occurring bacterium that is present in a wide variety of animal hosts and environmental sources. There are over 2000 strains of bacteria in the salmonella family. Ten strains are responsible for most of the reported salmonella infections. We will limit our discussions here to the four most common strains: *Salmonella* spp., *Salmonella typhi*, *Salmonella paratyphi*, and *Salmonella choleraesuis*. Some strains may make humans sick and not animals, while others make animals sick and not humans. The bacterium that causes salmonella infections is a single-celled organism that cannot be seen with the naked eye, touched, or tasted.

Salmonella bacteria occur naturally in the intestines and waste of poultry, dogs, cats, rats, and other warm-blooded animals. When live salmonella bacteria enter the body, which is usually the result of contaminated food, a salmonella infection occurs. Salmonellosis is the most common bacterial foodborne illness. Salmonella infections can be prevented by cooking food at the proper temperature and cleaning food utensils and preparation areas effectively. There are approximately 40,000 salmonella infections reported to public health officials each year in the United States. Experts believe that between 500,000 and 4,000,000 cases actually occur that go unreported. It is estimated that 2 of every 1000 cases result in the death of the patient, over 500 annually. Salmonella affects those that are young, those that are older, and those whose body has already been weakened by some other illness. Many of those who contract the illness believe they have the flu and never go to see a doctor.

While there are many strains of salmonella, with a few exceptions, they produce similar symptoms. Incubation periods range from 6 h to many weeks, depending on the strain and the amount of bacteria ingested. The infectious dose varies widely, and there is no rhyme or reason for why certain people are affected by the bacteria. Some become ill with ingestion of as little as 10 organisms, while others have ingested food contaminated with millions of bacteria and experienced no adverse effects. There is not any particular type of food that contains salmonella. Infection can occur from any undercooked food, eating raw foods, or eating food contaminated by people preparing the food.

Salmonella spp. (excluding *S. typhi*, *S. choleraesuis*, and *S. paratyphi*) causes acute gastroenteritis (inflammation of the GI tract). After exposure and the appropriate incubation period, the onset of symptoms is sudden, with abdominal pain, diarrhea, nausea, and vomiting. Dehydration can occur and be severe in infants. Hosts range from humans to domestic and wild animals. The infectious dose is 100–1000 organisms by ingestion. Symptoms can appear in 6–72 h and normally within 12–36 h. Transmission occurs from eating food made from infected animals or food contaminated from feces of an infected animal or person. Infected animal feeds and fertilizers prepared from contaminated meat scraps and fecal–oral transmission from person to person also occurs. Salmonella bacterium is sensitive to antibiotic therapy using ampicillin, amoxicillin, TMP-SMX, and chloramphenicol.

Salmonella choleraesuis produces acute gastroenteritis (infection of the GI tract), sudden headache, abdominal pain, diarrhea, nausea, and sometimes vomiting. This bacterial illness may develop into enteric fever, with septicemia or focal infection in any tissue of the body. The infectious dose is 1000 organisms by ingestion. Modes of transmission and incubation periods are similar to *Salmonella* spp. Carriers are contagious throughout the infection, which can last for several days to several weeks. Antibiotic treatment can prolong the period in which the disease is contagious.

Salmonella typhi is a generalized systemic infection, which produces the following symptoms: fever, headache, malaise, anorexia, enlarged spleen, rose spots on the trunk of the body, and constipation. These symptoms are followed by more serious ones, including ulceration of Peyer's patches in the ileum, which can produce hemorrhage or perforation.

Mild and atypical infections can also occur, with a death rate of 10% if not treated with antibiotics. Drug-resistant strains are appearing in several parts of the world. The infectious dose is 100,000 organisms by ingestion. Transmission occurs by ingesting food or water that has been contaminated with the feces or urine of a patient or carrier. Infection can also occur from food handlers who do not practice proper hygiene or flies contaminating foods. Incubation periods depend on the strength of the dose, but are usually 1–3 weeks. Antibiotic treatment with chloramphenicol, ampicillin, or amoxicillin is usually effective.

Salmonella paratyphi is a bacterial enteric (intestinal) infection with an abrupt outbreak, which produces the following symptoms: continued fever, headache, malaise, enlarged spleen, rose spots on the trunk of the body, and diarrhea. These symptoms are similar to those of typhoid fever, but the death rate is much lower. Mild and asymptomatic infections may also occur upon exposure. Outbreaks and locations are similar to those of the other salmonella bacteria. The infectious dose is 1000 organisms by ingestion. Transmission occurs by direct or indirect contact with feces or, in rare cases, urine of patients or carriers. It is spread by food, especially milk and dairy products, shellfish, and, in some isolated cases, water supplies. Incubation depends on the strength of the dose, but usually 1–3 weeks for enteric fever and 1–10 days for gastroenteritis. Antibiotic treatment with chloramphenicol, ampicillin, or TMP-SMX is usually effective.

Rickettsia are pleomorphic (come in varying sizes), parasitic microorganisms. A number of strains of these bacteria exist naturally. *Rickettsia Canada*, also known as louseborne typhus fever and classical typhus fever, occurs in areas of poor hygiene that are also louse-infected. Outbreaks generally occur in Central America, South America, Asia, and Africa. It is primarily a disease of humans and squirrels. The body louse, *Pediculus humanus*, is the primary carrier of the disease. It feeds on the blood of an infected patient with acute typhus fever and becomes infected. Once this occurs, the lice excrete rickettsiae in their feces, which is defecated as they feed. Infection occurs from feces left on the skin by the lice, which is rubbed in at the site of the bite or other breaks already existing in the skin, or through inhalation of infected dust. Squirrels become infected from the bite of the squirrel flea.

The incubation period ranges from 1 to 2 weeks, with an average of 12 days. Rickettsia Canada cannot be directly transmitted from human to human, but it is a bloodborne pathogen, and universal precautions should be practiced. Victims are infective for lice from the time in which the febrile (fever) illness is present and for 2–3 days after the body temperature returns to normal. Infection remains in the louse for 2–6 days after biting the source, although it may occur quicker if the louse is crushed. Symptoms include headache, chills, fever, prostration, and general pains. On the 5th or 6th day, a macular eruption (unraised spots on the skin) occurs on the upper trunk and spreads to the entire body (except for the face, palms of the hands, and soles of the feet). The illness lasts for approximately 2 weeks. Without treatment, the fatality rate is about 10%–40%. Treatment involves antibiotic therapy with tetracyclines and chloramphenicol.

Glanders, Burkholderia (formerly *Pseudomonas*) *mallei*, is a gram-negative bacillus found naturally in horses, mules, and donkeys. Man is rarely infected even during frequent and close contact with infected animals. No naturally occurring cases in man have been reported in the United States in over 60 years. Sporadic cases do continue to occur in Asia, Africa, the Middle East, and South America. The disease exists in four basic forms in man and horses. Acute forms occur primarily in mules and donkeys and result in death in 3–4 weeks. Chronic forms occur in horses and result in lymphadenopathy (enlargement of the lymph nodes), skin nodules that ulcerate and drain, and induration (increase in fibrous tissue). Transmission to man occurs through the nasal, oral, and conjunctival mucous membranes, by inhalation into the lungs, and breaks in the skin. Aerosols have been reported to be highly infectious in laboratory exposures and resulted in over 46% of the cases becoming severe. There is no vaccine or effective treatment available. Incubation periods are from 10 to 14 days. Symptoms include fever, rigors, sweats, myalgia, headache, pleuritic chest pain, cervical adenopathy, splenomegaly, and generalized papular/pustular eruptions. The disease is almost always fatal without treatment. Sulfadiazine may be an effective treatment in some cases. Ciprofloxacin, doxycycline, and rifampin have also been shown to be effective. Most antibiotic sensitivities are based upon animal studies because of the low incidence of human exposure.

Viruses

Viruses are the simplest type of microorganisms and the smallest of all living things. They are much smaller than bacteria and range in size from 0.02 to 1 m (m = 1000 mm). One drop of blood can contain over 6 billion viruses! Viruses were first discovered in 1898 and observed for the first time in 1939. Virus is a Latin word meaning "poisonous slime." Every living entity is composed of cells, except for viruses. They are, in fact, totally inert until they come in contact with a living host cell. Hosts can include human, animal, plants, or bacteria. The infection point created from the virus occurs at the cellular level. There must be an exact fit between the virus and the cell, or the invasion of the cell cannot occur. After a virus attaches to a cell, it begins to reproduce itself, resulting in an acute viral infection. Once a virus takes hold of the cell, it can cause the host cell to die. Common examples of viral agents include measles, mumps, meningitis, influenza, and the common cold.

Viruses that most people are familiar with today are HIV (the virus that causes AIDS), HBV (the virus that causes hepatitis B), and HCV (the virus that causes hepatitis C). These viruses are not airborne and are usually difficult to transmit. Very specific actions have to take place to transmit the viruses. HIV, HBV, and HCV are transmitted by contact with blood and body fluids. They are more commonly referred to as bloodborne pathogens. Viral hemorrhagic fevers (VHFs) are a group of viruses that include Ebola, Marburg, Arenaviridae, Lassa fever, Argentine and Bolivian, Congo-Crimean, Rift Valley, *Hantavirus*, yellow fever, and dengue (Figure 8.44).

HIV is caused by the human immunodeficiency virus. There are more than 1 million people in the United States today that are HIV-positive. Exposure to HIV occurs through direct contact with blood or blood components; cerebrospinal fluid; synovial fluid; peritoneal fluid; amniotic fluid; or blood in saliva, semen, vaginal secretions, or any other body fluid that contains blood. Infection may occur from unprotected sex with an infected person, sharing needles, needle and other sharp-object sticks, and contact with blood or blood products. The incubation period for HIV varies from 2 months to over 10 years. Treatments

- HIV
- Hepatitis
- Smallpox
- Venezuelan equine encephalitis (VEE)
- Viral hemorrhagic fevers

Smallpox virus

Figure 8.44 Examples of viral agents.

are available, which may extend the incubation period. Symptoms include an acute, self-limited mono-like illness lasting for 1–2 weeks. Those infected then remain symptom-free for many years. Full-blown AIDS then rears its ugly head. Symptoms are somewhat non-specific and vary from person to person. HIV destroys the body's immune system, allowing infections, some which are rare, to develop. Common infections include pneumocystis carinii pneumonia and cancers, such as Kaposi's sarcoma. Victims also may develop a wasting syndrome, extrapulmonary tuberculosis, or neurological diseases, like HIV dementia. Symptoms are insidious at onset. They may include swollen glands, anorexia, chronic diarrhea, weight loss, and fever. Treatments and vaccines are under development, which have offered the promise of reducing HIV to a chronic illness.

HBV and HCV are diseases of the liver, sometimes referred to as hepatitis or inflammation of the liver. Both HBV and HCV are bloodborne pathogens, like HIV. The virus is most concentrated in blood, serum, and wound exudates. The virus is also present in smaller concentrations in semen, vaginal fluid, and breast milk. Low concentrations exist in urine, feces, sweat, tears, and saliva. The incubation period for HVB is from 45 to 160 days, and for HCV 2–26 weeks, with a fatality rate of 1%–1.4%. Like HIV, symptoms are insidious at onset. They include anorexia, malaise, nausea, vomiting, abdominal pain, jaundice, skin rashes, arthralgias, and arthritis. A vaccine is available for HBV.

Smallpox is a lethal infection caused by the variola virus, which has at least two strains, variola major and variola minor. Cases of smallpox date back over 2000 years, and it is the oldest-known human pathogen. Naturally occurring smallpox was declared eradicated from the earth in 1980 by the World Health Organization (WHO), a branch of the UN. The last reported case in the world occurred in Somalia in 1977. Two laboratories in the world still hold the last-known stocks of variola virus: the Centers for Disease Control in Atlanta and VECTOR in Novosibirsk, Russia. Clandestine stocks could exist in other parts of the world, but are as yet unknown. The WHO's governing body recommended the total destruction of the remaining stockpiles by the year 1999. An effective vaccination is available for smallpox and has been used for years for the general population. Since it is primarily a children's disease, vaccinations were given during early childhood and were effective only for about 10 years. Vaccination of civilians in the United States was discontinued in the early 1980s. Children who are no longer vaccinated would be at great risk from exposure to smallpox (Figure 8.45).

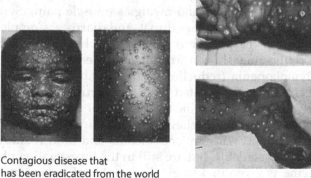

Contagious disease that
has been eradicated from the world
as a naturally occurring disease.
However, it is still a potential terrorist threat.

Figure 8.45 Smallpox.

Monkey pox and cowpox are closely related to variola and might be genetically manipulated to produce a smallpox-like virus. Once exposure to the smallpox virus occurs, the incubation period is approximately 12 days. Those who may have contacted exposed persons are quarantined for a minimum of 16–17 days following the exposure. Symptoms of smallpox include malaise, fever, rigors, vomiting, headache, and backache, and about 15% of the patients develop delirium (hallucinations). In approximately 2–3 days, an enanthem develops concomitantly with a particular rash on the face, hands, and forearms. This is followed by eruptions on the lower extremities and the trunk of the body, which occur over a week's time. Lesions progress from discolored spots flush with the surface of the skin, to raised spots on the skin, and finally to an inflamed swelling on the skin containing pus (skin blisters). Lesions are more abundant on the extremities and face, which is important in the diagnosis of the disease. Within 8–14 days, scabs form on the skin blisters. Once the scabs fall off, a discolored depression is left behind. As long as the scabs are in place, the patient is considered contagious and should be isolated. Transmission occurs from close person-to-person contact, and it is unknown if an airborne dispersion would be effective. Most antiviral drugs for smallpox are experimental at the present time and would not be available for large numbers of victims. Vaccinia-immune globulin has shown to be an effective prevention following an exposure to smallpox.

VEE is a virus that is naturally transmitted from horse to horse by mosquitoes. Humans can also get the virus from the infected mosquito. Each year, thousands of persons acquire the disease naturally from mosquito bites. Human outbreaks usually follow an epidemic among the horse population. Humans infected can infect mosquitoes for up to 72 h. Once a mosquito becomes infected, it remains so for life. VEE is considered a bloodborne pathogen. Universal precautions for bloodborne pathogens should be taken by emergency workers when around the VEE patient. Human-to-human transmission through inhalation of respiratory droplets can theoretically cause infection, but has not been proven. VEE is rarely fatal (less than 1%) and acts as an incapacitating agent. Nearly 100% of those exposed acquire the disease; however, only a small number actually develop encephalitis. Usually young children are the most vulnerable for developing encephalitis.

VEE is characterized by convulsions, coma, and paralysis. Encephalitis is characterized by inflammation of the meninges (surrounding membranes) of the brain and the brain itself, which produces central nervous system symptoms. Onset of symptoms is sudden, following an incubation period of 1–5 days. Symptoms are flu-like and may include malaise, spiking fevers, rigors (chills and severe shivering), severe headache, photophobia (sensitivity to light), and myalgias (muscle pain). Symptoms of VEE may include nausea, vomiting, cough, sore throat, and diarrhea. Complete recovery requires 1–2 weeks. Diagnosis of the disease is difficult, as physical symptoms are nonspecific. White blood cell counts may show a striking leukopenia (abnormally low leukocytes in the blood) and lymphopenia (reduction in the number of lymphocytes circulating in the blood). The virus can be isolated from the serum (fluid that forms blood clots). Because there are no drugs or specific treatments for the disease, treatment is supportive. Analgesics may be given to relieve headache and myalgia. Victims that develop encephalitis can be treated with anticonvulsants and fluid therapy for electrolyte balance. Two vaccines are available, but are still in the investigation phase of development. TC-83, a live vaccine, is given in a single dose of 0.5 mL subcutaneously. A second vaccine, C-84, is used to boost those who do not respond to the TC-83. It is given as 0.5 mL subcutaneously in three doses at 2–4 week intervals. Research is also underway using antiviral drugs that have shown some promise with laboratory animals. However, no human clinical data are available.

VHFs are an assorted group of human diseases originated by viruses from several different families. Included are the Filoviridae family, which includes Marburg and Ebola, and the Arenaviridae family, which encompasses Lassa fever and Argentine and Bolivian hemorrhagic fevers (BHFs). Additionally, the Bunyaviridae family, which involves various members of the *Hantavirus* genus, the Congo-Crimean hemorrhagic fever (CCHF) family from the *Nairovirus* genus, and the Rift Valley fever from the *Phlebovirus* genus are family members. The last family is the Flaviviridae, which includes yellow fever and the dengue hemorrhagic fever virus.

The virus spreads through close personal contact with a person who is infected with the disease and can also be spread through sexual contact. Universal bloodborne pathogen precautions should be practiced when treating victims. It is not known what the natural host is for the virus or how a person contracts the disease initially. The incubation period is 2–21 days.

Mortality rates in Africa from *Ebola* range from 50% to 90% of those infected. Another strain of Ebola, the *Reston*, was reported among monkeys in the Philippines in 1989. This strain is not yet known to transmit to humans. African strains, on the other hand, have caused severe disease and death. Why the disease has only shown sporadic outbreaks is unknown.

Marburg virus has been reported to cause infection in man on four previous occasions. Three occurrences were in Africa and one in Germany, where the virus was named. Outbreaks first occurred in Germany and Yugoslavia involving 31 reported cases, which occurred from exposure to African green monkeys. Seven people died. The incubation period for Marburg is 3–7 days. Methods of transmission of these diseases are not well known. The disease seems to be spread by direct contact with infected blood, secretions, organs, or semen.

Argentine hemorrhagic fever (AHF) is caused by the Junin virus, which first appeared in 1955 among corn harvesters in Argentina. The virus is spread naturally from contact with the infected excreta of rodents. Somewhere between 300 and 600 cases of AHF occur each year in the Pampas region of Argentina. A similar disease, BHF, which is caused by the Machupo virus, appeared in northeastern Bolivia following the appearance of AHF. Another closely related virus is Lassa, which occurs over much of western Africa.

CCHF is a disease carried by ticks and transmitted by bites to humans. Infections occur primarily in the Crimea and other parts of Africa and in Europe and Asia. The incubation period is 3–12 days. It is stable in blood up to 10 days at 105°F (40.5°C). *Rift Valley fever* also occurs only in Africa and results in sporadic, widespread epidemics of the disease.

Hantavirus was first identified prior to World War II in Manchuria along the Amur River. Incubation periods vary from 5 to 42 days, with 12–16 days being the average.

Yellow fever and *dengue* fever are two diseases that are transmitted by infected mosquitoes. Yellow fever has an incubation period of 3–6 days. Yellow fever is the only one of the VHFs that does have a vaccine available. Dengue fever virus has an incubation period from 3 to 14 days, with 7–10 being the average. The virus is stable in dried blood and exudates up to 2 days at room temperature. All VHFs, except for dengue fever, are capable of being spread by aerosolization and skin-crawling insects. Patients infected with VHFs other than *Hantavirus* will have the virus in the blood, and it can be transmitted through contact with blood or body fluids containing blood. Bloodborne pathogen precautions should be undertaken when exposure to blood is a possibility. Routes of infection for the filoviruses in humans are not well understood at this time.

VHFs are feverish illnesses that are complicated by easy bleeding, petechiae (bleeding under the skin), hypotension (abnormally low blood pressure), shock, flushing of the face and chest, and edema (excess fluid in tissues). Not all infected humans actually develop VHFs. The reason for this is not well understood. Congenital symptoms, such as malaise, myalgias, headache, vomiting, and diarrhea may appear in association with any of the hemorrhagic fevers. Treatment is supportive of the symptoms that are presented. Ribavirin may be an effective antiviral therapy for Lassa fever, Rift Valley fever, and CCHF viruses. During recovery, plasma may be effective in Argentine hemorrhagic fever.

Toxins

Biological toxins are defined as any toxic substance occurring in nature produced by an animal, plant, or microbes (pathogenic bacteria), such as bacteria, fungi, flowering plants, insects, fish, reptiles, or mammals. Under the DOT classification system, these materials would be classified as Class 6.1 Poisons. Unlike chemical agents, such as sarin, cyanide, or mustard, toxins are not manmade. Generally, toxins are not volatile and not considered a dermal exposure hazard (except for mycotoxins). Toxins as a family are much more toxic than any of the chemical agents, including VX and sarin. Ricin, a biological toxin produced from the castor bean plant, is 10,000 times more toxic than sarin; as little as a milligram (1/1000 of a gram) can kill an individual. Botulinum toxin is the most toxic material known to man. One ounce of botulinum toxin (0.12 μg or 1200-millionths of a gram) is enough to kill 60 million people. Mycotoxins are the least toxic of the toxins (thousands of times less than botulism).

Routes of exposure also have a bearing on the level of toxicity. Some toxins are much more lethal when aerosolized and inhaled than when taken orally. Ricin, saxitoxin, and T2 mycotoxins are examples of these types of materials. Botulism, for example, has a lower toxicity through aerosolization and inhalation than through ingestion. When looking at the potential toxicity of toxins, the lower the LD_{50}, in micrograms per kilogram of body weight, the less agent that would be required to be toxic. Conversely, some agents such as ricin would require great quantities (tons) for an aerosol attack out in the open. Toxins can be used either for their lethality or as an incapacitating agent. Some toxins are incapacitating at lower doses and would cause serious illness.

Toxins can be divided into groups based upon the mechanism by which they function. *Protein toxins* are created by bacteria. Protein toxins include botulinum (seven related toxins), diphtheria, tetanus, and staphylococcal enterotoxins (seven different toxins). They function by paralyzing the respiratory muscles. Staphylococcal enterotoxins can incapacitate at levels at least 100 times lower than the lethal level (Figure 8.46).

- Botulinum
- Staphylococcus enterotoxins
- Ricin
- Mycotoxins

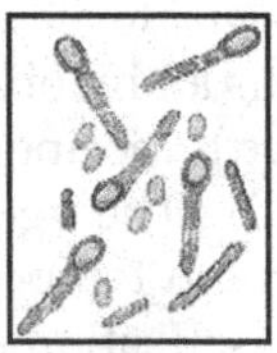

Botulinus
bacteria

Figure 8.46 Types of biological toxins.

Bacterial toxins can be classified as membrane-damaging. This group includes *Escherichia coli* (hemolysins), aeromonas, pseudomonas, and staphylococcus alpha (cytolysins and phospholipases). Many of these toxins function by interfering with bodily functions and kill by creating pores in cell membranes.

Marine toxins may be developed from marine organisms. Examples include saxitoxin, tetrodotoxin, palytoxin, brevetoxins, and microcystin. Saxitoxin is a sodium-channel blocker and is most toxic by inhalation compared to the other routes of exposure. Saxitoxin and tetrodotoxin are similar in mechanical action, toxicity, and physical attributes. They can be lethal within a few minutes when inhaled. It has not yet been chemically synthesized efficiently or easily created in large quantities from natural sources. Palytoxin is produced from soft coral and is highly toxic. It is, however, difficult to produce or harvest from nature.

Trichothecene mycotoxins are created by numerous species of fungi. Over 40 toxins are known to be produced by fungi. T2 is a stable toxin even when heated to high temperatures. However, unlike other toxins, the mycotoxins are dermally active. Once absorbed into the body, it would become a systemic toxin.

Plant toxins are derived from plants or plant seeds. One of the premier plant toxins is ricin. It is a protein taken from the castor bean. Approximately 1 million tons of castor beans are grown annually worldwide for the production of castor oil. Castor beans, to grow the plants, can be purchased through seed catalogs, from lawn and garden centers, and from agricultural co-ops. Waste mash resulting from the production of castor oil contains 3%–5% ricin by weight. Videos and books are available giving detailed step-by-step instructions for producing ricin.

Animal venoms can contain protein toxins as well as nontoxic proteins. Many venom toxins can be cloned, created by molecular biological procedures, and manufactured by simple chemical synthesis. Venoms can be divided into groups: *ion channel toxins*, like those found in rattlesnakes, scorpions, and cone snail; *presynaptic phospholipase A2 neurotoxins*, found in the banded krait, Mojave rattler, and Australian taipan snake; *postsynaptic neurotoxins*, found in the coral, mamba, cobra, sea snake, and cone snail; *membrane-damaging toxins* of the Formosan cobra and rattlesnake; and the *coagulation/anticoagulation toxins*, found in the Malayan pit viper and carpet viper. Toxins are unlike chemical agents in that they vary widely in their mechanism of action. Time from exposure to onset of symptoms also varies widely.

Saxitoxin acts quickly and can kill an individual within a few minutes of inhalation of a lethal dose. It acts by directly blocking nerve conduction and causes death by paralyzing muscles of respiration. At slightly less than the lethal dose, the victim may not experience any effects at all. Botulinum toxin needs to invade nerve terminals in order to block the release of neurotransmitters, which under normal conditions control muscle contraction. The symptoms from botulinum toxin are slow to develop (from hours to days), but are just as lethal, causing respiratory failure. This toxin blocks biochemical action in the nerves, which activate the muscles necessary for respiration, which leads to suffocation. Unlike saxitoxin, toxicity for botulinum is greater through ingestion than inhalation.

Neurotoxins are effective in stopping nerve and muscle function without producing microscopic injury to the tissues, whereas other toxins destroy or damage tissue directly. Microcystin is a toxin produced by blue-green algae. When it enters the body, it binds to an important enzyme inside the liver cells. No other cells in the body are affected. If microcystin is not blocked from reaching the liver within 15–60 min of receiving a lethal dose, irreversible damage to the liver will occur. Damage to the liver from this toxin is the same, regardless of the route of exposure. With other toxins, the damage that occurs after contact may vary greatly, depending on the route of exposure, even with the same toxin family. Death occurs from ricin because it blocks protein synthesis in many different cells within the body. However, no damage occurs to the lungs unless the route of exposure is inhalation.

For example, the staphylococcal enterotoxins can cause illness at low concentrations, but require large doses to be lethal. Trichothecene mycotoxins are the only biological toxins that are dermally active. Exposure results in skin lesions and systemic illness without being inhaled and absorbed through the respiratory system. Primary routes of exposure are through skin contact and ingestion. Nanogram (one-billionth-of-a-gram) quantities per square centimeter of skin can cause irritation. One-millionth-of-a-gram quantities per square centimeter of skin can cause destruction of cells. Microgram doses to the eyes can cause irreversible damage to the cornea. Because most biological agents are not skin-absorbent hazards, simple washing of contaminated skin surfaces with soap and water within 1–3 h of an exposure can greatly reduce the risk of illness or injury.

Botulinum clostridium (botulism) toxin is a deadly illness caused by any one of seven different, but related, neurotoxins (A through G). All seven types have similar mechanisms of action. They each produce similar symptoms and effects when inhaled or ingested, but the length of time to the development of symptoms may vary, depending on the route of exposure and dose received. Botulism is not spread from person to person. Botulinum toxins as a group are among the most toxic compounds known to man. Lethal doses in research animals are 0.001 µg/kg of body weight, which will kill 50% of the animals. Botulinum toxins are 15,000 times more toxic than lethal nerve agent VX, and 10,000 times more toxic than sarin. Botulism occurs naturally in improperly canned foods and infrequently in contaminated fish. Ingestion of the canned food or fish causes the illness. Low-dose inhalation may not produce symptoms for several days. Inhalation or ingestion of high doses would produce symptoms much quicker. Botulism bacterium is commonly found in the soil.

Two types of illness are associated with the botulinum toxin, infant and adult botulism. An adult becomes ill by eating spoiled food that contains the toxin. Infants become ill from eating the spores of the botulinum bacterium. One source of these spores comes from the ingestion of honey. Spores are not normally toxic to adults. Botulinum toxins work by binding to the presynaptic nerve terminal at the neuromuscular junction and at cholinergic autonomic sites. They then act to stop the release of acetylcholine presynaptically, thus blocking neurotransmission.

This function is quite unlike the action of the nerve agents, where there is too much acetylcholine due to the inhibition of acetylcholinesterase. What occurs with botulism is a lack of the neurotransmitter in the synapse. Therefore, using atropine as an antidote would not be helpful and could even provoke symptoms. When contaminated food is ingested by adults, the toxin is absorbed from the intestines and attaches to the nerves, causing the signs and symptoms of botulism poisoning. Symptoms include blurred vision, dry mouth or sore throat, difficulty in swallowing or speaking, impairment of the gag reflex, general muscular weakness, dilated pupils, and shortness of breath. Paralysis of the skeletal muscles follows, with a proportional, downward, and growing weakness, which may result in sudden respiratory failure. The time from the beginning of symptoms to respiratory failure can occur in as little as 24 h when the toxin is ingested. One-third of patients die within 3–5 days. An antitoxin for botulism can be effective if administered quickly after the onset of symptoms. The bacterium that produces botulism survives well in soil and agricultural products. Botulism toxin can be destroyed by boiling for 10 min.

SEB is a common cause of food poisoning. Cases have usually been isolated to a group of people exposed to contaminated food at some public event or through airline travel. While it can cause death, it is thought of as an incapacitating agent rather than a lethal agent. It could, however, make a large number of people ill for an extended period of time. The primary route of exposure is ingestion. Symptoms from inhalation exposure and ingestion are completely different. Either route of exposure may produce fatalities. Symptoms appear within 3–12 h after aerosol exposure. They include sudden onset of fever, chills, headache, myalgia, and cough. Some patients may also exhibit shortness of breath and retrosternal (behind the breastbone) chest pain. Fevers of between 103°F and 106°F generally last for 2–5 days, and cough may persist for several weeks. Ingestion produces nausea, vomiting, and diarrhea. Very large doses may result in pulmonary edema, septic shock, and possibly death. No antitoxin has been developed for this illness, so treatment remains supportive. No preventative vaccines are available. Naturally occurring food poisoning cases would not present with respiratory symptoms.

SEB infection has a tendency to develop quickly due to a somewhat unchanging clinical condition. Respiratory difficulties occur much later with SEB inhalation. Laboratory testing will provide limited data for diagnosing the disease. SEB toxin is difficult to detect in serum when symptoms develop; however, a baseline specimen for antibody detection should be drawn anyway as early as possible after exposure. Additional specimens should be drawn during recovery. SEB can be detected in the urine, and a sample should be taken and tested. Test results may be helpful retrospectively in developing a diagnosis. High concentrations inhibit kidney function. Disinfectant solutions include 0.5% sodium hypochlorite for 10–15 min or soap and water.

Ricin (*Ricinus communis*) is a protein toxin that is produced from the castor bean and functions as a cellular poison (Photo 8.11). Ricin is widely available in any part of the

Photo 8.11 Castor bean plants are the source of ricin toxin that has been used as a terrorist agent. Seeds are available at most lawn and garden centers and some castor bean plants grow wild.

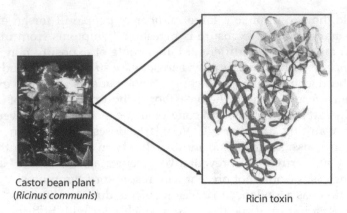

Castor bean plant
(*Ricinus communis*)

Ricin toxin

Figure 8.47 Ricin.

world. It is highly toxic and can enter the body by ingestion, inhalation, and injection. Inhalation from aerosol dispersion produces symptoms based upon the dose that was inhaled. During the 1940s, humans were accidentally exposed to sublethal doses, which produced fever, chest tightness, cough, dyspnea, nausea, and arthralgias within 4–8 h. After several hours, profuse sweating occurred, which signaled the end of the symptomatic phase (Figure 8.47).

Little data are available on human inhalation exposure, but victims would be expected to develop severe lung inflammation with a progressive cough, dyspnea, cyanosis, and pulmonary edema. When ricin enters the body through routes other than inhalation, it is not a direct lung irritant. Ingestion causes GI hemorrhage with hepatic, splenic, and renal necrosis. Intramuscular injection causes severe localized necrosis of muscle and regional lymph nodes, with moderate visceral organ involvement. The toxicity of ricin compared to botulinum and SEB, based upon LD_{50} values, is much less. Natural intoxication from ricin can occur by the ingestion of the castor bean. This produces severe GI symptoms, vascular collapse, and death. When exposure to ricin occurs through inhalation of small particles, pathogenic changes can occur in as little as 8 h. This is followed by severe respiratory symptoms and acute hypoxic respiratory failure in 36–72 h. Intravenous injection may result in disseminated intravascular coagulation, microcirculatory failure, and multiple organ failure. Ricin is toxic to the cells in the body and acts by inhibiting protein synthesis. During tests conducted on rodents, ricin was more toxic through inhalation than ingestion. A vaccine for ricin is under development, but not currently available. A vaccine would provide the best protection against ricin poisoning.

Trichothecene mycotoxins are toxins produced by several types of fungi (mold). They are the only group of biological agents that enter the body through skin absorption. Other routes of exposure include inhalation and ingestion. Most mycotoxins act by inhibiting protein synthesis and respiration. Fungi toxins most likely to be used by terrorists include diacetoxyscirpenol (DAS), nivalenol, 4-deoxynivalenol (DON), and T2. Of those listed, T2 is the most likely candidate for terrorist use because of its stability. T2 could be aerosolized or used to contaminate food supplies. Mycotoxins are fast-acting and may produce symptoms within minutes of exposure. Initial symptoms include burning skin pain, redness, tenderness, blistering, and progression to skin necrosis (tissue death), with leathery blackening and sloughing of large areas of skin in lethal cases. When inhaled, the symptoms include itching and pain, sneezing, epistaxis (bleeding

Table 8.15 Mycotoxins

Dangerous biological toxins
The only bioagents that are absorbed through the skin
Not sensitive to heat or ultraviolet light
Can be used effectively as a weapon

from the nose), and rhinorrhea. Other symptoms include pulmonary/tracheobronchial toxicity by dyspnea, wheezing, and cough. Mouth and throat exposures are characterized by pain and blood-tinged saliva and sputum. When the toxin reaches the GI tract, anorexia, nausea, vomiting, watery or bloody diarrhea, and abdominal cramp pain may occur (Table 8.15).

SEB and ricin can cause similar systemic symptoms; however, neither of them produces eye or skin symptoms. If the eyes are exposed, eye pain, tearing, redness, foreign-body sensation, and blurred vision may result. Irrespective of the route of exposure, when the toxin reaches the rest of the body's systems, it may cause weakness, prostration, dizziness, ataxia, and loss of coordination. When victims have been exposed to lethal doses, tachycardia, hypothermia, and hypotension follow. Death may occur in minutes, hours, or days. No antidotes are known for mycotoxins. Treatment is supportive and symptomatic.

Chemistry of clandestine drug labs

According to the DEA, "Clandestine drug labs are illicit operations consisting of chemicals and equipment necessary to manufacture controlled substances." Production of some substances, such as methamphetamine, PCP, MDMA, and methcathinone, requires little sophisticated equipment or knowledge of chemistry. Synthesis of other drugs, such as fentanyl and LSD, requires much higher levels of expertise and equipment. The clandestine drug problem continues to grow across the United States. No part of the country is spared as a potential site. Once-tranquil rural areas have become targets because of the remote locations. No emergency responders can safely say that they will not find a clandestine drug lab in their jurisdiction. Unlike other hazardous materials locations, drug labs, because of their clandestine nature, do not have the usual signs or hints to the presence of hazardous materials. Some hints to the presence of illicit drug activity include the following (Photo 8.12):

- Mixing of unusual chemicals in a house, garage, or barn by persons not involved in the chemical industry
- Late-night secretive activity in a rural/farm area
- The possession of chemical glassware by someone not involved in the chemical field
- Possession of unusual chemicals, such as large quantities of methyl ethyl ketone, Coleman Fuel 7, toluene/paint thinner, acetone, alcohol, benzene, freon, chloroform, starting fluid, anhydrous ammonia, "Heet," white gasoline, phenyl-2-propane, phenylacetone, phenylpropanolamine, iodine crystals, red phosphorous, black iodine, lye (Red Devil Lye), muriatic/hydrochloric acid, battery acid/sulfuric acid, Epsom salts, batteries/lithium, sodium metal, wooden matches, propane cylinders, bronchodilators, rock salt, diet aids, energy boosters, or cold/allergy medications

Photo 8.12 According to the DEA, "Clandestine drug labs are illicit operations consisting of chemicals and equipment necessary to manufacture controlled substances."

Response personnel need to be alert to strong unusual odors, such as cat urine, ether, ammonia, acetone, or other chemicals; fast-burning fires; chemical containers and apparatus in places they should not be expected; houses with windows blacked out; excessive trash, including large amounts of items such as antifreeze containers, lantern fuel cans, red chemically stained coffee filters, drain cleaner, and duct tape. Clandestine drug labs have been found in houses, apartments, trailers, motels, mountain cabins, rural farms, and other occupancies.

Under nonfire conditions, health effects of drug labs that responders could be exposed to can be varied. They can range from respiratory problems, skin and eye irritation, headaches, nausea, and dizziness. Little is known about long-term effects of exposure. Cleanup of an abandoned lab is a hazardous materials team function, usually by a private contractor, but emergency services hazmat teams could be called upon to assist in evidence preservation and collection.

Anhydrous ammonia is a key ingredient in the illegal production of methamphetamines. Drugmakers often steal ammonia from farms and agricultural supply companies. Leaks and releases of anhydrous ammonia have occurred as a result of the thefts from valves being left open. Ammonia has been transferred to inappropriate makeshift containers, such as propane tanks used on barbeque grills. These leaks add to the risks of emergency responders who are called upon to deal with the releases.

Summary

Class 6.1 Poisons, when spilled, usually do not affect large segments of the population, as do the Class 2.3 Poison Gases. However, they can still present serious dangers to emergency responders. Several protective measures may minimize the effects of toxic materials. Antidotes are available for a small number of toxic materials, but they must be administered immediately after exposure. Your body has the ability to filter out some toxic materials through the normal process of eliminating wastes. Subclass 6.2 Infectious Substances are transported, used, and stored in small quantities. They can, however, create a significant hazard to responders and the public if mishandled. Protect yourself from toxic materials by wearing protective clothing and avoiding contact with toxic materials. Practice contamination prevention. Establish zones, deny entry, and provide protection to responders and to the public.

Review questions

8.1 A one-time, short-duration exposure to a toxic material is referred to as which of the following?
   A. Chronic exposure
   B. Subacute exposure
   C. Acute exposure
   D. None of the above

8.2 Multiple exposures, including 8 h/day, 40 h/week, are referred to as which of the following?
   A. Chronic
   B. Subacute
   C. Acute
   D. Mutagenic

8.3 List the four routes of exposure by which toxic materials may enter the body.

8.4 Long-term effects of exposure to toxic materials may cause which of the following?
   A. Cancer
   B. Skin rash
   C. Eye irritation
   D. Birth defects

8.5 Etiologic agents are which of the following?
   A. Chemicals
   B. Pesticides
   C. Living organisms
   D. None of the above

8.6 Which of the following is a rate of exposure to toxic materials?
   A. RADs
   B. Isomers
   C. Curies
   D. TLV-TWA

8.7 Protective measures against toxic materials include all but which of the following?
   A. Internal
   B. Vaccinations
   C. External
   D. Antidotes

8.8 Provide the names, structures, and hazards for the following toxic hydrocarbon and hydrocarbon-derivative compounds.

$$C_7H_8 \quad C_2H_5F \quad CH_3OH \quad HCOOH \quad t\text{-}C_4H_9OH$$

8.9 What are the three "signal words" associated with pesticide labels?

8.10 List four terms that express concentrations of toxic materials.

chapter nine

Radioactive materials

History of radiation

Radiation is a phenomenon characterized more by its ability to cause biological effects than where it originates. Radiation cannot be detected by the human senses. For response personnel to determine radiation is present, they must use meters to detect it. Radiation was first discovered by German scientist Antoine Henri Becquerel, who received the Nobel Prize in Physics in 1903 for his work. Many of the terms associated with radioactivity come from those early pioneers in radiation physics: Wilhelm Conrad Roentgen (1845–1923) and Pierre (1859–1906) and Marie Curie (1867–1934), who also received the Nobel Prize in Physics in 1903 for their work on radiation. Ernest Rutherford (1871–1937) is considered the father of nuclear physics. He developed the language that describes the theoretical concepts of the atom and the phenomenon of radioactivity. Particles named and characterized by him include the alpha particle, beta particle, and proton. Rutherford won the Nobel Prize in Chemistry in 1909 for his work.

The effects of radiation have been studied for over 100 years. Scientists know a great deal about how to detect, monitor, and control even the smallest amounts of radiation. More is known about the health effects of radiation than any other chemical or biological agent. Radiation is naturally occurring and is a part of everyday life. It is present in the earth's crust, travels from outer space, and is in the air and rocks. Radiation can also be manmade. Our way of life is characterized by the many uses of manmade radiation. Radioisotopes are used in medicine, scientific research, energy production, manufacturing, mineral exploration, agriculture, and consumer products. Radioisotopes can come from three sources. They are naturally occurring, such as radon in the air or radium in the soil. Linear accelerators and cyclotrons can produce radioisotopes, as can a nuclear reactor. Thirty-one nuclear reactors are licensed in research facilities around the United States by the Nuclear Regulatory Commission (NRC). They are primarily located in colleges and universities.

Radioactivity is caused by changes in the nucleus of the atom. Radioactivity is not a chemical activity, but rather it is a nuclear event. Chemical activity involves electrons orbiting around the nucleus of an atom, particularly the outer-shell electrons. It is within these outer-shell electrons that chemical reactions and chemical bonding take place. Radioactivity, on the other hand, involves the nucleus of the atom. There is normally a "strong force" that holds the nucleus of an atom together. There are some nuclei of elements that the force cannot hold together and the nuclei begin to disintegrate. A basic law of nature says that unstable materials may not exist naturally for long. Unstable materials must do whatever they can to achieve stability. Radioactive elements throw off particles from the nucleus to reach stability. This throwing-off of particles is called radioactivity; the process is known as nuclear decay. This decay process is a random, spontaneous occurrence. There is no way that it can be shut off, nor is there any way to predict when a particular atom will begin to decay.

According to the U.S. Department of Transportation (DOT), a radioactive material (RAM) is "any material having a specific activity greater than 0.002 microcuries per gram." Specific activity of a radionuclide includes "the activity of the radionuclide per unit mass of that nuclide." Simply stated, a microcurie is a measurement of radioactivity. When a RAM emits more than 0.002 µCi/g of material, the material is then regulated in transportation by the DOT. Radiation is ionizing energy spontaneously emitted by a material or combination of materials. A RAM, then, is a material that spontaneously emits ionizing radiation. There are three types of DOT labels used to mark radioactive packages. They are Radioactive, Radioactive I, II, and III. These RAMs are determined by the radiation level at the package surface (Table 9.1). Radioactive III materials are the only radioactives that require placarding on a transportation vehicle.

Elements above lead (atomic numbers 83 and above) on the periodic table are radioactive (Figure 9.1). Other elements may have one or more radioactive isotopes. Isotopes are atoms of elements that have the same number of protons in the nucleus but differing numbers of neutrons than the most common type (Table 9.2). For example, Carbon-12 is

Table 9.1 Radioactive Placarding and Labeling

Radioactive I:	≤0.5 mrem/h
Radioactive II:	≥0.5–≤50 mrem/h
Radioactive III:	≥50–≤200 mrem/h
All measurements at the surface of the package	

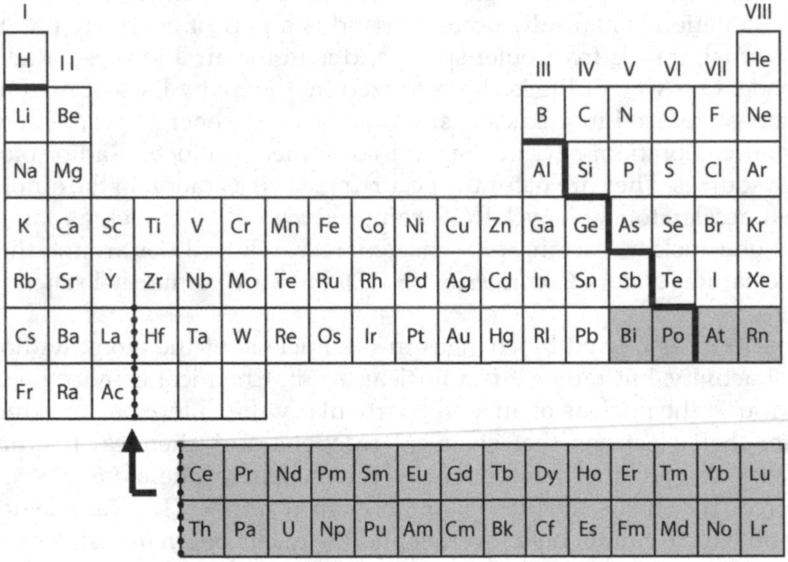

Figure 9.1 Radioactive elements.

Table 9.2 Isotopes of Selected Elements

Carbon-12 (normal)	Hydrogen 1	Protium
Carbon-13	Hydrogen 2 (heavy water)	Deuterium
Carbon-14	Hydrogen 3	Tritium

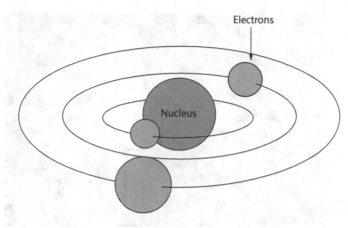

Figure 9.2 Atom.

the normal form of carbon as shown on the periodic table of elements, but carbon has a total of 16 isotopes. Some elements occur naturally, while others are man made. Each symbol on the periodic table represents one atom of that element. An atom is made up of a nucleus with varying numbers of electrons in orbits circling around the nucleus (Figure 9.2). Located inside the nucleus are protons and neutrons. Protons in the nucleus of an atom represent the atomic number of that element. Neutron numbers may vary within the same type of element or from one element to another, but the number of protons must stay the same. The sum of the protons and neutrons in the nucleus is the atomic weight of the element. For the purposes of this book, electrons do not have weight. The atom is the smallest part of an element that normally exists, so any particle of an element that is smaller than an atom is commonly referred to as a subatomic particle.

Types of radiation

There are two types of radiation: ionizing and nonionizing. Ionizing radiation involves particles and "waves of energy" emitted from the nucleus of the atom and traveling in a wave-like motion. Examples are alpha, beta, gamma and X radiation (Photo 9.1). Nonionizing radiation is also made up of "waves of energy." Examples include ultraviolet, radar, radio, microwave, visible light, and infrared light. While all radioactive waves travel in a wave-like motion, all radioactivity travels in a straight line.

There are three primary types of radioactive emissions from the nucleus of radioactive atoms (Figure 9.3). The first is the alpha particle, which looks much like an atom of helium stripped of its electrons with two protons and two neutrons remaining (Figure 9.4). An alpha particle is a positively charged particle. It is large in size and, therefore, will not penetrate as much or travel as far as beta or gamma radiation. Alpha particles travel 3–4 in. and will not penetrate the skin. Complete turnouts, including self-contained breathing apparatus (SCBA), hood, and gloves, will protect responders from external exposure. However, if alpha emitters are ingested or enter the body through a broken skin surface, they can cause a great deal of damage to internal organs.

Beta particles are negatively charged and smaller, travel faster, and penetrate farther than alpha particles. A beta particle is 1/1800 the size of a proton, or roughly equal to an electron in mass (Figure 9.5). Beta particles will penetrate the skin and travel from 3 to 100 ft. Full turnouts and SCBAs *will not* provide full protection from beta particles. Particulate

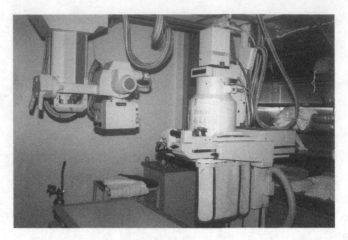

Photo 9.1 An x-ray machine at a hospital is a source of radiation only when the machine is turned on. There is no radioactive source when the power is off.

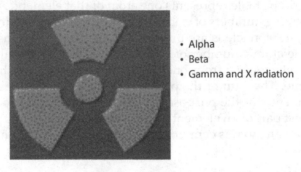

- Alpha
- Beta
- Gamma and X radiation

Figure 9.3 Three primary types of radioactive emissions.

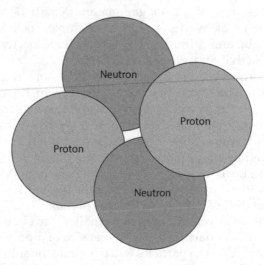

Figure 9.4 Alpha particle.

Figure 9.5 Beta particle.

radiation results in contamination of personnel and equipment where the particles come to rest. Electromagnetic energy waves, like gamma, do not cause contamination.

There is a third type of radioactive particle, but it does not occur naturally. The neutron particle is the result of splitting an atom in a nuclear reactor or accelerator, or it may occur in a thermonuclear explosion. When an atom is split, neutron particles are thrown out. You would have to be inside a nuclear reactor or experience a thermonuclear explosion to be exposed to neutron particles. If you were exposed to neutron particles, you would be quickly vaporized.

Gamma radiation is a naturally occurring, high-energy electromagnetic wave that is emitted from the nucleus of an atom. It is not particulate in nature and has high-penetrating power. Gamma rays have the highest energy level known and are the most dangerous of common forms of radiation. Gamma rays travel at the speed of light or more than 186,000 mps and will penetrate the skin, can injure internal organs, and pass through the body. No protective clothing can protect against gamma radiation. Shielding from gamma radiation requires several inches of lead, other dense metal, or several feet of concrete or earth. Gamma radiation does not result in contamination because there are no radioactive particles, only energy waves. Examples of other electromagnetic energy waves include ultraviolet, infrared, microwave, visible light, radio, and x-ray (Figure 9.6). There is little difference between gamma rays and x-rays; x-rays are produced by a cathode ray tube. To be exposed to x-rays, however, there has to be electrical power to the x-ray machine and the machine has to be turned on. If there is no electrical power, there is no radiation.

Isotopes

When the nucleus of an element contains more or less neutrons than a "normal" atom of that element, it is said to be an isotope of that element (Photo 9.2). All atoms have from 3 to 25 isotopes; the average is 10 per element. *All isotopes are not radioactive.* Hydrogen has three

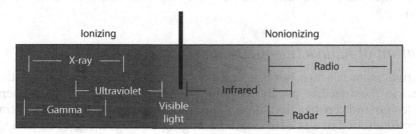

Figure 9.6 Gamma rays are a type of ionizing radiation and travel at the speed of light or more than 186,000 mps and their energy is greater than visible light.

Photo 9.2 Refrigerator with radioactive isotopes inside.

important isotopes: Hydrogen 1, sometimes called protonium, has 1 proton in the nucleus and no neutrons. Hydrogen 2 has 1 proton and 1 neutron in the nucleus and is called deuterium, or heavy water. Hydrogen 1 and Hydrogen 2 are not radioactive. Hydrogen 3 has 1 proton and 2 neutrons in the nucleus and is called tritium. Hydrogen 3 is radioactive. Tritium is used in some "exit" signs, which gives them the glow-in-the-dark ability without batteries or any other electrical source (Table 9.2). The signs last approximately 20 years before they have to be replaced. Radiation in the signs being replaced needs to be properly disposed of.

Carbon also has several isotopes. "Normal" carbon is known as carbon-12. Carbon-12 has six protons and six neutrons in the nucleus and is not radioactive. Carbon-13 has six protons and seven neutrons in the nucleus and is not radioactive. Carbon-12 makes up 999 out of 1000 carbon atoms; the other is carbon-13. Carbon-14 has six protons and eight neutrons and is naturally radioactive. Carbon-14 is a beta emitter that is produced in the atmosphere by the action of cosmic radiation on atmospheric nitrogen. In the process, a proton is forced from the nucleus of nitrogen, which then becomes carbon-14. The human body is made up of 1/10,000 of 1% of carbon-14. You inhale carbon-14 every time you take a breath. It decays and is replaced in the body, so there is a constant supply. Carbon-14 is sometimes used to determine the age of organic materials, as it takes years to disappear. The amount of carbon-14 can be used to determine how long a person has been dead, or the age of a body or other object. There are three other carbon isotopes that are all manmade: carbon-11 has 6 protons and 5 neutrons, carbon-15 has 6 protons and 9 neutrons, and carbon-16 has 6 protons and 10 neutrons; all three are radioactive.

Regulation of radioactive materials

Radioactive materials are heavily regulated during transportation by the DOT and NRC and in fixed-facility use by the NRC, which is part of the Department of Energy (DoE). Design and construction of packaging for Radioactive materials in transportation make the likelihood of a release small. Packaging undergoes rigorous testing before it is approved for use with Radioactive materials. There are four types of radioactive packaging approved for transport: Excepted, Industrial (IP), Type A, and Type B.

Excepted packaging is designed to survive normal conditions of transport. Excepted packaging is used for transportation of materials that are either low specific activity (LSA) or surface-contaminated objects (SCO), that are limited quantity shipments, instruments, or articles, and that are articles manufactured from natural or depleted uranium or natural thorium; empty packaging is also excepted (49CFR 173.421-428). Excepted packaging can be almost any packaging that meets the basic requirements, with any of the aforementioned contents. They are excepted from several labeling and documentation requirements.

Industrial packaging (IP) is designed to survive normal conditions of transport (IP-1) and at least the DROP test and stacking test for Type A packaging (IP-2 and IP-3). IP is used for transportation of materials with very small amounts of radioactivity (LSA or SCO). IP is usually metal boxes or drums.

Type A packaging is designed to survive normal transportation, handling, and minor accidents. They are used for the transportation of limited quantities of radioactive material that would not result in significant health effects if they were released.

Type B packaging is certified as Type A on the basis of performance requirements, which means it must survive certain tests. Type A packaging may be cardboard boxes, wooden crates, or drums. The shipper and carrier must have documentation of the certification of the packages being transported. Type B packaging must be able to survive severe accidents.

They are used for the transportation of large quantities of radioactive material. A Type B packaging may be a metal drum or a huge, massive shielded transport container. Type B packaging must meet severe accident performance standards that are considerably more rigorous than those required for Type A packages. Type B packaging either has a Certificate of Compliance (COC) by the NRC or Certificate of Competent Authority (COCA) by the DOT. Metal casks (Type B packaging) used for high-level Radioactive materials have never been involved in an accident in which a serious release occurred. Most releases of Radioactive materials are low level in nature and often involve radioactive isotopes used for medical purposes (Photo 9.3).

Photo 9.3 Casks are used to transport spent nuclear fuel rods. They are well constructed and there has never been a leak from a cask in an accident.

Intensity of radiation

Several terms are used to express the intensity of radiation (Table 9.3). "Radiation level" is a term often substituted for dose rate or exposure rate. It is generally referred to as the effect of radiation on matter; i.e., the amount of radiation that is imparted from the source and absorbed by matter due to emitted radiation per unit of time. Curie is a radiological term for the physical amount of a radioactive material. A curie consists of 37 billion disintegrations per second. It is a physical amount of material that is required to produce a specific amount of ionizing radiation: 1 mCi = 0.001 Ci and 1 μCi = 0.000001 Ci.

Several hundred pounds of one radioactive material may be required to produce the same amount of curies as one pound of another radioactive material. While a curie is a measure of the physical amount, the roentgen is a measure of the amount of ionization produced by a specific material. It is the amount of x-ray or gamma radiation that produces 2 billion ionizations in 1 cm^3 of dry air. A RAD is the radiation-absorbed dose (roughly equal to a roentgen). Radiation equivalent man (REM), also roughly equal to a roentgen, is a term for how much radiation has been absorbed, or the biological effect of the dose.

When radioactive materials are released, the human senses *cannot* detect radioactivity. The only way responders will know if radioactive materials are present is with the use of instruments specially designed to detect radioactivity. During the Cold War of the 1950s and 1960s, America's civil defense agencies distributed radiation meters around the country to measure radiation fallout from a thermonuclear detonation. Many of those meters have fallen into the hands of emergency responders because they were free and maintained by the civil defense agencies (a program that has since been abandoned).

There are two types of civil defense meters that have been widely available to emergency responders: the CD V-700 and the CD V-715. Neither of these instruments is able to detect alpha radiation. The CD V-700 survey meter has a range of 0–50 mR/h. An experienced meter operator can detect beta radiation with the CD V-700 through a process of elimination. Geiger–Mueller (GM) tubes are designed to survey for gamma radiation. If you check for radiation with the tube on the CD V-700 with the window closed, and no radiation exists beyond normal background, no gamma radiation is present. If the window is opened and another reading detects radiation, then it is beta radiation that is present. The CD V-715 survey meter has a range of 0.05–500 R/h or 50–500,000 mR/h. Radiation is detected through an ionization chamber much like the ionization chamber of a smoke detector (Photo 9.4).

Dosimeters are used in conjunction with survey meters to monitor the amount of radiation that personnel are exposed to over a given period of time. There are two types of civil defense dosimeters with different monitoring scales. The CD V138 is used for monitoring relatively low levels of exposure and has a minimum-scale reading of 200 mR. The CD V-742 has a range up to 200 R (200,000 mR) and is used for high levels

Table 9.3 Radiation Measurements

Curie = A physical amount of radioactive material
   1 Megacurie (MCi) = 1,000,000 Curies
   1 Kilocurie (kCi) = 1,000 Curies
   1 Millicurie (mCi) = 0.0001 Curies
   1 Microcurie (mCi) = 0.000001 Curies
Roentgen = Ionization per cm^3 of dry air
RAD = Radiation absorbed dose—dosage
REM = Biological effects

Photo 9.4 Civil defense radiological meters V-700 and the CD V-715 can detect gamma and beta radiation with a skilled operator.

of personnel exposure. Both meters should be worn by responders to ensure proper protection (Photo 9.5).

Similar commercial radiation instruments are available, including those sensitive enough to detect alpha radiation. Dosimeters are available in the form of a pen-type detector, a badge, or a digital readout. The pen or tube types record exposure and can be read by being inserted into a lighted socket on a charging unit. After the reading is taken and recorded, the pen is set to zero again for its next period of use. The badge types contain one or several rolls of photographic film that record the passage of ionizing radiation. The film must be removed and developed before the cumulative dose can be established. The digital readout dosimeters are usually handheld devices about the size of a pocket calculator. They constantly record radiation exposure and will sound an alarm if hazardous limits are reached.

Self-Indicating Radiation Alert Dosimeter (*SIRAD*®) SMART dosimeters have a sensor (a rectangle strip between the color bars) with 0, 20, 50, 100, and 250 mSv bars on its top and 500, 1,000, 2,000, 4,000, and 10,000 mSv bars on its bottom for triaging information

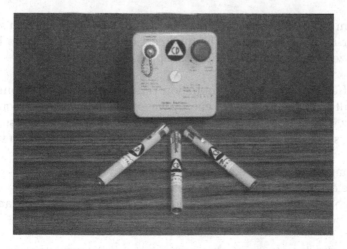

Photo 9.5 Dosimeters are used in conjunction with survey meters to monitor the amount of radiation that personnel are exposed to over a given period of time.

in emergencies. If during or after the incident, the color of sensor has not changed, the wearer has not received radiation exposure large enough to cause acute medical effects. If the sensor turns light green in color, a low radiation exposure is indicated. In this case, further exposure should be avoided. If a user is exposed to a dose between the 250 and 500 mSv, the user should seek medical advice and/or evaluation. The SIRAD® can be carried in your wallet or attached to other identification cards you carry at an incident scene.

Four general types of radiation monitors are available for emergency response:

Geiger counter

- You can use a portable Geiger counter to detect radiation coming from samples in many different settings, from laboratories to mine shafts. It contains a cigar-sized high-voltage vacuum tube called a GM tube that registers many kinds of ionizing radiation, including x-rays, gamma rays, and alpha particles. The Geiger counter indicates radiation by making a clicking sound and showing the intensity level on a meter or digital display.

Lithium fluoride crystal

- People who work around radiation wear badges called dosimeters. The dosimeter keeps track of radiation they encounter on the job. A lithium fluoride crystal in the badge stores energy whenever ionizing radiation hits it. Technicians test the exposed crystal to make certain a worker's radiation dose stays within safe limits.

Bubble detector

- Most radiation monitors cannot detect neutrons, so you need another device for this kind of radiation. Bubble detectors serve as inexpensive, simple-to-use dosimeters. Neutrons produce tiny vapor trails in a special gel. The vapor forms bubbles, and the bubbles accumulate. The number of bubbles relates to the number of neutrons, which helps determine exposure to neutron radiation.

Helium 3 detector

- A helium 3 detector can monitor levels of neutron radiation much as a Geiger counter does for ionizing radiation. The high cost of helium 3 detectors limits their use to specialized laboratory and industrial applications.

Health effects of exposure to radiation can vary (Table 9.4). Nonionizing radiation comes from ultraviolet and infrared energy waves. This type of radiation causes a sunburn type of injury. This is not a major concern for hazardous materials responders. Ionization damage occurs at the cellular level. Four types of short-term effects on the cells can occur:

1. No damage at all; the ionization passes through the cell
2. Damage occurs, but the damaged cells can be repaired
3. Irreparable damage to the cells that does not cause death, but the damage is permanent
4. Destruction of the cells

There are also long-term effects from ionizing radiation. Exposures can cause cancer and birth defects of a teratogenic or mutagenic nature. Teratogenic birth defects result

Table 9.4 Health Effects of Exposures

25 REM	Maximum single lifetime exposure
20–100 REM	Chromosomal damage, alteration of white blood count
100–200 REM	Nausea and vomiting, WBC reduction
200–400 REM	Severe WBC reduction, hair loss, some death from infection
600–1000 REM	50% death in 30 days
1000–2000 REM	Death within 4–14 days
2000 or more	Death (immediate)

from the fetus being exposed to and damaged by radiation. The child is then born with some kind of birth defect as a result of the exposure. Providing no further exposures occur during pregnancy, the mother can have normal children again. Mutagenic damage occurs when the DNA or other part of the reproductive system is damaged by exposure to radiation. The ability to produce normal children is lost; the damage is permanent.

Radiation exposure

Routes of entry for radioactive materials are much the same as for poisons. However, the radioactive source or material does not have to be directly contacted for radiation exposure to occur. Exposure occurs from the radiation being emitted from the radioactive source. Once a particulate radioactive material enters the body, it is dangerous because the source now becomes an internal source rather than an external one. You cannot protect yourself by time, distance, or shielding from a source that is inside your body. Contact with or ingestion of a radioactive material does not make you radioactive. Contamination occurs with radioactive particles, but with proper decontamination, these can be successfully removed. After they are removed, they cannot cause any further damage to the body.

Because radiation exposure can be cumulative, there are no truly safe levels of exposure to radioactive materials. Radiation does not cause any specific diseases. Symptoms of radiation exposure may be the same as those from exposure to cancer-causing materials. The tolerable limits for exposure to radiation that have been proposed by some scientists are arbitrary. Scientists concur that some radiation damage can be repaired by the human body. Therefore, tolerable limits are considered acceptable risks when the activity benefits outweigh the potential risks. The maximum annual radiation exposure for an individual person in the United States is 0.1 REM. Workers in the nuclear industry have a maximum exposure of 5 REMs per year. An emergency exposure of 25 REMs has been established by The United States Environmental Protection Agency for response personnel. This type of exposure should be attempted under only the most dire circumstances and should occur only once in a lifetime.

Effects of exposure to radiation on the human body depend on the amount of material the body was exposed to, the length of exposure, the type of radiation, the depth of penetration, and the frequency of exposure. Cells that are the most susceptible are rapidly dividing cells, such as in the bone marrow. Children are more susceptible than adults, and the fetus is the most susceptible. Radiation injuries frequently do not present themselves for quite a long time after exposure. It can be years, or even decades, before symptoms appear. Cancer is one of the main long-term effects of exposure to radiation. Leukemia may take from 5 to 15 years to develop. Lung, skin, and breast cancer may

Table 9.5 Radiation Sickness

25 REM	No detectable symptoms
50 REM	Temperature, blood count change
100 REM	Nausea, fatigue
200–250 REM	Fatal to some in 30 days, all sick
500 REM	1/2 dead in 30 days
600+ REM	All will die

take up to 40 years to develop. Table 9.5 shows exposure rates and resulting radiation-sickness effects. Varying levels of illness will occur from radiation exposure depending on the dose:

- No detectable symptoms are the result of up to 25 REM.
- Elevated temperature and changes in blood count occur from 50 REM.
- Nausea and fatigue result from 100 REM.
- Two hundred to two-hundred fifty REM results in sickness to all exposed and death to some within 30 days.
- Five hundred REM results in the death of half of those exposed in 30 days.
- Exposure to over 600 REM results in the death of all exposed.

Radiation burns are much like thermal burns, although they can be much more severe:

- First-degree radiation burns result from an exposure of 50–200 RADs.
- Second-degree burns result from 500 RADs.
- Third-degree burns result from 1000 RADs.

Because of the physical characteristics of radioactive materials, protection for emergency responders can be provided by taking a few simple, protective actions, commonly referred to as time, distance, and shielding. *Time* refers to the length of exposure to a radioactive source and the half-life of a radioactive material. A half-life is the length of time necessary for an unstable element or nuclide to lose one half of its radioactive intensity in the form of alpha, beta, and gamma radiation. Half-lives range from fractions of seconds to millions of years. In 10 half-lives, almost any radioactive source will no longer put out any more radiation than normal background.

Distance is the second protective measure against radiation. As previously mentioned, radiation travels in a straight line, but only for short distances. Therefore, the greater the distance from the radioactive source, the less the intensity of the exposure will be. There is a law in dealing with radioactivity known as the "inverse-square law" (Figure 9.7). This means that as the distance from the radioactive source is doubled, the radiation intensity drops off by one quarter. If the distance is increased 10 times, the intensity drops off to 1/100 of the original intensity.

Shielding is the third protective measure against radiation. Shielding simply means placing enough mass between personnel and the radiation, which will provide protection from the radiation (Figure 9.8). In the case of alpha particles, your skin or a sheet of paper will produce enough shielding. Turnouts will provide extra protection. Ingestion is the major hazard of radioactive particles, and wearing SCBA will prevent ingestion. Beta particles require more substantial protection from entering the body.

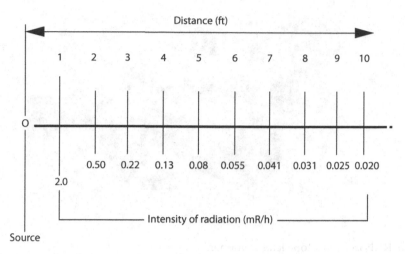

Figure 9.7 Inverse square law.

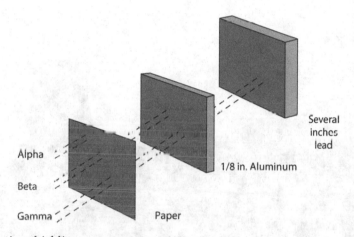

Figure 9.8 Radiation shielding.

A 1/4-in. thick piece of aluminum will stop beta radiation. Turnouts will not provide adequate protection. Gamma radiation requires 3–9 in. of lead or several feet of concrete or earth (Photo 9.6).

Radioactive elements and compounds

Uranium, U, is a radioactive metallic element. Uranium has three naturally occurring isotopes: uranium 234 (0.006%), uranium 235 (0.7%), and uranium 238 (99%). Uranium 234 has a half-life of 2.48×10^5 years; uranium 235 has a half-life of 7.13×10^8 years; uranium 238 has a half-life of 4.51×10^9 years. Uranium is a dense, silvery, solid material that is ductile and malleable; however, it is a poor conductor of electricity. As a powder, uranium is a dangerous fire risk and ignites spontaneously in air. It is highly toxic and a source of ionizing radiation. The TLV, including metal and all compounds, is 0.2 mg/m³ of air. The four-digit UN identification number for uranium is 2979. Uranium is used in nuclear reactors to produce electricity and in the production of nuclear weapons systems (Photo 9.7).

Photo 9.6 Radioactive isotope lead container.

Photo 9.7 A nuclear reactor at a power plant can be a source of highly radioactive materials including spent fuel rods.

Uranium compounds

Uranium compounds are primarily used in the nuclear industry. Uranium has been used over the years for a number of commercial ventures, some successful and others not. Uranium dioxide was employed as a filament in series with tungsten filaments for large incandescent lamps used in photography and motion pictures. Uranium dioxide has a tendency to eliminate the sudden surge of current through the bulbs when the light is turned on, which extends the life of the bulbs. Some alloys of uranium were used in the production of steel; however, they never proved commercially valuable. Sodium and ammonium diuranates have been used to produce colored glazes in the production of ceramics. Uranium carbide has been suggested as a good catalyst for the production of synthetic ammonia. Uranium salts in small quantities are claimed to stimulate plant growth; however, large quantities are clearly poisonous to plants.

Uranium carbide, UC_2, is a binary salt. It is a gray crystal that decomposes in water. It is highly toxic and a radiation risk. Uranium carbide is used as nuclear reactor fuel.

Uranium dioxide, also known as yellow cake, has a molecular formula of UO_2. It is a black crystal that is insoluble in water. It is a high radiation risk and ignites spontaneously in finely divided form. It is used to pack nuclear fuel rods for nuclear reactors.

Uranium hexafluoride has a molecular formula of UF_6. It is a colorless, volatile crystal that sublimes and reacts vigorously with water. It is highly corrosive and is a radiation risk. The four-digit UN identification number for fissile material containing more than 1% of uranium 235 is 2977; for lower specific activity, the number is 2978. Uranium hexafluoride is used in a gaseous diffusion process for separating isotopes of uranium.

Uranium hydride is a binary salt and has a molecular formula of UH_3 and is a brown-gray to black powder that conducts electricity. It is highly toxic and ignites spontaneously in air.

Uranium tetrafluoride, with the molecular formula of UF_4, is a green, nonvolatile, crystalline powder that is insoluble in water. It is highly corrosive and is also a radioactive poison.

Radium compounds

Radium, Ra, is a radioactive metallic element. There are 14 radioactive isotopes of radium; however, only radium 226, with a half-life of 1620 years, is usable. It is a brilliant, white solid that is luminescent and turns black upon exposure to air. Radium is water soluble and contact with water evolves hydrogen gas. It is in the alkaline-earth metal family, and, like calcium, it seeks the bones when it enters the body. It is highly toxic and emits ionizing radiation. Radium is destructive to living tissue. It is used in the medical treatment of malignant growths and industrial radiography. Compounds formed with radium all have the same hazards as radium itself. Most are used in the treatment of cancer and for radiography in the medical and industrial fields. The compounds are all solids, and the degree of water solubility varies.

Radium bromide is a binary salt and has a molecular formula of $RaBr_2$, it is composed of white crystals that turn yellow to pink. It sublimes at about 1650°F and is water soluble. The hazards are the same as that for radium. It is used in the medical treatment of cancers.

Radium carbonate, with the molecular formula of $RaCO_3$, is an amorphous, radioactive powder that is white when pure. Because of impurities, radium carbonate is sometimes yellow or pink. It is insoluble in water.

Radium chloride is a binary salt with the molecular formula of $RaCl_2$, which is a yellowish-white crystal that becomes yellow or pink upon standing. It is radioactive and soluble in water. It is used in cancer treatment and physical research. Radium is a common source of naturally occurring radioactive material (NORM) especially in the oil industry.

Cobalt

Cobalt, Co, is a metallic element. Cobalt 59 is the only stable isotope. Common isotopes are cobalt 57, cobalt 58, and the most common, cobalt 60. Cobalt is a steel-gray, shining, hard, ductile, and somewhat malleable metal. It has magnetic properties and corrodes readily in air. Cobalt dust is flammable and toxic by inhalation, with a TLV of 0.05 mg/m³ of air. It is an important trace element in soils and animal nutrition. Cobalt 57 is radioactive. It has a half-life of 267 days.

It is a radioactive poison and is used in biological research. Cobalt 58 is also radioactive and has a half-life of 72 days. It is a radioactive poison, and it is used in biological and medical research. Cobalt 60 is one of the most common radioisotopes. It has a half-life of

5.3 years, is available in larger quantities, and is cheaper than radium. It is a radioactive poison and is used in radiation therapy for cancer and radiographic testing of welds and castings in industry. Compounds of cobalt are not radioactive.

Iodine

Iodine, I, is a nonmetallic element of the halogen family. There is only one natural stable isotope: iodine 127. There are many artificial radioactive isotopes. Iodine is a heavy, grayish-black solid or granules having a metallic luster and characteristic odor. It is readily sublimed to a violet vapor and is insoluble in water. Iodine is toxic by ingestion and inhalation and is a strong irritant to the eyes and skin, with a TLV ceiling of 0.1 ppm in air. Iodine 131 is a radioactive isotope of iodine and has a half-life of 8 days. It is used in the treatment of goiter, hyperthyroidism, and other disorders. It is also used as an internal radiation therapy source. Most iodine compounds are not radioactive.

Krypton

Krypton, Kr, is an elemental, colorless, odorless inert gas. It is noncombustible, nontoxic, and nonreactive; however, it is an asphyxiant gas and will displace oxygen in the air. Krypton 85 is radioactive and has a half-life of 10.3 years. The four-digit UN identification number for krypton is 1056 as a compressed gas and 1970 as a cryogenic liquid. These forms of krypton are not radioactive. Radioactive isotopes of krypton are shipped under radioactive labels and placards as required. Its primary uses are in the activation of phosphors for self-luminous markers, detecting leaks, and in medicine to trace blood flow.

Radon

Radon, Rn, is a gaseous radioactive element from the noble gases in family eight on the periodic table. There are 18 radioactive isotopes of radon, all of which have short half-lives. For example, radon 222 has a half-life of 3.8 days. Radon is a colorless gas that is soluble in water. It can be condensed to a colorless transparent liquid and to an opaque, glowing solid. Radon is the heaviest gas known, with a density of 9.72 g/L at 32°F. Radon is derived from the radioactive decay of radium. It is highly toxic and emits ionizing radiation. Lead shielding must be used in handling and storage. Radon has appeared naturally in the basements of homes, causing some concern for the residents. The primary uses are as a cancer treatment, a tracer in leak detection, in radiography, and in chemical research.

Radioactive materials are often found in transportation. They are heavily regulated, and the containers are well constructed. Most radioactive incidents are not handled by local emergency responders. Agencies other than fire, police, and EMS are responsible for response and handling of radioactive emergencies. Emergency responders must, however, be aware of radioactive materials and know how to protect themselves. Each state has radiological response teams for radioactive emergencies. They may be a part of the emergency management agency, the health department, the department of environment, or some other agency. Federal interests are represented by the U.S. DoE and the NRC (301) 492–7000. Incidents involving weapons are handled by the Department of Defense Joint Nuclear Accident Center (703) 325–2102.

Medical uses of radioactive sources include sterilization, implants using radium, scans using iodine, and therapy using cobalt. X-rays are used in diagnostic medical procedures. In addition to medical facilities, RAMs may be found in research laboratories, educational institutions, industrial applications, and hazardous waste sites.

Department of Energy Nuclear Emergency Search Team

For response to emergencies involving nuclear incidents, the U.S. DoE has created a Nuclear Emergency Search Team (NEST) (Photo 9.8). It is headquartered on a remote corner of Nellis Air Force Base near Las Vegas, Nevada. It is, in reality, the nation's "nuclear fire department," poised to respond to terrorist incidents involving nuclear devices or materials. Its job would be to locate the nuclear device or materials, determine what it is, render it safe, and get rid of it. Resources available to the team include satellites and radiation-sensing planes and helicopters. It is reported that it can detect a radioactive particle or debris as small as a grain of salt. On the ground, resources include power generators, secure phone systems, and radiation monitors packed into nondescript vans. These vans can be loaded into a wide-body aircraft and flown anywhere in the country, or the world for that matter, within a moment's notice (Photo 9.9).

The team was organized in 1975, following a threat by an extortionist in Boston to set off a nuclear device. This team comprises over 200 searchers and is a joint effort of the

Photo 9.8 RAMs may be found in many types of transportation vehicles, including taxi cabs, delivery trucks, and private cars.

Photo 9.9 Storage areas for radioactive materials are marked with this type of sign.

FBI, DoD, and DoE. Since its inception, it has responded to dozens of incidents involving threats or actual use of radiological materials. Some of its responses included stolen plutonium from a Wilmington, North Carolina, plant, recovery of radioactive debris from a fallen Soviet satellite, radioactive monitoring after the Three Mile Island Incident in Pennsylvania, and nuclear extortion attempts against Union Oil Company in Los Angeles and Harrah's Club in Reno, Nevada. The team has responded to threats of actual nuclear devices but has never actually come face-to-face with one.

Review questions

9.1 Radioactivity takes place in which part of the atom?
   A. Outer-shell electrons
   B. Electrons next to the nucleus
   C. In the nucleus
   D. In the protons

9.2 Which of the following best describes an isotope of an element?
   A. It is located in the nucleus
   B. More or less neutrons than normal
   C. More or less protons than normal
   D. Extra electrons

9.3 List the three major types of radioactivity.

9.4 Name the types of radiation that are particulate and create contamination.

9.5 Radiation always travels in the following manner.
   A. Up
   B. Down
   C. East to west
   D. In a straight line

9.6 Name the two types of radiation.

9.7 Which of the following statements concerning types of radiation is correct?
   A. Beta high-energy wave
   B. Alpha small, travels faster, and penetrates farther
   C. Gamma large and travels short distances

9.8 Gamma radiation is considered to be which of the following?
   A. Limited hazard
   B. Particulate in nature
   C. Electromagnetic energy wave
   D. Easy to protect against

9.9 Which of the following radioactive-labeled materials require placarding on the transport vehicle regardless of the quantity?
   A. Radioactive II
   B. Radioactive I
   C. Radioactive III
   D. All of the above

9.10 List the three protective measures for radiation.

chapter ten

Corrosives

Corrosives are the largest class of chemicals used by industry, so it stands to reason that they would frequently be encountered in transportation and at fixed facilities (Photo 10.1). DOT Class 8 materials are corrosive liquids and solids. There are no DOT subclasses of corrosives. There are, however, two types of corrosive materials found in Class 8: acids and bases. Acids and bases are actually two different types of chemicals that are sometimes used to neutralize each other in a small spill. Corrosives are grouped together in Class 8 because the corrosive effects from both acids and bases are much the same on tissue and metals, if contacted. It should be noted, however, that the correct terminology for an acid is *corrosive* and for a base is *caustic*. DOT, however, does not differentiate between the two when placarding and labeling. DOT defines a corrosive material as "a liquid or solid that causes visible destruction or irreversible alterations in human skin tissue at the site of contact, or a liquid that has a severe corrosion rate on steel or aluminum. This corrosive rate on steel and aluminum is 0.246 inches per year at a test temperature of 131°F (55°C)."

A definition for an acid from the *Condensed Chemical Dictionary* is "a large class of chemical substances whose water solutions have one or more of the following properties: sour taste, ability to make litmus dye turn red and to cause other indicator dyes to change to characteristic colors, the ability to react with and dissolve certain metals to form salts, and the ability to react with bases or alkalis to form salts." It is important to note here that tasting any chemical is not an acceptable means of identification for obvious reasons. In addition to being corrosive, acids and bases can explode or polymerize; they can also be water reactive, toxic, flammable (applies to organic acids only, because inorganic acids do not burn), reactive, and unstable oxidizers.

There are two basic types of acids: organic and inorganic (Table 10.1). These can be further subdivided into families including binary and oxyacids. Inorganic acids are sometimes referred to as mineral acids. As a group, organic acids are generally not as strong as inorganic acids. The main difference between the two is the presence of carbon in the compound: inorganic acids do not contain carbon. Inorganic acids are corrosive, but they do not burn. They may, however, be oxidizers and support combustion, or may spontaneously combust with organic material. Organic acids on the other hand do burn. Inorganic acid molecular formulas begin with hydrogen (H). For example, H_2SO_4 is the molecular formula for sulfuric acid, HCl is hydrochloric acid, and HNO_3 is nitric acid. Organic acids are hydrocarbon derivatives; therefore, they have carbon in the compound, and the name begins with the prefix indicating the number of carbons. For example, the prefix for a one-carbon compound with the organic acids is "form," so a one-carbon acid is called formic acid; a two-carbon acid is acetic acid; and a three-carbon acid is propionic acid. Organic acids are corrosive, may polymerize, and may burn.

Inorganic acids

Acids are materials that release hydrogen ions (H^+) when placed in water. Inorganic acids can generally be identified by hydrogen at the beginning of the formula (Table 10.2), because few other compounds begin with hydrogen. The hydrogen ion, H^+, consists of just

Photo 10.1 Chemical protective clothing Level B, splash protection is adequate for dealing with most corrosive materials.

Table 10.1 Types of Acids

Inorganic	Organic
Sulfuric (H_2SO_4)	Formic (HCOOH)
Hydrochloric (HCl)	Acetic (C_2H_5COOH)
Nitric (HNO_3)	Propionic (C_2H_5COOH)
Phosphoric (HPO_4)	Acrylic (C_2H_3COOH)
Perchloric ($HClO_4$)	Butyric (C_3H_7COOH)

Table 10.2 Inorganic Acids Begin with Hydrogen in the Formula

Binary acids	Ternary acids
Hydrofluoric (HF)	Nitric (HNO_3)
Hydrochloric (HCl)	Perchloric ($HClO_4$)
Hydrobromic (HBr)	Sulfuric (H_2SO_4)
Hydroiodic (HI)	Phosphoric (H_3PO_4)
Hydrosulfuric (H_2S)	Carbonic (H_2CO_3)

a hydrogen nucleus, without electrons, and is composed of just one proton. Acids that supply just one H^+ are often referred to as monoprotic acids, e.g., HCl and HNO_3. Acids that supply more than one H^+ are referred to as polyprotic acids; more specifically, H_2CO_3 and H_2SO_4 with two H^+ are referred to as diprotic acids and H_3PO_4 as triprotic acids.

There are two general types of inorganic acids: binary and oxyacids. Binary acids are composed of just two elements: hydrogen and some other nonmetal, e.g., HCl and H_2S. Binary acids are named by placing the prefix "hydro" before and the suffix "ic" after the nonmetal element; the compound ends with the word "acid." For example, when hydrogen is combined with chlorine, the "ine" is dropped from chlorine and the prefix "hydro" and suffix "ic" are added: hydrochloric acid; hydrogen combined with sulfur is called hydrosulfuric acid.

Acids that contain hydrogen, oxygen, and some other nonmetal element are called oxyacids, e.g., H_2SO_4, HNO_3, and $HClO_4$ (note the similarities to the oxyradicals). Like the oxysalts, these acids are named according to the number of oxygen atoms in the compound. The acid with the largest number of oxygen atoms in a series ends with the suffix "ic," and the one with the fewest number of oxygen atoms takes the suffix "ous" (similar to the alternate naming of the transitional metal salts discussed in Chapter 2).

For example, when hydrogen is combined with sulfur, the base state of the compound is SO_4 and the acid, H_2SO_4, is called sulfur*ic* acid. If there is one less oxygen present in the compound, such as SO_3, the ending changes to "ous" and the acid, H_2SO_3, is called sulfur*ous* acid. HNO_3 is nitric acid, HNO_2 is nitrous acid, etc. When halogens are present in the acid, the compound with the most oxygen atoms in the base state ends in "ic," such as chloric acid, $HClO_3$. If the oxygen is increased by one to $HClO_4$, the prefix "per" is added, yielding the name *per*chloric acid. The acid compound with the least number of oxygen atoms ends with "ous," such as chlorous acid, $HClO_2$. If the oxygen is reduced by one, to $HClO$, the prefix "hypo" is added, yielding the name *hypo*chlorous acid.

Strength and concentration

Most inorganic acids are produced by dissolving a gas or liquid in water, e.g., hydrochloric acid is derived from dissolving hydrogen chloride gas in water (Photo 10.2). All inorganic acids contain hydrogen. Hydrogen is in the form of an ion (H^+) and can be measured by using the pH scale (Figure 10.1). In simple terms, the pH scale measures the hydrogen ion concentration of a solution. Concentrated acids and bases measure off the pH scale.

To determine if a concentrated material is an acid or a base, litmus paper is used; however, this does not yield a numerical value. As a group, acids have high hydrogen ion concentrations. Bases have low hydrogen ion concentrations and high hydroxyl (OH^-) concentrations. Strength or weakness of an acid or base is the amount of hydrogen ions or hydroxyl ions that are produced as the acid or base is created. If the hydrogen ion concentration of an acid is high, the acid is concentrated. If the hydroxyl concentration is high, it is a concentrated base. In both cases, there is almost total ionization of the material dissolved in water to make the strong acid and base (Table 10.3). For example, hydrochloric acid is a strong acid because practically all of the hydrogen chloride gas is ionized in the water. Acetic acid is a weak acid because only a few molecules ionize in producing the acid.

Another term associated with corrosives is *concentration*. Concentration has to do with the amount of acid that is mixed with water and is often expressed in terms of percentages.

Photo 10.2 Stainless steel "kegs" of nitric acid above 40% concentration, which is also an oxidizer.

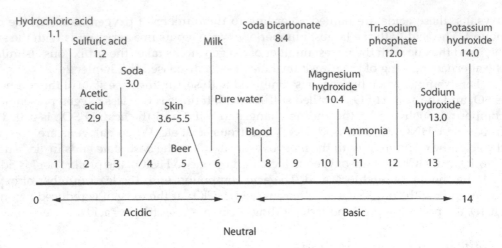

Figure 10.1 pH scale.

Table 10.3 Ionization of Common Acids and Bases

Completely ionized	Moderately ionized	Slightly ionized
Nitric	Oxalic	Hydrofluoric
Hydrochloric	Phosphoric	Acetic
Sulfuric	Sulfurous	Carbonic
Hydriodic		Hydrosulfuric
Hydrobromic		(most others)
Potassium hydroxide		Ammonium
Sodium hydroxide		hydroxide
Barium hydroxide		(all others)
Strontium hydroxide		
Calcium hydroxide		

A 98% concentration of sulfuric acid is 98% sulfuric acid and 2% water; a solution of 50% nitric acid is 50% nitric acid and 50% water. In the 50% concentration, the solution has only half the H^+ ions that the 100% concentration would have. A 50% concentration of nitric acid is a solution diluted to 50% of the original acid.

pH

The pH scale measures the acidity or alkalinity of a *solution* (Photo 10.3). It cannot measure some strong acids and bases that are full strength because they have values less than 0 or greater than 14. They would be off the scale. Acid solutions are considered acidic, and base solutions are considered alkaline. Acid solutions have a value on the pH scale from 1 to 6.9. Materials with a pH value of 7 are considered to be neutral (7 is also the pH of pure water), i.e., they are neither acidic nor basic. Base solutions have values on the scale from 7.1 to 14. It is not important for emergency responders to understand or know how the pH scale measures corrosivity or the specific values of any given acid or base. It is, however, important for responders to know that numerical values lower than 7 are acids and values higher than 7 are bases.

Usually, when dealing with numerical values, the higher the number, the greater the value that is being measured and the number 2 is twice the value of 1. When using the pH

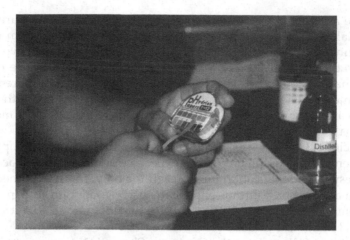

Photo 10.3 pH paper is an easy and cheap way to determine whether an unknown material is an acid or a base.

scale for acids, however, the lower the pH number, the more acidic an acid solution is and an acid solution with a pH of 1 is 10 times more acidic than an acid solution with a pH of 2, and so on (Table 10.4). The ratio and the intervals between the numbers are exponential, e.g., a pH of 5 is 10 times more acidic than a pH of 6. The result of this exponential ratio is that on the full scale, a solution with a pH of 1 is 1,000,000 times more acidic than an acid with a pH of 6.9. So the difference between individual values on the pH scale is great and is one of the reasons why dilution and neutralization are not as simple as they might sound. Those terms will be discussed further under "Dilution vs. neutralization" section of this chapter.

If the chemical name of a hazardous material is known and it is determined to be a corrosive, looking up the chemical name in reference sources will identify whether the material is an acid or a base. It will not be necessary for responders to get a pH measurement of the material unless it is to verify the reference information. The use of pH measurements can be useful when a material has not been positively identified. The pH measurement can be used to narrow the chemical family possibilities in the identification process.

Table 10.4 Exponential Logarithm of pH Values

pH value	H^+ and OH^- concentration
1	1,000,000
2	100,000
3	10,000
4	1,000
5	100
6	10
7	1
8	10
9	100
10	1,000
11	10,000
12	100,000
13	1,000,000

There are a number of ways for emergency responders to measure the pH of a corrosive material. First of all, the proper chemical protective clothing must be worn when working around corrosive materials. The simplest and least expensive method of determining pH is the use of pH paper, which changes color based on the type and strength of corrosive material that is present. The colored paper is then compared to a chart on the pH paper container. The chart indicates numerical pH values much the same way as do expensive measuring instruments. Although not as accurate as a pH meter, the numbers will give a "ballpark" measure of the pH of the material.

Commercial pH meters are also available, from handheld to sophisticated laboratory instruments. This equipment can be expensive and pH paper is accurate enough for emergency response identification purposes. If the only information needed is whether a material is an acid or a base, litmus paper can also be used. Litmus paper turns blue if the corrosive material is a base, red if the corrosive material is an acid. Litmus paper will not give actual pH numerical values.

The definition of a base from the *Condensed Chemical Dictionary* is "a large class of compounds with one or more of the following properties: bitter taste, slippery feeling in solution, ability to turn litmus blue and to cause other indicators to take on characteristic colors, and the ability to react with (neutralize) acids to form salts." It is important to note that while the definition of acid and base mentions the taste and feeling of the materials, these are dangerous chemicals and can cause damage to tissues upon contact. Therefore, it is *NOT* recommended that responders come in contact with these materials through taste or touch!

Ionization occurs with the bases, just as with the acids, as they are created. Most bases are produced by dissolving a solid, usually a salt, in water. However, with the bases, the ion produced is the hydroxyl ion (OH^-). A base is considered strong or weak depending on the number of hydroxyl ions produced as a corrosive material is dissolved in water. Large OH^- concentrations produce a strong base; a small OH^- concentration produces a weak base. Sodium and potassium hydroxide are strong bases; calcium hydroxide (hydrated lime) is a weak base. Bases will have a pH from 7.1 to 14 on the pH scale. Degree of alkalinity increases from 7.1 to 14, with 7.1 being the least basic and 14 the most basic. Amount of alkalinity between the numerical values on the pH scale is exponential, just as with the acids. A base with a pH of 9 is 10 times more basic than one with a pH of 8, and so on.

Corrosivity is not the only hazard of Class 8 materials. In addition to being corrosive, they may have other hazards, such as toxicity, flammability, or oxidation. Corrosives, especially acids, can be violently water reactive. Contact with water may cause splattering of the corrosive, produce toxic vapors, and evolve heat, which may ignite nearby combustible materials. Some of the water may be turned to steam by the heat produced in the reaction. This can cause overpressurization of a container. Corrosives may also be unstable, reactive, explode, polymerize, or decompose and produce poisons.

Picric acid, $C_6H_2(NO_2)_3OH$, e.g., becomes a high explosive when dried out and is sensitive to shock and heat. The hazard class for picric acid is 4.1 Flammable Solid. It is considered a wetted explosive. The name would indicate acid; however, the corrosivity of picric acid is far outweighed by its explosive dangers. The slightest movement of dry picric acid may cause an explosion. Picric acid, when shipped, is mixed with 12%–20% water to keep it stable. When this water evaporates in storage over time, the material becomes explosive.

Perchloric acid, $HClO_4$, is a colorless, volatile, fuming liquid that is unstable in its concentrated form. It is a strong oxidizing agent and will spontaneously ignite upon contact with organic materials. It is corrosive, with the highest concentration of 70%. Contact with water produces heat; when shocked or heated, it may detonate. The boiling point is 66°F (18°C), and it is soluble in water, with a specific gravity of 1.77, which is heavier than

water. The vapor density is 3.46, which is heavier than air. Perchloric acid is toxic by ingestion and inhalation. It is used in the manufacture of explosives and esters, in electropolish-
ing and analytical chemistry, and as a catalyst.

Hydrocyanic acid, HCN, is corrosive in addition to toxic. It is also a dangerous fire and
explosion risk. It has a wide flammable range of 6%–41% in air. The boiling point is 79°F
(26°C), the flash point is 0°F, and the ignition temperature is 1004°F (540°C). It is toxic by
inhalation and ingestion and through skin absorption. The TLV of hydrocyanic acid is
10 ppm in air. It is used in the manufacture of acrylonitrile, acrylates, cyanide salts, dyes,
rodenticides, and other pesticides.

Organic acids

Organic acids are hydrocarbon derivatives. They are flammable and corrosive and may
polymerize by exposure to heat or sudden shock. Organic acids are "super-duper" polar
materials; they are the most polar of the hydrocarbon derivatives. Organic acids have
hydrogen bonding and a carbonyl that gives them a double dose of polarity. The functional group is represented by a carbon atom, two oxygen atoms, and a hydrogen atom.
The general formula is *R-C-O-O-H.*

One radical is attached to the carbon atom of the functional group. Organic acids use
the alternate prefix for one- and two-carbon compounds. When naming them, all of the
carbons, including the one in the functional group, are counted to determine the hydrocarbon prefix name. To represent an acid, "ic" is added to the hydrocarbon prefix and the name
ends in "acid," e.g., a one-carbon acid uses the alternate prefix name "form"; "ic" is added
to "form," making it formic; and "acid" is added to the end: formic acid. A two-carbon acid
uses the alternate name for two carbons, which is "acet," plus "ic," and ends with "acid":
acetic acid. Naming three and four-carbon acids reverts back to the normal prefixes for
three- and four-carbon radicals, with some minor alterations to make the names flow more
smoothly. For example, a three-carbon acid uses the prefix "prop," indicating three carbons; the letters "ion" are then added to make the name flow smoothly. The ending "ic" is
added to the radical and the word "acid" for the compound name: propionic acid. A four-
carbon organic acid begins with the radical prefix "but", the filler letters "yr" are attached,
the radical ends with "ic", acid is added, and the name for a four-carbon organic acid is
butyric acid. Structures and molecular formulas for organic acids with one through four
carbons are shown in Figure 10.2. Note that the carbon in the functional group is counted
when determining which hydrocarbon radical is used in naming it.

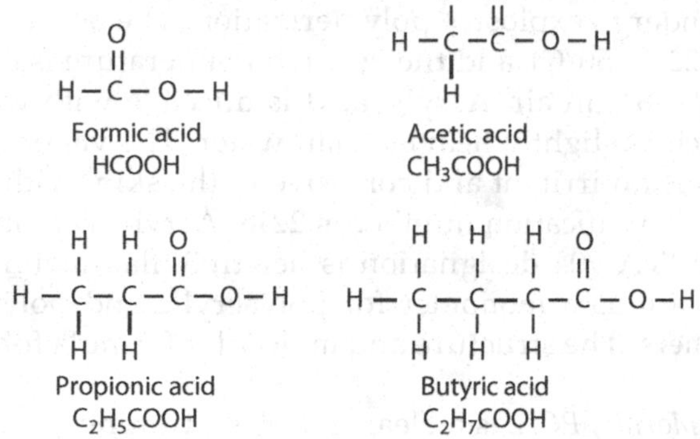

Figure 10.2 Organic acids.

It is also possible to add double-bonded radicals to the organic acid functional group. For example, when the vinyl radical is attached to the carbon atom in the functional group, a three-carbon, double-bonded radical is created. The acryl radical is used for three carbons with a double bond, the ending "ic" is added to the radical, and the word "acid" is added to the end. The compound formed is acrylic acid. The double bond between the carbons can come apart in a polymerization reaction. Generally, materials that have double bonds are reactive in some manner. If polymerization occurs inside a container, an explosion may occur that can produce heat, light, fragments, and a shock wave.

Formic acid, HCOOH, is a colorless, fuming liquid with a penetrating odor. The highest commercial concentration is 90%. Formic acid, as well as all organic acids, is a polar material that is soluble in water, and it has a specific gravity of 1.2, which is heavier than water. As with many of the organic acids, formic acid is flammable. The boiling point is 213°F (100°C), the flash point is 156°F (68°C), and the flammable range is 18%–57%. The ignition temperature is 1004°F (540°C), and the vapor density is 1.6, which is heavier than air. Formic acid is toxic, with a TLV of 5 ppm in air. The four-digit UN identification number is 1779. The NFPA 704 designation is health 3, flammability 2, and reactivity 0. It is used in the dyeing and finishing of textiles, the treatment of leather, and the manufacture of esters, fumigants, insecticides, refrigerants, etc.

Propionic acid, C_2H_5COOH, is a colorless, oily liquid with a rancid odor. It is a polar compound and soluble in water. Propionic acid is flammable, with a flammable range of 2.9%–12% in air, a boiling point of 286°F (141°C), a flash point of 126°F (52°C), and an ignition temperature of 955°F (512°C). Polar-solvent foam will have to be used to extinguish fires. It is toxic, with a TLV of 10 ppm in air.

The four-digit UN identification number for propionic acid is 1848. The NFPA 704 designation is health 3, flammability 2, and reactivity 0. It is used as a mold inhibitor in bread and as a fungicide, an herbicide, a preservative for grains, in artificial fruit flavors, pharmaceuticals, and others.

Butyric acid, also known as butanoic acid, is a colorless liquid with a penetrating, obnoxious odor. It is miscible with water, with a specific gravity of 0.96, which makes it slightly lighter than water. Butyric acid is flammable, with a boiling point of 326°F (163°C) and a flash point of 161°F (71°C). The flammable range is 2%–10% in air, and the ignition temperature is 846°F (452°C). Butyric acid is a strong irritant to the skin and eyes. The four-digit UN identification number is 2820. The NFPA 704 designation is health 3, flammability 2, and reactivity 0. The primary uses of butyric acid are in the manufacture of perfume, flavorings, pharmaceuticals, and disinfectants.

Acrylic acid, C_2H_3COOH, is a colorless liquid with an acrid odor. It polymerizes readily and may undergo explosive polymerization. The boiling point is 509°F (265°C), the flash point is 122°F (50°C), and the ignition temperature is 820°F (437°C). The flammable range is 2.4%–8% in air. Acrylic acid is miscible with water and has a specific gravity of 1.1, which is slightly heavier than water. The vapor density is 2.5, which is heavier than air. It is an irritant and corrosive to the skin, with a TLV of 2 ppm in air. The four-digit UN identification number is 2218. Acrylic acid must be inhibited when transported. The NFPA 704 designation is health 3, flammability 2, and reactivity 2. The primary uses are as a monomer for polyacrylic and polymethacrylic acids and other acrylic polymers. The structure and molecular formula for acrylic acid are shown in Figure 10.3.

Phosphorus trichloride, PCl_3, is a clear, colorless, fuming, corrosive liquid. It decomposes rapidly in moist air and has a boiling point of about 168°F (75°C). PCl_3 is corrosive

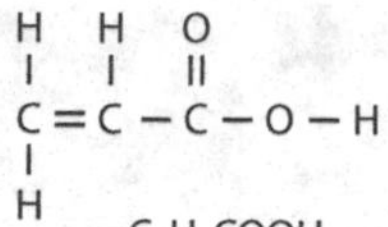

$$\begin{array}{ccc} H & H & O \\ | & | & \| \\ C = C & - C & - O - H \\ | & & \\ H & & C_2H_3COOH \end{array}$$

Figure 10.3 Acrylic acid.

to skin and tissue and reacts with water to form hydrochloric acid. The TLV is 0.2 ppm, and the IDLH is 50 ppm in air. The four-digit UN identification number is 1809. The NFPA 704 designation is health 4, flammability 0, and reactivity 2. The white section at the bottom of the diamond contains a W with a slash through it, indicating water reactivity. The primary uses are in the manufacture of organophosphate pesticides, gasoline additives, and dyestuffs; as a chlorinating agent; as a catalyst; and in textile finishing. Corrosives in contact with a poison may produce poison gases as the poison decomposes. In responding to an incident involving corrosives, the toxicity of the vapors could be much more of a concern for personnel than the corrosivity. When acids come in contact with cyanide, hydrogen cyanide gas, which is highly toxic, with a TLV of 10 ppm in air, is produced. The structure and molecular formula of phosphorous trichloride are shown in Figure 10.4.

When strong corrosives contact flammable liquids, the chemical reaction that occurs may produce heat. Heat produced will cause more vapor to be produced, and if an ignition source is present, combustion may occur. Corrosives may also be strong oxidizers. If they come in contact with particulate combustible solids, spontaneous combustion may occur. Once ignition has occurred, the corrosive will act as an oxidizer and accelerate the rate of combustion. Nitric acid in contact with combustible organic materials containing cellulose will produce a chemical reaction. This reaction will produce nitrocellulose, which is a dangerous fire and explosion risk. Toxic vapors may also be produced when the cellulose burns. After flammable liquids and gases, corrosives are the next most common hazardous material encountered by emergency responders.

Sulfuric acid, H_2SO_4, is a strong corrosive, with a solution pH of 1.2. It is a dense, oily liquid, colorless to dark brown, depending on purity (Photo 10.4). Sulfuric acid is miscible with water but is violently water reactive, producing heat and explosive splattering if water is added to the acid. The boiling point is 626°F (330°C), and the specific gravity is 1.84, which is heavier than water. Sulfuric acid is highly reactive and dissolves most metals. When in contact with metals, hydrogen gas is released. The vapors are toxic by inhalation, and the TLV is 1 ppm in air. Sulfuric acid is incompatible with potassium chlorate, potassium perchlorate, potassium permanganate, and similar compounds of other light metals. Sulfuric acid has a four-digit UN identification number of 1830. The NFPA 704 designation is health 3, flammability 0, and reactivity 2. The white section contains a W with a slash through it, indicating water reactivity. Sulfuric acid is used in batteries for cars and other vehicles. It is also used in the manufacture of fertilizers, chemicals,

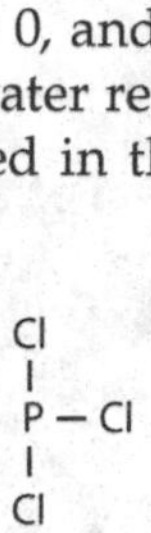

$$\begin{array}{c} Cl \\ | \\ P - Cl \\ | \\ Cl \\ PCl_3 \end{array}$$

Figure 10.4 Phosphorus trichloride.

Photo 10.4 Bulk container of sulfuric acid. All bulk containers are required to display the UN four-digit identification number in the center of the placard.

and dyes; as an etchant and a catalyst; in electroplating baths and explosives; in or for pigments; and many others (Photo 10.5).

Fuming sulfuric acid is also called oleum, which is a trade name. It is a heavy, oily liquid, colorless to dark brown depending on purity, and fumes strongly in moist air and is extremely hygroscopic. Fuming sulfuric acid is a solution of sulfur trioxide in sulfuric acid. Sulfur trioxide is forced into solution with sulfuric acid to the point that the solution cannot hold any more. As soon as the solution is exposed to air, the fuming begins, forming dense vapor clouds. It is violently water reactive, as are most acids. Oleum is also a strong irritant to tissue. The four-digit UN identification number is 1831 (Photo 10.6).

Lime, CaO (also known as calcium oxide, quicklime, hydrated lime, and hydraulic lime), a binary oxide salt, is a white or grayish-white material in the form of hard clumps.

Photo 10.5 Railcar with liquid sulfuric acid.

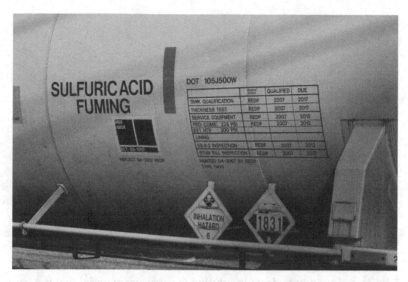

Photo 10.6 Fuming sulfuric acid is formed when a solution will no longer dissolve hydrogen sulfide gas. At that point, the gas is forced into the container in a solution with the liquid and will stay as long as the container remains closed. If opened, large amounts of hydrogen sulfide gas will be released.

It may have a yellowish or brownish tint due to the presence of iron. Lime is odorless and crumbles upon exposure to moist air. It is a corrosive caustic that yields heat and calcium hydroxide when mixed with water and is a strong irritant, with a TLV of 2 mg/m³ of air. The four-digit UN identification number for calcium oxide is 1910. The primary uses are in the manufacture of other chemicals, like calcium carbide; in pH control; and in the neutralization of acid waste, insecticides, and fungicides.

Sodium hydroxide, NaOH, also known as caustic soda, is a hydroxide salt. Sodium hydroxide is a strong base and is severely corrosive, with a solution pH of 13. It is the most important industrial caustic material. Sodium hydroxide is a white, deliquescent solid found in the form of beads or pellets. It is also found in solutions with water of 50% and 73%. Sodium hydroxide is water soluble and water reactive and absorbs water and carbon dioxide from the air. The specific gravity is 2.8, which is heavier than water. It is corrosive to tissues in the presence of moisture and is a strong irritant to the eyes, skin, and mucous membranes and is toxic by ingestion. The TLV ceiling is 2 mg/m³ of air. The four-digit UN identification number is 1823 for dry materials and 1824 for solutions. The NFPA 704 designation is health 3, flammability 0, and reactivity 1. It is used in the manufacture of chemicals, as a neutralizer in petroleum refining, in metal etching, in electroplating, and as a food additive (Photo 10.7).

Phosphoric acid, H₃PO₄, is a colorless, odorless, sparkling liquid or crystalline solid, depending on concentration and temperature. Phosphoric acid has a boiling point of 410°F (210°C); at 68°F (20°C), the 50% and 75% concentrations are mobile liquids. The 85% concentration has a syrupy consistency, and the 100% acid is in the form of crystals. Phosphoric acid is water soluble and absorbs oxygen readily, and the specific gravity is 1.89, which is heavier than water. It is toxic by ingestion and inhalation and an irritant to the skin and eyes, with a TLV of 1 mg/m³ of air. The four-digit UN identification number is 1805. The NFPA 704 designation is health 3, flammability 0, and reactivity 0. The primary use of phosphoric acid is in chemical analysis and as a reducing agent.

Photo 10.7 Railcar of sodium hydroxide solution, a corrosive material that is a base.

Sodium carbonate, Na₂CO₃, also known as soda ash and sodium bicarbonate, is an oxy-salt and is a base with a pH of 11.6. It is not particularly hazardous and is used to neutralize acid spills.

Nitric acid, HNO₃, an inorganic acid, is a colorless, transparent, or yellowish, fuming, suf-focating, corrosive liquid. Nitric acid will attack almost all metals. The yellow color results from the exposure of the nitric acid to light. Nitric acid is a strong oxidizer, is miscible with water, and has a specific gravity of 1.5, which is heavier than water. It may be found in solutions of 36, 38, 40, and 42 degrees B'e (specific gravity) and concentrations of 58%, 63%, and 95%. Nitric acid is a dangerous fire risk when in contact with organic materials. It is toxic by inhalation and is corrosive to tissue and mucous membranes. The TLV is 2 ppm in air. Nitric acid is incompatible with acetic acid, hydrogen sulfide, flammable liquids and gases, chromic acid, and aniline. The four-digit UN identification number for nitric acid at <40% concentration is 1760. The NFPA 704 designation for nitric acid at <40% concentration is health 3, flammability 0, and reactivity 0. There is not any information in the white area of the diamond for <40% concentrations. Below 40% concentration, nitric acid is not consid-ered an oxidizer. The four-digit UN identification number for nitric acid at >40% is 2031. It is placarded as a Class 8 Corrosive; however, individual containers are labeled corrosive, oxidizer, and poison. The NFPA 704 designation for nitric acid at >40% concentration is health 4, flammability 0, and reactivity 0. The prefix "oxy" appears in the white section of the diamond. Nitric acid >40% concentration is an oxidizer.

Nitric acid is used in the manufacture of ammonium nitrate fertilizer and explosives, in steel etching, and in reprocessing spent nuclear fuel. There are two types of fuming nitric acid. *White fuming nitric acid* is concentrated with 97.5% nitric acid and less than 2% water. It is a colorless to pale-yellow liquid that fumes strongly. It is decomposed by heat and exposure to light and becomes red in color from nitrogen dioxide. *Red fuming nitric acid* contains more than 85% nitric acid, 6%–15% nitrogen dioxide, and 5% water. The four-digit UN identification number for red fuming nitric acid is 2032. The NFPA 704 designa-tion is health 4, flammability 0, and reactivity 1. The prefix "oxy" appears in the white section of the diamond. Red fuming nitric acid is considered an oxidizer. Both white and red fuming acids are toxic by inhalation, strong corrosives, and dangerous fire risks that

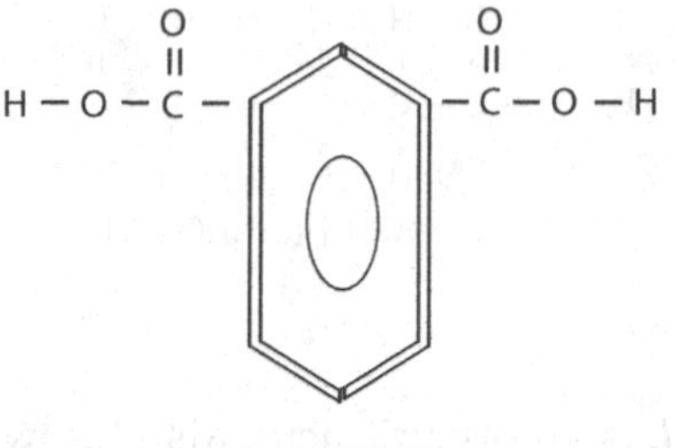

$$H-O-\overset{\overset{\displaystyle O}{\|}}{C}-\!\!\!\!\bigcirc\!\!\!\!-\overset{\overset{\displaystyle O}{\|}}{C}-O-H$$

C$_6$H$_4$(COOH)$_2$

Figure 10.5 Terephthalic acid.

may explode upon contact with reducing agents. They are used in the production of nitro compounds, rocket fuels, and as laboratory reagents.

Terephthalic acid (TPA), $C_6H_4(COOH)_2$, is an organic acid. It is a white crystalline or powdered material that is insoluble in water. It undergoes sublimation above 572°F (300°C). In addition to being corrosive, it is also combustible. The primary uses are in the production of polyester resins, fibers, and films; it is also an additive to poultry feeds. The structure and molecular formula for TPA are shown in Figure 10.5. The naming of TPA does not follow any of the rules of naming organic acids under the trivial naming system. However, the formula and the structure indicate an organic compound and the name indicates acid. The hazards of the acids, except for flammability, are similar. The fact that the name indicates acid should lead you to assume flammability and toxicity, in addition to corrosiveness, until other information is known.

Hydrochloric acid, HCl, an inorganic acid, is a colorless or slightly yellow, fuming, pungent liquid produced by dissolving hydrogen chloride gas in water. Hydrochloric acid in solution has a pH of 1.1. The specific gravity is 1.19, which is heavier than water. It is water soluble, a strong corrosive, and toxic by ingestion and inhalation. It is an irritant to the skin and eyes. The four-digit UN identification number is 1050 for anhydrous and 1789 for solution. Hydrochloric acid is used in food processing, pickling, and metal cleaning; as an alcohol denaturant; and as a laboratory reagent (Photo 10.8).

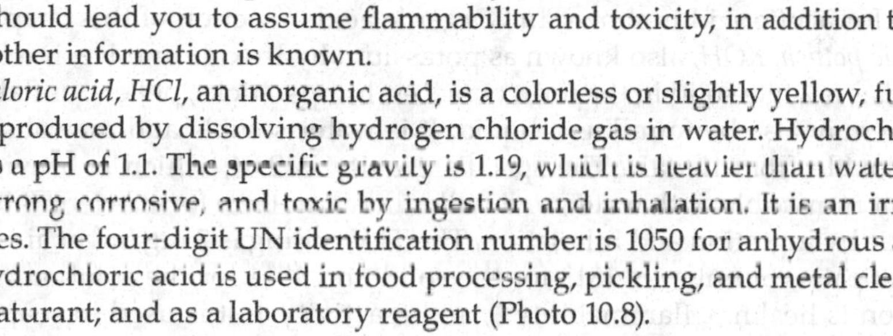

Photo 10.8 Muriatic acid is a weaker form of hydrochloric acid.

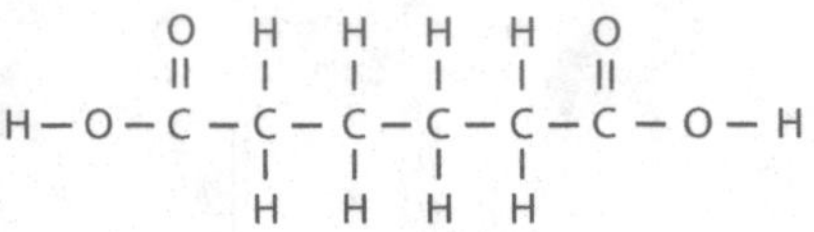

$$O \quad H \quad H \quad H \quad H \quad O$$
$$\parallel \quad | \quad | \quad | \quad | \quad \parallel$$
$$H-O-C-C-C-C-C-C-O-H$$
$$| \quad | \quad | \quad |$$
$$H \quad H \quad H \quad H$$

COOH(CH₂)₄COOH

Figure 10.6 Adipic acid.

Acetic acid, CH₃COOH, is an organic acid, also known as ethanoic acid and vinegar acid. Acetic acid is a clear, colorless, corrosive liquid with a pungent odor. In solution, acetic acid has a pH of 2.9. The glacial form is the pure form without water; it is 99.8% pure. Glacial acetic acid is a solid at normal temperatures. It is flammable, with a flash point of 110°F (43°C) and a flammable range of 4%–19.9%. The ignition temperature is 800°F (426°C). Acetic acid is "super-duper" polar and water soluble. It will require polar-solvent-type foam to extinguish fires. The specific gravity is 1.05, which is slightly heavier than water, but being miscible, it will mix rather than form layers. It is toxic by inhalation and ingestion, with a TLV of 10 ppm in air. Acetic acid is a strong irritant to the skin and eyes. It is incompatible with nitric acid, peroxides, permanganates, ethylene glycol, hydroxyl compounds, perchloric acid, and chromic acid. The four-digit UN identification number is 2789. The NFPA 704 designation is health 3, flammability 2, and reactivity 0. It is a food additive at lower concentrations; it is used in the production of plastics, pharmaceuticals, dyes, insecticides, and photographic chemicals. The structure for acetic acid is shown in "Organic acids" section of this chapter.

Caustic potash, KOH, also known as potassium hydroxide and lye, is a hydroxide salt. It is found as a white solid in the form of pieces, lumps, sticks, pellets, or flakes. Potassium hydroxide may also be found as a liquid. It is water soluble and may absorb water and carbon dioxide from the air. The specific gravity is 2.04, which is heavier than water; however, it is miscible with water, so it will mix rather than form layers. It is a strong base and is toxic by ingestion and inhalation. The TLV ceiling is 2 mg/m³ of air. The four-digit UN identification number is 1813 for the solid and 1814 for the solution. The NFPA 704 designation is health 3, flammability 0, and reactivity 1. It is used in soap manufacture, bleaching, as an electrolyte in alkaline storage batteries and some fuel cells, as an absorbent for carbon dioxide and hydrogen sulfide, and in fertilizers and herbicides.

Adipic acid, also known as hexanedioic acid, is an organic acid. It is a white, crystalline solid that is slightly soluble in water. In addition to being a corrosive, it is also flammable; however, it is a relatively stable compound. Adipic acid is used in the manufacture of nylon and polyurethane foams. It is also a food additive and adhesive. The structure and molecular formula are shown in Figure 10.6; notice that the structure has two organic acid functional groups attached. The naming of adipic acid does not follow any of the rules of naming organic acids under the trivial naming system. However, the formula and the structure indicate an organic acid, and the name indicates acid. The hazards of the acids, except for flammability, are similar. The fact that the name indicates acid should lead you to assume flammability and toxicity, in addition to corrosiveness, until other information is known.

Dilution vs. neutralization

Dilution and neutralization are often tactics considered when dealing with spills of corrosive materials. *Dilution* involves placing water into the acid to reduce the pH level. The addition of water to a corrosive can create a dangerous chemical reaction. Acids are highly water reactive, creating vapors, heat, and splattering. With dilution, you must consider the exponential values of the numbers on the pH scale. Just moving the pH from 1 to 2 on the scale will take an enormous amount of water. Dilution may not be a practical approach for large spills. For example, if a 3000 gal spill of concentrated hydrochloric acid occurs, applying enough water to dilute the material to a pH of 6 would require the following efforts: one 1000 gpm pumper, pumping 24 h a day, 7 days a week, 365 days a year, for 64 years. This would produce 1,440,000 gal of water per day! A large reservoir would be required to hold the water. As the process proceeds, it would become necessary to stir the mixture of water and acid to ensure uniformity in the dilution process. Dilution may work on small spills, but it will not work well on large spills.

Neutralization involves a chemical reaction that works well under laboratory conditions, using small amounts of acids and bases. However, in the field, facing a large spill of a corrosive material, neutralization may not be feasible. Neutralization requires a large amount of neutralizing agent. For the same spill of 3000 gal of concentrated hydrochloric acid mentioned in the previous example, it would require 8.7 tons of sodium bicarbonate, 5.5 tons of sodium carbonate, or 4.15 tons of sodium hydroxide to neutralize the spill. The latter would not be recommended, because sodium hydroxide is a strong base and would be dangerous to work with by itself without trying to add it to a concentrated acid. There would also be a need for a method to apply the neutralizing agent. The reaction that occurs will be a violent one, producing heat, vapor, and splattering of product. Neutralization may not work well for emergency responders at the scene of an incident with a large spill. The method of choice may turn out to be one of cleaning up the product by a hazardous waste contractor. They may use vacuum trucks; absorbent, gelling materials; or neutralization to accomplish the task.

The main danger of corrosive materials to responders is the contact of these materials with the body. Corrosive materials destroy living tissue. Destruction begins immediately upon contact. Many strong acids and bases will cause severe damage upon contact with the skin. Weaker corrosives may not cause noticeable damage for several hours after exposure. A chemical burn is nine times more damaging than a thermal burn. There are four basic methods of reducing the chemical action of corrosives on the skin: physical removal, neutralization, dilution, and flushing. Flushing is the method of choice. Removal of a corrosive material is difficult to accomplish and may leave a residue behind. Neutralization is a chemical reaction that may be violent and produce heat. This type of reaction on body tissues may cause more damage than it prevents. Neutralization should not be attempted on personnel wearing chemical suits, for the same reason as mentioned earlier. The layer of chemical protection is thin, and the heat from the neutralization may melt the suit and cause burns to the skin below the suit.

Dilution takes a large amount of water to lower the pH to a neutral position. While dilution may be similar to flushing, the intended outcome is different. With dilution, the goal is to reduce the pH number to as near neutral as possible. With flushing, the goal is to remove as much of the material as possible with a large volume of water. Flushing should be started as soon as possible to reduce the amount of chemical damage and should continue for a minimum of 15 min. This also applies to the eyes. Most corrosives are highly water soluble. Contact lenses should not be worn by personnel at HazMat incident scenes. Contact with acids can "weld" the contact to the eye, which almost always produces blindness. The person being treated may be in a great deal of pain and may have to be restrained during the flushing operation. Treatment after flushing involves standard first aid for burns.

Corrosives are transported in MC/DOT 312/412 tanker trucks (Photo 10.9). These trucks have a small-diameter tank with heavy reinforcing rings around the circumference of the

Photo 10.9 MC/DOT 312/412 tanker for heavy corrosive materials.

tank. The tank diameter is small because most corrosives are heavy. No other type of hazardous material is carried in this type of tanker. The 312/412 is a corrosive tanker regardless of how it is placarded. The placard may indicate a poison, an oxidizer, or a flammable; but do not forget the "hidden hazard": the tank identifies corrosives. Lighter corrosives may also be found in MC/DOT 307/407 tankers and may be placarded corrosive, flammable, poison, and oxidizer. Corrosives may also be found in tank cars, intermodal containers, and varying sizes of portable containers. Portable containers may range from pint and gallon glass bottles to stainless steel carboys and 55 gal drums. Some are also shipped in plastic containers (Photo 10.10).

Photo 10.10 Hydrochloric acid in 55 gal drums.

Incidents

Emergency responders should have a thorough knowledge of corrosive materials. After flammable liquids and gases, corrosives are the most frequently encountered hazardous material. Sulfuric acid has been the number one produced industrial chemical for many years. Responders should have proper chemical protective equipment and SCBA to deal safely with corrosive materials. Firefighter turnouts will not provide protection from corrosives. Most common exposures are contact with the hands and feet and inhalation of the vapors. Make sure that the chemical suits chosen for use are compatible with the corrosive material. No suit will protect you from chemicals indefinitely; they all have breakthrough times. Make sure personnel are rotated to avoid prolonged exposure and make sure they do not contact the material unless absolutely necessary. Safety should be your primary concern.

In California, an MC/DOT 312/412 tanker truck developed a leak along an interstate highway (Photo 10.11). On arrival, responders found a reddish-brown vapor cloud coming from the tank. The shipping papers indicated that the load was spent sulfuric acid; however, the color of the vapor coming from the trailer was in conflict with that information. As it turns out, the driver was hauling spent sulfuric acid but had room to pick up some nitric acid and put it in the same tank with the sulfuric acid. The nitric acid was not compatible with the tank and ate through it quickly. The entire load of acid was spilled onto the highway when the tank failed. Certain hazardous materials have specific colors, and responders should be aware of these colors.

A tank car placarded "empty," which contained an estimated 800 gal of anhydrous hydrogen fluoride, a corrosive liquid, was found leaking in a rail yard. "Empty" or "residue" placarded tank cars, as they are now called, can have as much as 3000 gal of product still in the tank if it has not been purged. Responders attempted to control the leak over a 4 h period. In the meantime, a vapor cloud formed and traveled approximately 2.5 miles downwind. This forced the evacuation of 1500 people from a 1.1 sq. mile radius around the leaking tank car for 9 h. Local hospitals treated approximately 75 people for minor skin and eye irritations.

Photo 10.11 On arrival, responders found a reddish-brown vapor cloud coming from the tank.

Review questions

10.1 List the two types of chemicals that make up the DOT corrosive class.

10.2 List the two types of acids.

10.3 The strength of an acid is a result of passing a gas through water. Which of the following terms reflects the name of the process?
  A. Radiation
  B. Sublimation
  C. Ionization
  D. Cationazation

10.4 Concentration of an acid is an expression of the relationship between acid and what other material?
  A. Alcohol
  B. Water
  C. Gas
  D. Dissolved solid

10.5 A corrosive hazardous material with a pH of 7.0 is considered to be
  A. Weak
  B. Acidic
  C. Basic
  D. Neutral

10.6 A corrosive hazardous material with a pH of 2.3 is considered to be
  A. Acidic
  B. Basic
  C. Neutral
  D. Weak

10.7 A corrosive hazardous material with a pH of 8.7 is considered to be
  A. Neutral
  B. Acidic
  C. Basic
  D. Weak

10.8 List two procedures that can be used to reduce the corrosive effects of an acid.

10.9 Provide the names and structures for the following organic acids.

$$CH_3COOH \quad C_2H_5COOH \quad C_3H_7COOH$$

10.10 List the formulas and names for the following organic acid structures.

chapter eleven

Miscellaneous hazardous materials

Miscellaneous hazardous materials in DOT/UN Class 9 are defined as "a material which presents a hazard during transportation, but which does not meet the definition of any other hazard class." Other hazards might include anesthetic, noxious (harmful to health), elevated temperature, hazardous substance, hazardous waste, or marine pollutant. They may be encountered as solids of varying configurations, gases, and liquids. Examples include asbestos, dry ice, molten sulfur, and lithium batteries. These materials would be labeled and placarded with the Class 9 Miscellaneous Hazardous Materials placard, which is white with seven vertical black stripes on the top half (Photo 11.1).

Also included in the Miscellaneous Hazardous Materials class is "Other Regulated Materials ORM-D, Consumer Commodities." They are "materials that present a limited hazard during transportation due to the form, quantity, and packaging." Some of these materials, if they were shipped in tank or box truck quantities, would fit into another hazard class. However, because the individual packaging quantities are so small, the DOT considers the hazard is limited and they are labeled ORM-D. Generally, these ORM-D materials are destined for use in the home, industry, and institutions. The materials are in small containers, including aerosol cans, with a quantity that is usually a gallon or less.

Caution should be observed if fire is involved in an incident, since many small containers can become projectiles as pressure builds up inside from the heat and the containers explode. Aerosol cans may be particularly dangerous because they are already pressurized, and exposure to heat can cause them to explode and rocket from the pressure. Those materials used in industry and institutions are usually service products used in cleaning and maintenance rather than in industrial chemical processes.

Examples of ORM-D materials include low-concentration acids, charcoal lighter, spray paint, disinfectants, and cartridges for small firearms. Even though the container sizes may be small, the products inside can still cause contamination of responders or death and injury if not handled properly.

There is no one specific hazard that can be attributed to Class 9 materials. The hazards will vary and may include all of the other eight hazard classes. The physical and chemical characteristics mentioned in the first nine chapters of this book may be encountered with Class 9 materials. The difference is that the quantities may be small, or the materials may be classified as hazardous wastes, which can be almost any of the other hazard classes (Photo 11.2). With miscellaneous hazardous materials, it is important to obtain more information about the shipment to determine the chemical names and the exact hazards of the materials involved.

Class 9 placards on transportation vehicles may include a four-digit UN identification number. The corresponding information in the *Emergency Response Guide (ERG)* may not give detailed names of the materials. There may be generalizations, such as "hazardous substance, n.o.s. (not otherwise specified)." When the material is not specifically identified in the *ERG*, the shipping papers or other sources will have to be consulted to determine the exact hazard of the shipment (Photo 11.3).

Photo 11.1 Intermodal container of miscellaneous hazardous materials.

Photo 11.2 Electrical transformers were once a primary source of PCBs, a miscellaneous hazardous material.

Elevated-temperature materials

In addition to the Class 9 placard, a second placard may appear next to it with the word HOT (Photo 11.4). The word may also appear outside of a placard by itself. It indicates that the material inside has an elevated temperature that may be a hazard to anyone who comes in contact with it. An elevated-temperature material is usually a solid that has been heated to the point that it melts and becomes a molten liquid. The change is in physical state only; the chemical characteristics of the material remain the same.

Photo 11.3 Box truck with overpack drum underneath, usually hauling hazardous waste.

Photo 11.4 MC/DOT 307/407 tanker of molten sulfur.

There may, however, be vapors produced from molten materials that are not present in the solid form. These vapors may be flammable or toxic. Water in contact with molten materials can cause a violent reaction and instantly turn to steam. If this happens inside a container, the pressure buildup from the steam can cause a boiler-type explosion that has nothing to do with the characteristics of the chemical inside. The steam, which is a gas, builds up pressure inside what is usually a non pressure container. When the container can no longer withstand the pressure, it fails. The molten material inside may be splattered around by the explosion.

There are two molten materials specifically listed in the Hazardous Materials Tables in 49 CFR: molten aluminum and molten sulfur (Photo 11.5). *Molten aluminum* has a four-digit UN identification number of 9260. The *ERG* refers to Guide 169 for hazards of the material.

Photo 11.5 Railcar with molten sulfur, a Class 9 Miscellaneous Hazardous Material. It is transported at 284°F (140°C).

Molten aluminum is the only material that refers to this guide. The guide indicates that the material is above 1300°F and will react violently with water, which may cause an explosion and release a flammable gas.

Molten materials in contact with combustible materials may cause ignition if the molten material is above the ignition temperature of the combustible. For example, gasoline has an average ignition temperature of around 800°F. Diesel fuel has an average ignition temperature around 400°F, depending on the blend and additives. In an accident, gasoline or diesel fuel could be spilled. The molten material could be an ignition source for the gasoline or diesel fuel with which it comes in contact. When contacting concrete on a roadway or at a fixed facility, molten materials could cause spalling and small pops. This could cause pieces of concrete to become projectiles. Contact with the skin would cause severe thermal burns. There is no personnel protective clothing that adequately protects responders from contact with molten materials.

Other molten materials are not as hot as molten aluminum. *Sulfur*, in the molten state, refers you to Guide 133 in the *ERG*. Molten sulfur, with the four-digit UN identification number 2448, may ignite combustible materials that it comes in contact with if it is above the ignition temperature of the material. Molten sulfur has a melting point of approximately 245°F. The molten sulfur in transportation would be above that temperature, but not as hot as molten aluminum. Contact would still cause severe thermal burns, and the vapor is toxic.

Hot asphalt in the liquid form can also cause combustion of combustible materials and severe thermal burns. Asphalt refers you to Guide 130 in the *ERG* for hazard information. Asphalt has a boiling point of >700°F and a flash point of >400°F. The ignition temperature is 905°F. Fires involving asphalt should be fought with care. Water may cause frothing, as it does with all combustible liquids with flash points above 212°F. This does not mean that water should not be used, but be aware that the frothing may be violent and the water contacting the molten material may also cause a reaction. Asphalt has a four-digit UN identification number of 1999 for all forms. The NFPA 704 designation for asphalt is health 0, flammability 1, and reactivity 0.

Other miscellaneous hazardous materials

White, gray, green, brown, and blue *asbestos* are impure magnesium silicate minerals that occur in fibrous form. Asbestos is noncombustible and was used extensively as a fire-retardant material until it was found to cause cancer. Asbestos is highly toxic by inhalation of dust particles. The four-digit UN identification number for white asbestos is 2590. The primary uses of asbestos are in fireproof fabrics, brake linings, gaskets, as a reinforcing agent in rubber and plastics, and as cement reinforcement. Many uses of asbestos are banned because of the cancer danger of the material.

Ammonium nitrate fertilizers that are not classified as oxidizers are classified as miscellaneous hazardous materials. This type of fertilizer has other materials in the mixture, and there are controlled amounts of combustible materials. Mixtures of ammonium nitrate, nitrogen, and potash that are not more than 70% ammonium nitrate and do not have more than 0.4% combustible material are included as a miscellaneous hazardous material. Additionally, ammonium nitrate mixtures with nitrogen and potash, with not more than 45% ammonium nitrate, may have combustible material that is unrestricted in quantity. The four-digit UN identification number for these mixtures of ammonium nitrate fertilizer is 2071.

Solid carbon dioxide, also known as dry ice, presents a danger in transport because carbon dioxide gas is produced as it warms. This warming is much like melting ice, although no liquid is formed in the case of dry ice. While carbon dioxide gas is nontoxic, it is an asphyxiant and can displace oxygen in the air or in a confined space. It is nonflammable; in fact, carbon dioxide gas is used as a fire-extinguishing agent. The four-digit UN identification number for solid carbon dioxide is 1845.

Solutions of formaldehyde, 30%–50%, such as those used in preservatives, are listed as miscellaneous hazardous materials. These solutions are nonflammable, and the toxicity is below the requirements for a poison liquid. However, the material may still be carcinogenic. Formaldehyde solutions usually contain up to 15% methanol to retard polymerization. The four-digit UN identification number for nonflammable solutions is 2209.

Polychlorinated biphenyls (PCBs) are composed of two benzene rings attached together with at least two chlorine atoms in the compound. PCBs were widely used in industry since 1930 because of their stability; however, it was this same stability that led to their downfall. They are highly toxic, colorless liquids with a specific gravity of 1.4–1.5, which is heavier than water. They are known carcinogens. In the human body, they tend to settle in the liver and fat cells, where they stay for a long period of time. They are not biodegraded and remain as an ecological hazard through water pollution. The only known way to remove PCBs from the environment is high-temperature incineration (at least 2200°F) for a proper length of time. The manufacture was discontinued in the United States in 1976. The material that remains is considered hazardous waste and is shipped as a miscellaneous hazardous material. The structure for PCB is shown in Figure 11.1.

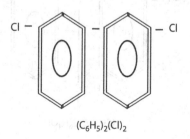

$(C_6H_5)_2(Cl)_2$

Figure 11.1 Polychlorinated biphenyl.

Batteries containing lithium are listed as miscellaneous hazardous materials. The storage batteries are composed of lithium, sulfur, selenium, tellurium, and chlorine. These batteries have four-digit UN identification numbers assigned depending on the use and composition of the battery. Lithium batteries contained in some kind of equipment have the four-digit number 3091. Batteries with liquid or solid cathodes, not in any kind of equipment, are given the number 3090.

Other miscellaneous hazardous materials listed in CFR 49 Hazardous Material Tables include solid materials and fish meal or fish scrap that has been stabilized. Fish meal is subject to spontaneous heating. These materials are given the four-digit UN identification number of 2216. Castor beans, meal, or flakes may also undergo spontaneous heating. The four-digit UN identification number is 2969. Additional materials include cotton (wet) 1365, polystyrene beads 2211, lifesaving appliances, self-inflating 2990, not self-inflating 3072, environmentally hazardous liquids or substances 3077, hazardous waste liquid 3082, and hazardous waste solid 3077. Additionally, self-propelled vehicles, including internal combustion engines or other apparatus containing internal combustion engines, or electric storage batteries are regulated. Self-propelled vehicles include electric wheelchairs with spillable or nonspillable batteries.

Titanium dioxide, TiO_2, is a white powder and has the greatest hiding power of all white pigments. It is noncombustible; however, it is a powder and, when suspended in air, may cause a dust explosion if an ignition source is present. It is not listed in the DOT Hazardous Materials Table, and the DOT does not consider it hazardous in transportation. The primary uses are as a white pigment in paints, paper, rubber, and plastics; in cosmetics; in welding rods; and in radioactive decontamination of the skin.

Sodium silicate, $2Na_2OSiO_2$, is the simplest form of glass. It is found as lumps of greenish glass soluble in steam under pressure, white powders of varying degrees of solubility, or liquids cloudy or clear. It is noncombustible; however, when the powdered form is suspended in air, it could cause a dust explosion if an ignition source is present. Breathing the dust may also cause health problems. The glass form could also create a hazard to responders in an accident. It is not listed as a hazardous material in the DOT Hazardous Materials Tables. The primary uses are as catalysts, soaps, adhesives, water treatment, bleaching, waterproofing, and flame retardant.

Bisphenol A, $(CH_3)_2C(C_6H_4OH)_2$, is made up of white flakes that have a mild phenolic odor. It is insoluble in water. Bisphenol A is combustible, with a flash point of 175°F. It is not listed in the DOT Hazardous Materials Tables. It is used in the manufacture of epoxy, polycarbonate, polysulfone, and polyester resins; as a flame retardant; and as a fungicide. The structure and molecular formula are shown in Figure 11.2.

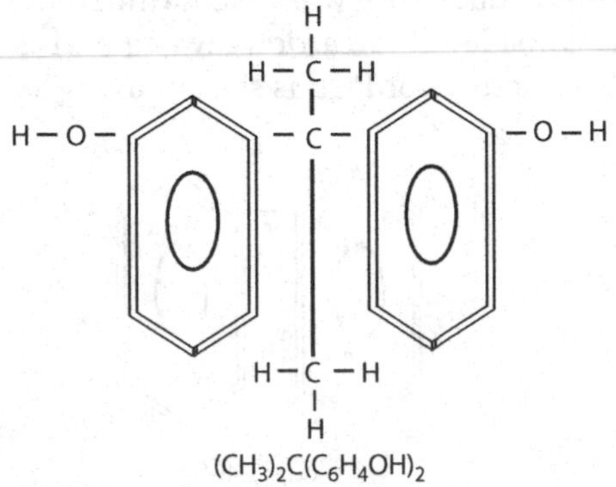

$$(CH_3)_2C(C_6H_4OH)_2$$

Figure 11.2 Bisphenol A.

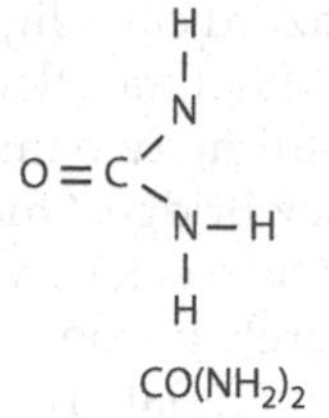

$$O=C \begin{array}{c} H \\ | \\ N \\ \diagup \\ \diagdown \\ N-H \\ | \\ H \end{array}$$

$$CO(NH_2)_2$$

Figure 11.3 Urea.

Urea *(carbamide)*, $CO(NH_2)_2$, is composed of white crystals or powder, almost odorless, with a saline taste. It is soluble in water and decomposes before reaching its boiling point. Urea is noncombustible. The primary uses of urea are in fertilizers, animal feed, plastics, cosmetics, flame-proofing agents, pharmaceuticals, and as a stabilizer in explosives. Urea appears to be both a ketone and an amine by structure and molecular formula; however, it is neither, nor does it have any of the characteristics of either family. The structure and molecular formula are shown in Figure 11.3.

Incidents

Hot materials, such as asphalt, can cause serious thermal burns if contacted with parts of the body (Photo 11.6). The television show *Rescue 911* highlighted a rescue operation that involved hot asphalt. A dump truck used to haul solid hot asphalt to patch holes in the road collided with a car. As a result of the collision, the load of hot asphalt was dumped into the car, covering the driver to the point that he could not escape. Before rescuers could remove the driver, he suffered severe second and third-degree thermal burns from the hot asphalt.

This same hazard exists with all elevated-temperature materials. Responders should work carefully around transportation vehicles that have the HOT placard or the word "HOT" on the container. Miscellaneous hazardous materials can expose emergency responders to a wide variety of hazards. The placard itself does not indicate what those

Photo 11.6 MC/DOT 307/407 highway tanker used to haul molten asphalt with "HOT" marking.

hazards may be. Do not treat this hazard class lightly, as there are some materials that can cause injury or death to responders if not handled properly.

In Benicia, California, a truck pulling two tank trailers of molten sulfur was involved in a collision on the Benicia–Martinez Bridge. One of the tanks ruptured, spilling the molten sulfur onto two other vehicles. The truck driver died, along with a passenger in one of the cars; another passenger was severely burned. At the time of the accident, molten sulfur was not regulated as a hazardous material. The molten sulfur produced sulfur dioxide vapors, which hampered visibility, along with fog.

Hazardous materials, regardless of class, almost always have multiple hazards. It is important for emergency responders to recognize that the hazard classes indicate only the most severe hazard of the materials as determined by the DOT. Research has to be conducted with all hazardous materials even if the hazard class is known. The correct chemical name must be identified and all of the associated hazards evaluated before tactics are determined. Responders should have a thorough understanding of the physical and chemical characteristics of hazardous materials, including parameters of combustion, water and air reactivity, incompatibilities with other materials, and the effects of temperature and pressure on hazardous materials. Emergency responders should have the same level of understanding of hazardous materials as they do for firefighting, EMS protocols, and law enforcement procedures.

All emergency response incidents have the potential to involve hazardous materials; your knowledge of the physical and chemical characteristics of hazardous materials will help ensure a safe outcome to incidents.

Review questions

11.1 What is the word that elevated-temperature materials may have on a placard on a vehicle?
   A. Dangerous when wet
   B. Infectious substance
   C. HOT
   D. Thermal hazard
11.2 ORM-D materials are also known as which of the following?
   A. Consumer commodities
   B. Household substances
   C. Other regulated materials
   D. Not otherwise specified
11.3 A Class 9 hazardous material is best defined as which of the following?
   A. Smaller amounts of other classes
   B. Unregulated materials
   C. Shipped by highway only
   D. Do not fit in other categories

chapter twelve

Incompatible and unstable chemicals

Chemical processing and mixing under controlled circumstances safely create many of the useful products we have come to depend on in our daily lives. However, when certain chemicals come in contact with each other during uncontrolled situations, such as accidents and spills, dangerous reactions can occur. This chapter will focus on incompatible chemicals, safe storage practices, water- and air-reactive materials, and some of the hazards to responders that may occur as a result of an uncontrolled chemical reaction.

Responders often ask what will happen when certain chemicals mix together. That question may not always have an answer. Many chemicals that are stored or shipped together may not normally be mixed together; therefore, there may be little knowledge of the outcome of mixing during an emergency. Other chemicals or families of chemicals are commonly used or stored together, and it is fairly easy to predict the consequences of mixing them.

Mixing of chemicals may cause a number of reactions, from mild to violent. Reactions can produce heat or cold; splattering of the material; spontaneous combustion; production of flammable, oxidizing, toxic, or corrosive vapors; and explosions. Some reactions occur when chemicals contact each other or when a chemical contacts air, oxidizers, or water. Reactions can also occur within a chemical without contact with other chemicals. Compounds that have double and triple bonds are highly reactive. When these bonds break, heat is produced and may result in spontaneous combustion, which may be slow or quite rapid. Resulting reactions, if they occur inside a container, may result in the container coming apart explosively. Table 12.1 provides a brief listing of chemicals and some of the things they are incompatible with. Chemical families may also be incompatible with each other (Table 12.2).

Acids and bases

One of the most common groups of chemicals that are violently incompatible is acids and bases. While grouped together in the same DOT hazard class of corrosives, these materials are generally not stored together because of the potential dangers.

Under controlled circumstances, acids and bases can be used as neutralizing agents for each other. Under controlled circumstances, weak acids are used to neutralize strong bases, and weak bases are used to neutralize strong acids. During an accident or spill, strong acids and strong bases could come in contact with each other, resulting in violent reactions and placing emergency responders in danger. Acids include sulfuric, hydrochloric, nitric, acetic, propionic, butyric, and others. Bases include sodium and potassium hydroxide, ammonium hydroxide, and others (Table 12.3).

Oxidizers and organic materials

As previously discussed in Chapter 5, in order for combustion to occur, there must be oxygen, fuel, heat, and a chemical chain reaction. When oxidizers (oxygen) encounter organic materials (fuel), the result of the reaction may produce enough heat for

Table 12.1 Common Chemical Incompatibilities

Chemical	Incompatible chemical(s)
Acetylene	Bromine, chlorine, fluorine, copper, silver, mercury
Ammonia	Mercury, hydrogen fluoride, calcium hypochlorite, chlorine
Flammable liquids	Ammonium nitrate, chromic acid, hydrogen peroxide, sodium peroxide, nitric acid, and halogens
Oxygen	Oils, grease, hydrogen, and flammable liquids, solids, gases
Nitric acid	Acetic acid, hydrogen sulfide, flammable liquids and gases
Phosphorus	Air, oxygen
Sulfuric acid	Potassium chlorate, perchlorate, permanganate
Alkali metals	Carbon tetrachloride, carbon dioxide, water, and halogens
Alkaline earth metals	Carbon tetrachloride, halogens, hydrocarbons

Table 12.2 Compatibility of Chemical Families

Chemical family	1	2	3	4	5	6	7	8	9	10	11	12	13	14	15	16
1. Inorganic acids																
2. Organic acids	×															
3. Caustics	×	×														
4. Amines and alkoholamines	×	×														
5. Halogenated compounds	×		×	×												
6. Alcohols, glycols	×															
7. Aldehydes	×	×	×	×		×										
8. Ketones	×		×	×			×									
9. Saturated hydrocarbons																
10. Aromatic hydrocarbons	×															
11. Olefins	×				×											
12. Esters	×		×	×												
13. Halogens			×			×	×	×	×	×		×				
14. Ethers	×															
15. Acid anhydrides	×	×			×	×										
16. Oxidizers	×	×	×	×	×	×	×	×	×	×	×	×	×	×	×	

Table 12.3 Chemical Compatibility Chart

Group	Name	Incompatible groups
Group 1	Inorganic acids	2, 3, 4, 5, 6, 7, 8, 10, 13, 14, 16, 17, 18, 19, 21, 22, 23
Group 2	Organic acids	1, 2, 4, 7, 14, 16, 17, 18, 19, 22
Group 3	Caustics	1, 2, 6, 7, 8, 13, 14, 15, 16, 17, 18, 20, 23
Group 4	Amines	1, 2, 5, 7, 8, 13, 14, 15, 16, 17, 18, 23
Group 5	Halogenated compounds	1, 3, 4, 11, 14, 17
Group 6	Alcohols	1, 7, 14, 16, 20, 23
Group 7	Aldehydes	1, 2, 3, 4, 6, 8, 15, 16, 17, 19, 20, 23
Group 8	Ketones	1, 3, 4, 7, 19, 20
Group 9	Saturated hydrocarbons	20
Group 10	Aromatic hydrocarbons	1, 20
Group 11	Olefins	1, 5, 20
Group 12	Petroleum oils	20
Group 13	Esters	1, 2, 4, 19, 20
Group 14	Monomers/esters	1, 2, 3, 4, 5, 6, 15, 16, 19, 20, 21, 23
Group 15	Phenols	3, 4, 7, 14, 16, 19, 20
Group 16	Alkylene oxides	1, 2, 3, 4, 6, 7, 14, 15, 17, 18, 19, 23
Group 17	Cyanohydrins	1, 2, 3, 4, 5, 7, 16, 19, 23
Group 18	Nitriles	1, 2, 3, 4, 16, 23
Group 19	Ammonia	1, 2, 7, 8, 13, 14, 15, 16, 17, 20, 23
Group 20	Halogens	3, 6, 7, 8, 9, 10, 11, 12, 13, 14, 15, 19, 21, 22
Group 21	Ethers	1, 14, 20
Group 22	Phosphorus	1, 2, 3, 20
Group 23	Acid anhydrides	1, 3, 4, 6, 7, 14, 16, 17, 18, 19

combustion to occur. Oxidizer sources include elemental oxygen compressed, cryogenic liquid oxygen, the halogen family of the periodic table, peroxide salts, oxysalts, and organic peroxide hydrocarbon derivatives. Reactions may vary, from just dissolving as with the oxysalts to violent contact explosions from liquid oxygen mixing with asphalt paving.

Aging chemicals

Incompatibility can also result from chemicals that are stored too long and become unstable. Organic peroxides are formed in the containers when exposed to the oxygen in the air. This forms an organic peroxide compound, which is highly explosive that can react from exposure to heat, shock, and friction. Compounds susceptible to peroxide formation include ether, formaldehyde, and potassium metal. All compounds subject to peroxide formation should be stored away from heat and light. Sunlight is a particularly good promoter of peroxidation. Protection from physical damage and ignition sources during storage is also important. Make sure enclosures of the chemicals are tightly closed. Loose or leaky closures may permit or enhance evaporation of the stored material, leaving a hazardous concentration of peroxides in the containers.

Some peroxide-forming compounds can be inhibited during storage using hydroquinone, alkyl phenols, aromatic amines, or similar materials (Table 12.4). The selection of a proper inhibitor should be made to avoid possible conflicts with use or purity requirements of the compound. The more volatile the peroxidizable compound, the easier it is to concentrate the peroxides. Pure compounds are subject to peroxide accumulation because impurities may inhibit peroxide formation or catalyze their slow decomposition. Compounds that are suspected of having high peroxide levels because of visual observation of unusual viscosity or crystal formation, or because of age, should be considered extremely dangerous. If crystals form in a peroxidizable liquid, or discoloration occurs in a peroxidizable solid, peroxidation may have occurred, and the product should be considered extremely dangerous and should be destroyed without opening the container. Testing procedures are available to determine if peroxides have formed by sampling outsides of containers; however, they should only be conducted by persons with a chemical background and experience with the test procedures. The precautions taken for disposal of these materials should be the same as for any material that can be detonated by friction or shock. In general, the material should be carefully removed, using explosive handling procedures.

Ethyl ether (diethyl ether), and other ethers, are organic compounds that form explosive peroxides when in contact with air (Photo 12.1). They are found in college, high school, research, and industrial laboratories. When a container of ether is opened, oxygen from the air bonds with the single oxygen in each ether molecule and forms an organic peroxide. These peroxides are very unstable and become sensitive to shock, heat, and friction. Moving or shaking a container can cause an explosion. Ethers are also very flammable, with wide flammable ranges. Fire is likely to follow an explosion of an ether container. Ethers were once used extensively as anesthetics in hospitals and, while not highly toxic, could injure or impair emergency responders. Ethers in laboratories should be dated when opened and discarded after 6 months in storage; otherwise they run the risk of peroxide formation.

Potassium metal, a metallic element from family one on the periodic table of elements, is soft and silvery in color and frequently found in high school and college laboratories. In transport, it is found in metal containers stored under kerosene or naphtha to keep it from contact with air. While not air reactive, potassium and other metals of family one can react to moisture in the air. When encountered in labs, potassium and other family one metals are often stored in improper containers, such as mayonnaise or canning jars. This is dangerous—during an emergency, the glass containers can break and expose the metal to the air and spill the flammable liquid also in the container. Like other members of family one, potassium is a dangerous fire risk and reacts violently with water to liberate hydrogen gas (which is highly flammable). The heat from the reaction of the water and potassium can be enough to ignite the hydrogen. When exposed to moist air, it can also ignite spontaneously. Potassium metal is closely related to sodium and lithium metals, which are in the same family on the periodic table of elements. However, the similarity ends with their water- and air-reactive characteristics. Potassium metal can form peroxides and superoxides at room temperature and may explode violently when handled. Simply cutting a piece of potassium metal with a knife to conduct an experiment could cause an explosion. Potassium metal's dangers far outweigh its usefulness in lab experiments in schools and should be replaced with sodium or lithium metals, which also react with water, but do not form explosive peroxides.

Table 12.4 Peroxide-Forming Compounds

Severity of storage hazard	Storage time limits
Severe storage hazard	
Butadiene	3 months
Chloroprene	
Divinyl acetylene	
Isopropyl ether	
Potassium amide	
Potassium metal	
Sodium amide	
Tetrafluoroethylene	
Vinylidene chloride	
Moderate storage hazard	1 year
Acetal	
Acetaldehyde	
Benzyl alcohol	
2-Butanol dioxanes	
Chlorofluoroethylene	
Cyclohexene	
2-Cyclohexen-1-ol	
Cyclopentene	
Decahydronaphthalene (decalin)	
Diacetylene (butaldiyne)	
Dicyclopentadiene	
Diethylene glycol dimethyl ether (diglyme)	
Diethyl ether	
Ethylene glycol ether acetates (cellosolves)	
Furan	
4-Heptanol	
2-Hexanol	
Methyl acetylene	
3-Methyl-1-butanol	
Methyl-isobutyl ketone	
4-Methyl-2-pentanol	
2-Pentanol	
4-Penten-1-ol	
1-Phenylethanol	
2-Phenylethanol	
Tetrahydrofuran	
Tetrahydronaphthalene	
Vinyl ethers	
Other secondary alcohols	
Subject to autopolymerization	6 months
Butadiene	
Chlorobutadiene	

(continued)

Table 12.4 (continued) Peroxide-Forming Compounds

Severity of storage hazard	Storage time limits
Chloroprene	
Chlorotrifluoroethylene	
Stryene	
Tetrafluoroethylene	
Vinyl acetate	
Vinyl acetylene	
Vinyl chloride	
Vinyl pyridine	
Vinyldiene chloride	

Photo 12.1 Compounds that are suspected of having high peroxide levels because of visual observation of unusual viscosity or crystal formation, or because of age, should be considered extremely dangerous.

Picric acid is a type of chemical that is shipped and stored with a minimum of 10%–20% water in its container. While it is a high explosive when dry, it is classified as a 4.1 Flammable Solid, Wetted Explosive, because of the moisture content in the container. It cannot be shipped when dry. As long as the moisture remains in the container, the compound is stable. Picric acid is a yellow crystal that becomes highly explosive when it dries out and is shocked or heated. The structural and molecular formulas of picric (the common name for trinitrophenol) and trinitrotoluene (TNT) are very similar. When dry, picric acid also closely resembles the explosive power of TNT pound for pound. Picric acid is another chemical found in high school, college, and research laboratories. As it ages, the moisture that keeps it stable evaporates, and it becomes an unstable high explosive.

Benzoyl peroxide is a white, granular, and crystalline solid that is highly flammable, explosive, and toxic by inhalation. It also may explode spontaneously when it becomes dry.

Phosphorus is a waxlike crystal, transparent solid material. White or yellow (it is the same material but can be called by either color) phosphorus is the most common form, and it is reactive and dangerous. Red and black phosphorus can also be found in laboratories, but do not possess the same dangers as white or yellow. White phosphorus is an air-reactive material that must be stored under water or other liquid to keep it from igniting spontaneously. Like potassium and other metals, it is shipped in metal containers. However, many times it can be found in laboratories, especially high schools, in glass containers that can prove a significant hazard during an emergency. In addition to being air reactive, it is also quite toxic at 0.1 mg/m^3 in air and is commonly used as a rat poison. It doesn't appreciably deteriorate with age, is very dangerous if not properly stored and handled, and can cause serious fire and burns (Table 12.5).

Unstable functional groups

Several hydrocarbon derivative functional groups are subject to instability (see Table 12.7). These include the organic peroxides, nitro, azo, nitrate ester, imino, azide, nitroso, and nitramine. As previously discussed, functional groups are families of chemicals that have similar hazards. While there may be some in a group that do not fit the family hazard rule, use caution with all of the chemicals in a family until positive identification of individual chemical compounds can be determined. Most of the instability with these functional groups involves fire, explosion, or violent decomposition. They all should be treated with extreme caution and handled only by trained personnel.

Water- and air-reactive materials

Some hazardous materials may release toxic gases, oxygen, and flammable gas when contacting water. The *Emergency Response Guidebook* contains a section at the end of the green pages that contains a listing of water-reactive materials and the toxic gases released when they contact water. Some water-reactive materials release toxic inhalation hazard (TIH) gases when in contact with water. The gases released are not the same chemicals that reacted with water. Look up the gases once again in the appropriate section of the ERG and find an orange guide page for the gas. Also look in the green water-reactive section to determine what gas is produced.

Phosphorus is a material that spontaneously combusts when exposed to air. Thus, it is shipped under water to keep it from contacting air. Fires involving phosphorus should be fought using large quantities of water, as it will not react with water. Care should be taken to ensure that these types of materials are not stored in breakable containers. Alkali metals, such as lithium, sodium, and potassium, are considered water reactive, and explosive reactions can occur when these metals contact water releasing flammable hydrogen gas.

Calcium carbide is a water-reactive salt, which releases acetylene gas when in contact with water. When water-reactive materials contact flammable solids or oxidizers and fire occurs, water may be the only thing available in a large enough quantity to extinguish the fire. Magnesium is an element that is water reactive but only when it is on fire. Putting water on a magnesium fire can cause serious explosions, which may produce shock waves. Personnel should be at a safe distance using unmanned monitors and aerial devices to put water on a magnesium fire.

Table 12.5 Incompatible Chemical Listing

Chemical	Incompatible chemical(s)
Acetic acid	Aldehyde, bases, carbonates, hydroxides, metals, oxidizers, peroxides, phosphates, xylene
Acetic anhydride	Chromic acid, nitric acid, hydroxyl containing compounds, ethylene glycol, perchloric acid, peroxides and permanganates
Acetylene	Halogens (chlorine, fluorine, etc.), mercury, potassium, oxidizers, silver
Acetone	Acids, amines, oxidizers, plastics
Alkali and alkaline earth metals	Acids, chromium, ethylene, halogens, hydrogen, mercury, nitrogen, oxidizers, plastics, sodium chloride, sulfur
Aluminum alkyls	Halogenated hydrocarbons, water
Ammonia	Acids, aldehydes, amides, halogens, heavy metals, oxidizers, plastics, sulfur
Ammonium nitrate	Acids, alkalis, chloride salts, combustible materials, metals, organic materials, phosphorous, reducing agents, urea
Aniline	Acids, aluminum, dibenzoyl peroxide, oxidizers, plastics
Arsenical materials	Any reducing agent
Azides	Acids, heavy metals, oxidizers
Benzoyl peroxide	Chloroform, organic materials
Bromine	Acetaldehyde, alcohols, alkalis, amines, combustible materials, ethylene, fluorine, hydrogen, ketones (acetone, carbonyls, etc.), metals, sulfur
Calcium carbide	Water
Calcium hypochlorite	Methyl carbitol, phenol, glycerol, nitromethane, iron oxide, ammonia, activated carbon
Calcium oxide	Acids, ethanol, fluorine, organic materials
Carbon (activated)	Alkali metals, calcium hypochlorite, halogens, oxidizers
Carbon tetrachloride	Benzoyl peroxide, ethylene, fluorine, metals, oxygen, plastics, silanes
Chlorates	Powdered metals, sulfur, finely divided organic or combustible materials
Chromic acid	Acetone, alcohols, alkalis, ammonia, bases
Chromium trioxide	Benzene, combustible materials, hydrocarbons, metals, organic materials, phosphorous, plastics
Chlorine	Alcohols, ammonia, benzene, combustible materials, flammable compounds (hydrazine), hydrocarbons (acetylene, ethylene, etc.), hydrogen peroxide, iodine, metals, nitrogen, oxygen, sodium hydroxide
Chlorine dioxide	Hydrogen, mercury, organic materials, phosphorous, potassium hydroxide, sulfur
Chlorosulfonic acid	Organic materials, water, powdered metals
Copper	Calcium, hydrocarbons, oxidizers
Cumene hydroperoxide	Acids, organic or mineral
Cyanide	Acids
Hydroperoxide	Reducing agents
Cyanides	Acids, alkaloids, aluminum, iodine, oxidizers, strong bases

Table 12.5 (continued) Incompatible Chemical Listing

Chemical	Incompatible chemical(s)
Ethylene oxide	Acids, bases, copper, magnesium perchlorate
Flammable liquids	Ammonium nitrate, chromic acid, hydrogen peroxide, nitric acid, sodium peroxide, halogens
Fluorine	Alcohols, aldehydes, ammonia, combustible materials, halocarbons, halogens, hydrocarbons, ketones, metals, organic acids
Hydrocarbons (such as butane, propane, benzene, turpentine, etc.)	Acids, bases, oxidizers, plastics
Hydrocyanic acid	Nitric acid, alkalis
Hydrides	Water, air, carbon dioxide, chlorinated hydrocarbons
Hydrofluoric acid	Metals, organic materials, plastics, silica (glass), (anhydrous) sodium
Hydrogen peroxide	Acetylaldehyde, acetic acid, acetone, alcohols, carboxylic acid, combustible materials, metals, nitric acid, organic compounds, phosphorous, sulfuric acid, sodium, aniline
Hydrogen sulfide	Acetylaldehyde, metals, oxidizers, sodium
Hypochlorites	Acids, activated carbon
Iodine	Acetylaldehyde, acetylene, ammonia, metals, sodium
Maleic anhydride	Sodium hydroxide, pyridine, and other tertiary amines
Mercury	Acetylene, aluminum, amines, ammonia, calcium, fulminic acid, lithium, oxidizers, sodium
Nitrates	Acids, nitrites, metals, sulfur, sulfuric acid
Nitric acid	Acetic acid, acetonitrile, alcohols, amines, (concentrated) ammonia, aniline, bases benzene, cumene, formic acid, ketones, metals, organic materials, plastics, sodium, toluene
Nitroparaffins	Inorganic bases, amines
Oxalic acid	Oxidizers, silver, sodium chlorite
Oxygen	Acetaldehyde, secondary alcohols, alkalis, and alkalines, ammonia, carbon monoxide, combustible materials, ethers, flammable materials, hydrocarbons, metals, phosphorous, polymers
Oxalic acid	Silver, mercury, organic peroxides
Perchlorates	Acids
Perchloric acid	Acetic acid, alcohols, aniline, combustible materials, dehydrating agents, ethyl benzene, hydriotic acid, hydrochloric acid, iodides, ketones, organic material, oxidizers, pyridine
Peroxides, organic	Acids (organic or mineral)
Phosphorus (white)	Oxygen (pure and in air), alkalis
Phosphorus pentoxide	Propargyl alcohol
Potassium	Acetylene, acids, alcohols, halogens, hydrazine, mercury, oxidizers, selenium, sulfur
Potassium chlorate	Acids, ammonia, combustible materials, fluorine, hydrocarbons, metals, organic materials, sugars

(continued)

Table 12.5 (continued) Incompatible Chemical Listing

Chemical	Incompatible chemical(s)
Potassium perchlorate	Alcohols, combustible materials, fluorine, hydrazine, metals, organic matter, reducing agents, sulfuric acid
Potassium permanganate	Benzaldehyde, ethylene glycol, glycerol, sulfuric acid
Selenides	Reducing agents
Silver	Acetylene, ammonia, oxidizers, ozonides, peroxyformic acid
Sodium	Acids, hydrazine, metals, oxidizers, water
Sodium amide	Air, water
Sodium nitrate	Ammonium nitrate and other ammonium salts
Sodium oxide	Water, any free acid
Sodium peroxide	Acitic acid, benzene, hydrogen sulfide metals, oxidizers, peroxyformic acid, phosphorous, reducers, sugars, water
Sulfides	Acids
Sulfuric acid	Potassium chlorates, potassium perchlorate, potassium permanganate
Tellurides	Reducing agents
UDMH (1, 1-Dimethylhydrazine)	Oxidizing agents such as hydrogen peroxide and furning nitric acid
Zirconium	Prohibit water, carbon tetrachloride, foam, and any dry chemical on zirconium fires

Basic chemical storage segregation (Tables 12.6 and 12.7)

There are likely to be few places where firefighters and other emergency response personnel will encounter a wider variety of dangerous chemicals than in a chemical laboratory (Photo 12.2). Laboratories can be found in a variety of locations, including industrial, research facilities, high school, and college labs to name a few. High school and college laboratories are particularly dangerous because the variety of chemicals used and stored there may be greater than any other lab (Photo 12.3). While the quantities of individual chemicals are usually relatively small, collectively, and in some cases individually, the danger to response personnel can be significant. Chemicals can be toxic in very small amounts when absorbed through the skin, inhaled, or ingested. They can also cause eye damage, skin burns, illness, and cancer. Every category of Department of Transportation Hazard Class can be identified among the chemicals in a high school or college laboratory setting. Other labs are more specialized, and the range of chemicals is limited to research projects or analytical needs of the facility. When stored with caution, most chemicals do not pose an unreasonable threat under normal conditions. However, chemicals in laboratories are often stored in improper locations and alphabetical order, which may place totally incompatible chemicals on the same shelf with each other (Photo 12.4). Storing chemicals alphabetically can place dangerous materials such as nitric acid, which is a strong oxidizer, on the same shelf with flammable liquids. Mixing the two can result in an explosive compound. A more appropriate storage system might be one in which organic and inorganic chemical families are stored together. Many chemical supply companies include proper storage system information in their chemical catalogs. Fisher Scientific Company has a very good one in their catalog. Material safety data sheet (MSDS) also may contain information about chemical compatibility. MSDS should be maintained on file

Table 12.6 Basic Chemical Storage Segregation

Class of chemicals	Recommended storage method	Examples	Incompatibilities
Compressed gases flammable	Store in a cool, dry area, away from oxidizing gases. Securely strap or chain cylinders to prevent falling over	Methane, acetylene, propane, hydrogen	Oxidizing and toxic compressed gases, oxidizing solids
Compressed gases oxidizing	Store in a cool, dry area, away from flammable gases and liquids. Securely strap or chain cylinders to prevent falling over	Oxygen, fluorine, chlorine, bromine	Flammable gases
Compressed gases poisonous	Store in a cool, dry area, away from flammable gases and liquids. Securely strap or chain cylinders to prevent falling over	Carbon monoxide, hydrogen sulfide	Flammable and/or oxidizing gases
Corrosives-acids	Store in separate acid storage cabinet or area	Mineral acids— hydrochloric, sulfuric, nitric, perchloric, chromic, chromerge	Flammable liquids, flammable solids, bases, oxidizers
Corrosives-bases	Store in separate storage area	Ammonium hydroxide, sodium hydroxide	Flammable liquids, oxidizers, poisons, and acids
Explosives	Store in secure location away from all other chemicals	Ammonium nitrate, nitro urea, picric acid, trinitroaniline, trinitroanisole, trinitrobenzene, trinitrobenzenesulfonic acid	Flammable liquids, oxidizers, poisons, acids, and bases
Flammable liquids	In grounded flammable liquid storage cabinets or specially constructed rooms.	Acetone, benzene, diethyl ether, methanol, ethanol, toluene, glacial acetic acid	Acids, bases, oxidizers, and poisons
Flammable solids	Store in a cool, dry area away from oxidizers, corrosives, and flammable liquids	Phosphorus	Acids, bases, oxidizers, and poisons

(continued)

Table 12.6 (continued) Basic Chemical Storage Segregation

Class of chemicals	Recommended storage method	Examples	Incompatibilities
General chemicals nonreactive	Store on general laboratory benches or shelving, in chemical storage rooms or warehouse locations	Agar, sodium chloride, sodium bicarbonate, and most nonreactive salts	See MSDS
Oxidizers	Store in a spill tray inside a noncombustible cabinet, separate from flammable and combustible materials	Sodium hypochlorite, benzoyl peroxide, potassium permanganate, potassium chlorate, potassium dichromate. The following are generally considered oxidizing substances: peroxides, perchlorates, chlorates, nitrates, bromates, and superoxides	Separate from reducing agents, flammables, and combustibles
Poisons	Store separately in a vented, cool, dry area, in unbreakable chemically resistant secondary containers	Cyanides, heavy metals compounds, i.e., cadmium, mercury, osmium	Flammable liquids, acids, bases, and oxidizers
Water-reactive chemicals	Store in cool, dry area, and protect from water from fire sprinkler	Sodium metal, potassium metal, lithium metal, lithium aluminum hydride	Separate from all aqueous solutions, and oxidizers

Table 12.7 Unstable Functional Groups

$O-O$	(peroxide)	$-N$	(imino)
$-NO_2$	(Nitro)	$-N_3$	(azide)
$-N=N-$	(azo)	$-N=O$	(nitroso)
$-ONO$	(nitrate ester)	$-NHNO_2$	(nitramine)

Note: Generally, chemicals within these functional groups are prone to instability.

for each chemical in the laboratory. Manufacturers or chemical suppliers can provide the MSDS upon request. In many cases, they will be shipped automatically with the chemicals. Chemicals that require refrigeration and storage in flammable liquid or acid cabinets are often stored on open shelves in the lab or classroom. Acids and bases, while categorized by the DOT as corrosives, react violently with each other and should be stored in separate locations and cabinets. Other chemicals can be dangerous under normal conditions,

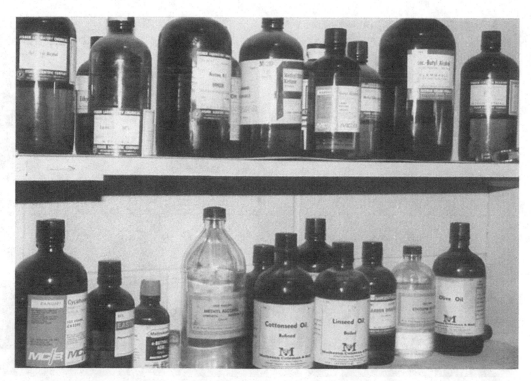

Photo 12.2 Chemicals not stored in proper hazard groups can be a hazard during a fire or accident.

Photo 12.3 Aging and no-longer-used chemicals removed from high schools and colleges during a cleanup program.

Photo 12.4 Corrosion of a flammable liquid container from an improperly stored acid.

when exposed to heat, shock, friction, water, or air. Care must be taken to store these chemicals in safe locations. To make ordinary storage conditions worse, chemicals can degrade, dehydrate, or form other dangerous compounds as they age. Many compounds that are normally safe can be turned into shock- or heat-sensitive explosive materials as they age. Many of these factors can create risk for emergency response personnel if they are unaware of the dangers. Oftentimes, laboratory personnel and teachers are unfamiliar with the hazards of aging chemicals. Fire inspectors need to be aware of these dangers so they can point them out during laboratory inspections (Photo 12.5).

In many industrial and educational occupancies, laboratory operations take up only a small portion of the space in the facility. Plating labs are used to maintain the quality of plating solutions within the plant. Chemicals found there are fairly limited and include cyanide, sodium hydroxide, and many different types of acids. Contact between acids and cyanide can produce deadly hydrogen cyanide gas. Other types of industrial labs that are used to maintain quality control of manufactured products and chemicals may vary from one facility to the next. Preplanning is an important step in determining the hazards of these facilities. National Fire Protection Association (NFPA) Standard 45, as well as NFPA 30, 49, 325, 491M, and 704 are good sources of information concerning fire code issues in laboratory occupancies. Research facilities located in industrial plants or universities also use limited but very specific chemicals based upon their research. Some of these facilities also have an additional biological hazard over and above the chemical hazards present. Once again, preplanning should ease the minds of response personnel when fires or other emergencies occur at these locations. Marking systems such as NFPA 704 or Hazard Communication should be used to identify areas

Photo 12.5 Unstable, flammable, and reactive chemicals removed from schools being disposed of by explosives personnel.

within buildings where dangerous chemical or biological materials may be used and stored along with emergency contact information for use by response personnel during an emergency.

Firefighting operations usually involve the application of water through a hose line. Small fires in laboratories can be extinguished safely by using portable dry chemical fire extinguishers. When metals such as potassium are involved, Class D dry powder fire extinguishers should be used. Any other type will not be effective. Inserting a hose line into a laboratory chemical storage area can cause glass containers to break and mix chemicals together. Even a chemist will not be able to foretell the potential outcome if chemicals that are not normally placed together are mixed. Firefighters and other rescue personnel may encounter highly toxic and carcinogenic chemicals and mixtures along with flammable, water and air reactive, and explosive materials. Great care should be taken when fires occur in laboratories. Runoff from firefighting can be very toxic and cause environmental damage, or at the very least danger to personnel and contamination of personal protective equipment. Some fires may be better left to the sprinkler system to extinguish or allowed to burn out. To ensure safety of personnel, some important steps should be taken in handling chemical fires and emergencies:

- Firefighters and other rescue personnel should never approach a chemical emergency scene or fire without SCBA and proper protective clothing. Care should be taken not to contact any chemicals or runoff with turnouts.
- During overhaul, extreme care should be taken to ensure firefighters are not exposed to chemicals. Also, runoff water, which may have become contaminated during firefighting operations, should be controlled.

Prevention is always crucial. Some effective preventive measures include the following:

- Preplan schools, colleges, universities, and other locations where laboratories are located. Note locations of hazardous chemicals and use the NFPA 704 or other types of marking systems to identify the locations of chemical use and storage.
- Conduct regular fire inspections of these facilities.
- Obtain inventories of chemicals stored in the laboratory areas and make them available to emergency response personnel during emergencies.
- Remove explosive, reactive, severely flammable, toxic, or no-longer-used chemicals from high school and college laboratories through the use of qualified chemical and explosives experts.
- Encourage teachers, researchers, and laboratory technicians to order small amounts of chemicals instead of large supplies that present a danger to response personnel or may become dangerous through long-term storage.
- Instruct laboratory workers and instructors to date substances when purchased and again when opened that may become dangerous with age. Unused portions should be properly disposed of within 6 months to 1 year.
- Encourage lab managers to set up proper storage systems so the chance of incompatible chemicals combining is reduced. Flammables should be stored in approved cabinets or 1 h fire rated storage rooms, with proper ventilation and sprinkler protection. Acids should be stored in approved acid cabinets away from flammable liquids and solids. Nitric acid should always be stored away from other acids. Bases should be stored separately from acids.
- Provide familiarization tours for fire and other emergency response companies so that personnel will have a better understanding of the hazards in laboratories.

In addition to removing dangerous chemicals from laboratory settings, it is important to educate teachers, laboratory workers, and research personnel on proper storage, use, and purchasing practices through training classes and literature. High school and college instructors should also be given assistance to develop experiments using safer chemicals that produce similar results but reduce the danger to students, teachers, and emergency responders. Through the cooperative efforts of school administrations, industry leaders, research organizations, and emergency response agencies, schools and laboratory settings can be made safer for occupants and response personnel. With proper training and preparation, emergencies involving hazardous chemicals in schools and laboratories can be safely and successfully handled.

Appendix

List of acronyms and recognized abbreviations

AAR/BOE	Association of American Railroads/Bureau of Explosives
ACC	American Chemical Council (Formerly CMA: Chemical Manufacturers Association, parent organization for CHEMTREC)
AIChE	American Institute of Chemical Engineers
ASCS	Agricultural Stabilization and Conservation Service
ASME	American Society of Mechanical Engineers
ASSE	American Society of Safety Engineers
ATSDR	Agency for Toxic Substances and Disease Registry (HHS)
CAER	Community Awareness and Emergency Response (ACC)
CDC	Centers for Disease Control (HHS)
CEPP	Chemical Emergency Preparedness Program (EPA)
CERCLA	Comprehensive Environmental Response, Compensation, and Liability Act of 1980 (PL 96–510)
CFR	Code of Federal Regulations
Chemnet	A mutual-aid network of chemical shippers and contractors
CHEMTREC	Chemical Transportation Emergency Center
CHLOREP	A mutual-aid group comprised of shippers and carriers of chlorine
CHRIS/HACS	Chemical Hazards Response Information System/Hazard Assessment Computer System
CSEPP	Chemical Stockpile Emergency Preparedness Program
CWA	Clean Water Act
DOC	U.S. Department of Commerce
DOD	U.S. Department of Defense
DOE	U.S. Department of Energy
DOI	U.S. Department of Interior
DOJ	U.S. Department of Justice
DOL	U.S. Department of Labor
DOS	U.S. Department of State
DOT	U.S. Department of Transportation
EENET	Emergency Education Network (FEMA)
EMA	Emergency Management Agency
EMI	Emergency Management Institute
EOC	Emergency Operating Center
EOP	Emergency Operations Plan
EPA	U.S. Environmental Protection Agency
ERD	Emergency Response Division (EPA)

477

FEMA   Federal Emergency Management Agency
FWPCA   Federal Water Pollution Control Act
HazMat   Hazardous material
HazOp   Hazard and Operability Study
HHS   U.S. Department of Health and Human Services
ICS   incident command system
IEMS   integrated emergency management system
LEPC   Local Emergency Planning Committee
MSDS   material safety data sheet
NACA   National Agricultural Chemical Association
NCP   national contingency plan
NCRIC   National Chemical Response and Information Center (CMA)
NETC   National Emergency Training Center
NFA   National Fire Academy
NFPA   National Fire Protection Association
NIOSH   National Institute of Occupational Safety and Health
NOAA   National Oceanic and Atmospheric Administration
NRC   National Response Center
NRT   National Response Team
OHMTADS   Oil and Hazardous Materials Technical Assistance Data System
OSC   on-scene coordinator
OSHA   Occupational Safety and Health Administration (DOL)
PSTM   Pesticide Safety Team Network
RCRA   Resource Conservation and Recovery Act
RQs   reportable quantities
RRT   Regional Response Team
RSPA   Research and Special Programs Administration (DOT)
SARA   Superfund Amendments and Reauthorization Act of 1986 (PL 99–499)
SCBA   self-contained breathing apparatus
SERC   State Emergency Response Commission
SPCC   spill prevention control and countermeasures
TSD   treatment, storage, and disposal facilities
USCG   U.S. Coast Guard (DOT)
USDA   U.S. Department of Agriculture
USGS   U.S. Geological Survey
USNRC   U.S. Nuclear Regulatory Commission

Resource guide

Chemical Hazard Information Response System (CHRIS)
Condensed Chemical Dictionary (CCD)
Cross-handling Guide for Potentially Hazardous Materials (CROSS)
Dangerous Properties of Hazardous Materials (SAX)
Emergency Response Guidebook (ERG)
Emergency Handling of Hazardous Materials in Surface Transportation (EHHM)
NFPA Fire Protection Guide on Hazardous Materials (FPG)
NIOSH/OSHA Pocket Guide to Chemical Hazards (NIOSH)
TLV Guide (American Conference of Governmental Industrial Hygienists) (TLV)

Sources of specific information

Chemical name to four-digit UN number ERG, EHHM
Four-digit UN number to chemical name ERG, EHHM
Chemical name to STCC EHHM
STCC to chemical name EHHM
STCC to four-digit UN number EHHM
Four-digit UN number to STCC EHHM
Chemical name to synonym CHRIS(MAN 2), NIOSH, CROSS, CCD (Limited), Sax (Limited)
Synonym to chemical name CHRIS(MAN 2), CCD, Sax, MNFC
(From trade name) FPG (Flash Point Index), CCD
(From chemical name) CHRIS(MAN 2)

Product uses

(From trade name) (Flash Point Index) FPG, CCD
(From chemical name) CCD, CROSS
Product trade name to chemical composition (contact manufacturer)
NFPA 704 designation FPG (Sec. 325M, Sec. 49)
Chemical formula NIOSH, CCDCHRIS(MAN 2), FPG (Sec. 325M, Sec. 49), Sax
Reactions FPG (Sec. 491M), NIOSH, CHRIS, CROSS, EHHM
IDLH CHRIS (MAN 2), NIOSH
TLV, TLV
PEL, NIOSH
LD50, CHRIS(MAN 2), Sax
LC50, CHRIS(MAN 2), Sax
Odor threshold, CHRIS(MAN 2)
Physical, chemical properties. FPG (Sec.49, Sec. 325), CHRIS(MAN 2), CROSS, EHHM, CCD, NIOSH, ERG
(BASIC) medical information FPG (Sec. 49), CHRIS(MAN 2), CROSS, EHHM, CCD, NIOSH, ERG
(BASIC) fire protection, personal protection FPG(Sec.49), CHRIS(MAN 2), CROSS, EHHM, CCD, NIOSH, ERG

Selected technical references

Computer-Aided Management of Emergency Operations (CAMEO), National Safety Council, Washington, DC.
Chemical Hazard Response Information System (CHRIS), Superintendent of Documents, U.S. Government Printing Office, Washington, DC.
Condensed Chemical Dictionary, Van Nostrand and Reinhold Co., New York.
Dangerous Properties of Industrial Chemicals (Sax), Van Nostrand and Reinhold Co., New York.
Dictionary of Chemical Names and Synonyms, CRC/Lewis Publishers, Boca Raton, FL.
Emergency Action Guides, Bureau of Explosives, Association of American Railroads, Washington, DC.
Emergency Handling of Hazardous Materials in Surface Transportation, Bureau of Explosives, Association of American Railroads, Washington, DC.
Emergency Response Guidebook (ERG), U.S. Department of Transportation, RSPA, Washington, DC.
Farm Chemical Handbook, Meister Publishing, Willoughby, OH.
Fire Protection Guide on Hazardous Materials, National Fire Protection Association (NFPA), Quincy, MA.
Handbook of Chemistry and Physics, CRC/Lewis Publishers, Boca Raton, FL.
Merck Index, Merck & Company, Rahway, NJ.
NIOSH Pocket Guide to Chemical Hazards, Superintendent of Documents, U.S. Government Printing Office, Washington, DC.

IUPAC rules of nomenclature

Hydrocarbons with more than 10 carbons

$C_{11}H_{24}$	Undecane
$C_{20}H_{42}$	Eicosane
$C_{12}H_{26}$	Dodecane
$C_{21}H_{42}$	Heneicosane
$C_{13}H_{28}$	Tridecane
$C_{22}H_{46}$	Docosane
$C_{14}H_{30}$	Tetradecane
$C_{23}H_{48}$	Tricosane
$C_{15}H_{32}$	Pentadecane
$C_{26}H_{54}$	Hexacosane
$C_{16}H_{34}$	Hexadecane
$C_{30}H_{62}$	Triacontane
$C_{17}H_{36}$	Heptadecane
$C_{31}H_{64}$	Hentriacontane
$C_{18}H_{38}$	Octadecane
$C_{32}H_{66}$	Dotriacontane
$C_{19}H_{40}$	Nonadecane
$C_{33}H_{68}$	Tritriacontane
$C_{40}H_{82}$	Tetracontane
$C_{49}H_{100}$	Nonatetracontane
$C_{50}H_{102}$	Pentacontane
$C_{60}H_{122}$	Hexacontane
$C_{70}H_{142}$	Heptacontane
$C_{80}H_{162}$	Octacontane
$C_{90}H_{182}$	Nonacontane
$C_{100}H_{202}$	Hectane
$C_{132}H_{266}$	Dotriacontahectane

There are four types of Formula:
Molecular:
C_2H_{10}
Structural:

```
        H   H   H   H   H
        |   |   |   |   |
   H  - C - C - C - C - C - H
        |   |   |   |   |
        H   H   H   H   H
```

Condensed structural:
$CH_3CH_2CH_2CH_3$

Skeleton:
C–C–C–C

The naming of all the alkanes is based upon the number of carbon atoms in the longest continuous chain of carbon atoms. If, for example, the longest chain contains four carbon atoms, the compound would be called butane. If it has five carbon atoms, it is pentane, and so on.

Names of branched-chain hydrocarbons and hydrocarbon derivatives using the IUPAC system are based on the name of the longest continuous carbon chain in the molecule, with a number indicating the location of a branch or substituent.

In order to locate the position of a branch or substituent, the carbon chain is numbered consecutively from one end to the other, starting at that end that gives the lowest numbers to the substituents. For example,

```
        1   2   3   4   5
        H   H   H   H   H
        |   |   |   |   |
   H — C — C — C — C — C — H   is 2-Methylpentane
        |   |   |   |   |
        H   |   H   H   H
        H — C — H
            |
            O
            |
            H
```

```
        1   2   3   4
        H   Cl  H   H
        |   |   |   |
   H — C — C — C — C — H   Is 2,2-Dichlorobutane
        |   |   |   |
        H   Cl  H   H
```

The prefixes "di," "tri," "tetra," "penta," "hexa," etc., indicate how many of each substituent is in the molecule. A cyclic (ring) hydrocarbon is designated by the prefix "cyclo."

Double bonds in hydrocarbons are indicated by changing the suffix "ane" to "ene," and triple bonds by changing to "yne." The position of the multiple bond within the structure is indicated by the number of the first or lowest-numbered carbon atom attached to the multiple bond. For example,

```
        1   2   3   4   5
        H   H   H   H   H
        |   |   |   |   |
   H — C — C — C = C — C — H      is 2-Pentene
        |   |           |
        H   H           H
```

```
        5   4   3   2   1
        H   H   H
        |   |   |
   H — C — C — C — C ≡ C — H   is 1-Pentyne
        |   |   |
        H   H   H
```

```
        4   3   2   1
        H   H   H   H
        |   |   |   |
   H — C = C — C = C — H   is 1,3,-Butadiene
```

Most of the hydrocarbon-derivative functional groups in organic compounds are designated by either a suffix or a prefix, as shown in Table B. Rules regarding whether a prefix or a suffix designation is used are as follows:

1. When one such group is present, the suffix will be used.
2. When more than one such group is present, only one will be designated by a suffix; the others are designated by prefixes.

3. The order of precedence for deciding which group takes the suffix designation is the same as the order in Table B.

Examples are as follows:

$$H-\overset{\displaystyle H}{\underset{\displaystyle H}{C}}-\overset{\displaystyle H}{\underset{\displaystyle H}{C}}-O-H \quad \text{is ethanol, not hydroxyethane}$$

$$H-O-\overset{3}{\underset{\displaystyle H}{\overset{\displaystyle H}{C}}}-\overset{2}{\underset{\displaystyle H}{\overset{\displaystyle H}{C}}}-\overset{1}{\overset{\displaystyle O}{C}}-O-H \quad \text{is 3-Hydroxypropanoic acid}$$

In numbering the carbon chain, the lowest numbers will be given preference to

1. Groups in Table B named by suffixes
2. Double bonds
3. Triple bonds
4. Groups named by prefixes (groups named by prefixes are listed in alphabetical order)

Nomenclature of Aromatic Compounds

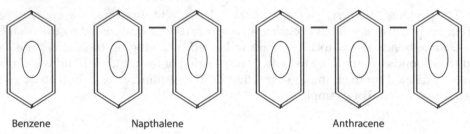

Benzene   Napthalene   Anthracene

Monosubstituted compounds

Names are derived using prefixes from Table B and Table C, followed by the name "benzene."

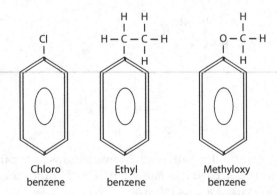

Chloro   Ethyl   Methyloxy
benzene   benzene   benzene

Names can be indicated by commonly accepted names.

Disubstituted compounds

Names are derived using prefixes (including commonly accepted names) and numbers or words.

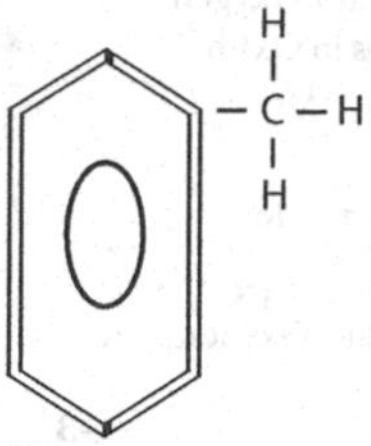

Toluene

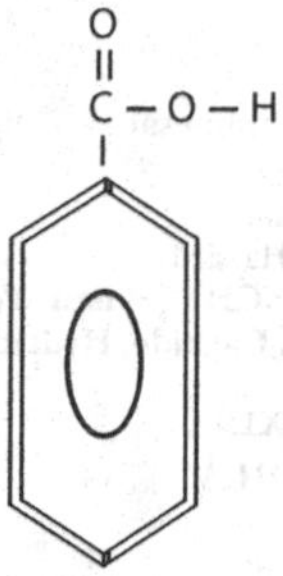

Benzoic acid

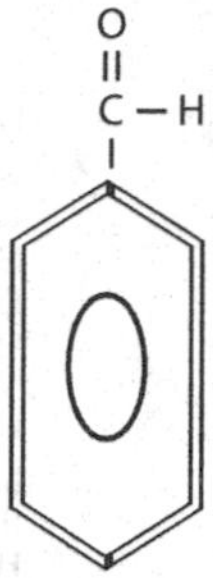

Benzaldehyde

Phenol

Analine

Benzene sulfonic acid

Ortho or 1, 2

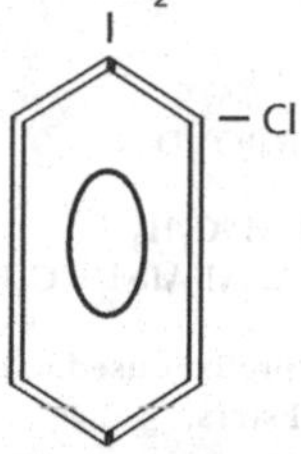

Meta or 1,3

Para or 1,4

2-Chloroaniline
orthochloroaniline

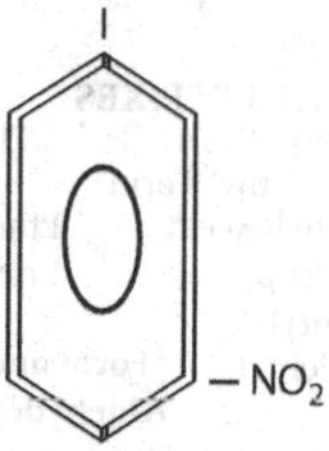

3-Nitrophenol
metanitrophenol

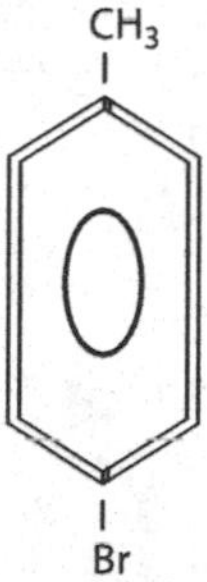

4-Bromotoluene
parabromotoluene

CHEMISTRY OF HAZMAT STUDY SHEET

PERIODIC TABLE
Family 1-Column 1-Alkali Metals
Family 2-Column-2-Alkaline Earth Metals
Family 3-Between Towers-Transitional Metals
Family 4-Column 7-Halogens
Family 5-Column 8-Inert/Noble Gases

ATOMIC STRUCTURE
Atomic Number = Number of Protons
Atomic weight = Protons + Neutrons
Number of Protons = #Electrons
Octet-8 electrons outer shell
Duet-2 electrons outer shell

HAZMAT ELEMENTS

Hydrogen	H
Lithium	Li
Sodium	Na
Potassium	K
Magnesium	Mg
Calcium	Ca
Chromium	Cr
Iron	Fe
Cobalt	Co
Copper	Cu
Mercury	Hg
Tin	Sn
Lead	Pb
Aluminum	Al
Carbon	C
Nitrogen	N
Phosphorus	P
Oxygen	O
Sulfur	S
Fluorine	F
Chlorine	Cl
Bromine	Br
Iodine	I
Boron	B
Krypton	Kr
Uranium	U
Xenon	Xe
Gold	Au
Arsenic	As
Plutonium	Pu
Zinc	Zn
Silicon	Si
Titanium	Ti
Radium	Ra
Silver	Ag
Manganese	Mn
Beryllium	Be
Helium	He
Argon	Ar

BINARY
Metal + **Nonmetal**
Not Oxygen
Ends in **ide**
General Hazard
Except *NCHP (When Wet)
*Nitride, Carbide, Hydride, Phosphide**

BINARY OXIDE
Metal + **Oxygen**
Ends in **Oxide**
CL, RH

PEROXIDE
Metal + O_2^{-2}
Ends in
Peroxide
CL, RH, RO

HYDROXIDE
Metal + OH^{-1}
Ends in
Hydroxide
CL, RH

AMMONIUM
$(NH_4)^{+1}$ + NM
ends in **NM**
Various

OXYSALTS
Metal + **Oxy Radical**

-1	-2	-3
FO_3	CO_3	PO_4
ClO_3	SO_4	BO_3(Borate)
BrO_3	CrO_4(Chromate)	AsO_4(Arsenate)
IO_3		
MnO_3		
NO_3		

+1 Oxygen Per___ate
Base State____ate
−1 Oxygen_____ite
−2 Oxygen Hypo__ite

CYANIDE
Metal + $(CN)^{-1}$
ends in **Cyanide**
(Highly Toxic)

(Cation = Positive Ion, Anion = Negative Ion)

H/C PREFIXES		**GREEK NUMBERS**		**ALKANES**	**ALKENES**
Meth	1	Mono	1	Single Bond	Double Bond
Eth	2	Di (bis)	2	Saturated	Unsaturated
Prop	3	Tri (tris)	3	Ends in **ANE**	Ends in **ENE**
But	4	Tetra	4	C_nH_{2n+2}	C_nH_{2n}
Pent	5	Penta	5		
Hex	6	Hexa	6	**ALKYNES**	**AROMATICS**
Hept	7	Hepta	7	Triple Bond	Resonant Bond
Oct	8	Octa	8	Unsaturated	Acts Saturated
Non	9	Nona	9	Ends in **YNE**	"BTX" BENZENE
Dec/Dek	10	Deca	10	C_nH_{2n-2}	TOLUENE
				Know (Acetylene)	XYLENE

(diene-2 double bonds, triene-3 double bonds)
Iso—Branched, **Neo**—Center Carbon surrounded by Carbons
Cyclo—All Carbons hooked together in a circle or end to end.

Transitional Metal Charges

Mercury (Hg)	I or II
Chromium (Cr)	II or III
Iron (Fe)	II or III
Copper (Cu)	I or II
Manganese (Mn)	II or III
Tin (Sn)	II or IV
Cobalt (Co)	II or III
Zinc (Zn)	II
Lead (Pb)	II or IV

"ous" ending for lower charge
"ic" ending for higher charge

RADICAL PREFIXES
SINGLE BOND
Meth = Methyl/Form*
Eth = Ethyl/Acet*
Prop = Propyl
But = Butyl
Pent = Pentyl

DOUBLE BOND
Ethene=**Vinyl**=C_2H_3
Propene = **Acryl(Allyl)** = C_3H_5

*Form and Acet prefixes used with Acids
Aldehydes, and Esters.

AROMATIC
Benzene=**Phenyl**=C_6H_5
Toluene = Benzyl = $C_6H_5CH_2$

Hydrocarbon derivatives

Func Grp	Gen form	Structure	Hazard
Alkyl Halide	R-X	Cl F Br I	Toxic/flam
Nitro	R-NO$_2$	$-N\begin{smallmatrix}O\\\\O\end{smallmatrix}$	Explosive
Amine	R-NH$_2$, R$_2$NH, R$_3$N	$-N-H-N-H-N-$ (each N with H above)	Toxic/flam
Ether	R-O-R	$-O-$	Anest/WFR[b]
Organic peroxide	R-O-O-R	$-O-O-$	Explosive/oxy
Alcohol	R-O-H	$-O-H$	Toxic/WFR
Ketone	R-C-O-R	$-\overset{\overset{O}{\|\|}}{C}-$	Narc/flammable
Ester[a]	R-C-O-O-R, R-C-O$_2$-R	$-\overset{\overset{O}{\|\|}}{C}-O-$	Polymerize/flam
Aldehyde[a]	R-C-H-O	$-\overset{\overset{O}{\|\|}}{C}-H$	Toxic/WFR
Organic acid[a]	R-C-O-O-H	$-\overset{\overset{O}{\|\|}}{C}-O-H$	Toxic/flam/cor
Nitrile[a]	R-CN	$-C\equiv N$	Flammable/toxic
Isocyanate	R-NCO	$-N=C=O$	Toxic/flammable
Sulfide	R-S-R	$-S-$	Toxic/flammable
Thiol	R-S-H	$-S-H$	Toxic/flammable

[a] When naming the structure count all carbons hooked together, and attached to the carbon in the functional group.
[b] WFR, Wide flammable range.

Naming rules for hydrocarbon derivatives

Alkyl halides: Name functional group first, such as Chlorine, which would become *Chloro*, Fluorine would become *Fluoro* and so on. Then name the hydrocarbon backbone, one carbon would be methane, two carbons ethane and so on, such as *Chloro Methane*. Or name the hydrocarbon backbone first, one carbon would become *Methyl*, two carbons *Ethyl*, and so on. Then name the functional group and add ide for an ending, such as chlorine which would become *chloride*, fluorine would become *fluoride* and so on, such as *Methyl Chloride*.

Nitros: The word Nitro comes first followed by the hydrocarbon radical(s), such as *Nitro Methane*.

Amines: The hydrocarbon radical(s) come first in the name followed by the word Amine. When there is more than one radical, start with the smallest and go to the largest and then end with the word *Amine*, such as *Vinyl Amine*, or *Methyl Vinyl Amine*, or *Methyl Ethyl Vinyl Amine*.

Ethers: Start with the smallest hydrocarbon radical and name it, then the other hydrocarbon radical and name it, then end with the word *Ether*, such as *Methyl Ethyl Ether*.

Organic peroxides: Start with the smallest hydrocarbon radical and name it, then find the other hydrocarbon radical and name it, then end with the word *Peroxide*, such as *Ethyl Propyl Peroxide*.

Alcohols: Name the hydrocarbon radical, then end with the word *Alcohol*, or name the hydrocarbon radical and end with *ol*, such as *Ethanol*.

Ketones: Start with the smallest hydrocarbon radical and name it, then find the other hydrocarbon radical and name it, then end with the word *Ketone*, such as *Methyl Ethyl Ketone*.

Esters: First of all, nothing is named *ESTER!* If there is a *Vinyl Radical to the left* of the ester functional group it is an *Acrylate*, if there is a *Methyl Radical to the left* of the ester functional group it is an *Acetate*. (You can also count all the carbons.) Then name the radical on the right and name the compound, such as *Ethyl Acetate*, or *Propyl Acrylate*.

Aldehydes: Name the hydrocarbon radical first, counting the carbon in the aldehyde functional group, and end with the word *Aldehyde*. When naming aldehydes, you use the prefix *Form for one carbon* and *Acet for two carbons*, such as *Formaldehyde* for a one carbon aldehyde.

Organic acids: Name the hydrocarbon radical first, adding *"ic"* to the radical, and end with the word *Acid*. As with the aldehydes, when naming the organic acids, count the carbon in the functional group and use *Form* and *Acet*, such as *Acetic Acid or Formic Acid*.

Nitriles: Count carbon in functional group plus "o nitrile" or hydrocarbon radical plus "cyanide."

Isocyanate: Hydrocarbon radical plus "isocyanage."

Sulfides: Hydrocarbon radical plus "sulfide" also referred to as "thioethers."

Thiols: Hydrocarbon radical plus "mercaptans" or hydrocarbon backbone plus "thiol."

Branching hydrocarbon derivatives

n-Normal-Straight Chained.

```
        H   H   H   H
        |   |   |   |
    H - C - C - C - C - O - H
        |   |   |   |
        H   H   H   H
```

Iso-Carbon attached to functional group is touching one other Carbon.*

```
        H
        |
    H - C
        | \   H   H
        H  \ |   |
        H   C - C - O - H
        | /     |
    H - C       H
        |
        H
```

Secondary-Carbon attached to functional group is touching two other Carbons.

```
                H
                |
        H   H   O   H
        |   |   |   |
    H — C — C — C — C — H
        |   |   |   |
        H   H   H   H
```

Tertiary-Carbon attached to functional group is touching three other Carbons.

```
                H
                |
        H       O   H
        |       |   |
    H — C   —   C — C — H
        |       |   |
        H   H — C — H   H
                |
                H
```

*Propane has only one possible branch which is Iso.

Polarity:

Organic acid	Super duper polar
Alcohol	Super polar
Ketone	Polar
Aldehyde	Polar
Ester	Polar
Amine	Slightly polar
Ether	Nonpolar
Nitro	Nonpolar
Alkyl halide	Nonpolar
Organic peroxide	Nonpolar
Hydrocarbons	Nonpolar

Things that affect boiling point are as follows:

1. Weight
2. Polarity
3. Branching

Boiling Point Relationships

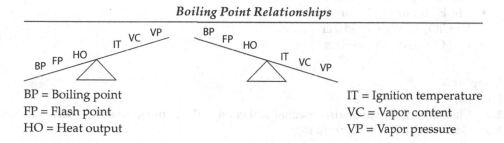

BP = Boiling point   IT = Ignition temperature
FP = Flash point   VC = Vapor content
HO = Heat output   VP = Vapor pressure

First determine boiling point and then compare boiling point to the other components.

Weak Bonds within a Structure

Carbon to carbon double bond	Alkenes	$C=C$
Carbon to carbon triple bond	Alkynes	$C\equiv C$
Oxygen to oxygen single bond	Peroxide/nitro	$-O-O-$
Carbonyl when on the last carbon	Aldehyde	$\overset{\displaystyle O}{\underset{\displaystyle -C-H}{\|}}$

Answers to review questions

Chapter 2

2.1 B

2.2 C

2.3 Noble Gases VIII, Bromine not a family, Alkaline Earth I, Alkali II, Halogens VII

2.4 B

2.5 A

2.6 C

2.7 B and C

2.8 B

2.9 NaCl, sodium chloride, binary salt, varying
- $Ca_3(PO_4)_2$, calcium phosphate, oxysalt oxidizer
- $Al_2(O_2)_3$, aluminum peroxide, peroxide RH, CL, RO
- $CuBr_2$, copper II, bromide, binary salt, varying
- KOH, potassium hydroxide, hydroxide RH, CL
- Li_2O, lithium oxide, metal oxide RH, CL
- $Mg(ClO)_2$, magnesium hypochlorite, oxysalt, oxidizer
- HgO_2, mercury II, peroxide, peroxide RH, CL, RO
- NaF, sodium fluoride, binary salt, varying
- $FeCO_3$, iron II, carbonate, oxysalt, oxidizer

2.10 $Ca(ClO)_2$, oxysalt, oxidizer
- $AlCl_3$, binary salt, varying
- LiOH, hydroxide RH, CL
- CuO_2, peroxide RH, CL, RO
- Na_2O, metal oxide RH, CL
- KI, binary salt, varying
- Mg_3P_2, binary salt, varying
- $HgClO_4$, oxysalt, oxidizer
- $Fe(FO_3)_3$, oxysalt oxidizer

Chapter 3

3.1 C

3.2 Mechanical overpressure, mechanical chemical, chemical reaction, dust, nuclear

3.3 Detonation and deflagration

3.4

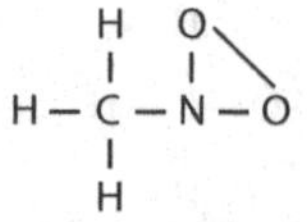

Nitro methane

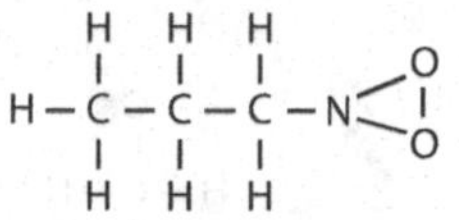

$C_3H_7NO_2$

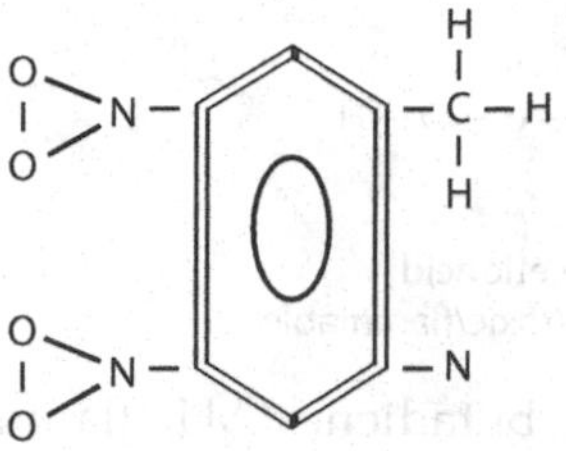

Tri-nitro toluene

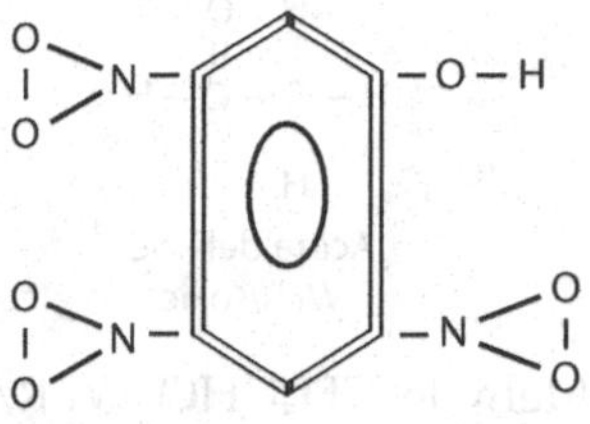

$C_6H_2(NO_2)_3OH$

3.5 Confinement, fuel, chemical oxidizer, heat
3.6 B
3.7 A
3.8 C
3.9 C
3.10 B
3.11 1.1, 1.2, 1.3, 1.4, 1.5, 1.6
3.12 B
3.13 C
3.14 C
3.15 A

Chapter 4
4.1 B
4.2 A
4.3 D
4.4 A
4.5 D
4.6 B
4.7 A
4.8 Isobutane C_4H_9, ethane C_2H_6, methane CH_4, ethene C_2H_4
4.9 Alkane, alkyne, alkene, alkane, alkene
4.10 Alkane saturated, alkene unsaturated, alkyne unsaturated

Chapter 5
5.1 B
5.2 B
5.3 A
5.4 C
5.5 Weight, polarity, branching
5.6 B, C, B, A, D, B
5.7 B
5.8 Alkane, aromatic, alkene, aromatic, alkene

5.9

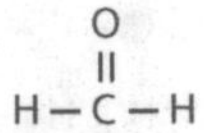

$$\begin{array}{c} O \\ \| \\ H-C-H \end{array}$$

Formaldehyde
WFR/toxic

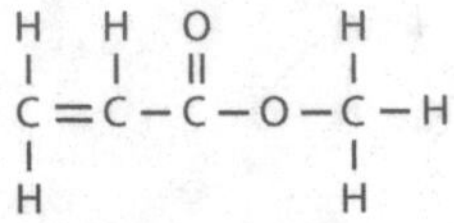

$$\begin{array}{c} H \quad H \quad O \qquad\quad H \\ | \quad | \quad \| \qquad\quad | \\ C=C-C-O-C-H \\ | \qquad\qquad\qquad | \\ H \qquad\qquad\qquad H \end{array}$$

Methyl acrylate
polymerize

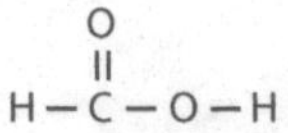

$$\begin{array}{c} O \\ \| \\ H-C-O-H \end{array}$$

Formic acid
corrosive/toxic/flammable

$$\begin{array}{c} H \quad O \\ | \quad \| \\ H-C-C-H \\ | \\ H \end{array}$$

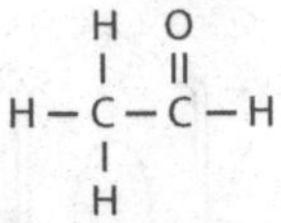

Acetaldehyde
WFR/toxic

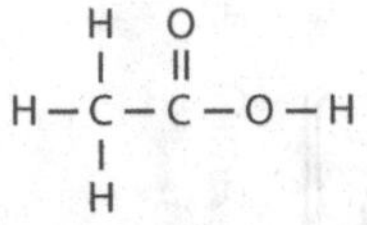

$$\begin{array}{c} H \quad O \\ | \quad \| \\ H-C-C-O-H \\ | \\ H \end{array}$$

Acetic acid
corrosive/toxic/flammable

5.10 Acetaldehyde CH_3CHO WFR/toxic, butadiene C_4H_6 flammable, isopropyl alcohol C_3H_7OH WFR/toxic, methyl amine CH_3NH_2 toxic/flammable

Chapter 6

6.1 Flammable solid, spontaneously combustible, dangerous when wet

6.2 D

6.3 B

6.4 D

6.5 A, C, D

6.6 D

6.7 B

6.8 D

6.9 D

6.10 D

Chapter 7

7.1 Fluorine, chlorine, bromine, oxygen

7.2 Halogens, peroxide salts, oxysalts, organic peroxides

7.3 D

7.4 Aluminum persulfate $Al_2(SO_5)_3$
Lithium chlorite, balanced
Sodium fluorate, balanced
Magnesium hypophosphite $Mg_3(PO_2)_2$
Copper I chlorate, balanced

7.5

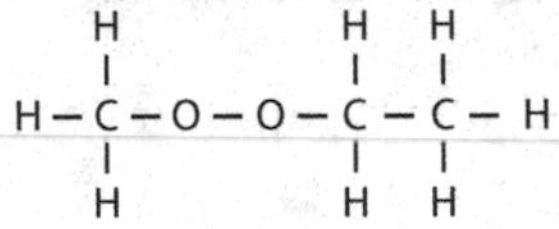

$$\begin{array}{c} H \qquad\quad H \quad H \\ | \qquad\quad | \quad | \\ H-C-O-O-C-C-H \\ | \qquad\quad | \quad | \\ H \qquad\quad H \quad H \end{array}$$

$CH_3OOC_2H_5$

$$\begin{array}{c} H \quad H \qquad\qquad H \quad H \\ | \quad | \qquad\qquad | \quad | \\ C=C-O-O-C=C \\ | \qquad\qquad\qquad\quad | \\ H \qquad\qquad\qquad\quad H \end{array}$$

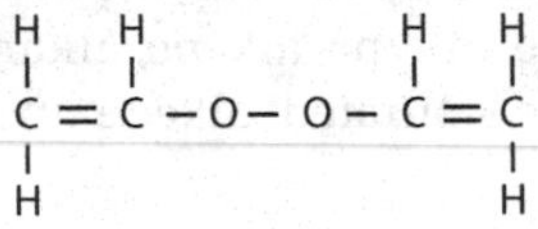

$C_2H_3OOC_2H_3$

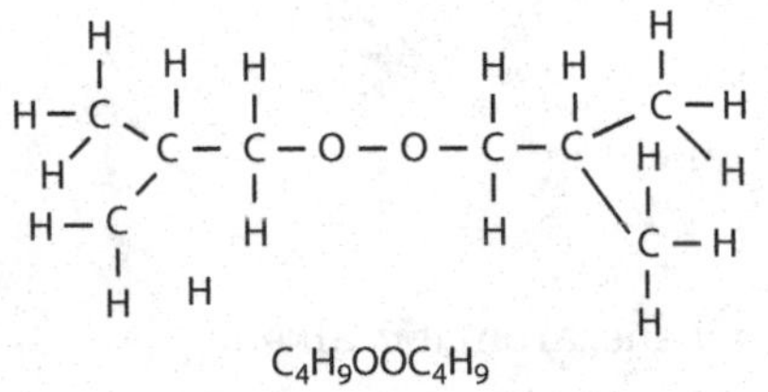

$C_4H_9OOC_4H_9$

7.6 C
7.7 D
7.8 B

Chapter 8
8.1 C
8.2 A
8.3 Inhalation, ingestion, absorption, injection
8.4 A
8.5 C
8.6 D
8.7 B
8.8

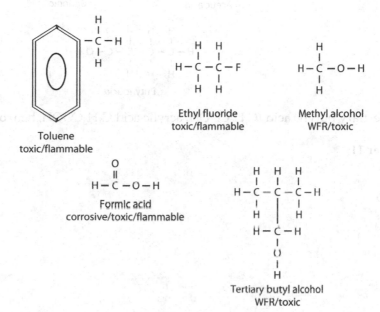

Toluene
toxic/flammable

Ethyl fluoride
toxic/flammable

Methyl alcohol
WFR/toxic

Formic acid
corrosive/toxic/flammable

Tertiary butyl alcohol
WFR/toxic

8.9 Caution, danger, warning
8.10 STEL, TLV-TWA, PEL, IDLH

Chapter 9
9.1 C
9.2 B
9.3 Alpha, beta, gamma
9.4 Alpha, beta
9.5 D
9.6 Ionizing, nonionizing
9.7 C, A, B
9.8 C
9.9 C
9.10 Time, distance, shielding

Chapter 10
10.1 Acids, bases
10.2 Inorganic, organic
10.3 C
10.4 B
10.5 D
10.6 A
10.7 C
10.8 Neutralization, dilution

```
      H  O                              H  H  O
      |  ||                             |  |  ||
  H – C – C – O – H               H – C – C – C – O – H
      |                                 |  |
      H                                 H  H
```
      Acetic acid   Proprionic acid

```
          H  H  H  O
          |  |  |  ||
      H – C – C – C – C – O – H
          |  |  |
          H  H  H
```
              Butyric acid

10.10 Tertiary butyric acid tC_4H_9COOH, acrylic acid C_2H_3COOH, benzoic acid C_6H_5COOH

Chapter 11
11.1 C
11.2 A
11.3 D

Glossary

Absorption: A route of exposure. It occurs when a toxic material contacts the skin and enters the bloodstream by passing through the skin.

Accidental explosion: An unplanned or premature detonation/ignition of explosive/incendiary material or material possessing explosive properties. The activity leading to the detonation/ignition has no criminal intent and is primarily associated with legal, industrial, or commercial activities.

Acetylcholine: A chemical compound formed from an acid and an alcohol that causes muscle to contract (neurotransmitter). It is rapidly broken down by an enzyme, cholinesterase.

Acetylcholinesterase: An enzyme present in nerve tissue, muscles, and red blood cells that catalyzes the hydrolysis of acetylcholine to choline and acetic acid, allowing neural transmission across synapses to occur; true cholinesterase.

Acid: (1) Any of a class of chemical compounds whose aqueous solutions turn litmus paper red (has a pH less than 7) or react with and dissolve certain metals or react with bases to form salts. (2) A compound capable of transferring a hydrogen ion in solution. (3) A molecule or ion that combines with another molecule or ion by forming a covalent bond with two electrons from other species.

Acid, corrosive: A material that usually contains an H^+ ion and is capable of dehydrating other materials.

Acute exposure: The adverse effects resulting from a single dose or exposure to a material. Ordinarily used to denote effects observed in experimental animals.

Acute toxicity: Any harmful effect produced by a single short-term exposure that may result in severe biological harm or death.

Aerosol: The dispersion of very fine particles of a solid or liquid in a gas, fog, foam, or mist.

Agent dosage: The concentration of a toxic vapor in the air multiplied by the time that the concentration is present or the time that an individual is exposed (mg-min/m^3).

Alcohol foam: A type of foam developed to suppress ignitable vapors on polar solvents (those miscible in water). Examples of polar flammable liquids are alcohols and ketones.

Alkaline: Any compound having the qualities of a base. Simplified, a substance that readily ionizes in aqueous solution to yield hydroxyl (OH$^-$) anions. Alkalis have a pH greater than 7 and turn litmus paper blue.

Alpha particle: A form of ionizing radiation that consists of two protons and neutrons.

Ambient temperature: The normal temperature of the environment.

ANFO: An ammonium nitrate and fuel oil mixture, commonly used as a blasting agent. The proportions are determined by the manufacturer or user. It is commonly mixed with the addition of an "enhancer," such as magnesium or aluminum, to increase the rate of burn.

Anhydrous: Describes a material that contains no water (water-free).

Anion: A negatively charged ion that moves toward the anode (+ terminal) during electrolysis. Oxidation occurs at the anode.

Anticholinergic: An agent or chemical that blocks or impedes the action of acetylcholine, such as the (also cholinolytic) antidote atropine.

Anticholinesterase: A substance that blocks the action of cholinesterase (acetylcholinesterase), such as nerve agents.

Antidote: A material administered to an individual who has been exposed to a poison in order to counteract its toxic effects.

Arsenical: Pertaining to or containing arsenic; a reference to the vesicant lewisite.

Asphyxia: Lack of oxygen and interference with oxygenation of the blood. Can lead to unconsciousness.

Asphyxiant: A vapor or gas that can cause unconsciousness or death by suffocation (lack of oxygen). Most simple asphyxiants are harmful to the body when they become so concentrated that they reduce (displace) the available oxygen in air (normally about 21%) to dangerous levels (18% or lower). Chemical asphyxiants, like carbon monoxide (CO), reduce the blood's ability to carry oxygen or, like cyanide, interfere with the body's utilization of oxygen.

Asphyxiation: Asphyxia or suffocation. Asphyxiation is one of the principal potential hazards of working in confined spaces.

Atmospheric container: A type of container that holds products at atmospheric pressure (760 mm).

Atom: The smallest unit into which a material may be broken by chemical means. In order to be broken into any smaller units, a material must be subjected to a nuclear reaction.

Atomic weight (at. wt.): The relative mass of an atom. Basically, it equals the number of protons plus neutrons.

Atropine: An anticholinergic used as an antidote for nerve agents to counteract excessive amounts of acetylcholine. It also has other medical uses.

Autoignition: A process in which a material ignites without any apparent outside ignition source. In the process, the temperature of the material is raised to its ignition temperature by the heat transferred by radiation, convection, combustion, or some combination of all three.

Autoignition temperature: *See* Ignition temperature.

B-NICE: The acronym developed by the National Fire Academy for identifying the five categories of terrorist incidents: *B*iological, *N*uclear, *I*ncendiary, *C*hemical, and *E*xplosive.

Bacteria: Single-celled organisms that multiply by cell division and can cause disease in humans, plants, or animals. Examples include anthrax, cholera, plague, tularemia, and Q fever.

Base: A chemical compound that reacts with an acid to form a salt. The term is applied to the hydroxides of the metals, to certain metallic oxides, and to groups of atoms containing one or more hydroxyl groups (OH^-) in which hydrogen is replaceable by an acid radical. *See* Alkaline.

Beta particle: A form of ionizing radiation that consists of either electrons or positrons.

Biohazard: Those organisms that have a pathogenic effect on life and the environment, and can exist in normal ambient environments. These hazards can represent themselves as disease germs and viruses.

Biological agent: Living organism, or the materials derived from it, that cause disease in or harm humans, animals, or plants, or cause deterioration of material. Biological agents may be found as liquid droplets, aerosols, or dry powders. A biological agent can be adapted and used as a terrorist weapon, such as anthrax, tularemia, cholera, encephalitis, plague, and botulism. There are three different types of biological agents: bacteria, viruses, and toxins.

Blasting agent: A material designed for blasting that has been tested in accordance with Sec. 173.114a (49 CFR). It must be so insensitive that there is little probability of accidental explosion or going from burning to detonation.

Blepharospasm: A twitching or spasmodic contraction of the orbicular oculi muscle around the eye.

BLEVE: *See* Boiling liquid, expanding vapor, and explosion.

Blister agent: A chemical agent, also called a vesicant that causes severe blistering and burns to eyes, skin, and tissues of the respiratory tract. Exposure is through liquid or vapor contact. Also referred to as mustard agent. Examples include mustard and lewisite.

Blood agent: A chemical agent that interferes with the ability of blood to transport oxygen and causes asphyxiation. These substances injure a person by interfering with cell respiration (the exchange of oxygen and carbon dioxide between blood and tissues). Common examples are hydrogen cyanide and cyanogens chloride.

Blood asphyxiant: A chemical that is absorbed by the blood and changes or prevents the blood from flowing or carrying oxygen to cells. An example is carbon monoxide poisoning.

Boiling liquid, expanding vapor, explosion (BLEVE): The explosion and rupture of a container caused by the expanding vapor pressure as liquids in the container become overheated.

Boiling point: At this temperature, vapor pressure of a liquid now equals the surrounding atmospheric pressure (14.7 psi at sea level).

BTU (British Thermal Unit): Amount of heat required to raise 1 lb of H_2O 1°F at sea level.

Carcinogen: A material that either causes cancer in humans or, because it causes cancer in animals, is considered capable of causing cancer in humans.

Cation: A positively charged ion that moves toward the cathode during electrolysis. Reduction occurs at the cathode.

Caustic: (1) Burning or corrosive. (2) A hydroxide of a light metal. Broadly, any compound having highly basic properties. A compound that readily ionizes in aqueous solution to yield OH^- anions, with a pH above 7, and turns litmus paper blue. *See* Alkaline; Base.

Cellular asphyxiant: A material that, upon entering the body, inhibits the normal function of cells. Examples are CO, hydrogen cyanide, or hydrogen sulfide poisoning.

Central nervous system (CNS): In humans, the brain and spinal cord, as opposed to the peripheral nerves found in the fingers, etc.

Chemical agent: There are five classes of chemical agents, all of which produce incapacitation, serious injury, or death: nerve agents, blister agents, blood agents, choking agents, and irritating agents. A chemical substance used in military operations intended to kill, seriously injure, or incapacitate people through its physiological effects.

Chemical burn: A burn that occurs when the skin comes into contact with strong acids, strong alkalis, or other corrosive materials. These agents literally eat through the skin and, in many cases, continue to do damage as long as they remain in contact with the skin.

Chemical properties: A property of matter that describes how it reacts with other substances.

Chemical reaction: A process that involves the bonding, unbonding, or rebonding of atoms. A chemical change takes place that actually changes substances into other substances.

Chemical reactivity: The process whereby substances are changed into other substances by the rearrangement, or recombination, of atoms.

CHEMTREC: Chemical Transportation Emergency Center operated by the Chemical Manufacturers Association. Provides information or assistance to emergency responders. CHEMTREC contacts the shipper or producer of the material for more detailed information, including on-scene assistance when feasible. Can be reached 24 h a day by calling 1-800-424-9300.

CHLOREP: Chlorine Emergency Plan operated by the Chlorine Institute. A 24-h mutual-aid program. Response is activated by a CHEMTREC call to the designated CHLOREP's geographical-sector assignments for teams.

Choking agents: These agents exert their effects solely on the lungs and result in the irritation of the alveoli of the lungs. Agents cause the alveoli to constantly secrete watery fluid into the air sacs, which is called pulmonary edema. When a lethal amount of a choking agent is received, the air sacs become so flooded that the air cannot enter and the victim dies of anoxia (oxygen deficiency); also known as dry drowning.

Cholinergic: Resembling acetylcholine, especially in physiological action. Cholinergic symptoms include nausea, vomiting, headache, and sweating.

Cholinesterase (Ache): Acetylcholinesterase is the enzyme that breaks down the neurotransmitter acetylcholine after it has transmitted a signal from a nerve ending to another nerve, muscle, or gland. Organophosphate pesticides and military nerve agents block the normal activity of Ache, which results in the accumulation of excess acetylcholine at nerve endings.

Chronic: Applies to long periods of action, such as weeks, months, or years.

Chronic effects: An adverse health effect on a human or animal body with symptoms that develop slowly or that recur frequently due to the exposure of hazardous chemicals.

Chronic exposure: Repeated doses or exposure to a material over a relatively prolonged period of time.

Closed-cup tester: A device for determining flash points of flammable and combustible liquids, utilizing an enclosed cup or container for the liquid. Recognized types are the Tagliabue (Tag) Closed Tester, the Pensky–Martens Closed Tester, and the Setaflash Closed-Cup Tester.

CNS: *See* Central nervous system.

Combustibility: The ability of a substance to undergo rapid chemical combination with oxygen, with the evolution of heat.

Combustible dust: Particulate material that, when mixed in air, will burn or explode.

Combustible liquid: The term commonly used for liquids that emit burnable vapors or mists. Technically, a liquid whose vapors will ignite at a temperature of 100°F or above.

Compound: A substance composed of two or more elements that have chemically reacted. The compound that results from the chemical reaction is unique in its chemical and physical properties.

Compressed gas: Any material or mixture having in the container an absolute pressure exceeding 40 psi at 70°F or, regardless of the pressure at 70°F, having an absolute pressure exceeding 104 psi at 130°F; or any liquid flammable material having a vapor pressure exceeding 40 psi absolute at 100°F as determined by testing. Also includes cryogenic or "refrigerated liquids" (DOT) with boiling points lower than −130°F at 1 atm.

Concentration: The amount of a material that is mixed with another material.

Concentration (corrosives): In corrosives, the amount of acid or base compared to the amount of water present. Corrosives have "strength" and "concentration." *See* Strength.

Contaminant: (1) A toxic substance that is potentially harmful to people, animals, and the environment. (2) A substance not in pure form.

Corrosive: A chemical that causes visible destruction of or irreversible alterations in living tissue by chemical action at the site of contact; a liquid that causes a severe corrosion rate in steel. A corrosive is either an acid or a caustic (a material that reads at either end of the pH scale).

Corrosive material (DOT): A material that causes the destruction of living tissue and metals.

Covalent bond: A chemical bond in which atoms share electrons in order to form a molecule.

Critical pressure: The pressure required to liquefy a gas at its critical temperature.

Critical temperature: The temperature above which a gas cannot be liquefied by pressure.

Cryogenic burn: Frostbite; damage to tissues as a result of exposure to low temperatures. It may involve only the skin, extend to the tissue immediately beneath it, or lead to gangrene and loss of affected parts.

Cryogenic cylinder: An insulated metal cylinder contained within an outer protective metal jacket. The area between the cylinder and the jacket is normally under vacuum. The cylinders range in size from a Dewier (similar to a small thermos) up to 24 in. in diameter and 5 ft in length. Examples of materials found in these types of cylinders are argon, helium, nitrogen, and oxygen.

Cryogenic liquid: A liquid with a boiling point below −130°F.

Cylinder: A container for liquids, gases, or solids under pressure. Ranges in size from aerosol containers found at home, such as spray deodorant, to the cryogenic (insulated) cylinders for nitrogen that can be approximately 24 in. in diameter and 5 ft in length. Pressure ranges from a few pounds to 6000 lb per in.^{2}

Dangerous when wet: Materials that when exposed to water allow a chemical reaction to take place and often produce flammable or poisonous gases, heat, and a caustic solution. An example is sodium.

Decomposition: Separation of larger molecules into separate constituent and smaller parts.

Decomposition (chemical): A reaction in which the molecules of a chemical break down to its basic elements, such as carbon, hydrogen, or nitrogen, or to more simple compounds. This often occurs spontaneously, liberating considerable heat and often large volumes of gas.

Decontamination: The physical or chemical process of reducing and preventing the spread of contamination from persons and equipment used at a hazardous materials incident.

Deflagration: Explosion, with rapid combustion, up to 1250 ft/s.

Detonating cord: A flexible cord containing a center cord of high explosives used to detonate other explosives with which it comes in contact.

Detonation: An explosion at speeds above 1250 ft/s and many times over 3300 ft/s.

Detonator: Any device containing a detonating charge that is used for initiating detonation in an explosive. This term includes, but is not limited to, electric and nonelectric detonators (either instantaneous or delayed) and detonating connectors.

Dewier container: Small (less than 25 gal) container used for temporary storage or handling of cryogenic liquids.

Dilution: The application of water to water-miscible hazardous materials. The goal is to reduce the hazard of a material to safe levels by reducing its concentration.

Dose: The accumulated amount of a chemical to which a person is exposed.

DOT: U.S. Department of Transportation. Regulates transportation of materials to protect the public as well as fire, law, and other emergency response personnel.

Dry bulk: A type of container used to carry large amounts of solid materials (more than 882 lb, or 400 kg). It can either be placed on or in a transport vehicle or vessel constructed as an integral part of the transport vehicle.

Dyspnea: Shortness of breath, a subjective difficulty or distress in breathing, usually associated with disease of the heart or lungs; occurs normally during intense physical exertion or at high altitudes.

Element: A substance that cannot be broken down into any other substance by chemical means.

Empirical formula: Describes the *ratio* of the number of each element in the molecule, but not the exact number of atoms in the molecule.

Emulsification: The process of dispersing one liquid in a second immiscible liquid. The largest group of emulsifying agents are soaps, detergents, and other compounds whose basic structure is a paraffin chain terminating in a polar group.

Encephalitis, pl. encephalitides: Inflammation of the brain.

Endothermic: A process or chemical reaction that is accompanied by absorption of heat.

Enterotoxin: A cytotoxin specific for the cells of the intestinal mucosa.

Erythema: Red area of the skin caused by heat or cold injury, trauma, or inflammation.

Etiologic agent: Those living organisms or their toxins that contribute to the cause of infection, disease, or other abnormal condition.

Evaporation: The process in which liquid becomes vapor as more molecules leave the vapor than return.

Exothermic reaction: A chemical reaction that liberates heat during the reaction.

Expansion ratio: The amount of gas produced from a given volume of liquid escaping from a container at a given temperature.

Explosion: The sudden and rapid production of gas, heat, noise, and many times a shock wave, within a confined space.

Explosive (DOT): Any chemical compound or mixture whose primary function is to produce an explosion.

Explosives, high: Explosive materials that can be used to detonate by means of a detonator when unconfined (e.g., dynamite).

Explosives, low: Explosive materials that deflagrate rather than detonate (e.g., black powder, safety fuses, and "special fireworks" as defined by Class 1.3 Explosives).

Explosive limits: *See* Flammable limits.

Febrile: Denoting or relating to fever.

Fire point: The lowest temperature at which the vapor above the liquid will ignite and continue to burn, usually a few degrees above the flash point.

Flammable gas: A gas that at ambient temperature and pressure forms a flammable mixture with air at a concentration of 13% by volume or less; or a gas that at ambient temperature and pressure forms a range of flammable mixtures with air greater than 12% by volume, regardless of the lower explosive limit.

Flammable limits: The range of the percentages of vapor mixed with air that are capable of ignition, as opposed to those mixtures that have too much or too little vapor to be ignited. Also called explosive limits.

Flammable liquid: A liquid that gives off readily ignitable vapors. Defined by the NFPA and DOT as a liquid with a flash point below 100°F (38°C).

Flammable range: The percentage of fuel vapors in air where ignition can occur. Flammable range has an upper and lower limit.

Flammable solid: A solid (other than an explosive) that ignites readily and continues to burn. It is liable to cause fires under ordinary conditions or during transportation through friction or retained heat from manufacturing or processing. It burns so vigorously and persistently as to create a serious transportation hazard. Included in this class are spontaneously combustible and water-reactive materials. An example is white phosphorus.

Flash back: The ignition of vapors and the travel of the flame back to the liquid/vapor-release source.

Flash point: The minimum temperature at which a liquid gives off vapor within a test vessel in sufficient concentration to form an ignitable mixture with air near the surface of the liquid.

Foam: Firefighting material consisting of small bubbles of air, water, and concentrating agents. Chemically, the air in the bubbles is suspended in the fluid. The foam clings to vertical and horizontal surfaces and flows freely over burning or vaporizing materials. Foam puts out a fire by blanketing it, excluding air, and blocking the escape of volatile vapor. Its flowing properties resist mechanical interruption and reseal the burning material.

Formula: A combination of the symbols for atoms or ions that are held together chemically.

Freezing point: The temperature at which a material changes its physical state from a liquid to a solid.

Frothing: A foaming action caused when water, turning to steam in contact with a liquid at a temperature higher than the boiling point (212°F), picks up a part of a viscous liquid.

Gas: A formless fluid that occupies the space of its enclosure. It can settle to the bottom or top of an enclosure when mixed with other chemicals. It can be changed to its liquid or solid state only by increased pressure and decreased temperature.

Gastrointestinal (GI) tract: The entire digestive canal from mouth to anus.

Halogens: A chemical family that includes fluorine, chlorine, bromine, and iodine.

Halon: Halogenated hydrocarbons (containing the elements F, Cl, Br, or I) used to suppress or prevent combustion.

Hazard class: One of nine classes of hazardous materials as categorized and defined by the DOT in 49 CFR.

Hazmat foam: A special vapor-suppressing mix that can be applied to liquids or solids to prevent off-gassing.

Hematemesis: Vomiting of blood.

Hemolytic anemia: Anemia caused by increased destruction of red blood cells where the bone marrow is not able to compensate for it.

Hemoptysis: The spitting of blood derived from the lungs or bronchial tubes as a result of pulmonary or bronchial hemorrhage.

Hepatoxin: A chemical that is injurious to the liver.

High-expansion foam: A detergent-based foam (low water content) that expands at ratios of 1000–1.

High-order explosion: Materials that require moderate heat and reducing agents to initiate combustion.

Hypergolic materials: Materials that ignite upon contact with one another.

Hypergolic reaction: The immediate spontaneous ignition when two or more materials are mixed.

Hypotension: Subnormal arterial blood pressure.

Hypovolemia: A decreased amount of blood in the body.

Hypoxemia: Subnormal oxygenation of arterial blood, short of anoxia.

IDLH (Immediately Dangerous to Life and Health): The maximum levels to which a healthy worker can be exposed for 30 min to a chemical and escape without suffering irreversible health effects or escape impairing symptoms.

Ignition temperature: The minimum temperature at which a material will ignite without a spark or flame present. This is also the temperature the ignition source must be.

Immiscible: Matter that cannot be mixed. For example, water and gasoline are immiscible.

Incapacitating agent: An agent that produces physiological or mental effects or both that may persist for hours or days after exposure, rendering an individual incapable of performing his or her assigned duties.

Incompatibility: The inability to function or exist in the presence of something else, such as when a chemical will destroy the container.

Inert: A material that under normal temperatures and pressures does not react with other materials.

Inhibited: A substance that has had another substance added to prevent or deter its reaction either with other materials or itself (polymerization). This is usually used to deter polymerization.

Inhibitor: A substance that is capable of stopping or retarding a chemical reaction. To be technically useful, it must be effective in low concentration (i.e., to stop polymerization).

Initiator: The substance or molecule (other than reactant) that initiates a chain reaction, as in polymerization.

Inorganic: Pertaining to or composed of chemical compounds that do not contain carbon as the principal element (except carbonates, cyanides, and cyanates). Matter other than plant or animal.

Inorganic peroxides: Inorganic compounds containing an element at its highest state of oxidation (such as sodium peroxide), or having the peroxy group –O–O– (such as perchloric acid).

Ion: An atom that possesses an electrical charge, either (+) positive or (–) negative.

Ionic bond: A chemical bond in which atoms of different elements transfer (exchange) electrons. As the electrons are exchanged, charged particles known as ions are formed.

Ionizing radiation: High-energy radiation, such as an x-ray, that causes the formation of ions in substances through which it passes (gamma rays). Excessive amounts of ionizing radiation will cause permanent genetic or bodily damage.

Irritant: A noncorrosive material that causes a reversible inflammatory effect on living tissue by chemical action at the site of contact.

Lacrimation: Secretion and discharge of tears.

Latent period: Specifically in the case of mustard, the period between exposure and onset of signs and symptoms; otherwise, an incubation period.

LC_{50}: Lethal concentration 50, median lethal concentration. The concentration of a material in air that on the basis of laboratory tests (respiratory route) is expected to kill 50% of a group of test animals when administered as a single exposure in a specific time period.

LD_{50}: Lethal dose 50. The single dose of a substance that causes death of 50% of an animal population from exposure to the substance by any route other than inhalation.

Lethal chemical agent: An agent that may be used effectively in a field concentration to produce death.

Liquid: A substance that is neither a solid nor a gas; a substance that flows freely, like water.

Lower explosive limit: The lowest concentration of gas or vapor (% by volume in air) that burns or explodes if an ignition source is present at ambient temperatures.

Low-order explosion: Materials that require excessive heat and reducing agents to initiate combustion.

Low-pressure container: A container designed to withstand pressures from 5 to 100 psi.

LOX: Liquid oxygen.

Mechanical foam: A substance introduced into the water line by various means at a 6% concentration. Air is then introduced to yield foam consisting generally of 90 volumes air, 9.4 volumes water, and 0.6 volumes foam liquid. It uses hydrolyzed soybean, fish scales, hoof and horn meal, and peanut or corn protein as a base.

Median incapacitating dosage (ICT_{50}): The volume of a chemical agent vapor or aerosol inhaled that is sufficient to disable 50% of exposed, unprotected people (expressed as mg-min/m^3).

Median incapacitating dosage (LD_{50}): The volume of a liquid chemical agent expected to incapacitate 50% of a group of exposed, unprotected individuals.

Median lethal dosage (LCT_{50}): The dosage of a chemical-agent vapor or aerosol inhaled that is lethal to 50% of exposed, unprotected people (expressed as mg-min/m^3).

Median lethal dosage (LD_{50}): The amount of liquid chemical agent expected to kill 50% of a group of exposed, unprotected individuals.

Melting point: The degree of temperature at which a solid substance becomes a liquid, especially under the pressure of one atmosphere.

Miosis: A condition where the pupil of the eye becomes contracted (pinpointed), impairing night vision.

Miscible: Mixable in any and all proportions to form a uniform mixture. Water and alcohol are miscible; water and oil are not.

Molecular formula: Shows the *exact number* of each atom in the molecule.

Molecular weight: The sum of the atomic weights of all the atoms in a molecule.

Molecule: The smallest possible particle of a chemical compound that can exist in the free state and still retain the characteristics of the substance. Molecules are made up of atoms of various elements that form the compound.

Monomer: A simple molecule capable of combining with a number of like or unlike molecules to form a polymer. It is a repeating structure unit within a polymer.

Mutagen: A material that induces genetic changes (mutations) in the DNA of chromosomes. Chromosomes are the "blueprints" of life within individual cells.

Myalgia: Muscular pain.

Mydriasis: Dilation of the pupil.

Narcosis: General and nonspecific reversible depression of neuronal excitability, produced by a number of physical and chemical agents, usually resulting in stupor rather than in anesthesia.

Necrosis: Cell or tissue death due to disease or injury.

Nerve agent: A substance that interferes with the central nervous system. Exposure is primarily through contact with the liquid (skin and eyes) and secondarily through inhalation of the vapor.

Neutralization: A chemical reaction used to remove H^+ ions from acidic solutions and OH^- ions from basic solutions. The reaction can be violent and usually produces water, a salt, heat, and many times, a gas.

Nonpersistent agent: An agent that remains in the target areas for a relatively short period of time. The hazard, predominantly vapor, will exist for minutes or, in exceptional cases, hours after dissemination of the agent. As a general rule, nonpersistent agent duration will be less than 12 h.

Normal: A solution that contains one equivalent of solute per liter of solution.

NRC: National Response Center, a communications center for activities related to response actions, is located at Coast Guard headquarters in Washington, DC. The toll-free number, (800) 424-8802, can be reached 24 h a day for reporting actual or potential pollution incidents.

NRT: National Response Team, consisting of representatives of 14 government agencies (DOD, DOI, DOT/RSPA, DOT/USCG, EPA, DOC, FEMA, DOS, USDA, DOJ, HHS, DOL, NRC, and DOE), is the principal organization for implementing the NCP. When the NRT is not activated for a response action, it serves as a standing committee to develop and maintain preparedness, to evaluate methods of responding to discharges or releases, to recommend needed changes in the response organization, and to recommend revisions to the NCP.

Odor: A quality of something that affects the sense of smell; fragrance.

Odor threshold: The greatest dilution of a sample with odor-free water to yield the least definitely perceptible odor.

OHMTADS: Oil and Hazardous Materials Technical Assistance Data System, a computerized database containing chemical, biological, and toxicological information about hazardous substances. OSCs use OHMTADS to identify unknown chemicals and to learn how to best handle known chemicals.

Open-cup tester: A device for determining flash points of flammable and combustible liquids, utilizing an open cup or container for the liquid. Recognized types are the Tagliabue (Tag) Open-Cup Apparatus and the Cleveland Open-Cup Apparatus.

Organic: A material that comes from living plants or animals, such as waste or decay products. Distinguished from mineral matter. Organic chemistry deals with materials that contain the element carbon (C).

Organic peroxides: Any organic compound containing oxygen (O) in the bivalent –O–O– structure and that may be considered a derivative of hydrogen peroxide, where one or more of the hydrogen atoms have been replaced by organic radicals.

Organophosphate: A compound with a specific phosphate group that inhibits acetylcholinesterase. Used in chemical warfare and as an insecticide.

Oxidizer: A material that gives up oxygen easily, removes hydrogen from another compound, or attracts negative electrons (such as chlorine or fluorine), thus enhancing the combustion of other materials.

Oxidizing agent: A material that gains electrons from the fuel during combustion.

Oxime: A compound that blocks acetylcholinesterase from combining with organophosphates, formed by the action of hydroxylamine upon an aldehyde or a ketone.

Oxygen deficient: Defined by OSHA as ambient air containing less than 19.5% oxygen concentration.

Oxygen enriched: Defined by OSHA as ambient air containing above 24% oxygen concentration.

2-PAM CL: Pralidoxime chloride, Protopam®, is an antidote to organophosphate poisoning such as might result from exposure to nerve agents or some insecticides. The drug, which helps restore an enzyme called acetylcholinesterase, must be used in conjunction with atropine to be effective. Restores normal control of skeletal muscle contraction (relieves twitching and paralysis).

Papule: A small, circumscribed, solid elevation on the skin.

PEL (permissible exposure limit): Term used by OSHA for its health standards covering exposures to hazardous chemicals. PEL generally relates to legally enforceable TLV limits.

Persistency: An expression of the duration of effectiveness of a chemical agent, dependent on physical and chemical properties of the agent, weather, method of dissemination, and terrain conditions.

Persistent agent: An agent that remains in the target area for longer periods of time. Hazards from both vapor and liquids may exist for hours, days, or in exceptional cases, weeks or months after dissemination of the agent. As a general rule, persistent-agent duration will be greater than 12 h.

pH: The "power of hydrogen." A measure of the acidity or basicity of a solution, that is, of the concentration of H^+ or OH^- ions in solution. Scale ranges from 0 to 14, where a reading of 7 is neutral.

Photophobia: Morbid dread and avoidance of light. Photosensitivity, or pain in the eyes with exposure to light, can be a cause.

Physical properties: A property of matter that describes only its condition, not the way it reacts with other substances. Examples are size, density, color, and electrical conductivity.

Poison: Any substance (solid, liquid, or gas) that, by reason of an inherent deleterious property, tends to destroy life or impair health.

Polar-solvent liquids: Those liquids that mix (are miscible with water).

Polymer: A long chain of molecules having extremely high molecular weights made up of many repeating smaller units called monomers or comonomers.

Polymerization: A chemical reaction in which small molecules combine to form larger molecules. A hazardous polymerization is a reaction that takes place at a rate that releases large amounts of energy that can cause fires or explosions or burst containers. Materials that can polymerize usually contain inhibitors that can delay the reaction.

Powder: A solid reduced to dust by pounding, crushing, or grinding.

ppm (parts per million): Parts of vapor or gas per million parts of contaminated air by volume at 25°C and 1 torr pressure.

Pressure vessel: A tank or other container constructed so as to withstand interior pressure greater than that of the atmosphere.

Presynaptic: Pertaining to the area on the proximal side of a synaptic cleft.

Pruritus: Syn: itching.

psi: Pounds per square inch.

Ptosis, pl. ptoses: In reference to the eyes, drooping of the eyelids.

Pustule: A small, circumscribed elevation of the skin containing pus and having an inflamed base.

Pyrogenic: Causing fever.

Pyrolysis: A chemical decomposition or breaking apart of molecules produced by heating in the absence of air.

Pyrophoric: Material that ignites spontaneously in air below 130°F (54°C). Occasionally caused by friction.

Pyrophoric gas: Gaseous materials that spontaneously ignite when exposed to air under ambient conditions. An example is trimethyl aluminum.

Pyrophoric liquid: Liquid materials that spontaneously ignite when exposed to air under ambient conditions.

Pyrophoric solid: Solid materials that spontaneously ignite when exposed to air under ambient conditions. An example is phosphorus.

RAD: Radiation-absorbed dose.

Radiation: Ionizing energy, either particulate or wave, that is spontaneously emitted by a material or combination of materials.

Radioactive material (DOT): Materials that emit ionizing radiation.

Radioactivity: Any process by which unstable nuclei increase their stability by emitting particles (alpha or beta) or gamma rays.

Rate of explosion: Rate of decomposition measured in feet per second in relation to the speed of sound. If subsonic, the rate is described as a deflagration. If supersonic, the rate of decomposition is defined as a detonation.

RCRA: Resource Conservation and Recovery Act (of 1976), which established a framework for the proper management and disposal of all wastes.

Reducing agent: A substance that gives electrons to (and thereby reduces) another substance.

Respiratory asphyxiant: A material that prevents or reduces the available oxygen necessary for normal breathing. Divided into simple and chemical asphyxiants.

Respiratory dosage: This is equal to the time in minutes an individual is unmasked in an agent cloud multiplied by the concentration of the cloud.

Retinitis: Inflammation of the retina.

Rhinorrhea: A runny nose.

Roentgen: The amount of ionization that occurs per cubic centimeter of air.

Routes of exposure: Ways in which chemicals get in contact with or enter the body. These are inhalation, absorption, ingestion, or injection.

RRT: Regional response teams composed of representatives of federal agencies and a representative from each state in the federal region.

SADT: *See* Self-accelerating decomposition temperature.

Safety fuse: A flexible cord containing an internal burning medium by which fire or flame is conveyed at a uniform rate from point of ignition to point of use, usually a detonator.

Safety-relief valve: A safety-relief device containing an operating part that is held normally in a position closing a relief channel by spring force, and is intended to open and close at a predetermined pressure.

SARA: The Superfund Amendments and Reauthorization Act of 1986. Title III of SARA includes detailed provisions for community planning.

Self-accelerating decomposition temperature (SADT): Organic peroxides or other synthetic chemicals that decompose at ambient temperature, or react to light or heat, resulting in a chemical breakdown. This releases oxygen, energy, and fuel in the form of rapid fire or explosion. To ensure stabilization, these materials must be kept in a dark or refrigerated environment.

Sensitizer: A substance that on first exposure causes little or no reaction in humans or test animals, but that on repeated exposure may cause a marked response not necessarily limited to the contact site.

Septic shock: Shock associated with septicemia caused by gram-negative bacteria.

Shigellosis: Bacillary dysentery caused by bacteria of the genus *Shigella*, often occurring in epidemic patterns.

Simple asphyxiant: A material that replaces the amount of oxygen admitted into the body without further damage to tissue or poisoning. Examples are nitrogen and carbon dioxide.

Slurry: A pourable mixture of solid and liquid.

Solid: The state of matter having definite volume and rigid shapes. Its atoms or molecules are restricted to vibration only.

Solubility: The ability of a substance to form a solution with another substance.

Solution: The even dispersion (mixing) of molecules of two or more substances. The most commonly encountered solutions involve mixing of liquids and liquids or solids and liquids.

Solvent: A substance, usually a liquid, capable of absorbing another liquid, gas, or solid to form a homogeneous mixture.

Specific gravity: The weight of a solid or liquid substance as compared to the weight of an equal volume of water; specific gravity of water equals 1.

Spontaneous combustion: A process by which heat is generated within material by either a slow oxidation reaction or by microorganisms.

Spontaneous ignition: Ignition that can occur when certain materials, such as tung oil, are stored in bulk, resulting from the generation of heat, which cannot be readily dissipated; often heat is generated by microbial action.

Spontaneous ignition temperature: *See* Ignition temperature.

States of matter: Any of three physical forms of matter: solid, liquid, or gas.

Strength (acid/base): The amount of ionization that occurs when an acid or a base is dissolved in a liquid.

Stridor: A high-pitched, noisy respiration, like the blowing of the wind; a sign of respiratory obstruction, especially in the trachea or larynx.

Subacute exposure: (1) Less than acute. (2) Of or pertaining to a disease or other abnormal condition present in a person who appears to be clinically well. The condition may be identified or discovered by means of a laboratory test or by radiologic examination.

Sublimation: The direct change of state from solid to vapor.

Subscripts: Identify the number of atomic weights of the element present in the molecule.

Symbol: Letters used to identify each element. The symbol for an element represents a definite weight (1 atomic weight) of that element.

Systemic toxicity: Poisoning of the whole system or organism, rather than poisoning that affects, for example, a single organ.

Target organ: The primary organ to which specific chemicals cause harm. Examples are the lungs, liver, or kidneys.

Temperature: Measure of the vibratory rate of a molecule.

Teratogen: Material that affects the offspring when a developing embryo or fetus is exposed to that material.

Thermal burn: Pertaining to or characterized by heat.

TLV: Threshold limit value, *estimated* exposure value, below which no ill health effects *should* occur to the individual.

Toxemia: A condition in which toxins produced by cells at a local source of infection or derived from the growth of microorganisms are contained in the blood.

Toxic: Harmful, poisonous.

Toxicity: The ability of a substance to cause damage to living tissue, impairment of the central nervous system, severe illness, or death when ingested, inhaled, or absorbed by the skin.

Toxins: Toxic substance of natural origin produced by an animal, plant, or microbe. They differ from chemical substances in that they are not manmade. Toxins may include botulism, ricin, and mycotoxins.

TTL: Threshold toxic limit, *estimated* exposure value, below which no ill health effects *should* occur to the individual.

Upper explosive limit: The maximum fuel-to-air mixture in which combustion can occur.

Uricant: A chemical agent that produces irritation at the point of contact, resembling a stinging sensation, such as a bee sting. For example, the initial physiological effects of phosgene oxime (CX) upon contact with a person's skin.

Vapor density: The weight of a vapor or gas compared to the weight of an equal volume of air; an expression of the density of the vapor or gas calculated as the ratio of the molecule weight of the gas to the average molecule weight of air, which is 29. Materials lighter than air have vapor densities less than 1.

Vapor pressure: The pressure exerted by a saturated vapor above its own liquid in a closed container. Vapor pressure reported on MSDS is in millimeters of mercury at 68°F (20°C), unless stated otherwise.

Vapors: Molecules of liquid in air; moisture, such as steam, fog, mist, etc., often forming a cloud suspended or floating in the air, usually due to the effect of heat upon a liquid.

Variola: Syn: smallpox.

Vesicants: Chemical agents also called blister agents, that cause severe burns to eyes, skin, and tissues of the respiratory tract. Also referred to as mustard agents, examples include mustard and lewisite.

Vesicles: Blisters on the skin.

Violent reaction: The action by which a chemical changes its composition near or exceeding the speed of sound, often releasing heat and gases.

Virus: The simplest type of microorganism, lacking a system for its own metabolism. It depends on living cells to multiply and cannot live long outside of a host. Types of viruses are smallpox, Ebola, Marburg, and Lassa fever.

Viscosity: The measurement of the flow properties of a material expressed as its resistance to flow. Unit of measurement and temperature are included.

Vomiting agent: Compounds that cause irritation of the upper respiratory tract and involuntary vomiting.

Water-reactive material: A material that will decompose or react when exposed to moisture or water.

Water solubility: The ability of a substance to mix with water.

Zoonosis: An infection or infestation shared in nature by humans and other animals that are the normal host; a disease of humans acquired from an animal source.

Sources
http://www.webelements.com
http://www.airproducts.com
http://www.chemicool.com
http://www.ptable.com/
http://www.chemicalelements.com/
http://phmsa.dot.gov/pipeline/library/data-stats
http://www.cas.org/
http://periodic.lanl.gov/elements/3.html
http://www.lenntech.com/periodic/elements/li.htm
http://environmentalchemistry.com/yogi/periodic/Li.html

References

1. Killen, W.D., *Eight Die in Chlorine Tanker Derailment, Fire Command*, National Fire Protection Association, Quincy, MA.
2. National Fire Protection Association, *NFPA Quarterly*, National Fire Protection Association, Quincy, MA, October 1964.
3. Sullivan, B., Burns illustrate need for bunker gear, *Firehouse Magazine*, July 1994.
4. Chisholm, J. and Johnson, M., *Introduction to Chemistry*, Usborn Publishing, London, England.
5. *Milwaukee Fire Department Basic Training Manual*, 1986.
6. Harte, J. et al., *Toxics A-Z*, University of California Press, Berkley, CA, 1991.
7. National Transportation Safety Board, Anhydrous hydrogen fluoride release from NATX 9408, Elkhart, IN, February 4, 1985.
8. Dektar, C., Peroxide blast shatters chemical plant, *Fire Engineering*, January 1979.
9. Klem, T.J., High-rise fire claims three Philadelphia fire fighters, *NFPA Journal*, National Fire Protection Association, Quincy, MA, September/October 1991.
10. *JEMS*, Proper Use of Highway Flares, August 1982.
11. Hill, I., *Crescent City Remembers*, Scheiwe's Print Shop, Crescent City, IL, 1995.
12. Ryczkowski, J.J., Kingman revisited, *American Fire Journal*, July 1993.
13. U.S. Department of Transportation, *North American Emergency Response Guidebook 1996*, U.S. Government Printing Office, Washington, DC, 1996.
14. American Trucking Association, Code of Federal Regulations (CFR) 49, Parts 100–177, American Trucking Association, Arlington, VA, October 1, 1994.
15. Lewis, R.J., Sr., *Hawley's Condensed Chemical Dictionary*, 12th edn., Van Nostrand Reinhold, New York, 1993.
16. U.S. Government Printing Office, *National Fire Academy Chemistry of Hazardous Materials Instructor Guide*, U.S. Government Printing Office, Washington, DC, 1994.
17. U.S. Government Printing Office, *National Fire Academy Chemistry of Hazardous Materials, Student Manual*, U.S. Government Printing Office, Washington, DC, 1994.
18. U.S. Government Printing Office, *National Fire Academy Initial Response to Hazardous Materials Incidents: Basic Concepts, Student Manual*, U.S. Government Printing Office, Washington, DC, 1992.
19. U.S. Government Printing Office, *National Fire Academy Initial Response to Hazardous Materials Incidents: Concept Implementation*, U.S. Government Printing Office, Washington, DC, 1992.
20. National Fire Protection Association, *NFPA Fire Protection Handbook*, 17th edn., National Fire Protection Association, Quincy, MA, 1992.
21. NIOSH, *NIOSH Pocket Guide to Chemical Hazards*, U.S. Government Printing Office, Washington, DC, June 1997.
22. National Fire Protection Association, *Fire Protection Guide to Hazardous Materials*, 11th edn., National Fire Protection Association, Quincy, MA, 1994.
23. Kamrin, M.A., *Toxicology*, Lewis Publishers, Chelsea, MI, 1988.
24. Brady, J.E. and Holm, J.R., *Fundamentals of Chemistry*, 3rd edn., John Wiley & Sons, 1988.
25. Cashman, J.R., *Hazardous Materials Emergencies Response and Control*, 1st edn., Technomic Publishing, Lancaster, PA, 1983.
26. California Department of Justice Web page, Stop Drugs.org-Resources, August 17, 2001.

27. Burke, R., *Counter-Terrorism for Emergency Responders*, 2nd edn., Lewis Publishers, Boca Raton, FL, 2007.
28. United States Chemical Safety and Hazard Investigation Board web site, www.chemsafety.gov, March 26, 2002.
29. National Safety Council, Recognition and handling of peroxidizable compounds, Data sheet, I-655, National Safety Council, Chicago, IL, 1982.
30. Basic Chemical Segregation, www.acs.ucalgary.ca
31. Chemical Compatibility, *CRC Press Laboratory Handbook*, CRC Press, Boca Raton, FL, www.lab-safety.org
32. United States Department of Transportation, *2012 Emergency Response Guidebook*, United States Department of Transportation, Washington, DC.
33. Los Alamos National Labs Periodic Table of Elements, http//:pearl1.lanl.gov/periodic/default.htm
34. Burke, R., *Hazmat Teams across America*, 1st edn., MT Publishing, Evansville, IN, 2009.
35. U.S. Government Printing Office, *National Fire Academy Chemistry for Emergency Response (CES)*, *Instructor Guide*, 4th edn., 1st Printing, U.S. Government Printing Office, Washington, DC, July 2009.
36. WebElements: the periodic table on the WWW, http://www.webelements.com/
37. Exchange Club of Waverly, TN, *Explosion in Waverly*, 2nd edn., Taylor Publishing Company, Dallas, TX, 2009.

Index

fire hazard, 51–52
health hazard, 52
history, 50
isotopes, 51
properties, 50
sources, 50–51
uses, 51
uranium
chemical hazards, 101
compounds, 100
fire hazards, 101
health hazards, 101
history, 99
isotopes, 101
properties, 99
reactions, 101
sources, 99–100
uses, 100–101
xenon
compounds, 108
health hazards, 108
history, 107
isotopes, 108
properties, 107
reactions, 108
sources, 108
uses, 108
zinc
compounds, 66
flammability, 67
health hazards, 67
history, 65
isotopes, 66
properties, 65
reactions, 66
sources, 66
uses, 66
Helium
compounds, 102
hazards, 103
history, 102
isotopes, 103
properties, 102
reactions, 103
sources, 102
uses, 102–103
Helium 3 detector, 426
Homemade/terrorist explosives
acetone peroxide, 170–171
ANFO mixture, 168
Atlanta bombings, 167
family planning clinic bombing, 167
Olympic Park bombing, 167
pentaerythritol tetranitrate, 170–171
pipe bombs, 167
step-by-step instructions and diagrams, 170
World Trade Center bombing, 167–168
Human body, chemical makeup, 1–2
Hydrocarbon derivatives

compressed gases
dimethylamine, 192–193
families, 191–192
formaldehyde, 193
functional groups, 191
methyl chloride, 192
vinyl chloride, 193
flammability, 263
alcohol (*see* Alcohol)
aldehydes, 283–285
alkane, 263
alkyl halide, 264–265
amine, 266–267
double bonds, 264
ester, 285–288
ether, 267–271
isomers, 279–281
ketone, 281–283
organic acids, 288–289
sulfur compounds, 287–288
vinyl and acryl radical, 264
MRI and NMR scanners
anhydrous ammonia, 214–218
carbon dioxide, 218
nitrogen, 213
oxygen, 213–214
tetrafluoromethane, 213
trifluoromethane, 213
Hydrogen
compounds, 34
hazards, 35–36
history, 33–34
isotopes, 35
properties, 32–33
reactions, 35
sources, 34
uses, 34–35

I

Immediately dangerous to life and health (IDLH), 220
Incident commander, 22
Incompatible and unstable chemicals
acids and bases, 461
aging chemicals (*see* Peroxide-forming
compounds)
chemical incompatibilities, 462, 468–470
chemical storage segregation, 470–472, 474–476
oxidizers and organic materials, 461, 463
water- and air-reactive materials, 467
Infectious substances
bacterial agents
anthrax spores, 396–397
Brucella (brucellosis), 401
Cholera (*Vibrio cholerae*), 400
Glanders, Burkholderia, 403–404
inhalation exposure, 397–398
Plague (*Yersinia pestis*), 398–399
Q fever (*Coxiella burnetii*), 401

Printed in the United States
by Baker & Taylor Publisher Services